AF906122

Advanced Structured Materials

Volume 5

Series Editors:
Prof. Dr. Andreas Öchsner
Technical University of Malaysia, Skudai, Johor, Malaysia
Prof. Dr. Holm Altenbach
University of Halle-Wittenberg, Halle, Germany
Prof. Dr. Lucas Filipe Martins da Silva
University of Porto, Porto, Portugal

For further volumes:
http://www.springer.com/series/8611

Noorhana Yahya

Carbon and Oxide Nanostructures

Synthesis, Characterisation and Applications

Springer

Assoc. Prof. Dr. Noorhana Yahya
Department of Fundamental and Applied Sciences
Universiti Teknologi PETRONAS
Bandar Seri Iskandar
31750 Tronoh, Perak
Malaysia
noorhana_yahya@petronas.com.my

ISSN 1869-8433
ISBN 978-3-642-14672-5 e-ISBN 978-3-642-14673-2
DOI 10.1007/978-3-642-14673-2
Springer Heidelberg Dordrecht London New York

Library of Congress Control Number: 2010937766

© Springer-Verlag Berlin Heidelberg 2010
This work is subject to copyright. All rights are reserved, whether the whole or part of the material is concerned, specifically the rights of translation, reprinting, reuse of illustrations, recitation, broadcasting, reproduction on microfilm or in any other way, and storage in data banks. Duplication of this publication or parts thereof is permitted only under the provisions of the German Copyright Law of September 9, 1965, in its current version, and permission for use must always be obtained from Springer. Violations are liable to prosecution under the German Copyright Law.
The use of general descriptive names, registered names, trademarks, etc. in this publication does not imply, even in the absence of a specific statement, that such names are exempt from the relevant protective laws and regulations and therefore free for general use.

Cover design: WMXDesign GmbH, Heidelberg, Germany

Printed on acid-free paper

Springer is part of Springer Science+Business Media (www.springer.com)

Preface

It is my privilege as the Editor-in-Chief to present to you an effort of our team of prominent contributors to this monograph on Carbon and Oxide Nanostructures. Over the past 20 years, carbon and oxide nanostructures evolved into one of the most studied objects and are presently entering in the transition phase from nanoscience to nanotechnology. Carbon and oxide nanostructures constitute an enormous topic which may only be described in a simplified manner, which in essence is the intent of this book. It is hoped that this book would provide valuable resources for researchers as well as postgraduate students of physics, chemistry and engineering. Related carbon-based materials such as fullerenes, carbon fiber, glassy carbon, carbon black, amorphous carbon, diamond, graphite, buckminsterfullerene, and carbon nanotubes (CNTs) are discussed. CNTs which have attracted the attention of the scientific community due to their fundamental and technical importance are elaborated. It also presents a review of the applications of fullerene and its derivatives as electron beam resists, as well as outlining the effects of catalyst on the morphology of the carbon nanotubes. Structural and optical properties of hydrogenated amorphous carbon (a-C:H) thin films prepared in a DC-plasma-enhanced chemical vapor deposition reactor is discussed in greater detail. Some of the works done on polymer-CNTs-based solar cells with a variety of device architecture and band diagram are summarized. Several irregular configurations of carbon nanofibers (CNF) such as coiled, regular helical, and twisted coil are elaborated. This book also includes the molecular modeling of carbon-based nanomaterials including discussions on some aspects of the issues related to the synthesis and characterization of diamond prepared via CVD techniques using the hot filaments and plasma. Oxide-based materials related to fuel synthesis and solar hydrogen production are also presented. The versatility of ZnO nanostructures and some of the novel applications such as solar cells and light-emitting devices are being highlighted. A brief introduction of Fe–FeO nanocomposites and some superparamagnetism studies in the form of particles and thin films are included. The benefits and drawbacks of the properties of some nanomaterials used in optical sensing applications are given, and the recently developed optical chemical sensors and probes based on photoluminescence are also rigorously overviewed. Aspects of nanocatalytic reactions, the types of catalyst, and also the preparation and characterization of the active catalyst for ammonia synthesis are scrutinized.

I am grateful to all authors who have contributed to the chapters of this book. All merits on overview of such an enormous topic as Carbon and Oxide Nanostructures in this concise monograph should be credited to all contributing authors, but any shortcomings to be attributed to the Editor-in-Chief. The book is dedicated with all sincerity to all whose work has not received due reference and recognition.

Universiti Teknologi PETRONAS Assoc. Prof. Dr. Noorhana Yahya
Malaysia

Contents

Carbon Nanotubes: The Minuscule Wizards

Noorhana Yahya and Krzysztof Koziol

Abstract Carbon Nanotubes (CNTs) have attracted the attention of scientific community due to their fundamental and technical importance. The structural diversities and the related diverse physical properties with large aspect ratio, small diameter and low density, are extremely fascinating. CNTs can behave as metallic conductors, semiconductors or insulators depending on their chirality, diameter and presence of defects. Their nano-scale dimension can be exploited as they have high accessible surface areas that make them not only exhibit high electronic conductivity but also useful mechanical properties. This chapter discusses on the production of CNTs, both single wall nanotubes and multiwall nanotubes giving emphasis on pulsed laser technique and microwave assisted chemical vapor deposition technique. The word wizard is coined due to their remarkable properties leading to their potential applications which are likely to stretch across different areas of industry.

1 Introduction

Carbon nanotube (CNT) is a graphitic sheet consisting of covalently bonded carbon atoms in hexagonal-type arrangement. The sheet is rolled up into a cylinder with the ends closed by hemispherical graphitic domes. Carbon nanotubes (CNTs) have extraordinary structural, electrical, and mechanical properties, which are derived from their unique 1-D nature [1]. This feature is of great interest to physicists as it permits the exploration and application of quantum effects. The mechanical

N. Yahya (✉)
Fundamental and Applied Sciences Department, Universiti Teknologi PETRONAS, Bandar Seri Iskandar, 31750 Tronoh, Perak, Malaysia
e-mail: noorhana_yahya@petronas.com.my

K. Koziol
Department of Materials Science and Metallurgy, University of Cambridge, Pembroke Street, Cambridge CB2 3QZ UK

N. Yahya (ed.), *Carbon and Oxide Nanostructures*, Adv Struct Mater 5,
DOI 10.1007/8611_2010_27, © Springer-Verlag Berlin Heidelberg 2010,
Published online: 31 July 2010

[2, 3], (stiffness, strength and toughness), thermal [4] (heat dissipation) and electrical [5] (conductor and semi-conductor) properties of CNTs enable enormous potential applications from batteries and fuel cells, fibers and cables to pharmaceuticals and bio-medical materials. Scores of other applications, which open possibilities of generating new 1D structures emerge when the hollow/cavity of CNTs were filled with other.

Carbon nanotubes can be divided into two basic classes; single wall nanotubes (SWNTs) and multiwalled nanotubes (MWNTs). SWNTs are formed from a single graphitic layers with typical diameter in the range of 0.4–2 nm [2, 3, 6, 7]. There is no restriction in length of the SWNTs and researchers are working towards achieving as high as possible aspect ratios [8] which are currently limited by the activity of the catalyst particles, used for the CNT growth and other synthesis conditions. MWNTs include structures formed in coaxial arrangement of several (2–100) graphitic cylinders and their external diameter ranges from 10 to 100 nm [9].

Carbon nanotubes can be divided into two basic classes; single wall nanotubes (SWNTs) and multiwalled nanotubes (MWNTs). SWNTs are formed from a single graphitic layers with typical diameter in the range of 0.4–2 nm [2, 3, 6, 7]. There is no restriction in length of the SWNTs and researchers are working towards achieving as high as possible aspect ratios [8] which are currently limited by the activity of the catalyst particles, used for the CNT growth and other synthesis conditions. MWNTs include structures formed in coaxial arrangement of several (2–100) graphitic cylinders and their external diameter ranges from 10 to 100 nm [9].

2 Synthesis of Carbon Nanotubes

Various structures of CNTs were formed by using different techniques, among others are arc discharge [10], laser ablation, gas-phase pyrolysis [11, 12], plasma enhanced [13, 14] or thermal enhanced chemical vapor deposition (CVD) [15, 16]. Regardless of the techniques, metal catalysts are generally required to assist the growth of the CNTs. Stringent control on the growth of CNTs is required and reasonable cost for large scale production still remains the challenge. Some aims at having large reactors while others are working towards miniaturization of equipments.

2.1 Laser Ablation Technique

Laser ablation is a very successful technique to produce high yield and high quality CNTs [17–20]. Some of the important parameters which governed the growth of CNTs need to be scrutinized. Web-like structures of CNTs were found by using this technique [19]. It was also reported that by using a simple vertical evaporation chamber without a furnace around the graphite/metal-composite (Co, Ni, Fe and Y) rods target and with a laser power 250 W and 400 Torr of argon gas flowing, web like soot containing high densities of bundles of Single Wall Carbon Nanotubes (SWCNTs) could be produced [19]. Also, aligned MWCNTs are highly desirable and the dependency of the diameter of catalyst used during the synthesis is very important [21].

Pulsed laser ablation deposition (PLAD) systems as well as continuous laser ablation systems and components with low start-up cost must also reach the market to ensure total cost of CNTs production can be reduced. In order to meet the demands of R&D using the laser system, there is high need to incorporate this technique into the main stream of solid-state-device technology. In short, the feasibility of large scaled-up methods needs to be demonstrated. In developing the PLAD systems for CNTs production, various constraints need to be addressed in getting the right conditions for higher quality and higher yield of the CNTs. The target-substrate distance, rotation of the target and substrate, catalyst, laser power, heating element, type of gas (inert environment), time of ablation, position of sample and substrate are amongst the conditions that must be closely studied and fully understood.

Yahya et al. [22] have designed and developed an inexpensive new chamber for Pulsed Laser Ablation Deposition (PLAD) system to synthesis carbon nanotubes (CNTs) (Fig. 1). CNTs were formed by ablating a graphite pellet mixed with catalysts using pulsed laser. Hot vapor plume (Fig. 2a) is formed and expands then cools rapidly during the ablation process. A T-shape stainless steel vacuum

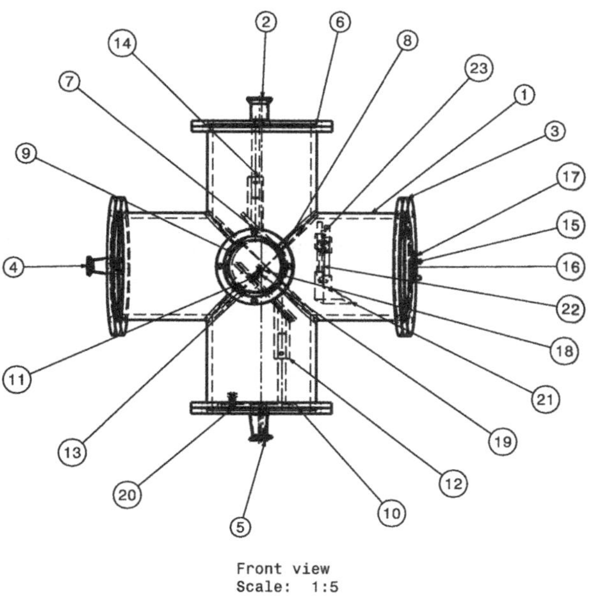

Bill of Material: Product1

Number	Part Number	Quantity
1	T-Chamber	1
2	Outlet Cap	1
3	Laser inlet Cap	1
4	Pressure Cap	1
5	Gas Inlet Cap	1
6	O-ring	4
7	Observation Cap	1
8	O-ring OB glass	1
9	Observation glass	1
10	Target Base	1
11	Target Connector	1
12	Target Holder	1
13	Motor	1
14	Substrat Holder	1
15	Quartz Window	1
16	Small O-ring	2
17	Quartz mount	1
18	Target Mounting	1
19	Small Screw	2
20	Mounting screw	2
21	Lense Holder	1
22	Concave lense	1
23	Lense Holder2	1

Fig. 1 A schematic diagram of the locally developed chamber for pulsed laser deposition (PLAD) system

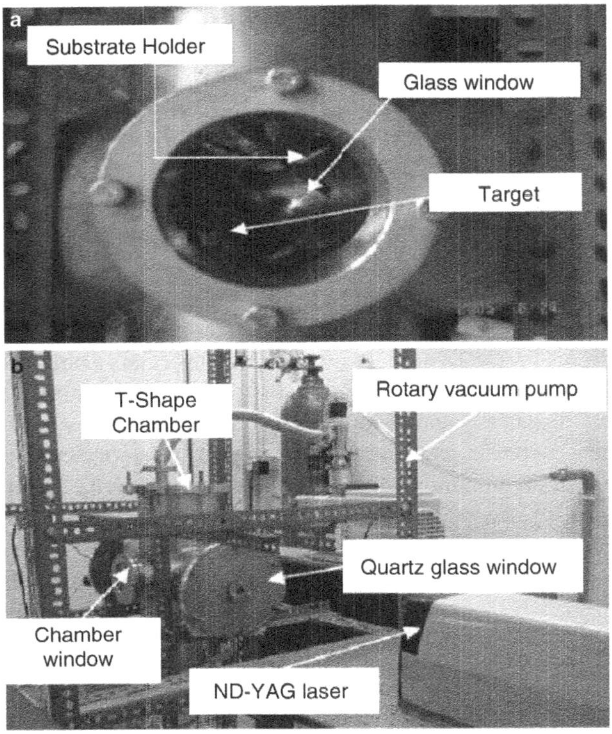

Fig. 2 (**a**) Formation of plume during the ablation process. (**b**) T-shape stainless steel chamber (designed in-house) ND: YAG laser and the rotary pump (adapted from [22])

chamber which has cylindrical shape, with diameter of about 15 and 45 cm length (Fig. 2b) was developed. An Nd: YAG laser (model SHG-LP-05) with a laser wavelength 532 nm was used as the evaporation source. An Edward RV Dual-Mode Vacuum pumps (model: RV5) was used to pump out the unwanted particle and to keep the chamber at vacuum condition. Maximum pumping speed of the vacuum pump was about 6.2 m^3/h and the pressure employed is about 2×10^{-6} bar. Argon gas was flowed through the chamber by using Concoa 65 mm flowmeter 565 series to ensure the inert environment. The vaporized small carbon molecules were condensed on a glass substrate to form CNTs. Quenching process can also be done whenever required.

Surface morphology of the CNTs collected from the PLAD process by using graphite-NiCo pellet was shown in Fig. 3. It can be seen that large quantities of web-like CNTs were formed on a glass substrate. The diameter of the web-like CNTs observed ranges from 35 to 100 nm. Bundles of CNTs can also be observed.

Transmission Electron Microscope (TEM) image reveals a bamboo-like CNTs with diameter of about 40 nm and a thick wall approximately 15 nm (Fig. 4a, b).

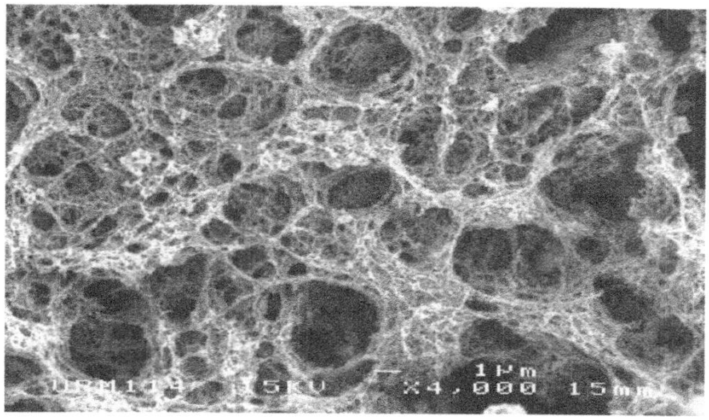

Fig. 3 Scanning electron microscope image of web-like CNTs collected from the PLAD method employing graphite-NiCo pellet

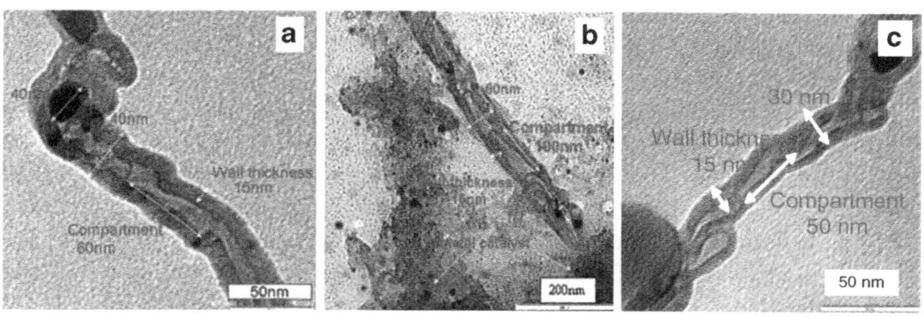

Fig. 4 (**a**) TEM image of 100 kX of bamboo-like CNTs and (**b**) TEM image of 80 kX magnification of bamboo-like CNTs with 10 weight% of NiCo as bi-catalyst in Argon gas. (**c**) TEM image of 100 kX bamboo like CNTs using 11 weight% of NiCo as bi-catalysts

The CNTs was formed using 10 and 11 weight percentage of NiCo bi-catalysts to initiate the catalytic activities. There are a few metal nanoparticles (catalysts) that can be observed in the TEM image. The metal particles ranging from 40 to 50 nm in diameter are seen as dark spots. The CNTs consist of hollow compartments and the distance between the adjacent compartments inside the tube is approximately 60 nm. The wide distribution of particle size of catalyst had probably caused the different diameter of CNTs. It should be noted that the CNTs that was formed using higher weight percentage (11%) of NiCo bi-catalyst does not give much significant effect on the length of the compartments and the diameter of the CNTs (Fig. 4c). Figure 5a–c give the image of CNTs when Fe_2O_3 was used as catalyst. It should be noted that the formation of bamboo-like structure can be also observed, however the inner diameter and the compartment of the structure are slightly less comparing to those prepared using NiCo catalysts. Further investigation on the effect of using Fe_2O_3 as the catalyst to the growth of the CNTs is currently been carried out.

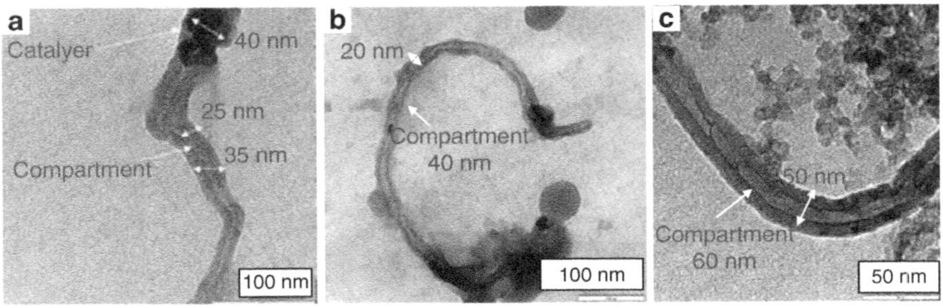

Fig. 5 (**a**) TEM image of 80 kX of bamboo-liked CNTs and (**b**) TEM image of 80 kX magnification of bamboo-like CNTs with 10 weight% of Fe_2O_3 catalyst in Argon gas. (**c**) TEM image of 100 kX bamboo like CNTs using 11 weight% of Fe_2O_3 catalysts

It is speculated that the bi-metal (NiCo) and the Fe_2O_3 catalysts and laser pulsed width (120 ns), had significantly resulted the bamboo-like structures. Carbon atoms that were produced from evaporation of target materials in this experiment a graphite powder by the pulsed laser diffused on the surface of the catalyst and formed the graphite sheets as a cap on the catalyst. The growth of compartment in the tube are attributed to the 120 ns pulsed width. The carbons and the catalysts had undergone temperature fall between the pulse-to-pulse laser ablation process and tended to grow towards horizontal rather than vertical. This continuous process had formed the hollow tube and had produced the compartments [18, 23].

A unique bamboo-like CNTs structure was formed by using in-house designed chamber for the PLAD system [22]. In short the bamboo-like structures of CNTs were formed by the PLAD system due to the precise conditions, such as laser power (10.24 W), pressure (4 Torr), catalyst (NiCo and Fe_2O_3), inert environment (argon gas) and the wave length (532 nm) of the laser and the pulse to pulse width time (about 120 ns) during the ablation process.

Klanwan et al. [24] reported high quality CNTs by using laser ablation method (Fig. 6). They employed Nd: YAG pulsed laser with 355 nm wavelength, 0.6 W and 10 ns pulse width at 10 Hz with C/Ni/Co rod target as the feed stock under 1.5 L/min nitrogen flow [24]. The experimental set up consists of electric field quartz tube furnace with outer diameter, inner diameter and length of 28, 25 and 700 mm, respectively with a rotating motor with 7 r.p.m. It should be noted that the feedstock/catalyst were heated at 1,000 and 1,080°C before ablation process was done.

It was found that web-like CNTs were produced using this system and the fibrous CNTs products have the average diameter of about 20 nm. Raman spectroscopy analysis was done to study the crystallinity of the single walled CNTs (SWCNTs). The Raman shift band in the range of 100–300 cm^{-1} known as the radial breathing mode (RBM) is the signature of SWCNTs [24]. At low Raman shift range, distinctive RBM signal at 220–240 cm^{-1}, was observed indicating the presence of SWCNTs when heated at 1,000 and 1,080°C. At higher Raman shift (1,300–1,600 cm^{-1})

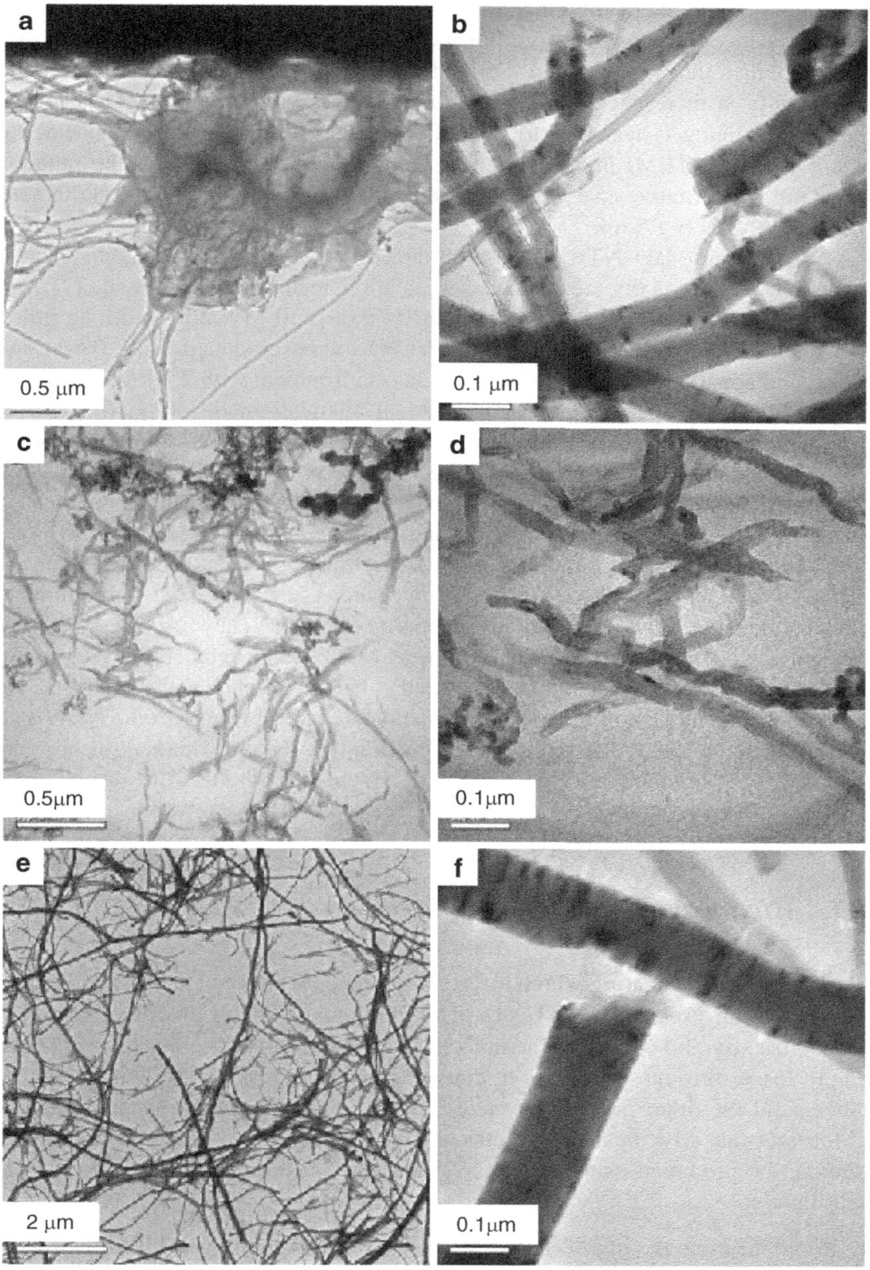

Fig. 6 TEM images of (**a**, **b**) MWNTs irradiated with microwave. (**c**, **d**) acid treated samples. (**e**, **f**) Thermally treated MWNTs (adapted from [25])

a strong G-band which strengthen the graphitic bonding and D-band which shows the presence of the defective carbonaceous constituents were observed. It was observed that this method had resulted in high G/D ratio indicating the high crystallinity with respect to the amorphous carbon components particularly for the samples that was heated up to 1,080°C. High Resolution Transmission Electron Microscopy (HRTEM) images revealed the CNTs with diameter in the range of 1–1.2 nm. Quantitative analysis was done and it was found that the mean diameter of the SWCNTs is 1.2 nm.

SWCNTs and MWCNTs were also produced by Zhu et al. [7] and Sabbaghzadeh et al. [26], respectively by using laser ablation technique. Zhu et al. [7] used Nd:YAG laser which has high power density 0.05–530 MW/cm^2 and long pulses 3–5 ms at 15 pulses per minute. The SWCNTs were produced using 0.6 at% of nickel and cobalt metal catalysts in argon gas environment with 2.2 L/min flow rate to give the inert environment. A cylindrical chamber made of quartz with 15 and 80 mm inner diameter and length, respectively was used for this process. Sabbaghzadeh et al. used copper vapor laser (CVL) with extremely high frequency, 10 kHz, in pulsed mode [26]. They found MWCNTs was produced when with short pulses with less than 10 ns duration and the wavelength used was 510.5 and 578.2 nm. High purity graphite target was used as the feedstock and cobalt and metals were used as the catalysts. Kusuba and Tsunawaki [27] had produced SWCNTs using XeCl excimer laser ablation method. They used a laser with 308 nm wavelength and pulse width of 16 ns which was irradiated onto a graphite target as the feedstock. The target contained cobalt and nickel as the catalysts. The Raman spectra at low frequency region indicated the presence of SWCNTs at 180 cm^{-1} due to the RBM and it was calculated that the tube diameter is approximately 1.3 nm in diameter. This is in good agreement with the TEM results.

2.2 Microwave Irradiation Method

Microwave (MW) heating which differs from the conventional heating had gained interest in the production of CNTs. This method heats the precursor materials volumetrically and causes the sample surface temperature slightly lower. This is due to losses through evaporation, convection, conduction and radiation. On the other hand the other part of the materials will have good heat dissipation. Due to the fact that MW heats volumetrically, materials with a uniform microstructure can be produced using microwave heating. Some other advantages of MW heating are:

1. Rapid heating is simply achieved due to direct coupling of the microwave energy to the materials
2. Energy is accumulated in the materials
3. Fast and clean
4. No direct contact between the energy source and the material [28]

Microwave irradiation method to functionalize CNTs has been reported. Also, numerous research works reported the application of microwave plasma enhanced chemical vapor deposition (MW-PECVD) which fulfills the need to synthesize CNTs at low temperature [29–32].

Talemi et al. [25] reported carbon nanotubes could be treated using deionised water as a reactive reagent. They used 10 mg of CNTs which was dispersed in 100 mL of deionized water. They employed 2.45 GHz, 900 W for 10 min at 100% power. This is a very promising way for low temperature synthesis of carbon nanostructures. They observed cloth-like amorphous carbon with some impurities when MWNTs were irradiated with microwave. Some of the CNTs were treated with acid and had resulted to shorter length but sharper tips. Some of the CNTs were thermally treated and had resulted in less amorphous carbon between the nanotubes while the catalysts still remain in the tubes (Fig. 6a–f).

Yoon et al. reported a method to transform solid carbon to CNTs using direct microwave irradiation of catalyst particles on the surface of solid carbon [33]. In this work cloth form of activated carbon fiber (ACF) as feedstock and $FeCl_3$ as catalyst were impregnated on the specimen (Fig. 7). Samples were directly irradiated with microwave of power up to 2,000 W and 2.45 GHz frequency in a quartz tube reactor under flowing argon gas. Fu et al. reported on microwave-CVD technique to synthesize CNTs [34]. A Y-junction CNTs were observed due to the microwave field and the methane gas flow fluctuation during the synthesis process [34]. It should be noted that the CNTs were produced without the presence of catalysts. Aligned MWCNTs can also be synthesized using microwave assisted

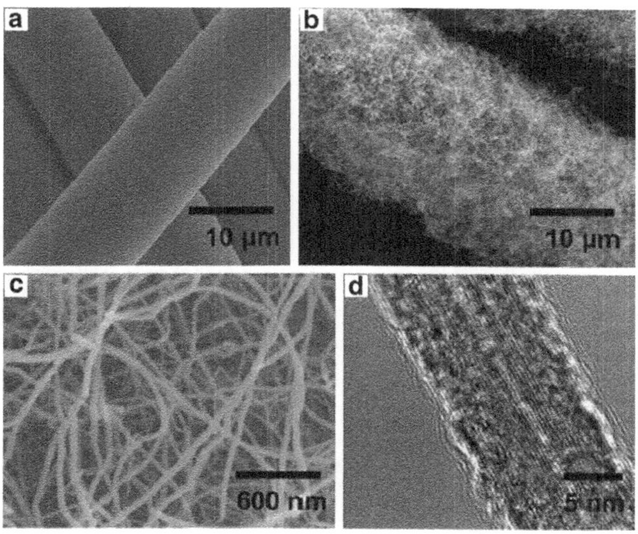

Fig. 7 Morphology of the ACF specimen (**a**) before microwave irradiation, (**b**) after microwave irradiation, (**c**) magnified image of (b), (**d**) microstructure of a fibrous carbon in (c) observed by HRTEM (adapted from [33])

CVD technique. Layers of aligned bamboo-like MWCNTs were observed using Co layers substrate [21].

Mendez et al. [35] also reported that when they heated graphite powder in a quartz capsule via microwave of 800 W power, 2.45 GHz of frequency and vacuum atmosphere (10^{-5} Torr), MWCNTs can be observed even without the presence of catalysts. They used silica target and the optimal process time was 60 min. The TEM image revealed MWCNTs with bamboo-like structures when the graphite feedstock was heated with boric acid.

Srivastava et al. had also prepared CNTs using microwave assisted PECVD by controlling the growth time and the power using C_2H_2 and ammonia gas composition and Fe as catalyst [2]. They were able to produce regular conical compartments with high crystallinity and with many open edges at the outer surface of the tubes (Fig. 8).

Generally, for the microwave assisted CVD, judicious control of process parameter namely, wavelength, power, time, substrate, catalyst and feedstock will result to different morphology and structures on the CNTs.

2.2.1 Vertically Aligned CNTs

Vertically aligned and high quality CNTs are highly sought [36] as they have potential applications in the microelectronic industries [37]. This part discusses on the production of vertically aligned CNTs produced by microwave assisted technique. Vertically-aligned MWCNTs have been synthesized using microwave plasma CVD technique [37]. The well vertically aligned CNTs were produced a

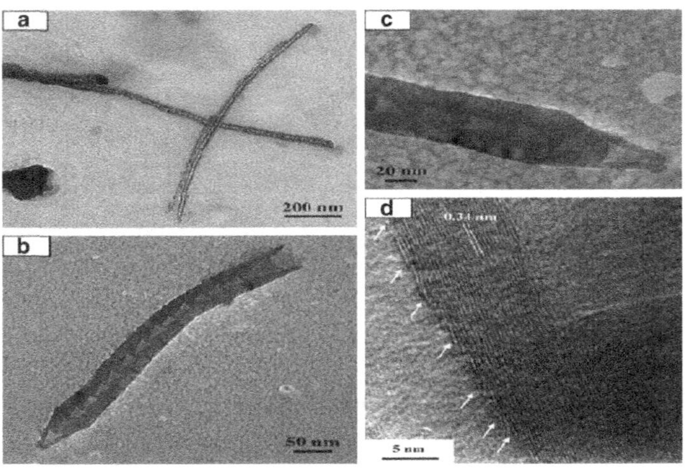

Fig. 8 TEM micrographs of short conical CNTs (**a**) Low magnification images, (**b**) magnified view of the shortest conical CNTs, (**c**) highly magnified view of tip and (**d**) typical HRTEM image of a conical CNT (adapted from [2])

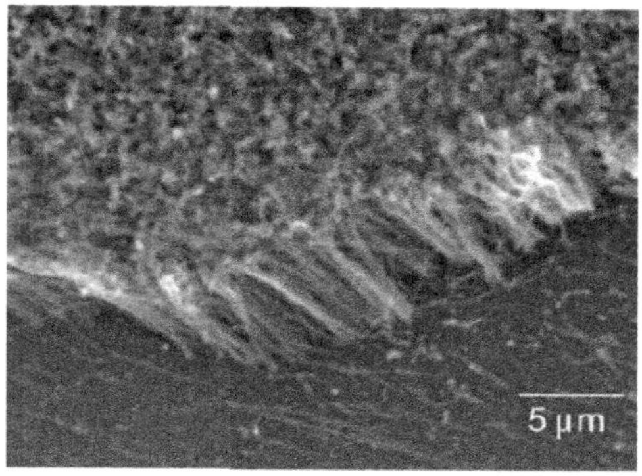

Fig. 9 SEM micrograph of vertically aligned CNTs grown on thermally oxidized silicon (001) substrate (adapted from [38])

low temperature (330°C) using gas mixtures of H_2–CH_4, H_2–C_2H_2 and H_2–C_6–H_6. A few experimental conditions were studied, namely, the microwave power, pressure, substrate temperature, CH_4 flow rate, CO_2 flow rate, DC bias, and deposition time leading to the quality of the CNTs. It should also be noted that Fe particles were deposited on an n-type silicon wafer substrate. In short, high quality and vertically aligned CNTs were produce using CO_2–CH_4 gas mixture, with the flow rate of 29.5 sccm/30 sccm, microwave power of 300 W, gas pressure of 15 Torr, DC bias of −150 V, and substrate temperature of 330°C.

Turq et al. [38] had successfully synthesis vertically aligned CNTs (Fig. 9) using microwave plasma enhanced CVD method. They used 2.45 GHz and 500 W for the ignition which was applied to silicon based substrate [38]. Prior to this process, thin iron catalyst was deposited on the silicon substrate. The feed gas is methane (CH_4)/H_2 with flow rate of 100 and 10 sccm respectively, at a total pressure of 2.7 kPa.

3 Carbon Nanotubes (CNTs) Properties

The superior properties of CNTs had made us coined the word *wizard* which may potentially have wide industrial applications and consumer products in the near future. This is due to the fact that CNTs exhibit excellent properties that are very attractive for many technology applications. CNTs have Young modulus ~10 times of steel and electrical conductivity up to 1,000 times that of copper [39]. It is thus far the strongest (in terms of tensile strength) and stiffest (elastic modulus) material. MWCNTs were tested to have tensile strength of 63 GPa [40]. The electronic properties of MWCNTs are quite similar to those of the SWCNTs. Their electronic transport in metallic CNTs occurs ballistically and this enables high currents carrying capacity with no heating [41]. CNTs are good thermal conductor element [42]. Single MWCNTs of a certain diameter and aligned SWCNTs exhibit thermal

Table 1 Single walled and multi walled CNTs properties

Properties	Values
Mechanical properties	
Young's modulus of multi walled CNTS [39]	~1–1.2 TPa
Young's modulus of multi walled CNTS [4]	~0.45 TPa
Young's modulus of multi walled CNTS ropes [39]	~1 TPa
Tensile strength of single walled CNTs ropes [39]	~60 GPa
Tensile strength of single walled CNTs [3]	Mean 30 GPa
Tensile strength of multi walled CNTs [4]	3.6 GPa
Thermal properties at room temperature	
Thermal conductivity of single walled CNTs [39]	1,750–5,800 W/mK
Thermal conductivity of single walled CNTs [3]	3,000 W/mK
Thermal conductivity of single walled CNTs [36]	200 W/mK
Electrical properties	
Typical resistivity of single and multi walled CNTs [39]	10^{-6} Ωm
Typical maximum current density of CNTs [39]	10^7–10^8 A cm^2
Quantized conductance (theoretical and measured) of CNTs [39]	6.5 kΩ^{-1} and 12.9 kΩ^{-1}
Electronic properties	
Single walled CNTs band gap [39]	
Whose n-m is divisible by 3 [39]	0 eV (metallic)
Whose n-m is non divisible by 3 [39]	0.4–0.7 eV (semiconducting)
Multi walled CNTs [39]	~0 eV (non-semiconducting

conductivity of 0–30 W/mK (4–300 K) and 200 W/mK (room temperature), respectively [36]. Pop et al. reported that the thermal conductivity of SWCNTs is about 3,500 W/mK measured at room temperature [43]. The thermal conductivity and the specific heat of CNTs are determined by phonon. At low temperature, the phonon contribution is primarily determined by the acoustic phonons [3]. Low thermal conductivity indicates the presence of impurities [3, 4]. Properties of CNTs are presented in Table 1

4 Potential Applications of Carbon Nanotubes (CNTs)

Carbon nanotubes are endowed as materials for the future particularly in the emerging technology. Due to the remarkable properties mentioned above, CNTs are largely perceived as the key research areas that will change the technology architecture. Current and long term applications are sought. However large scale synthesis of high quality and high crystallinity SWCNTs and MWCNTs is still the major drawback. The knowledge transfer from the academia to the industry is rather slow. Researchers are working towards the following areas to ensure their applications can be realized in the near future [44]:

1. High purity and defect free large scale CNTs must be produced
2. Establishment of useful techniques to quantify the defect structures

3. Development of effective purification techniques
4. Achieving homogeneous CNTs dispersion in polymer composite

Kohler et al. reported on the expected application areas of CNTs [45]. CNTs have been used as fillers for polymer composite to enhance the electrical, strength, elasticity, toughness, durability and conductivity of the material [45]. CNTs were used as microelectrodes in polyvinyldine fluoride (PDVF) composite materials [46]. They can also potentially be used as electromagnetic interference (EMI) shielding [47, 48], synthetic muscle [49], superconductors [50], supercapacitors [51], hydrogen storage [52], fuel cell [53], fire retardant [54], field emitter [55, 56] etc.

4.1 Field Emission Devices and Field Effect Transistors

Flat panel displays, nanotube lamps having long lifespan (>8,000 h) [41] emitters for microscopy [39], and field effect transistor (FET) [57–59] are among the lucrative applications of CNTs. CNTs surfaces are excellent field emitters due to (1) long lifetimes, (2) high current densities (3) stable emission [41].

Yu et al. [59] had investigated on field emission of CNTs using hydrogen-ion implantation on PE-CVD technique (Fig. 10). Vertically aligned CNTs were grown using this method. The electric-current density intensity was higher due to this treatment. The turn on-field of CNTs films were decreased due to the H-ion implantation However, the emission site density and the brightness from the H-implanted CNTs sample exceeded those of the unimplanted CNTs samples. They found that

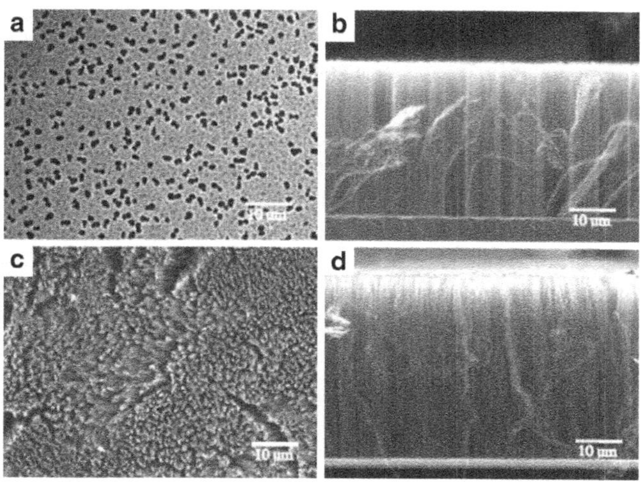

Fig. 10 SEM micrograph of vertically aligned CNTs grown on silicon (001) substrate with hydrogen-ion implantation treatment (**a**) Plan view (**b**) side view and without hydrogen-ion implantation (**c**) Plan view (**d**) side view (adapted from [59])

the treatment had resulted in increase of holes that had affectively increase the electron-emission intensity and the emission-current density of the CNTs.

CNTs as tips have several advantages over tungsten and molybdenum tip arrays as they can act as cold cathode source, which runs at 300–400°C [39] for microscopy tip. CNTs are added to the standard tungsten emitter by carbon glue [39] or electrophoresis [56]. It should be noted that microscopy demands bright, stable, low noise electron source with low kinetic energy which would maximize the resolution as well as the contrast [39]. The use of CNTs can potentially increase the high brightness and low energy spread requirements for transmission electron microscope (TEM) [39]. CNTs can also be potentially used as the X-Ray source due to their compact geometry. This allows improved quality images for biological samples and probably endoscopes for medical exploration [41].

CNTs give the right combinations of properties for field emission devices which are (1) nanometer size diameter, (2) structural integrity (3) high electrical conductivity (4) and good chemical stability [60]. It was reported that potential applied between a nanotube coated surface and an anode will create high electric field due to the diameter of the tip [60]. Figure 11 shows the CNTs based field emission display fabricated by Samsung. The electron emission from the CNTs extremely narrow tips has high density of state leading to much higher resolution. It should be noted that the major constraint is to have aligned CNTs which is pre-requisite for consistent and good field emission [61–64].

Sohn and Lee [64] had also reported on fabrication of CNTs as field emitter arrays (Fig. 12). They fabricated micropatterned vertically aligned CNTs which was grown on planar silicon surface using CVD, photolithography, pulsed laser deposition, reactive ion etching and lift off method. The well aligned vertically CNTs could be used as field emitters for cold cathodes and it was speculated that this method could revolutionize field-emitting electronic devices. The silicon based industry will come to its technology limit. CNT has been considered as one of the major constituent in the future microelectronic industry [59]. CNTs based electronic can show quantum effects at low temperature. Recently, CNT based FET which are generally p-type [58, 59] is expected to come in a diverse way. The common SWNT field effect transistors (Fig. 12) fabricated to date has Schottky barrier at the nanotube metal junction. The CNT is said to be OPEN Quantum Dot (QD) when an unpaired electron occupies the CNT state. The CNT connects the source (S) and drain (D) electrodes (Fig. 13) [57]. It was reported that electron transport is only feasible when Fermi energy, E_F is in resonance with the CNT levels, otherwise the current is blocked [59].

4.2 Catalyst Support

Ammonia production is capital-intensive industry as it requires high temperature (400–500°C) and also high pressure (150–300 bar) for its daily operations. Parameters such as catalysts and their support are important aspect which will determine

Fig. 11 A prototype 4.5_field emission display fabricated by Samsung using CNTs) (Adapted from [60])

the yield of the ammonia. Serp et al. reported potassium promoted ruthenium catalyst supported on MWNTs for ammonia production which was found to be much more active than their counterparts, deposited on graphite [65]. This is attributed to the much higher surface area of the nanotubes which enable better dispersion of the metallic catalyst. It was also reported that the electronic properties of the CNTs could enhance the electron transfer from potassium to ruthenium thereby increasing the ammonia yield [65]. Chen et al. reported alkali-promoted ruthenium, supported on MWNT for the production of ammonia [66]. They found that the MWNTs support comparing to other carbon-based support were able to produce much higher ammonia yield at atmospheric pressure.

Yahya et al. had designed and developed a microreactor for ammonia synthesis [67]. Iron particles were used as the catalyst for the ammonia production. They reported utilization of MWNTs as support for the ammonia synthesis in electromagnetic field in room temperature and ambient condition (Fig. 14). More work is

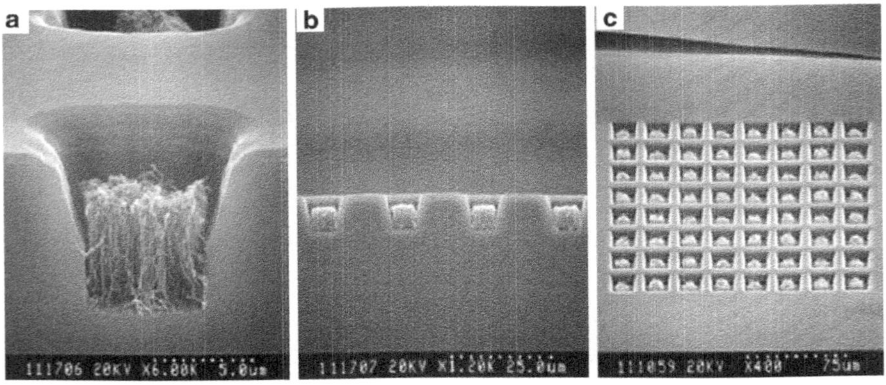

Fig. 12 SEM micrograph of vertically aligned CNTs arrays grown on planar silicon surface (**a**) regular arrays of trunches 10 μm deep with CNTs (**b**) side view (**c**) top view (adapted from [64])

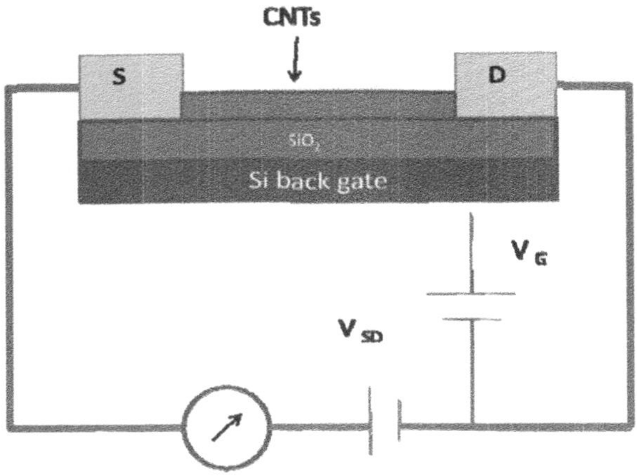

Fig. 13 Schematic diagram of single nanotube channel field effect transistor (NTFET) with CNTs conducting channels between (S) and drains (D) (adapted from [57])

currently being carried out which includes dispersion of iron nanoparticles on the MWNTs. Research activity focusing on development of Fischer–Tropsch catalysis is expanding and now covering support materials. Gusci et al. [68] investigated CO hydrogenation over Fe or Co catalysts supported on MWNTs (Simple impregnation method was studied and denoted as "I" was characterized using TEM for the morphology and Temperature Program Reduction (TPR) for the reduction process. They found the "I"-Fe supported on MWNTs gives highest higher catalytic activity and higher selectivity toward C_2–C_4 and C_{5+} fractions as compared to the "I"-Co [68].

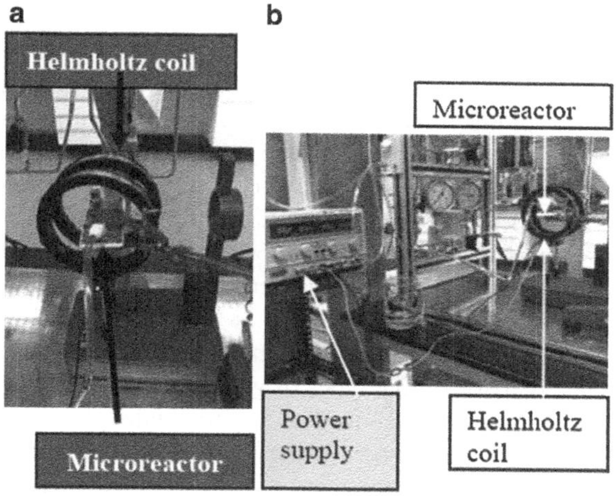

Fig. 14 (**a**) Helmholtz and microreactor connected (**b**) Helmholtz coil and microreactor with power supply (adapted from [67])

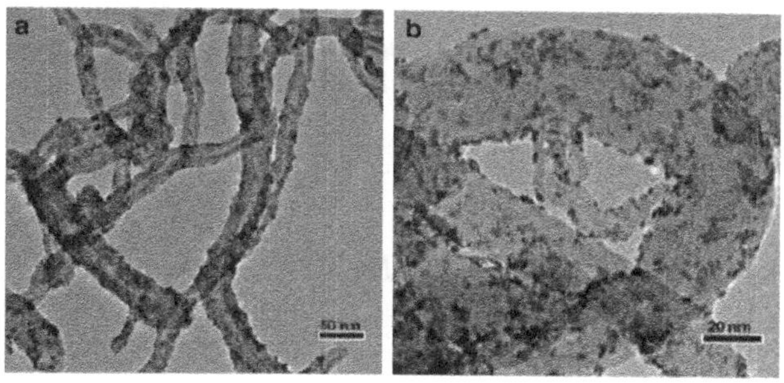

Fig. 15 TEM images of Pt/ CNT composites: (**a**) Low magnification (**b**) High magnification (adapted from [69])

Pt/MWNTs catalyst is to exhibit catalytic activity for methanol oxidation. Wang et al. [69] had developed a novel chemical method base d on ultrasonic assisted polyol synthesis to disperse Pt nanoparticles on MWNTs (Fig. 15). It should be noted that despite the nanoparticles dimension of the Pt, agglomeration was rare. The ultrasonic technique had effectively dispersed the Pt nanoparticles uniformly on the MWNTs.

4.3 Sensors

Chemical gas sensors usually utilize electrical or optical response by adsorption of gas molecules on an active surface layer [69]. ZnO, TiO_2 and SnO_2 metal oxide gas sensors are active layers where gas adsorption would lead to large electrical change [70]. Kim reported CNT array as electron emitters for the purpose of detecting inert gas sensor. The CNT array was prepared using thermal CVD technique. Electron is emitted from the tips of the vertically align CNTs array under high applied voltage. They are then accelerated towards the anode by the electric field and collide with the inert gas, in this experiment, Argon, which led to anodization of the gas. The I–V characteristics depend on the amount of electron and positive ions generated through the collision as well as the drift velocity of the electrons. Performance of the CNT' array as electron emitters for gas sensor applications were conducted and reported [70].

CNT Film Cathode (CNTFC) with different structures was studied [71] by using discharge I–V characteristics. Five types of gas, namely air, C_2H_2, Ar, H_2 and N_2 was used in a mixing system in a controlled flux to obtain the correct concentration of the gasses. The CNTs were grown using CVD technique and was grown on a silicone substrate. The CNTFC was found to be a good chemical sensor. It should also be noted that CNTs have been studied for other types of sensors, namely, EM sensors, capacitive sensors [57] and biological sensors [72] due to their remarkable properties. It should be recalled that MWNTs has good conducting properties. This in hand favor interface enzymatic hydrolysis reactions. Cai and Ju had developed a convenient and sensitive three electrode system as a portable sensor for fast determination of carbaryl pesticide [73]. In addition, CNTs were chosen due to its biocompatibility and lack of toxicity.

4.4 Dye Sensitized Solar Cells (DCS)

Solar technologies can be characterized into active solar or passive solar depending on the way they capture, distribute or convert sunlight into the other forms of energy, particularly electric energy. Solar technology has been dominated by solid-state devices usually made of silicon or germanium. To date, Dye Sensitized Solar Cell (DSC) is one of the solar families which have recently emerged as a promising approach to efficient energy conversion yet with low production cost. DSC also has major advantage over other solar cells because it can work when small amount of light falls on the cells. The light absorption by dye monolayer in DSC is low which limits the photocurrent efficiency with respect to incident light below 1% [74]. In DSC, the dye absorbs incident photons and uses this energy to make electrochemical charge carriers (e.g. electrons and holes) and this has resulted efficiencies up to 11% [75].

In DSC light is absorbed by a sensitizer which resides on the surface of a semiconductor (namely, TiO_2 or ZnO) that has large band gap. The semiconductor is used for charge separation. However, the charge separation is not provided by the semiconductor, but works in concert with a third element of the cell, an electrolyte which is in contact with both the semiconductor and the dye (Fig. 16). Recently, multiwalled CNTs nanocomposite were incorporated onto the TiO_2

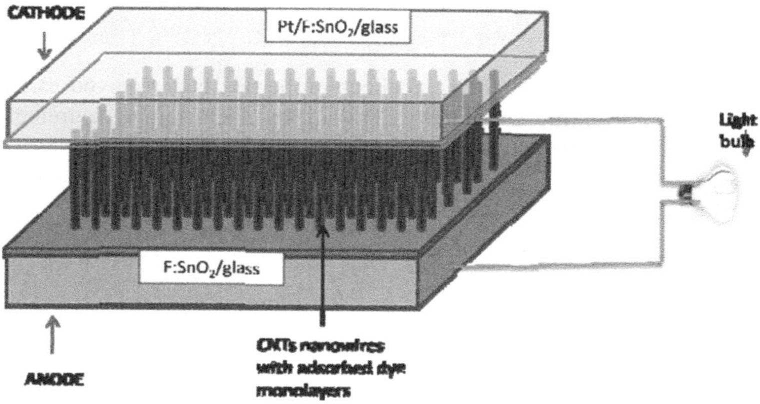

Fig. 16 Schematic diagram of dye sensitized solar cells (DSC) (adapted from [76])

based electrode to improve the roughness factor [75] and the recombination factor [77].

CNTs were also used as the counter electrode for DSC [77] for higher efficiency and better stability. The CNTs was added to the electrolyte and the counter electrode to increase the energy conversion efficiency of DCS [78]. CNTs which have high surface area and high electron conductivity were able to increase the efficiency up to 10% [77].

Single Walled Carbon Nanotube is speculated to be having suitable properties to increase the efficiency of DSC which is as per stated below [78]:

1. CNT is able to improve stability
2. CNT provides large surface area hence provide exciton diffusion length
3. CNT has suitable exciton binging energy
4. CNT has low energy gaps

4.4.1 Light Scattering Phenomena Effect

In DSC the dynamic competition between the generation and recombination of the photo-excited carriers is the major drawback that restricts the development of higher efficiencies. The thickness of the films has to be larger than the light absorption length to capture more photons. However, the thickness is constrained to be smaller than the electron-diffusion length. This is to avoid the recombination. A series of approaches was done to address the generation of photo-excited carriers by combining nanostructured films with optical effects. By adding a range of particle sizes of the sensitizer the overall energy conversion can be improved [76]. Hence CNTs can act as the light scattering element for light to electrical energy conversion scheme.

5 Conclusion

The versatility of CNT and their excellent properties had received exceptional attention by the scientific community. This makes them having extremely high commercial expectations and vast business opportunities.

References

1. Bernholc, J., Roland, C., Yakobson, B.I.: Modelling and simulation of solids. 706–715 (1997)
2. Srivastava SK, Vankar VD, Kumar V (2008) Nanoscale Res Lett 3:25–30
3. Popov VN (2004) Mater Sci Eng R 43:61–102
4. Xie S, Li W, Pan Z, Chang B, Sun L (2000) J Phys Chem Solids 61:1153–1158
5. Qiao, Y.J., Cao, M., Zhang, L.: In: Proceedings of the 1st IEEE International Conference on Nano/Micro Engineered and Molecular Systems, pp. 1331–1334 (2006)
6. Tian R, Wang X, Xu Y, Li S, Wan L, Li M, Cheng J (2009) J Nanopart Res 11:1201–1208
7. Zhu, X., Gupta, M.G., Lu, G.Q.: In: Proceedings of the 1st IEEE International Conference on Nano/Micro Engineered and Molecular Systems, pp. 450–454 (2006)
8. Saito Y, Uemura S (2000) Carbon 38:169–182
9. Ugarte D, Stokli T, Bornard JM, Chatelain A, de Heer WA (1998) Appl Phys A 67:101–105
10. Bethune DS, Kiang CH, de Vries MS, Gorman G, Savoy R, Vázquez J, Beyers R (1993) Nature 363:605
11. Thess A, Lee R, Nicolaev P, Dai H, Petit P, Robert J, Xu C, Lee YH, Kim SG, Rinzler AG, Colbert DT, Scuseria GE, Tomanek D, Fisher JE, Smalley RE (1996) Science 273:483
12. Terrones M, Grobert N, Zhang JP, Terrones H, Olivares J, Hsu WK, Hare JP, Cheetham AK, Kroto HW, Walton DRM (1998) Chem Phys Lett 285:299
13. Nath M, Satishkumar BC, Govindaraj A, Vinod CP, Rao CNR (2000) Chem Phys Lett 322:33
14. Cheol, J.L., Jeunghee, P.: School of Electrical Engineering, Kunsan National University, Kunsan, pp. 573–701 (2000)
15. Byszewsri P, Klusek Z (2001) Optoelectron Rev 9:203–210
16. Ma X, Wang EG (2001) Appl Phys Lett 78:978
17. Qin LC, Iijima S (1997) Chem Phys Lett 269:65–71
18. Scott CD, Arepalli S, Nikolaev P, Smalley RE (2001) J Appl Phys A 72:573–580
19. Maser MK, Munoz E, Benito AM, Martinez MT, De La Fuente GF, Maniette Y, Anglaret E, Sauvajol JL (1998) Chem Phys Lett 292:587–593
20. Park YS, Kim KS, Jeong JH, Kim WS, Moon JM, An KH, Bae DJ, Lee YS, Park GS, Lee YH (2002) Synth Met 126:245–251
21. Bartsch K, Leonhardt A (2004) Carbon 42:1731–1736
22. Yahya N, Beh HG, Mansor H (2005) Am J Appl Sci 2(11):1509–1515
23. Munoz E, Maser WK, Benito AM, Martinez MT, De La Fuente GF, Righi A, Sauvajol JL, Anglaret E, Maniette Y (2000) J Appl Phys A 70:145
24. Klanwan, J., Seto, T., Furukawa, T., Otani, Y., Charinpanitkul, T., Kohno, M., Hirasawa, M.: J. Nanopart. Res. **123** (2010)
25. Talemi PJ, Cervini R, Simon GP (2001) J Nanopart Res 12:393–403
26. Sabbaghzadeh J, Jafarkhani P, Dadras S, Torkamany MJ (2009) Appl Phys A 94:293–297
27. Kusaba M, Tsunawaki Y (2006) Thin Solid Films 506–507:255–258
28. Kappe CO (2008) Angew Chem Int Ed 43:6250–6284
29. Wang, X., Hu, Z., Chen, X., Chen, Y.: Microwave plasma-enhanced. **44**, 1567–1570 (2000)
30. Choi YC, Shin YM, Lee YH, Lee BS, Park GS, Choi WB, Lee NS, Kim JM (2000) Appl Phys Lett 76:2367

31. Bower C, Zhou O, Zhu W, Werder DJ, Jin S (2000) Appl Phys Lett 77:2767
32. Qin LC, Zhou D, Krauss AR, Gruen DM (1998) Appl Phys Lett 72:3437
33. Yoon DM, Yoon BJ, Lee KH, Kim HS, Park CG (2006) Carbon 44:1298–1352
34. Fu D, Zeng X, Zou J, Qian H, Li X, Xiong X (2009) Mater Chem Phys 118:501–505
35. Mendez UO, Kharissova KV, Rodriguez M (2006) Rev Adv Mater Sci 5:398–402
36. Gong Q, Li Z, Li D, Bai X, Liang J (2004) Solid State Commun 131:399–404
37. Chen M, Chen CM, Chen CF (2002) J Mater Sci 37:3561–3567
38. Turq V, Ohmae N, Martin JM, Fontaine J, Kinoshita H, Loubet JL (2005) Tribol Lett 19 (1):23–28
39. Mann M, Zhang Y, Teo KBK, Wells T, El Gomati MM, Milne WI (2010) Microelectron Eng 87:1491–1493
40. Yu MF, Lourie O, Dyer MJ, Moloni K, Kelly TF, Ruoff RS (2000) Science 287(5453): 637–640
41. Baughman RH, Zakhidov AA, de Heer WA (2002) Science 297:787–792
42. Berber S, Kwon YK, Tomanek D (2000) Phys Rev Lett 84(20):4613–4616
43. Pop E, Mann D, Wang Q, Goodson K, Dai H (2006) Nano Lett 6(1):96–100
44. Endo M, Hayashi T, Yoong AK, Terrones M, Dresselhaus MS (2004) Philos Trans R Soc Lond A 362:2223–2238
45. Kohler AR, Som C, Helland A, Gottschalk F (2008) J Cleaner Prod 16:927–937
46. Yahya N, Chen GZ, Andrew G (2009) J Adv Manuf Technol 3(1):109–122
47. Karpovichl, V., Rodionova, V., Rakov, A., Anoshkin, I., Philipenok, O.: In: MSMW'07 Symposium Proceedings, vol. 25, pp. 825–826 (2007)
48. Huang Y, Li N, Ma Y, Du F, Li F, He X, Lin X, Gao H, Chen Y (2007) Carbon 45:1614–1621
49. Maden JDW (2009) Science 323(5921):1571
50. Tang, Z.K., Zhang, L., Wang, N., Zhang, X.X., Wen, G.H., Li, G.D., Wang, J.N., Chan, C.T., Sheng, P.: **292** (5526), 2462–2465 (2001)
51. Chen GZ, Shaffer MSP, Dixon DCG, Zhou W, Fray DJ, Windle AH (2002) Adv Mater 12(7):522–526
52. Zidan, R.: In: Proceedings of the 2002 U.S. DOE Hydrogen Program Review NREL/CP-610-32405, pp. 1–3 (2002)
53. Chu H, Wei L, Cui R, Wang J, Li Y (2010) Coord Chem Rev 254:1117–1134
54. Zhou Z, Gou J (2009) Sci Technol Adv Mater 10:015001
55. de Heer WA, Bonard J-M, Stöckli T, Châtelain A, Forró L, Ugarte D (1997) Physica D 40:418–420
56. Collini, C., Guamieri, V., Marchion, R., Gottardi, G., Morganti, E., Lorenzelli, L.: In: 9th IEEE Conference on Nanotechnology, pp. 575–578 (2009)
57. Gruner G (2006) Anal Bioanal Chem 384:322–335
58. Endo, M., Strano, S., Ajayan, P.M.: Springer Berlin, **(111)**, 13-61 (2008)
59. Yu WDM, Zhang JH, Wang X, Li WM, Gao XD (2005) Appl Phys A 81:169–172
60. Ajayan PM, Zhou OZ (2001) Carbon Nanotubes: Synthesis, Structure, Properties and Application (Topics in Applied Physics) 80:391–425
61. Wang YY, Gupta S, Garguilo JM, Liu ZJ, Qin LC, Nemanicha RJ (2005) Diam Relat Mater 14:714–718
62. Hayashi N, Honda S, Tsuji K, Ikuno T, Fujimoto K, Ohkura S, Katayama M, Oura K, Hirao T (2003) Appl Surf Sci 212–213:393–396
63. Hyung SU, Lee SM, Jeona PG, Byung HK, Sang SP, Sang JK, Choc ES, Sung WK, Lee JD, Lee CG (2004) Thin Solid Films 462–463:19–23
64. Sohn JI, Lee S (2009) Appl Phys A 74:287–290
65. Serp P, Corrias M, Kalek P (2003) Appl Catal A 253:337–358
66. Chen HB, Lin JD, Cai Y, Wang XY, Yi J, Wang J, Wei G, Lin YZ, Liao DW (2001) Appl Surf Sci 180:328–335
67. Yahya N, Puspitasari P, Koziol K, Zabidi NAM, Othman MF (2009) Int J Basic Appl Sci 10:95–100

68. Guczi L, Stefler G, Gesti O, Kopanny Z, Konya Z, Molnar E, Urban M, Kiricsi I (2006) J Catal 244:24–32
69. Wang ZC, Zhou DD, Zhou GY, Li HL (2009) J Solid State Electrochem 13:371–376
70. Kim S (2006) Sensor 6:503–513
71. Yong Z, Junhua L, Xin L, Changchun Z (2006) Sens Actuators A 128:278–289
72. Pekarek, P., Ficek, R., Vrba, R., Magat, M.: In: 15th International Symposium for Design and Technology of Electronics Packages, pp. 323–326 (2009)
73. Cai J, Du D (2008) J Appl Electrochem 38:1217–1222
74. Grätzel M (2001) Nature 414:338
75. Bach U, Lupo D, Comte P, Moser JE, Weissortel F, Salbeck J, Spreitzer H, Gratzel M (1998) Nature 395:583
76. Zhang Q, Dandeneua S, Zhou X (2009) Adv Mater 21:4087–4108
77. Lee UL, Choi WS, Hong B (2010) Sol Energy Mat Sol Cells 95:680–685
78. Subash, S., Chowdury, M.H.: In: IEEE X'plore, pp. 240–243 (2010)

Synthesis of Carbon Nanostructures by CVD Method

Krzysztof Koziol, Bojan Obrad Boskovic, and Noorhana Yahya

Abstract The field of nanotechnology continues to develop. Carbon based materials with different structure and dimensions become increasingly important in the field. Carbon nanotubes (CNTs) are particularly promising due to their anisotropic extraordinary electrical, thermal and mechanical properties that have captured the imagination of researchers worldwide. However, the complexity involved in synthesis of nanotubes in a predictable manner has held back the development of real-world carbon nanotube based applications. In this chapter the structure and synthesis methods will be discussed of CNTs and other forms of nanostructures of carbons. Furthermore, their structuring into macroscopic assemblies, like mats and fibres will be presented as it has important role in future industrial applications of these materials.

1 Introduction to Carbon Nanomaterials

In 1985 chemists created a new allotrope of carbon [1] by heating graphite to very high temperatures. They named the allotrope buckminsterfullerene, after American architect Richard Buckminster Fuller. The buckminsterfullerene is a molecule

K. Koziol (✉)
Department of Materials Science and Metallurgy, University of Cambridge, Pembroke Street, Cambridge, CB2 3QZ, UK
e-mail: kk292@cam.ac.uk

B.O. Boskovic
Cambridge Nanomaterials Technology Ltd, 14 Orchard Way, Cambourne Cambridge CB23 5BN, UK
e-mail: Bojan.Boskovic@cnt-ltd.co.uk

N. Yahya
Fundamental and Applied Sciences Department, Universiti Teknologi PETRONAS, Bandar Seri Iskandar, 31750 Tronoh, Perak Malaysia
e-mail: noorhana_yahya@petronas.com.my

N. Yahya (ed.), *Carbon and Oxide Nanostructures,* Adv Struct Mater 5,
DOI 10.1007/8611_2010_12, © Springer-Verlag Berlin Heidelberg 2010,
Published online: 13 July 2010

consisting of 60 carbon atoms only (with a molecular formula of C_{60}). The molecules are shaped like tiny soccer balls (therefore sometimes referred to as buckyballs), with an atom at each point where the lines on a soccer ball would normally meet. The 60 carbon atoms bond in 20 six-membered rings and 12 five-membered rings. The discovery revolutionised the carbon field as researchers became interested in this new allotropic form of carbon. The carbon field expanded again in 1991 with Iijima's report on the observation of carbon nanotubes [2], an elongated version of buckminsterfullerenes. Carbon nanotubes, in particular, attract attention of hundreds of research groups around the world (Fig. 1) and their research still continues to grow.

The history of carbon nanotubes is much longer than 2 decades. In the 1950s and 1970s at least two groups synthesised and characterised carbon based nanotubes, but their discoveries went largely unnoticed [3, 4]. The field of carbon nanotubes has grown considerably with new, controllable production routes being developed, unusual properties predicted and measured, and many intriguing applications suggested.

The basic structure of a carbon nanotube is a hollow cylindrical tube of graphitic carbon capped by fullerene hemispheres with nanometer size diameters and macroscopic size lengths. The nanotubes may consist of one to hundreds of concentric graphitic shells of carbons. According to Saito et al. [5] the inter-sheet distance in multi-sheet nanotube is 0.344 nm. It is close to the distance between two layers in graphite, which equals to 0.335 nm [6]. The carbon network of each shell can be directly related to the hexagonal lattice of an individual layer of graphite. Nanotubes made of one hollow graphitic shell are called single wall nanotubes (SWNTs) and have diameters typically 0.6–3 nm. Nanotubes made of two or more concentric shells are called multi-walled nanotubes (MWNTs) [7] (shown in Fig. 2). In reality

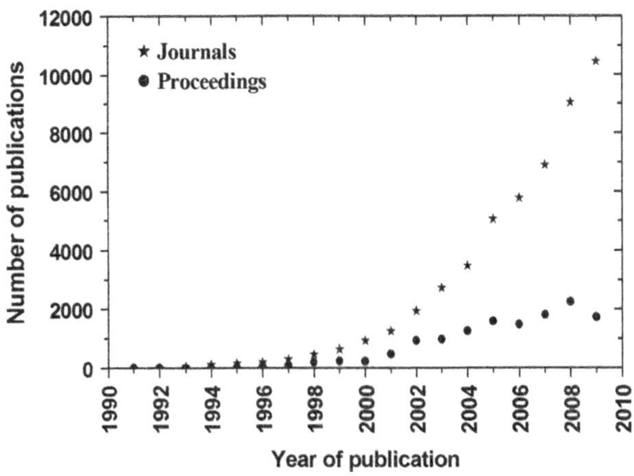

Fig. 1 Number of papers and proceedings on nanotubes per year.
Source: ISI (Institute for Scientific Information) Web of Knowledge. In the search window a term of "nanotub*" was used

multi-walled nanotubes have different lattice orientations (described with chiral vectors and angles) and defect concentration.

2 Structure of Carbon

There are several allotropes of carbon known in nature. The allotropes of carbon differ in the way the atoms bond with each other and arrange themselves into a structure (as shown in Fig. 3). As the structures of allotropes vary, they also have different physical and chemical properties [8].

In the most commonly used form, *graphite*, atoms of carbon form planar layers (graphene layers). Each layer is made up of rings containing six carbon atoms. The

Fig. 2 Examples of ideal, defect-free nanotube structures: (**a**) side view & end on view of a single wall carbon nanotube, (**b**) end on view of a multi-walled carbon nanotube

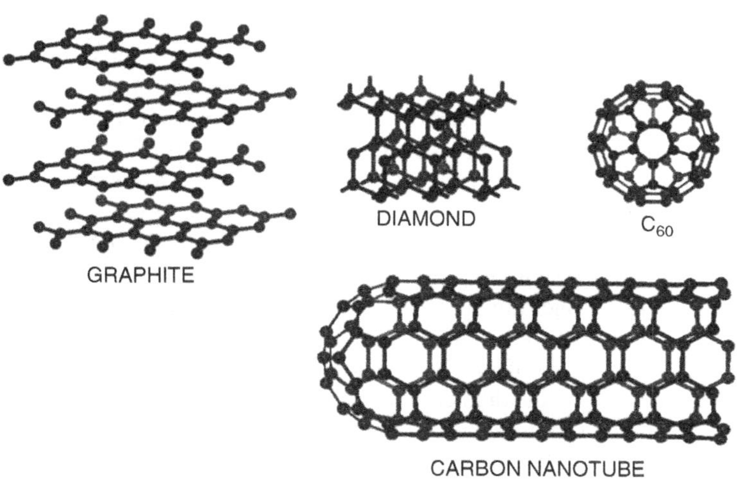

Fig. 3 Three main naturally occurring allotropes of carbon: graphite, diamond and fullerene

rings are linked to each other in a hexagonal structure. Each atom has three sigma bonds (with an angle of 120° between any two of the bonds) and belongs to three neighbouring rings. The fourth electron of each atom becomes part of an extensive π bond structure. Graphite conducts electricity due to the electrons in the π bond structure, which can move around throughout the graphite. Bonds between atoms within a graphene layer are strong, but the forces between the layers are weak, van der Waals forces [9]. The graphene layers can slip past each other, a property of graphite used in lubrication. Although graphite occur naturally, most commercial graphite is produced by treating petroleum coke, a black tar residue remaining after the refinement of crude oil, in an oxygen-free oven. Naturally existing graphite occurs in two forms, alpha (hexagonal) and beta (rhombohedral). These two forms have identical physical properties but different crystal structures. The alpha form can be converted to the beta by mechanical treatment, and the beta form reverts to the alpha on heating it above 1,000°C. All artificially produced graphite is of the alpha type.

Diamond, is one of the hardest substances known and naturally occurring form of carbon. In diamond structure, each carbon atom bonds tetrahedrally to four other carbon atoms to form a three-dimensional lattice. The shared electron pairs are held tightly in sigma bonds between adjacent atoms. Pure diamond is an electrical insulator. Due to its hardness, it is used in industrial cutting tools. The naturally occurring diamond is typically used for jewellery. However most commercial quality diamonds are artificially produced from graphite by applying extremely high pressure (more than 100,000 times the atmospheric pressure) and temperature (about 3,000°C). High temperatures break the strong bonds in graphite so that the atoms can rearrange themselves into a diamond lattice [10].

There are also amorphous forms of carbon containing varying proportions of sp^2 and sp^3 bonded carbon atoms. *Amorphous carbon* is formed when a material containing carbon is burned without enough oxygen for it to burn completely. This black soot is known as lampblack, gas black or channel black [10] and may, in fact, contain other elemental impurities. Amorphous carbon is not generally considered a third allotrope because its structure is poorly defined.

Fullerenes (buckyballs and carbon nanotubes) can be considered as a closed, zero and one dimensional carbon structure. They are the only allotrope of carbon existing in the pure form (without hydrogen terminations). Treated with hydrostatic pressure (at a level of 25 GPa) they can be converted into a hard and transparent form of amorphous carbon [11]. In comparison to atomistic crystals of graphite or diamond, fullerenes form molecular crystals. Due to the high aspect ratio of carbon nanotubes, the quasi-one-dimensional structure, and the graphite-like arrangement of the carbon atoms in the shells, nanotubes exhibit very broad range of unique properties. The properties of nanotubes can change depending on the different kinds of nanotube (defined by the diameter, length, and chiral angle) and quality (defined by defect concentration). Large increases in strength, toughness, superior electrical/thermal properties and their combination, are potential benefits of using nanotubes as the filler material in polymer-based composites as compared to traditional carbon, glass or metal fibres.

3 Synthesis Methods of Carbon Nanotubes

There is a huge demand for quality nanotubes both as research materials and for large scale industrial applications. The main problem with the currently available nanotubes is the heterogeneity of the sample, in terms of dimensions, chiral angles and purity. The nanotubes examined by Iijima in 1991 were synthesized by arc-discharge method [2], but since then several other production methods have been developed. A group led by Smalley [12] has used oven laser evaporation to produce carbon vapour, with nanotubes again observed in the condensed soot. Both arc-discharge and laser ablation techniques have the advantage of producing high quality nanotubes but at the same time relatively high amount of impurities (around 30%). Unfortunately, evaporation of carbon atoms from solid targets at temperatures above 3,000°C is neither economical and nor convenient. Synthesised CNTs may also be entangled, hindering purification steps and further application of the samples. Baker and co-workers [13, 14] demonstrated in early seventies growth of nanotubes, described at that time as carbonaceous deposits, from decomposition of acetylene. In 1976 Endo and co-workers [15–18] have also shown that CNTs can be synthesised by pyrolysis of benzene, followed by subsequent heat treatment.

Currently, the common method widely accepted in the synthesis of nanotubes, due to its simplicity and low cost, is the chemical vapour deposition (CVD) method. This method was originally developed in the 1960s and 1970s and has been successfully used in the production of carbon fibres and carbon nanofibres for more than 20 years [19–25]. Using this method, CNTs are produced from the carbon containing source (usually gaseous form) as it decomposes at elevated temperature and passes over a transition metal catalyst (typically Fe, Co or Ni) [26, 27]. A high yield of nanotubes can be achieved by this method, but the nanotubes are more structurally defective than those produced by arc or laser evaporation methods. There are several advantages of the CVD method, which make it preferred to other available synthesis methods. Firstly, the product tends to be purer (far fewer impurities in the form of nanoparticles of graphite or metal). Secondly, the growth occurs at a lower temperature (550–1,000°C) [26, 27], making the process both cheaper and more accessible for lab applications. Finally, the metal catalyst can be held on a substrate, which can lead to the growth of aligned nanotubes in a desired direction with respect to the substrate.

There are two basic mechanisms proposed for the growth of nanotubes by CVD method related to substrate bound catalyst (shown in Fig. 4), which are now widely recognised [9, 13, 14].

Top carbon diffusion through catalytic particle (tip growth model).

The decomposition of the carbon source on the exposed surface of the metal catalyst results in the formation of hydrogen and carbon species. The carbon dissolves in the particle and diffuses through it until it precipitates at the end in the form of graphene filaments. The catalytic particle sits always on the top of the growing nanotube.

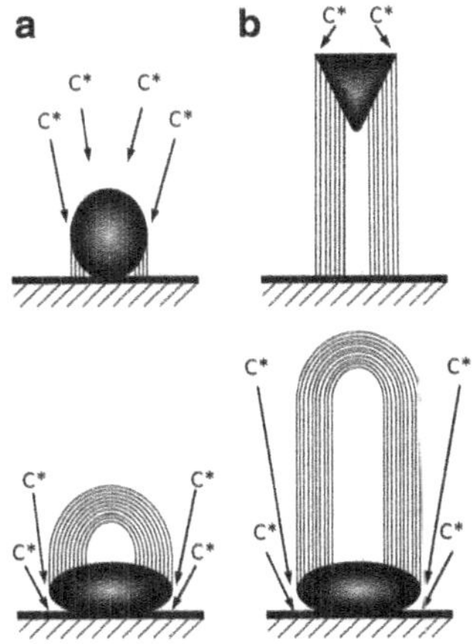

Fig. 4 Schematic diagram representing top carbon diffusion (*upper row*) and bottom carbon diffusion (*lower row*) growth mechanisms. (**a**) Pyrolysis of the hydrocarbon gas into carbon species which then dissolve in the catalyst metal particle, (**b**) precipitation of carbon in form of carbon filament

Bottom carbon diffusion through catalytic particle (base growth model). In this model, the catalytic particle stays on the growth substrate. The carbon species dissolve in the particle and diffuses through it until they precipitate on top of the metal particle in the form of graphene filaments. The carbon diffusion parameter depends on the dimensions of the particles, the characteristics of the metal used as a catalyst, the temperature and the hydrocarbons and gases involved in the process.

When the substrate-catalyst interaction is strong, a CNT grows up with the catalyst particle rooted at its base (base growth model). When the substrate-catalyst interaction is weak, the catalyst particle is lifted up by the growing nanotube and continues to promote CNT growth at its tip (tip growth model) [23]. Formation of SWNTs or MWNTs is governed by the size of the catalyst particle. If the particle size is a few nanometers, SWNTs form, whereas particles a few tens of nanometers wide favour MWNTs formation.

The growth mechanism suggested above is quite similar to the one proposed for the vapour grown carbon fibres (VGCF), again dating 20 years back (shown in Fig. 4). Growth of these fibres occurs by a dehydrogenation reaction of a hydrocarbon gas in several steps. In this mechanism, pyrolysis of the hydrocarbon gas occurs on the surface of the catalyst particle, releasing hydrogen gas and carbon, the later dissolving into the catalyst. The dissolved carbon then diffuses through the catalyst particle and is precipitated at the trailing edge of the particle. This step possibly relies on the presence of a temperature gradient across the particle, which is often created by the exothermic nature of the hydrocarbon decomposition. This gradient causes carbon to be precipitated at the cooler trailing edge of the catalyst particle, and therefore causing the elongation of the fibre. Below is a brief summary of three main methods, by which nanotubes are produced: arc-discharge, laser ablation and chemical vapour deposition (CVD).

3.1 Arc-Discharge

The arc-discharge method is the one by which CNTs were produced by Iijima [2]. CNTs can be synthesized in the arc-discharge AC/DC system (Fig. 5). DC provides higher yields of CNTs, which are deposited on the cathode. One important condition of stabilization of arc-discharge is maintaining a constant distance between the graphite electrodes, of around 1 mm [28]. Grams scale synthesis of MWNTs by arc discharge has been achieved in He gas [29, 30]. When a graphite rod containing a metal catalyst (Fe, Co, etc.) is used as the anode with a pure graphite cathode, single-walled carbon nanotubes (SWNTs) are generated in the form of soot [31, 32].

It was found that presence of hydrogen gas in the growth region gives the optimum synthesis of MWNTs with high crystallinity (having regular graphene sheets at an interlayer spacing of 0.34 nm) and few coexisting carbon nanoparticles [2, 33–39]. In contrast, fullerenes could not be produced in gas atmosphere which included hydrogen atoms, essential difference between CNT and fullerene production [40].

3.2 Laser Ablation

The laser vaporization method was developed for fullerene and CNT production by Smalley's group [41]. First used for fullerene synthesis [1] and further applied to produce CNTs [42] in 1996, especially SWNTs. The synthesis system consists of a furnace, quartz reactor tube and laser beam source (Fig. 6). It can also consist of a reactor chamber and a laser source. A laser beam (typically a YAG or CO_2 laser) is focused onto the graphite rod target located inside the reactor tube. The target is vaporized in high-temperature argon buffer gas and carried to the copper collector cooled down with coater. The deposit is rich in SWNTs and MWNTs (Fig. 7a, b). The method has several advantages, such as high-quality SWNT production,

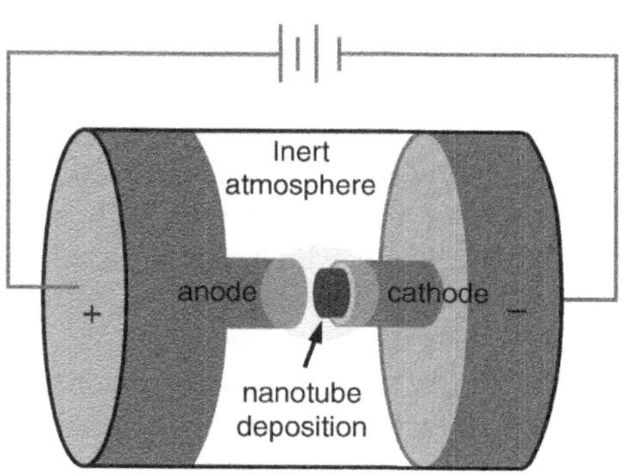

Fig. 5 Schematic diagram of the arc discharge apparatus

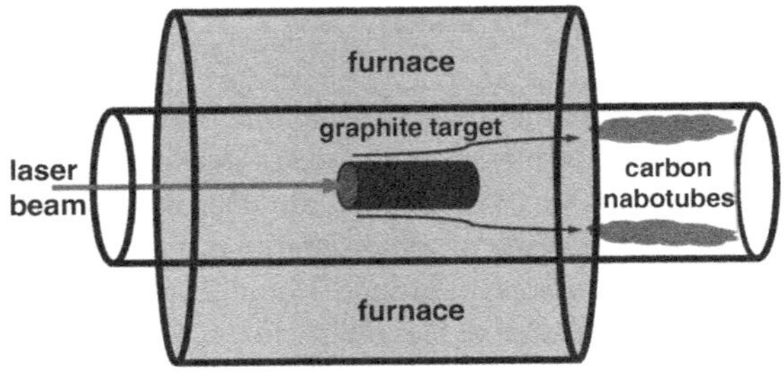

Fig. 6 Schematic diagram of the laser ablation method

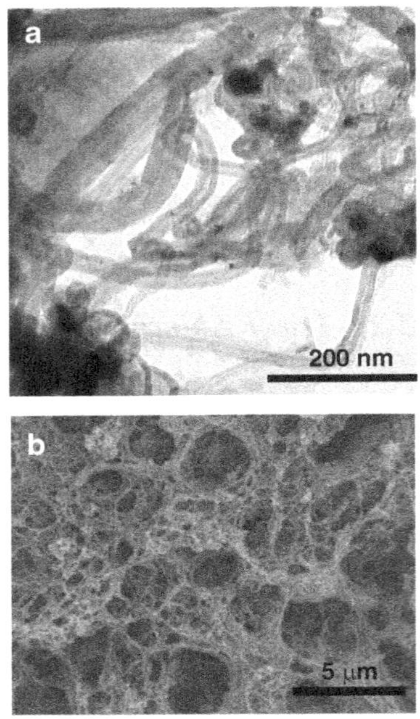

Fig. 7 (**a**) Transmission electron microscopy (TEM) image of CNTs (**b**) Scanning electron microscopy (SEM) image of carbon nanotube web structures. Both images show CNTs produced by pulsed laser ablation method (Nd:YAG laser with 532 nm wavelength was employed in this work)

diameter control, investigation of growth dynamics, and the production of new materials. High-quality SWNTs with minimal defects and contaminants, such as amorphous carbon and catalytic metals, can be synthesized using the laser-furnace method followed by suitable purification processes [43–45].

The laser has sufficiently high energy to vaporise the graphite target at the atomic level, which is then used as the material for synthesis of SWNTs [46–48]. SWNT diameter can be controlled by changing the furnace temperature, catalytic metals, and flow rate [47, 49, 50]. Raising the furnace temperature results in SWNTs with larger diameters [49]. Depending on the choice of the catalytic metals, the diameter of the SWNTs can either be increased or reduced [50, 51].

3.3 Thermal Catalytic Chemical Vapour Deposition

This method involves pyrolysis of hydrocarbons (acetylene, ethylene, propylene, methane, benzene, toluene etc.) or other carbon feedstock (polymers, carbon monoxide) diluted in the stream of inert gas in the furnace system over the surface of metal catalysts [15, 52–55]. The evaporation of a solid hydrocarbon can be conveniently achieved in another furnace at low temperature before the main, high-temperature reaction furnace [56–61]. The catalyst material may be solid, liquid, or gas and can be placed inside the furnace or fed in continuously from outside. Decomposed carbon species dissolve in the metal nanoparticles but, due to a finite solubility of carbon in the metallic particles, supersaturation will be reached followed by carbon precipitation out in the form of a fullerene dome extending into a carbon cylinder [19, 62]. Typical temperature range for the synthesis is 500–1,200°C at atmospheric pressure [6, 52].

Typical system used in the thermal CVD method of making carbon nanotubes, with horizontally positioned reaction tube is shown in Fig. 8.

The CVD method allows CNT growth in a variety of forms, such as powder, thin or thick films, aligned or entangled, straight or coiled, or even a desired architecture of nanotubes at predefined sites on a patterned substrate. It also offers better control over growth parameters in comparison to other synthesis methods. The three main parameters for CNT growth in CVD are the atmosphere, carbon source, catalyst, and growth temperature. Low-temperature (600–900°C) yields MWNTs, whereas a higher temperature (900–1,200°C) reaction favours SWNTs growth [63–68].

The most commonly used catalysts for CNT growth are the transition metals (Fe, Co, Ni) from sources like organometallocenes (ferrocene, cobaltocene, nickelocene), nitrates and others [69, 70]. A correlation was found between the size of catalyst particles and the nanotube diameter. Hence, metal nanoparticles of controlled size can be used to grow CNTs of controlled diameter [71].

The CVD process has been scaled up onto a large scale commercially, especially for MWNTs [72–74]. Smalley's lab developed a mass production of SWNTs by the so-called high pressure carbon monoxide (HiPco) technique [75]. Currently also

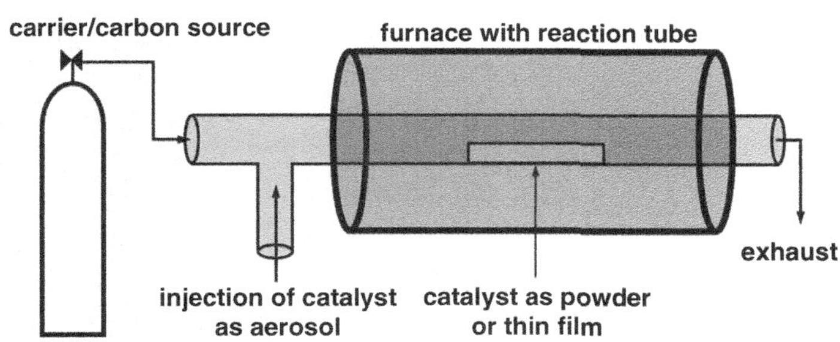

Fig. 8 Schematic design of a thermal CVD system with a tube furnace

kilograms scale of MWNTs per hour can be produced [76, 77] even with the control over the diameter of nanotubes.

3.3.1 Synthesis of Aligned Carbon Nanotubes

Generally, it is hard to grow aligned CNTs (SWNTs or MWNTs) by arc discharge, although partial alignment of the nanotubes can be achieved by convection [78] or directed arc plasma [79]. The CVD method is ideally suited to grow aligned CNTs on desired substrates for specific applications. Li et al. [80] have grown dense MWNTs arrays on iron-impregnated mesoporous silica prepared by a sol-gel process, Terrones et al. [81] have produced CNTs on Co-coated quartz substrates, while Pan et al. [82] have reported the growth of aligned CNTs of more than 2 mm in length over mesoporous substrates from acetylene. Depending on the preferred application highly aligned nanotubes were synthesised with different catalysts [83] or on different substrates [73, 84–86]. Using the CVD method it is also possible to grow aligned nanotubes in a desired direction with respect to the growth substrate. It was also found that not all materials can be active in the growth of aligned nanotubes. Metal, graphite or silicon used in the process would not yield any nanotubes. Substrates made of silica or alumina would generate nanotubes. Additionally it has been demonstrated that the growth of CNTs depends on the thickness of the oxide layer on silicon wafer surface [84]. Below 6 nm no detectable growth of the nanotubes was observed. Above 50 nm thick oxide layer gives saturation and growth dependence only on CVD time. However between 6 and 50 nm the growth of aligned nanotubes seems to be depended on both CVD time and SiO_2 layer thickness.

It has been shown that full control over the length of CNTs could be achieved and aligned, densely pack nanotubes produced (as in Fig. 9). The inhibition of CNTs growth at low SiO_2 thickness is explained by partial deactivation of catalyst

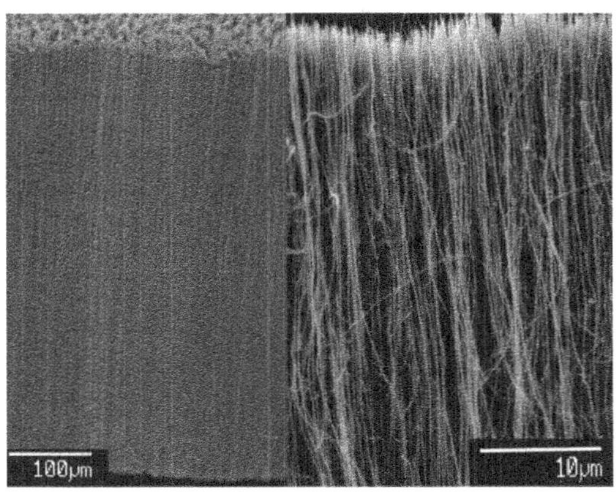

Fig. 9 Electron microscope images of highly aligned carbon nanotube car pets, at low and high magnifications

particles due to their reaction with the silicon substrate. Iron from ferrocene (source of metal) diffuses through SiO_2 layers thinner than ~5 nm and reacts with the silicon substrate, leading to formation of $FeSi_2$ and $FeSiO_4$, neither of which catalyses CNTs growth. The layer of SiO_2 with thickness above 5 nm is sufficient enough to keep the active metal particle and promote the suitable metal structure conducive to CNTs growth.

3.3.2 Synthesis of Nitrogen Doped Nanotubes

Shortly after the synthesis of carbon nanotubes, a quest of substitution of carbon atoms in the graphene network with heteroatoms such as boron, nitrogen, sulphur, phosphor and silicon begun. The intensive work on heteroatomic doping was aiming to alter some of the important properties of nanotubes, including electrical (electron density and semiconducting character), mechanical (improvement of Young's modulus), and chemical (change of reactivity, creation of catalytically active centres etc.) [87].

There are three basic ways that nitrogen can be incorporated into the graphene CNTs structure. (1) Substitution, where N is coordinated to three C atoms in sp^2 like fashion, which induces sharp localized states above the Fermi level associated with the injection of additional electrons into the structure. (2) Pyridine-like substitution, where N is arranged around a vacancy, since the valency of the nitrogen can be satisfied by two sp^2 bonds, a delocalised p-orbital, and a lone pair in the remaining sp^2 orbital, pointing at the vacancy. (3) Chemical adsorption of N_2 molecules.

Nitrogen contains one electron more than carbon; therefore, substitutional doping of nitrogen within graphene will n-dope the structure, enhancing the number of electronic states at the Fermi level depending on the location and concentration of dopant. Hernandez et al. calculated the mechanical properties of nitrogen and boron doped nanotubes [88, 89], demonstrating that high concentrations of N within SWNTs lower the Young's modulus. Nevertheless, the Young's modulus values still remain on the order of 0.5–0.8 TPa. This behaviour has been experimentally confirmed in pristine and N-doped MWNTs [90]. Unfortunately, the Young's modulus for pristine and N-doped MWNTs were 0.8–1 TPa and ~30 GPa, respectively. The decrease in mechanical strength of N-doped nanotubes could be explained by the nitrogen induced defects due to the relatively high N concentration (2–5%) within the tubes. If the N concentration is below 0.5%, it is expected that the mechanical properties will not be substantially altered [91].

Results from other theoretical studies demonstrated that relative position of nitrogen and carbon affects not only electronic properties but also their thermodynamic stability [92].

Studies using ab initio density functional theory have shown that the nitrogen substitution into zigzag and armchair SWNTs can cause a junction of separate tubes by the formation of covalent bonds [93]. If two neighbouring tubes have their nitrogen impurities facing one another, inter-tube covalent bonds could potentially be formed. If the density of inter-tube bond is high enough, a highly packed bundle of interlinked single-walled nanotubes can form, substantially enhancing the mechanical properties.

There are two main routes used to synthesis the N-CNTs: (1) direct delivery of heteroatoms with the carbon source stream, during the growth of the nanotubes (2) substitution of carbon atoms by heating the nitrogen containing compound with CNTs. The most common is the first route.

Similar methods as in the case of pure carbon nanotubes are used in the synthesis of nitrogen-doped nanotubes. In the arc-discharge method the atmosphere surrounding electrodes must contain nitrogen. Depending on the percentage of nitrogen in the growth atmosphere, different nitrogen doping levels have been recorded [94]. The doping level was up to 14 %wt (as determined by XPS) when 50 %vol of atmosphere was substituted by nitrogen. The resulting N-CNTs had diameters of about 20 nm and were coated with a thick layer of amorphous carbon. Computational calculations showed that incorporation of nitrogen atoms lead to distortion in graphite plane [94].

Arc experiments using pure graphite electrodes in an NH_3 atmosphere indicated that it was difficult to produce N-doped SWNTs and MWNTs, possibly because N_2 molecules are easily created and do not react with carbon [91]. N-doped SWNTs could be produced by arcing composite anodes containing graphite, melamine, Ni, and Y [95].

The laser ablation method was not fully explored in the synthesis of doped nanotubes. In 1997, Zhang et al. [96] reported that sandwich-like C-B-N nanotubes could be produced by laser vaporisation of graphite-BN targets. However it is likely that a large N content will result in the inhibition of SWNT growth. More energetic lasers were proposed in order to generate N- or B-doped SWNTs.

In the CVD method the usual approach relied on the pyrolysis of hydrocarbons or other carbon feedstock with the addition of a nitrogen source (e.g. nitrogen, ammonia, amines, nitriles) diluted in the stream of the inert gas in the furnace system over the surface of metallic catalyst particles (such as Fe, Co or Ni). The catalyst can be provided with the stream of starting materials or deposited directly onto the growth substrates. The differences between the reported processes arise from the application of different nitrogen sources, catalysts and pressures. Depending on the conditions and parameters of the synthesis, different quality of growth products was reported. It has been suggested that only small concentrations of nitrogen (below 15%) can be introduced into MWNTs [97]. The results demonstrated that it is extremely difficult to generate crystalline and highly ordered structures containing large concentrations of N within the hexagonal carbon network. The doped nanotubes with low N concentrations have been subsequently generated via pyrolysis of pyridine and methylpyrimidine [98]. Unfortunately, these nanotubes are easily oxidized in air. The degree of perfection within graphene sheets changes rapidly with different N concentration used. Keskar et al. prepared isolated N-doped SWNTs from thermal decomposition of a xylene-acetonitrile mixture over nanosized iron catalyst particles. The N dopant concentration was controlled by the amount of acetonitrile in the mixture [99].

Liang et al. reported that using ferrocene and ethylenediamine as a source of catalyst and nitrogen resulted in increased of the diameter of nanotubes with increasing growth temperature. The majority of the material containing nitrogen was formed as MWNTs in the bamboo-like structure. The N-doping level also was dependent on growth temperature. With increasing temperature from 780 to 1,080°C the amount of nitrogen decreased from 24 to 18 %wt. N-doped CNTs grown at lower temperatures have shown much higher degree of disorder and higher N-incorporation [100].Wang et al. shown that the longer the time of synthesis, the higher the length and diameter of nanotubes produced, which was suggested to correlate with the grain size of catalyst particles (the longer the growth time, the larger the iron catalyst particles). The bamboo-like morphology of nanotubes was again observed. The doping levels of nitrogen were estimated by EELS at 9% [101]. Lee et al. used acetylene and ammonia in argon and varied the growth temperature from 750 to 950°C. When increasing the amount of nitrogen source an increase in doping level from 2.8 to 6.6 %wt was observed by elemental analysis [102]. Again bamboo-shaped morphology of nanotubes was present. Additionally using ammonia as a source of nitrogen caused decrease in the growth rate of N-CNTs.

Two different bamboo-type morphologies of nanotubes were reported by Glerup et al. One type with a very frequent, regular compartments and another with irregular structure with fewer, longer and uneven compartments. Chemical analysis showed presence of molecular nitrogen trapped inside the nanotubes. It is not clear if the nitrogen is homogeneously distributed along the length of the nanotube or whether it is segregated into regions with higher and lower concentrations [103]. Jang et al. demonstrated that an increase in the flow rate of nitrogen yielded in more defective graphene sheets and higher doping levels [104]. Lee et al. used acetylene and ammonia as a source for synthesis and presented microscopy and spectroscopy evidence revealing consistently that as the nitrogen source increases the degree of crystallinity (nanotube structure perfection) decreases. Again the N-content varied in the range 2–6 %wt depending on the ammonia flow rate. It was found that the higher the nitrogen incorporation the more curved and thicker bamboo-like compartments appear [105].

In 2005, Koziol et al. demonstrated completely different outcome, to what was already reported, by using specific nitrogen precursors in CVD synthesis of nanotubes. In this case hydrocarbon feedstock containing diazine, aromatic compound with nitrogen, at a critical level, was injected to the reactor at 760°C. The nanotubes, which they synthesised, were multiwalled but found to be extremely straight and had unprecedented degrees of internal order [106]. Furthermore, electron diffraction patterns from individual nanotubes, revealed that all of the walls had the same chiral angle, which is not possible in concentric cylindrical nanotubes, due to a geometric constrains but possible in conical nanotubes (Fig. 10). The adjacent nanotube walls in these nanotubes were in crystallographic register with one another, with ABAB stacking sequences of layers [106]. Finally, and most importantly, the chiral angles seen in electron diffraction patterns were of the simple achiral forms and nanotubes were consistently either armchair or zigzag, as seen in Fig. 10 (middle and left) [106]. Very low conical angle was measured in these nanotubes, between 0.5° and 5° and

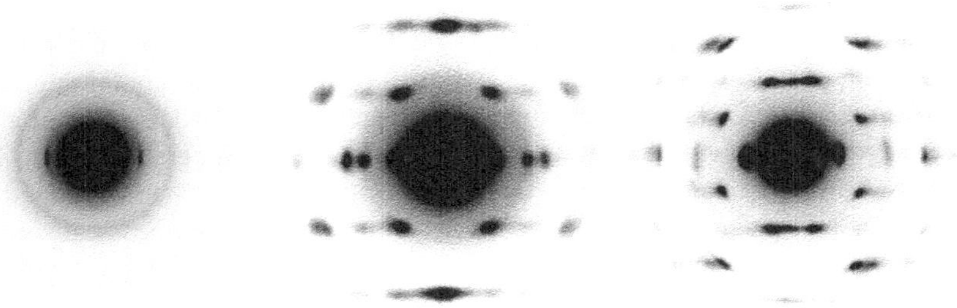

Fig. 10 Electron diffraction patterns from individual multiwalled nanotubes. Standard mix chirality (*left*), armchair (*middle*), zigzag (*right*) [106]

nitrogen was detected in two forms, as substitution in the lattice and as N_2 gas in the core of every tube [107–109]. Higher diazine concentrations in the feedstock seemed to allow the formation of shallower cones [108].

3.4 Plasma Enhanced Chemical Vapour Deposition

Carbon nanotubes and nanofibres can be synthesised using plasma enhanced CVD (PECVD) where the hydrocarbon gas is in an ionised state over the transition metal catalyst (nickel, iron, cobalt, etc.). The carbon nanotube and nanofibre aligned growth perpendicular to the substrate can be achieved using the electrical self-bias field from plasma (Fig. 11). PECVD systems are characterised primarily by the plasma energy sources used, and the most commonly used include: hot filament PECVD, direct current PECVD, radio-frequency PECVD, microwave PECVD.

Hot filament PECVD uses thermal energy for plasma creation and has been used successfully for carbon nanotube production by Ren and co-workers [110]. Microwave PECVD, widely used for the preparation of diamond films, has also been successfully used in the production of carbon nanotubes and nanofibres [111–115]. Synthesis of vertically aligned CNTs and CNFs requires electric field normal to the substrate, and dc PECVD is the most suitable method to achieve this [116, 117]. Inductively coupled plasma PECVD [118, 119] and radio frequency PECVD [120, 121] methods have also been used successfully for carbon nanotubes and nanofibres synthesis. Ren et al. in 1998 [110] reported first successful growth of large-scale well-aligned carbon nanofibres on nickel foils and nickel-coated glass at temperatures below 666°C. Bower et al. [114] have grown well-aligned carbon nanotubes using microwave PECVD with additional radio frequency graphite heater. They found that switching the plasma source off effectively turns the alignment mechanism off leading to the thermal growth of curly nanotubes. Merkulov et al. [116] reported synthesis of vertically aligned CNFs on patterned catalyst using dc PECVD. The catalyst patterns were fabricated using conventional electron beam lithography. The shape of CNFs depends on how much growth occurs at the tip by catalysis and now much by

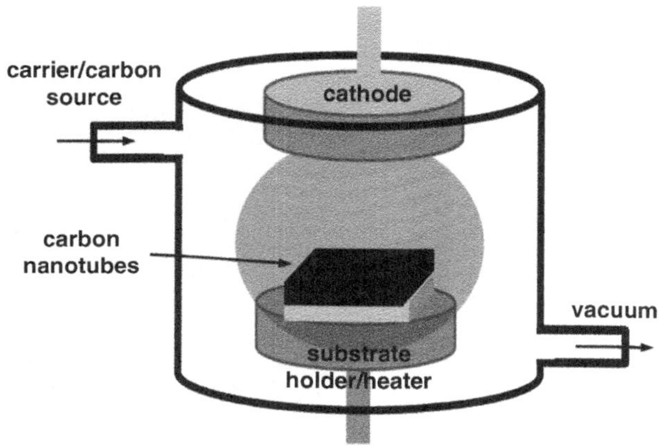

Fig. 11 Schematic design of a parallel plate PECVD system

deposition of a-C from the plasma along the sidewalls [122]. This ratio is controlled by the catalyst activity and by the balance of deposition and etching of a-C. The balance between deposition and etching depends on the plasma and the etchant (NH_3) and hydrocarbon gas (C_2H_2). This balance has been studied by Merkulov et al. [116] and Teo et al. [123].

In plasma enhanced CVD systems, plasma energy sources substitute for the thermal energy in a furnace, and provide the energy required for decomposition of hydrocarbon feedstock and allow growth of carbon nanostructures at much lower temperatures.

The PECVD method allows growth of carbon nanotubes and nanofibres at low temperatures suitable for use of temperature sensitive substrates. A radio frequency PECVD carbon nanofibres synthesis at room temperature has been reported by Boskovic et al. [121]. A room temperature growth of carbon nanofibers using PECVD was subsequently demonstrated by Minea et al. [124]. Using dc PECVD Hofmann et al. [125] demonstrated synthesis of aligned carbon nanofibres at temperatures as low as 120 °C and on plastic substrates [126].

Although MWCNT and nanofibers synthesis have been achieved through PECVD at low temperature [121], SWCNT synthesis still remains largely a high temperature process (800–1,200°C) produced in arc-discharge, laser ablation, or tube furnace. Cantoro et al. [127] recently reported thermal CVD synthesis of SWCNT at temperature as low as 350°C in very low pressure (10^{-3}–10^{-2} mbar) of pure acetylene in a cold-walled system.

4 Other Forms of Carbon Nanostructures

Besides the carbon nanotubes, other interesting carbon nanostructures have been sythesised using CVD. The carbon nanohorns, carbon nanowalls and graphene have received considerable interests. The radial packing of single-walled tubular carbon

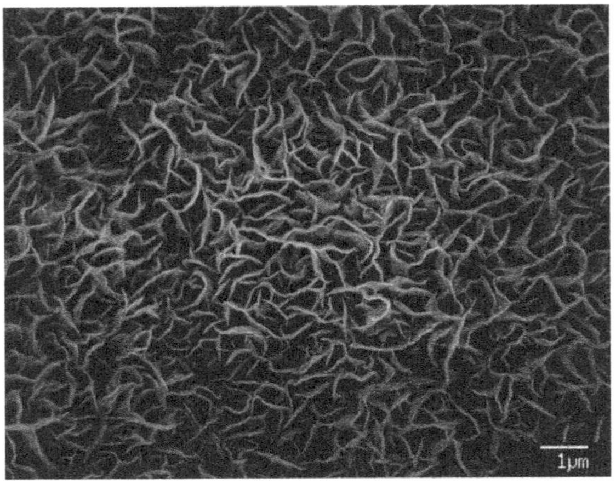

Fig. 12 Carbon nanowalls grown in the MW PECVD as described by Chuang et al. [138, 139]

nanohorns resembles a dahlia flower. Iijima et al. [128] described the growth mechanism of carbon nanohorns: In a high energy and low diffusion rate condition carbon species forms graphene sheets, and collide to form horn structures as predicted by tight-binding molecular-dynamics simulations [129].

Carbon nanowalls (CNWs) are networks of vertically aligned graphitic walls. They share similar morphology with other carbon nanomaterials such as carbon nanoflakes [36, 130, 131], and nanosheets [132, 133], and nanoflowers [134]. Two-dimensional CNWs, first reported by Wu et al. [135], are promising materials for a number of applications, and have been demonstrated as an efficient material for backlights of liquid crystal displays by field emission in the form of a nanodiamond/carbon nanowalls composite [136], also as high-brightness lamps based on CNW-coated nickel wires [137]. High surface area also makes CNW suitable for electrochemical applications, such as batteries and fuel cells.

Carbon nanowalls was first reported as a surface-bound material, by Wu et al. [135], synthesized in an attempt to produce CNT in PECVD environment. Chuang et al. [138, 139] reported the first successful synthesis non-surface bound free-standing macroscopic structure of CNW aggregates by microwave PECVD in various ammonia/acetylene gas mixtures (Fig. 12). This process is extremely efficient, and neither catalyst nor a flat substrate was needed. Carbon nanowall aggregates extrude from plasma sites induced by a growth stage and grow freely into three-dimensional space. The overall length can reach centimeters in 10 min of deposition time.

4.1 Carbon Nanotube Fibres

Significant attention was devoted into development of methods for manufacture of carbon nanotube based fibres. CNTs were used as the main constituent material in fibres or in combination with a polymeric matrix. In each case the aim was to take advantage of the spectacular axial properties of nanotubes. Carbon nanotube fibres would be an ideal system to translate the fabulous properties of individual nanotubes into real macroscopic use. One challenge in the fibre system is to achieve nanotube-nanotube bonding to get good load transfer and contact free flow of electrons. Second challenge is to find a convenient and economical way to manufacture CNT fibres.

First CNT fibre with a polymeric matrix was reported by Vigolo et al. [140]. Single wall nanotube dispersion was co-extruded with polyvinyl alcohol (PVA)/ water through a long syringe into a rotating water/PVA coagulation bath. The coagulation method used produced long fibres and ribbons. The diameter of the fibres could be adjusted by changing the injection rate, flow conditions, and dimensions of the capillary tube that affect the thickness of the ribbons. Authors have demonstrated the flexibility of the carbon nanotube fibres by making knots and they showed the fiber can be curved through 360° without breaking. The elastic modulus of SWNTs fibres was an order of magnitude higher than the modulus of high-quality bucky paper.

With long-range directional order, liquid crystals have long been used as precursor solutions for spinning high performance fibres. With lengths on the order of nanometers, and typical lengths in microns, CNTs have approximately the same shape as small molecules like tobacco mosaic virus, which readily form liquid crystalline phases. Liquid crystalline behaviour in CNTs was predicted by Somoza et al. in 2001, based on a computational model using continuum-based density-functional theory [141]. Somoza analyzed the different possibilities for tailored liquid crystalline CNT phases, predicting the formation of a columnar liquid crystalline phase. However liquid crystallinity in aqueous carbon nanotube suspension was first reported by Song et al. [142]. It opened a possible route for drawing fibres from liquid crystalline suspensions of carbon nanotubes.

Davis et al. at Rice University announced realization of nematic phases of SWNTs in superacid solutions. The SWNTs were produced using their high-pressure carbon monoxide (HiPco) process [143, 144]. Up to 10 wt% of SWNTs were dispersed in a superacid solution of sulphuric acid, chlorosulfonic acid, and triflic acid. Such a high concentration represents a tenfold increase over previous dispersions of SWNTs, and is due to the protonation of the nanotubes and the formation of an electrostatic double layer of protons and counter ions [145]. This charged layer surrounding individual nanotubes both encourages solubility in water, as well as preventing aggregation due to the repulsive force felt by like-charged nanotubes. Ericson et al. used sulfuric acid to promote the alignment of SWNTs and extruded fibres consisting entirely of SWNTs [146]. The purified SWNTs were mixed with 102% sulphuric acid and the mixture was extruded through a small capillary tube into a coagulation bath after its viscosity has reached a steady state. Fibres were obtained under different conditions, such as coagulants, different dope temperatures and coagulation bath temperatures. These fibres showed good alignment, with XRD analysis showing a mosaic angle of 31° at full width at half maximum (FWHM), and Raman spectroscopy showing a Raman ratio greater than 20:1. Additionally, fibres coagulated in water had a density that was 77% or the theoretical close packing density for 1.0 nm nanotubes. These fibres possess good mechanical properties, with a Yong's modulus of 120 ± 10 GPa and a tensile strength of 116 ± 10 MPa [146].

A simple and alternative route to spin CNT fibres directly from their lyotropic liquid crystalline phase consisting of multiwalled carbon nanotubes was shown by Zhang et al. [147]. The nanotubes were highly aligned within the fibres due to the combination of shear forces and the liquid crystalline phase. Fibres spun with carbon nanotubes and nitrogen-doped nanotubes (N-MWNTs) were both examined. High resolution transmission electron microscope shows N-MWNTs were much straighter than the MWNTs. Ethylene-glycol was used as a matrix to disperse nanotubes, with the concentrations between 1 and 3 wt%. A low power ultrasonic bath was used to assist the nanotubes dispersion process. The dispersion went from isotropic to biphasic to nematic phase with increasing concentration. The dispersions were then extruded out of the syringe through a needle with diameter less than 130 μm and transfer directly into a bath containing diethyl-ether. A syringe pump was used to control the extrusion rate of the dispersions and they were collected on a spindle outside the bath at the rate of 0.03–0.3 m/min. Young's modulus of MWNT fibres was found to be 69 ± 41 GPa. On the other hand, N-MWNT fibres had much higher stiffness of 142 ± 70 GPa, more than twice of the MWNT fibres [147]. The different mechanical properties between two types of fibres were believed to be the different interaction between individual nanotubes. The straighter N-MWNTs were thought to have less defects and a higher packing density, i.e. better interactions between the tubes. The electrical properties were measured by the two-probe method and both fibers were found to have ohmic behaviour, but N-MWNTs showed higher conductivity.

Direct spinning of CNTs into fibres is one method that can offer advantages over post-processing methods. Fewer processing steps lead to simpler and cheaper synthesis, and ease of scaling and commercialization. Jiang et al. have spun fibres directly from dense forests of MWNTs [148]. These CNT forests, grown by chemical vapour deposition (CVD), enable the continuous drawing of nanotubes due to van der Waal interactions between the nanotubes. Zhang et al. [149] introduced twist during spinning of multiwalled carbon nanotubes from nanotube forests to make multi-ply, torque-stabilized yarns. The yarn diameter was set by controlling the width of the forest sidewall that was used to generate an initial wedge-shaped ribbon and they have made singles (unplied), two-ply and four-ply MWNT yarns. The unplied yarn had diameters between 1 and 10 μm. The twist was typically 80,000 turns/m, versus 1,000 turns/m for conventional textiles (with much higher diameter). Single twisted fibres showed tensile strengths between 150 and 300 MPa. These single fibres were then spun into multi-ply yarns, with the two-ply having tensile strengths between 250 and 460 MPa. Later Zhang et al. made carbon nanotube sheets by rotating carbon nanotubes in vertically oriented nanotube arrays [150]. This method combines the dry-state spinning of nanotube yarns from forests and the introduction of twist. They demonstrated the thickness of the sheet depended on the forest size and increased with increasing the forest height. These transparent sheets have been used for the planar sources of polarized broad-band radiation and flexible organic light-emitting diodes. Zhang et al. at Los Alamos National Laboratory demonstrated the spinning of fibres from CNT arrays of 300, 500, and 650 μm in length and they found the tensile strengths for those as-spun fibres were 0.32, 0.56, and 0.85 GPa, respectively [151]. The work indicated that the fibre strength increased with increasing CNT length.

The most direct technique for spinning of CNT fibres was developed by Windle's group at University of Cambridge. This method relies on drawing carbon nanotube fibres directly and continuously from the CVD synthesis zone of a furnace [152]. Any type of hydrocarbon can be used as a source of carbon, injected at one end of the furnace together with thiophene (used as synthesis enhancer) and organometallic precursor, typically ferrocene, which after the decomposition forms iron nanoparticles allowing the formation of CNTs. These CNTs form an aerogel in the furnace hot zone, and due to their intermolecular interactions, as "elastic smoke" can be drawn from the furnace (as shown in Fig. 13) and wound onto a rotating spool [152]. There appears to be no limit to the length of the fibres drawn, presenting a truly continuous process. The continuous spinning process relies on two critical factors. One is to have sufficient high-purity nanotubes to form an aerogel in the furnace hot zone and the other is the forcible removal of the material from reaction by continuous wind-up. Different carbon sources and furnace temperature will produce CNT fibres with varies structures and properties. The composition of the fibres, in terms of double walled or multiwalled nanotubes could be controlled by changing the reaction parameters.

Additionally, Koziol et al. developed a controlled method for continuous spinning of fibres from the CVD reactor with different nanotube orientation based on the liquid condensation and drawing from the CVD reactor [153]. The mechanical data obtained demonstrate a considerable potential of carbon nanotube assemblies in the quest for maximal mechanical performance. The strength values measured in these fibres up to 10 GPa exceed any known available high performance material.

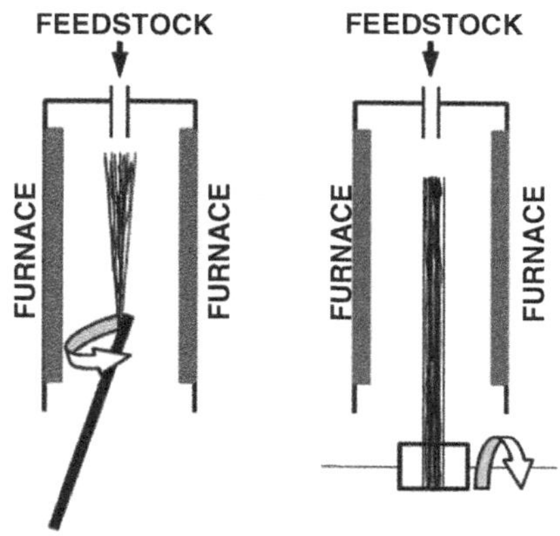

Fig. 13 Schematic of the direct aerosol spinning process (*left*); The wind-up procedure is operated outside the furnace hot zone at room temperature (*right*)

The development of continuous fibre drawing methods represents an enormous leap forward in the attempt to scale CNT properties for use in macroscopic applications. Now that researchers have realized success in spinning such fibres, attention must turn to designing processes that will provide increased tensile strength and modulus, approaching that of individual nanotubes. Better control of the underlying chemistry will allow experimentalists to fine-tune the nanotube properties, including length, axial alignment and surface functionalization.

4.2 3D Carbon-Carbon Nanomaterials

Three-dimensional (3D) nano-carbon structures that can transfer exceptional properties of carbon nanomaterials to meso- and micro-scale engineering materials are essential for development of many applications [154]. Tennent et al. [155] at Hyperion Catalysis in 1998 patented a method of preparing 3D microscopic structures by dispersing carbon fibrils (nanotubes or nanofibers) in a medium and separating them from the medium by filtration and evaporation to form a porous mat or sheet. Carbon nanotubes and nanofibers synthesized using CVD are usually in the form of a powder or a thin film on a flat substrate and direct synthesis of 3D carbon nanotube and nanofiber macroscopic structures are still challenging.

Well known engineering materials like carbon, ceramic or glass fibres could be exploited as a support for the formation of 3D nano-structures. Growth of CNTs and CNFs on the surface of carbon fibres was first reported to improve composite shear strength [156, 157] and load transfer at the fibre/matrix interface [158]. The high surface area of carbon and ceramic fibres coated with nanotubes and nanofibres is important for use in electrochemical applications [159–161]. Jo et al. [162] reported excellent field emission properties of CNTs grown on the surface of carbon fibres in carbon cloth, which could potentially be used in flat panel displays. Boskovic et al. [163] reported low temperature DC PECVD synthesis of carbon nanofibres on the surface of carbon fibres (Fig. 14) using Co colloid catalyst. It was also demonstrated that using the same Co colloid catalyst and the same PECVD method it is possible to grow carbon nanotubes and nanofibres on arbitrary micro-machined silicon three-dimensional "micro-grass" surfaces [164]. Hart et al. [164] demonstrated that conventional metal deposition techniques can be used to obtain uniform SWCNT and DWCNT film growth by atmospheric pressure thermal CVD on arbitrarily micro-structured silicon "micro-grass" surfaces, where the surfaces face the deposition source in any orientation from vertical to horizontal. These principles can be applied to grow a wide variety of nanostructures on microstructures having arbitrary 3D topography, extending the fabrication capability for hierarchically micro-structured and nano-structured substrates. Carbon fibres bundles, woven and non-woven carbon fibre cloth can be used as a three-dimensional scaffold for carbon nanotube synthesis on surface of carbon fibres and in the empty space between them. Boskovic has found [165] that when the catalyst is

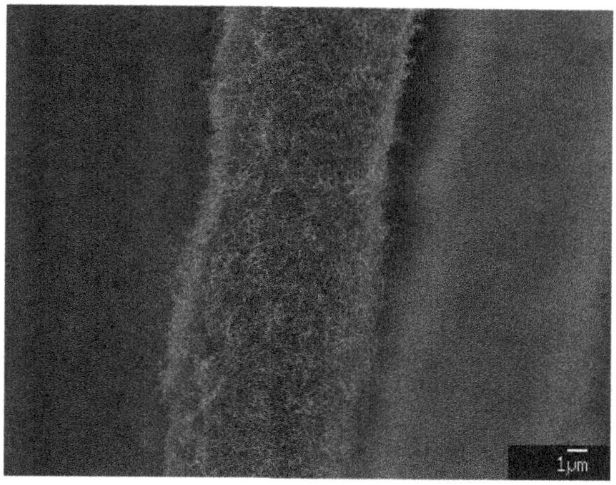

Fig. 14 Carbon nanotubes synthesised on the carbon fibre surface using thermal CVD

impregnated and dispersed within a fibrous matrix (carbon or ceramic fibre cloth or felt), rather than being left on the surface, a more efficient deposition of nanofibres and/or nanotubes results. Fine iron powder catalyst dispersed in isopropanol was impregnated within a 2.5 mm thick VCL N carbon cloth, obtained from Morgan Specialty Graphite, Fostoria, OH, USA using an ultrasonic bath. The samples were then dried producing a fibrous matrix with an impregnated finely dispersed metal powder. Carbon nanotubes and nanofibres were grown using an ethylene and hydrogen mixture at 650°C. The nanotubes/nanofibres are produced in clumps originating from the surface of the catalyst particles. The amount of produced carbon nanomaterials could be controlled using variation of catalyst loading.

Veedu et al. [166] reported that well-aligned CNTs grown perpendicular to 2D woven fabric of SiC fibres improved significantly the mechanical and thermal properties. Interlaminar fracture-toughness of the resulting 3D composite has shown an improvement of 348% compared with the base composite without CNTs. The interlaminar shear sliding fracture toughness was improved by about 54%. It is also reported that addition of carbon nanotubes has significantly improved dissipation of vibration energy under cyclic loading – damping (514%). The coefficient of thermal expansion was reduced to 38% of the original value and thermal conductivity was improved by 51%. Three-dimensional composite materials containing carbon nanotubes and carbon fibres are good candidate for many potential applications. High thermal conductivity of these materials may be of use in automotive and aerospace applications and for heat distribution or hot spot control. Recently, Boskovic patented use for aircraft brake applications [167]. The high electrical conductivity of these materials could be used for example in electronic components packaging, as gas diffusion layers in fuel cells or in electromagnetic shielding. The carbon fabric impregnated with carbon nanotubes could be used for lightweight structures and for bulletproof vests.

5 Conclusions

In this chapter we presented different carbon nanostructures but the focus was particularly on carbon nanotubes, their methods of synthesis, heteroatomic doping and exquisite properties. The processing of nanotubes and macroscopic realisation of the properties through the fabrication of fibres and 3D structures is further presented and compared.

Acknowledgement Dr Krzysztof Koziol thanks The Royal Society for financial support at the University of Cambridge.

References

1. Kroto, H.W., Heath, J.R., O'Brien, S.C., Curl, R.F., Smalley, R.E.: Nature **318**, 162–163 (1985)
2. Iijima, S.: Nature **354**, 56–58 (1991)
3. Ball, P.: Nature **414**, 142–144 (2001)
4. Radushkevich, L.V., Lukyanovich, V.M.: J. Phys. Chem. **26**, 88–95 (1952)
5. Saito, Y., Yoshikawa, T., Bandow, S., Tomita, M., Hayashi, T.: Phys. Rev. B **48**, 1907–1909 (1993)
6. Popov, V.N.: Mater. Sci. Eng. R **43**, 61–102 (2004)
7. Ebbesen, T.W.: Carbon Nanotubes: Preparation and Properties, 1st edn. CRC, Boca Raton (1997)
8. Encarta: Online Encyclopedia (2004)
9. Terrones, M.: Int. Mater. Rev. **49**, 325–377 (2004)
10. Pierson, H.O.: Handbook of Carbon, Graphite, Diamond and Fullerenes. William Andrew Publishing, Norwich (1993)
11. Moshary, F., Chen, N.H., Silvera, I.F., Brown, C.A., Dorn, H.C., Devries, M.S., Bethune, D.S.: Phys. Rev. Lett. **69**, 466–469 (1992)
12. Guo, T., Nikolaev, P., Rinzler, A.G., Tomanek, D., Colbert, D.T., Smalley, R.E.: J. Phys. Chem. **99**, 10694–10697 (1995)
13. Baker, R.T.K., Barber, M.A., Harris, P.S., Feates, F.S., Waite, R.J.: J. Catal. **26**, 51–62 (1972)
14. Baker, R.T.K., Waite, R.J.: J. Catal. **37**, 101–105 (1975)
15. Endo, M., Takeuchi, K., Igarashi, S., Kobori, K., Shiraishi, M., Kroto, H.W.: J. Phys. Chem. Solids **54**, 1841–1848 (1993)
16. Sarkar, A., Kroto, H.W., Endo, M.: Carbon **33**, 51–55 (1995)
17. Endo, M., Takeuchi, K., Kobori, K., Takahashi, K., Kroto, H.W., Sarkar, A.: Carbon **33**, 873–881 (1995)
18. Endo, M.: Chemtech **18**, 568–576 (1988)
19. Tibbetts, G.G.: J. Cryst. Growth **66**, 632–638 (1984)
20. Tibbetts, G.G.: Carbon **27**, 745–747 (1989)
21. Tibbetts, G.G.: J. Cryst. Growth **73**, 431–438 (1985)
22. Tibbetts, G.G., Devour, M.G., Rodda, E.J.: Carbon **25**, 367–375 (1987)
23. Baker, R.T.K.: Carbon **27**, 315–323 (1989)
24. Walker, P.L., Rakszawski, J.F., Imperial, G.R.: J. Phys. Chem. **63**, 133–140 (1959)
25. Dresselhaus, M.S., Dresselhaus, G., Surihara, K., Spain, I.L., Goldberg, H.A.: Graphite Fibers and Filaments. Springer, Berlin (1988)

26. Ci, L.J., Zhao, Z.G., Dai, J.B.: Carbon **43**, 883–886 (2005)
27. Dresselhaus, M.S., Dresselhaus, G., Avouris, Ph.: Springer-Verlag, Germany (2001)
28. Ebbesen, T.W.: Phys. Today **49**, 26–32 (1996)
29. Ebbesen, T.W., Ajayan, P.M.: Nature **358**, 220–222 (1992)
30. Colbert, D.T., Zhang, J., Mcclure, S.M., Nikolaev, P., Chen, Z., Hafner, J.H., Owens, D.W., Kotula, P.G., Carter, C.B., Weaver, J.H., Rinzler, A.G., Smalley, R.E.: Science **266**, 1218–1222 (1994)
31. Iijima, S., Ichihashi, T.: Nature **363**, 603–605 (1993)
32. Bethune, D.S., Kiang, C.H., Devries, M.S., Gorman, G., Savoy, R., Vazquez, J., Beyers, R.: Nature **363**, 605–607 (1993)
33. Ando, Y., Iijima, S.: Jpn. J. Appl. Phys. Part 2 **32**, L107–L109 (1993)
34. Ando, Y.: Fullerene Sci Technol **2**, 173–180 (1994)
35. Wang, M., Zhao, X.L., Ohkohchi, M., Ando, Y.: Fullerene Sci Technol **4**, 1027–1039 (1996)
36. Ando, Y., Zhao, X., Ohkohchi, M.: Carbon **35**, 153–158 (1997)
37. Zhao, X., Ohkohchi, M., Wang, M., Iijima, S., Ichihashi, T., Ando, Y.: Carbon **35**, 775–781 (1997)
38. Wang, X.K., Lin, X.W., Dravid, V.P., Ketterson, J.B., Chang, R.P.H.: Appl. Phys. Lett. **66**, 2430–2432 (1995)
39. Ando, Y., Zhao, X.L., Ohkohchi, M.: Jpn. J. Appl. Phys. Part 2 **37**, L61–L63 (1998)
40. Tai, Y., Inukai, K., Osaki, T., Tazawa, M., Murakami, J., Tanemura, S., Ando, Y.: Chem. Phys. Lett. **224**, 118–122 (1994)
41. Guo, T., Diener, M.D., Chai, Y., Alford, M.J., Haufler, R.E., Mcclure, S.M., Ohno, T., Weaver, J.H., Scuseria, G.E., Smalley, R.E.: Science **257**, 1661–1664 (1992)
42. Thess, A., Lee, R., Nikolaev, P., Dai, H.J., Petit, P., Robert, J., Xu, C., Lee, Y.H., Kim, S.G., Rinzler, A.G., Colbert, D.T., Scuseria, G.E., Tomanek, D., Fischer, J.E., Smalley, R.E.: Science **273**, 483–487 (1996)
43. Bandow, S., Rao, A.M., Williams, K.A., Thess, A., Smalley, R.E., Eklund, P.C.: J. Phys. Chem. B **101**, 8839–8842 (1997)
44. Chiang, I.W., Brinson, B.E., Huang, A.Y., Willis, P.A., Bronikowski, M.J., Margrave, J.L., Smalley, R.E., Hauge, R.H.: J. Phys. Chem. B **105**, 8297–8301 (2001)
45. Ishii, H., Kataura, H., Shiozawa, H., Yoshioka, H., Otsubo, H., Takayama, Y., Miyahara, T., Suzuki, S., Achiba, Y., Nakatake, M., Narimura, T., Higashiguchi, M., Shimada, K., Namatame, H., Taniguchi, M.: Nature **426**, 540–544 (2003)
46. Puretzky, A.A., Geohegan, D.B., Fan, X., Pennycook, S.J.: Appl. Phys. A **70**, 153–160 (2000)
47. Sen, R., Ohtsuka, Y., Ishigaki, T., Kasuya, D., Suzuki, S., Kataura, H., Achiba, Y.: Chem. Phys. Lett. **332**, 467–473 (2000)
48. Kokai, F., Takahashi, K., Yudasaka, M., Iijima, S.: J. Phys. Chem. B **104**, 6777–6784 (2000)
49. Bandow, S., Asaka, S., Saito, Y., Rao, A.M., Grigorian, L., Richter, E., Eklund, P.C.: Phys. Rev. Lett. **80**, 3779–3782 (1998)
50. Kataura, H., Kumazawa, Y., Maniwa, Y., Ohtsuka, Y., Sen, R., Suzuki, S.: Carbon **38**, 1691–1697 (2000)
51. Kataura, H., Kimura, A., Ohtsuka, Y., Suzuki, S., Maniwa, Y., Hanyu, T., Achiba, Y.: Jpn. J. Appl. Phys. Part 2 **37**, L616–L618 (1998)
52. Journet, C., Bernier, P.: Appl. Phys. A Mater. Sci. Process. **67**, 1–9 (1998)
53. Yacaman, J.M., Yoshida, M.M., Rendon, L.: Appl. Phys. Lett. **62**, 657–659 (1993)
54. Satishkumar, B.C., Govindaraj, A., Rao, C.N.R.: Chem. Phys. Lett. **307**, 158–162 (1999)
55. Hernadi, K., Fonseca, A., Nagy, J.B., Bernaerts, D., Lucas, A.A.: Carbon **34**, 1249–1257 (1996)
56. Maruyama, S., Kojima, R., Miyauchi, Y., Chiashi, S., Kohno, M.: Chem. Phys. Lett. **360**, 229–234 (2002)
57. Murakami, Y., Miyauchi, Y., Chiashi, S., Maruyama, S.: Chem. Phys. Lett. **377**, 49–54 (2003)

58. Murakami, Y., Chiashi, S., Miyauchi, Y., Hu, M.H., Ogura, M., Okubo, T., Maruyama, S.: Chem. Phys. Lett. **385**, 298–303 (2004)
59. Kumar, M., Ando, Y.: Diam. Relat. Mater. **12**, 998–1002 (2003)
60. Kumar, M., Ando, Y.: Chem. Phys. Lett. **374**, 521–526 (2003)
61. Kumar, M., Kakamu, K., Okazaki, T., Ando, Y.: Chem. Phys. Lett. **385**, 161–165 (2004)
62. Baker, R.T.K., Harris, P.S. Chemistry and physics of carbon. In: Walker, P.L., Thrower, P.A. (eds.) Dekker, New York (1978)
63. Dai, H., Rinzler, A.G., Nikolaev, P., Thess, A., Colbert, D.T., Smalley, R.E.: Chem. Phys. Lett. **260**, 471–475 (1996)
64. Cheng, H.M., Li, F., Sun, X., Brown, S.D.M., Pimenta, M.A., Marucci, A., Dreselhaus, G., Dresselhaus, M.S.: Chem. Phys. Lett. **289**, 602–610 (1998)
65. Satishkumar, B.C., Govindaraj, A., Sen, R., Rao, C.N.R.: Chem. Phys. Lett. **293**, 47–52 (1998)
66. Hafner, J.H., Bronikowski, M.J., Azamian, B.R., Nikolaev, P., Rinzler, A.G., Colbert, D.T., Smith, K.A., Smalley, R.E.: Chem. Phys. Lett. **296**, 195–202 (1998)
67. Kong, J., Cassell, A.M., Dai, H.J.: Chem. Phys. Lett. **292**, 567–574 (1998)
68. Flahaut, E., Govindaraj, A., Peigney, A., Laurent, C., Rousset, A., Rao, C.N.R.: Chem. Phys. Lett. **300**, 236–242 (1999)
69. Yudasaka, M., Kikuchi, R., Ohki, Y., Yoshimura, S.: Carbon **35**, 195–201 (1997)
70. Rao, C.N.R., Sen, R., Satishkumar, B.C., Govindaraj, A.: Chem. Commun. **15**, 1525–1526 (1998)
71. Ago, H., Komatsu, T., Ohshima, S., Kuriki, Y., Yumura, M.: Appl. Phys. Lett. **77**, 79–81 (2000)
72. Sen, R., Govindaraj, A., Rao, C.N.R.: Chem. Phys. Lett. **267**, 276–280 (1997)
73. Andrews, R., Jacques, D., Rao, A.M., Derbyshire, F., Qian, D., Fan, X., Dickey, E.C., Chen, J.: Chem. Phys. Lett. **303**, 467–474 (1999)
74. Nikolaev, P., Brownikowsky, M.J., Bradley, R.K., Rohmund, F., Colbert, D.T., Smith, K.A., Smalley, R.E.: Chem. Phys. Lett. **313**, 91–97 (1999)
75. Bronikowski, M.J., Willis, P.A., Colbert, D.T., Smith, K.A., Smalley, R.E.: J. Vac. Sci. Technol. A **19**, 1800–1805 (2001)
76. Couteau, E., Hernadi, K., Seo, J.W., Thien-Nga, L., Miko, C., Gaal, R., Forro, L.: Chem. Phys. Lett. **378**, 9–17 (2003)
77. Wang, Y., Wei, F., Luo, G.H., Yu, H., Gu, G.S.: Chem. Phys. Lett. **364**, 568–572 (2002)
78. Zhao, X., Inoue, S., Jinno, M., Suzuki, T., Ando, Y.: Chem. Phys. Lett. **373**, 266–271 (2003)
79. Huang, H., Kajiura, H., Tsutsui, S., Hirano, Y., Miyakoshi, M., Yamada, A., Ata, M.: Chem. Phys. Lett. **343**, 7–14 (2001)
80. Li, W.Z., Xie, S.S., Qian, L.X., Chang, B.H., Zou, B.S., Zhou, W.Y., Zhao, R.A., Wang, G.: Science **274**, 1701–1703 (1996)
81. Terrones, M., Grobert, N., Olivares, J., Zhang, J.P., Terrones, H., Kordatos, K., Hsu, W.K., Hare, J.P., Townsend, P.D., Prassides, K., Cheetham, A.K., Kroto, H.W., Walton, D.R.M.: Nature **388**, 52–55 (1997)
82. Pan, Z.W., Xie, S.S., Chang, B.H., Wang, C.Y., Lu, L., Liu, W., Zhou, M.Y., Li, W.Z.: Nature **394**, 631–632 (1998)
83. Li, J., Papadopoulos, C., Xu, J.M., Moskovits, M.: Appl. Phys. Lett. **75**, 367–369 (1999)
84. Cao, A.Y., Ajayan, P.M., Ramanath, G., Baskaran, R., Turner, K.: Appl. Phys. Lett. **84**, 109–111 (2004)
85. Fan, S.S., Chapline, M.G., Franklin, N.R., Tombler, T.W., Cassell, A.M., Dai, H.J.: Science **283**, 512–514 (1999)
86. Wei, B.Q., Vajtai, R., Jung, Y., Ward, J., Zhang, R., Ramanath, G., Ajayan, P.M.: Nature **416**, 495–496 (2002)
87. Zhao, M.W., Xia, Y.Y., Lewis, J.P., Zhang, R.Q.: J. Appl. Phys. **94**, 2398–2402 (2003)
88. Hernandez, E., Goze, C., Bernier, P., Rubio, A.: Phys. Rev. Lett. **80**, 4502–4505 (1998)
89. Hernandez, E., Goze, C., Bernier, P., Rubio, A.: Appl. Phys. A. Mater. **68**, 287–292 (1999)

90. Gao, R.P., Wang, Z.L., Bai, Z.G., de Heer, W.A., Dai, L.M., Gao, M.: Phys. Rev. Lett. **85**, 622–625 (2000)
91. Terrones, M., Jorio, A., Endo, M., Rao, A.M., Kim, Y.A., Hayashi, T., Terrones, H., Charlier, J.C., Dresselhaus, G., Dresselhaus, M.S.: Mater. Today **10**, 30–45 (2004)
92. Huang, Y.H., Gao, J.P., Liu, R.Z.: Synthetic Met. **113**, 251–255 (2000)
93. Nevidomskyy, A.H., Csanyi, G., Payne, M.C.: Phys. Rev. Lett. **91**, 105502 (2003)
94. Droppa, R., Hammer, P., Carvalho, A.C.M., dos Santos, M.C., Alvarez, F.: Journal of Non-Crystalline Solids **299**, 874–879 (2002). Part B
95. Glerup, M., Steinmetz, J., Samaille, D., Stephan, O., Enouz, S., Loiseau, A., Roth, S., Bernier, P.: Chem. Phys. Lett. **387**, 193–197 (2004)
96. Zhang, Y., Gu, H., Suenaga, K., Iijima, S.: Chem. Phys. Lett. **279**, 264–269 (1997)
97. Terrones, M., Hsu, W.K., Terrones, H., Zhang, J.P., Ramos, S., Hare, J.P., Castillo, R., Prassides, K., Cheetham, A.K., Kroto, H.W., Walton, D.R.M.: Chem. Phys. Lett. **259**, 568–573 (1996)
98. Sen, R., Satishkumar, B.C., Govindaraj, S., Harikumar, K.R., Renganathan, M.K., Rao, C.N. R.: J. Mater. Chem. **7**, 2335–2337 (1997)
99. Keskar, G., Rao, R., Luo, J., Hudson, J., Chen, J., Rao, A.M.: Chem. Phys. Lett. **412**, 269–273 (2004)
100. Liang, E.J., Ding, P., Zhang, H.R., Guo, X.Y., Du, Z.L.: Diam. Relat. Mater. **13**, 69–73 (2004)
101. Wang, X.B., Liu, Y.Q., Zhang, L., Ma, H.Z., Yao, N., Zhang, B.L.: J. Phys. Chem. B **106**, 2186–2190 (2002)
102. Lee, C.J., Lyu, S.C., Kim, H.W., Lee, J.H., Cho, K.I.: Chem. Phys. Lett. **359**, 115–120 (2002)
103. Glerup, M., Castignolles, M., Holzinger, M., Hug, M., Loiseau, A., Bernier, P.: Chem. Commun. **20**, 2542–2543 (2003)
104. Jang, J.W., Lee, C.E., Lyu, S.C., Lee, T.J., Lee, C.J.: Appl. Phys. Lett. **84**, 2877–2879 (2004)
105. Lee, Y.T., Kim, N.S., Bae, S.Y., Park, J., Yu, S.C., Ryu, H.: J. Phys. Chem. B **107**, 12958–12963 (2003)
106. Koziol, K., Shaffer, M., Windle, A.: Adv Mater **17**, 760–763 (2005)
107. Friedrichs, S., Windle, A.H., Koziol, K., Ducati, C., Midgley, P.A.: Microsc. Microanal. **11**, 1536–1537 (2005)
108. Ducati, C., Koziol, K., Friedrichs, S., Yates, T.J.V., Shaffer, M.S., Midley, P.A., Windle, A. H.: Small **2**, 774–784 (2006)
109. Ducati, C., Koziol, K., Stavrinadis, A., Friedrichs, S., Windle, A.H., Midgley, P.A.: J. Phys. Conf. Ser. **26**, 199–202 (2006)
110. Ren, Z.F., Huang, Z.P., Xu, J.W., Wang, J.H., Bush, P., Siegal, M.P., Provencio, P.N.: Science **282**, 1105 (1998)
111. Qin, L.C., Zhou, D., Krauss, A.R., Gruen, D.M.: Appl. Phys. Lett. **72**, 3437 (1998)
112. Tsai, S.H., Chao, C.W., Lee, C.L., Shih, H.C.: Appl. Phys. Lett. **74**, 3462 (1999)
113. Choi, Y.C., Shin, Y.M., Lee, Y.H., Lee, B.S., Park, G.S., Choi, W.B., Lee, N.S., Kim, J.M.: Appl. Phys. Lett. **76**, 2367 (2000)
114. Bower, C., Zhu, W., Jin, S., Zhou, O.: Appl. Phys. Lett. **77**, 830 (2000)
115. Okai, M., Muneyoshi, T., Yaguchi, T., Sasaki, S.: Appl. Phys. Lett. **77**, 3468 (2000)
116. Merkulov, V.I., Lowndes, D.H., Wei, Y.Y., Eres, G., Voelkl, E.: Appl. Phys. Lett **76**, 1534 (2000)
117. Chhowalla, M., Teo, K.B.K., Ducati, C., Rupesinghe, N.L., Amaratunga, G.A.J., Ferrary, A. C., Roy, D., Robertson, J., Milne, W.I.: J. Appl. Phys. **90**, 5308 (2001)
118. Li, J., Stevens, R., Delzeit, L., Ng, H.T., Cassell, A., Han, J., Meyyappan, M.: Appl. Phys. Lett. **81**, 910 (2002)
119. Delzeit, L., McAninch, I., Cruden, B.A., Hash, D., Chen, B., Han, J., Meyyappan, M.: J. Appl. Phys. **91**, 6027 (2002)
120. Wang, Y.H., Lin, J., Huan, C.H.A., Chen, G.S.: Appl. Phys. Lett. **79**, 680 (2001)
121. Boskovic, B.O., Stolojan, V., Khan, R.U., Haq, S., Silva, S.R.P.: Nat. Mater. **1**, 165 (2002)

122. Lee, O.J., Lee, K.H.: Appl. Phys. Lett. **82**, 3770 (2003)
123. Teo, K.B.K., Chhowalla, M., Amaratunga, G.A.J., Hasko, D.G., Pirio, G., Legagneux, P., Wyczisk, F., Pribat, D.: Appl. Phys. Lett. **79**, 1534 (2001)
124. Minea, T.M., Point, S., Granier, A., Touzeau, M.: Appl. Phys. Lett. **85**, 1244 (2004)
125. Hofmann, S., Ducati, C., Kleinsorge, B., Robertson, J.: Appl. Phys. Lett. **83**, 135 (2003)
126. Hofmann, S., Ducati, C., Kleinsorge, B., Robertson, J.: Appl. Phys. Lett **83**, 4661 (2003)
127. Cantoro, M., Hofmann, S., Pisana, S., Scardaci, V., Parvez, A., Ducati, C., Ferrari, A.C., Blackburn, A.M., Wang, K.Y., Robertson, J.: Nano Lett. **6**, 1107 (2006)
128. Kasuya, D., Yudasaka, M., Takahashi, K., Kokai, F., Iijima, S.: J. Phys. Chem. B **106**, 4947 (2002)
129. Kawai, T., Miyamoto, Y., Sugino, O., Koga, Y.: Phys. Rev. B **66**, 033404 (2002)
130. Chen, C.C., Chen, C.F., Lee, I.H., Lin, C.L.: Diam. Relat. Mater. **14**, 1897 (2005)
131. Shang, N.G., Au, F.C.K., Meng, X.M., Lee, C.S., Bello, I., Lee, S.T.: Chem. Phys. Lett. **358**, 187 (2002)
132. Wang, J.J., Zhu, M.Y., Outlaw, R.A., Zhao, X., Manos, D.M., Holloway, B.C.: Carbon **42**, 2867 (2004)
133. Lin, C.H., Chang, H.L., Tsai, M.H., Kuo, C.T.: Diam. Relat. Mater. **11**, 922 (2002)
134. Du, J.M., Liu, Z.M., Li, Z.H., Han, B.X., Sun, Z.Y., Huang, Y.: Mater. Lett. **59**, 456 (2005)
135. Wu, Y.H., Qiao, P.W., Chong, T.C., Shen, Z.X.: Adv. Mater. **14**, 64 (2002)
136. Hiraki, H., Jiang, N., Wang, H.X., Hiraki, A.: J. Phys. IV. **132**, 111 (2006)
137. Nishimura, K., Jiang, N., Hiraki, A.: IEICE Trans. Electron. **E86C**, 821 (2003)
138. Chuang, A.T.H., Boskovic, B.O., Robertson, J.: Diam. Relat. Mater. **15**, 1103 (2006)
139. Chuang, A.T.H., Robertson, J., Boskovic, B.O., Koziol, K.K.K.: Appl. Phys. Lett. **90**, 123107 (2007)
140. Vigolo, B., Penicaud, A., Coulon, C., Sauder, C., Pailler, R., Journet, C., Bernier, P., Poulin, P.: Science **290**, 1331 (2000)
141. Somoza, A.M., Sagui, C., Roland, C.: Phys. Rev. B **63**, 081403 (2001)
142. Song, W., Kinloch, I.A., Windle, A.H.: Science **302**, 1363 (2003)
143. Nikolaev, P., Bronikowski, M.L., Bradley, R.K., Rohmund, F., Colbert, D.T., Smith, K.A., Smalley, R.E.: Chem. Phys. Lett. **313**, 91 (1999)
144. Bronikowski, M.J., Willis, P.A., Colbert, D.T., Smith, K.A., Smalley, R.E.: J. Vac. Sci. Technol. A Vac. Surf. Films **19**, 1800 (2001)
145. Ramesh, S., Ericson, L.M., Davis, V.A., Saini, R.K., Kittrell, C., Pasquali, M., Billups, W.E., Adams, W.W., Hauge, R.H., Smalley, R.E.: J. Phys. Chem. B **108**, 8794 (2004)
146. Ericson, L.M., Fan, H., Peng, H., Davis, V.A., Zhou, W., Sulpizio, J., Wang, Y., Booker, R., Vavro, J., Guthy, C., Nicholas, A., Parra-Vasquez, G., Kim, M.J., Ramesh, S., Saini, R.K., Kittrell, C., Lavin, G., Schmidt, H., Adams, W.W., Billups, W.E., Pasquali, M., Hwang, W. F., Hauge, R.H., Fischer, J.E., Smalley, R.E.: Science **305**, 1447 (2004)
147. Zhang, S., Koziol, K.K.K., Kinloch, I.A., Windle, A.H.: Small **4**, 1217–1222 (2008)
148. Jiang, K., Li, Q., Fan, S.: Nature **419**, 801 (2002)
149. Zhang, M., Atkinson, K.R., Baughman, R.H.: Science **306**, 1358 (2004)
150. Zhang, M., Fang, S., Zakhidov, A.A., Lee, S.B., Aliev, A.E., Williams, C.D., Atkinson, K.R., Baughman, R.H.: Science **309**, 1215 (2005)
151. Zhang, X., Li, Q., Tu, Y., Li, Y., Coulter, J.Y., Zheng, L., Zhao, Y., Jia, Q., Peterson, D.E., Zhu, Y.: Small **3**, 244–248 (2007)
152. Li, Y.L., Kinloch, I.A., Windle, A.H.: Science **304**, 276 (2004)
153. Koziol, K., Vilatela, J., Moisala, A., Motta, M., Cunniff, P., Sennett, M., Windle, A.: Science **318**, 1892–1895 (2007)
154. Baughman, R.H., Zakhidov, A.A., de Heer, W.A.: Science **297**, 787 (2002)
155. Tennent, H., Hausslein, R.W., Leventis, N., Moy, D.: Hyperion Catalysis International, Inc. Patent US 5,846,658, 8 December 1998
156. Downs, W.B., Baker, R.T.K.: Carbon **29**, 1173 (1991)
157. Downs, W.B., Baker, R.T.K.: J. Mater. Res. **10**, 625 (1995)

158. Thostenson, E.T., Li, W.Z., Wang, D.Z., Ren, Z.F., Chou, T.W.: J. Appl. Phys. **91**, 6034 (2002)
159. Sun, X., Li, R., Villers, D., Dodelet, J.P., Desilets, S.: Chem. Phys. Lett. **379**, 99 (2003)
160. Wang, C., Waje, M., Wang, X., Tang, J., Haddon, R.C., Yan, Y.: Nano Lett. **4**, 345 (2004)
161. Marphy, M.A., Wilcox, G.D., Dahm, R.H., Marken, F.: Electrochem. Commun. **5**, 51 (2003)
162. Jo, S.H., Wang, D.Z., Huang, J.Y., Li, W.Z., Kempa, K., Ren, Z.F.: Appl. Phys. Lett. **85**, 810 (2004)
163. Boskovic, B.O., Golovko, V., Cantoro, M., Kleinsorge, B., Ducati, C., Chuang, A.T.H., Hofmann, S., Robertson, J., Johnson, B.F.G.: Carbon **43**, 2643 (2005)
164. Hart, A.J., Boskovic, B.O., Chuang, A.T.H., Golovko, V.B., Robertson, J., Johnson, B.F.G., Slocum, A.H.: Nanotechnology **17**, 1397 (2006)
165. Boskovic, B.O.: The Morgan Crucible Company Plc. Synthesis of carbon nanotubes and/or nanofibres on a porous fibrous matrix. WO 2004078649, 16 September 2004
166. Veedu, V.P., Cao, A., Li, X., Ma, K., Soldano, C., Kar, S., Ajayan, P.M., Ghasemi-Nejhad, M.N.: Nat. Mater. **5**, 458 (2006)
167. Boskovic, B.O.: Meggitt Aerospace Ltd. Carbon-carbon composite. Patent Int. Pub. No. WO 2009/004346 A1, 7 March 2007

Fullerene (C60) and its Derivatives as Resists for Electron Beam Lithography

Hasnah Mohd Zaid

Abstract The application of fullerene as a negative resist was first studied by Tada and Kanayama who verified that this material could be used as a negative electron beam resist. Its small molecule enables the resist to have a resolution of at least 20 nm. Robinson et al. demonstrated that chemical modification of C60 by adding functional groups to the C60 cage can significantly enhance the resist properties. Chemical amplification of the fullerene derivatives improves their sensitivities while maintaining their high resolution. In this chapter, the concepts of lithography and lithography techniques which include electron beam lithography technology systems are described. Current electron beam resists and their characteristics are discussed. A review of the application of fullerene and its derivatives as electron beam resists is presented. Finally, concepts of chemical amplification and current chemically amplified resists are discussed.

Device density of modern computer components has grown exponentially as predicted by Moore's Law [1] with a decrease in components sizes. Smaller devices mean a reduced interconnect length, reducing the distance electrons have to travel and thus signal delay. Although photolithography has been the technique of choice for the fabrication of microdevices for many years, electron beam lithography is a very promising lithographic technique for nanoscale patterning due to its flexibility and nearly unlimited resolution capability, able to fabricate sub-50 nm features. A factor that influences its resolution is the electron beam resists. The application of fullerene as a negative resist was first studied by Tada and Kanayama [2] who verified that this material could be used as a negative electron beam resist. Its small molecule enables the resist to have a resolution of at least 20 nm. Robinson et al. [3–5] demonstrated that chemical modification of C60 by adding functional groups to the C60 cage can significantly enhance the resist properties. Chemical amplification of the fullerene derivatives improves their sensitivities while

H.M. Zaid
Fundamental and Applied Sciences Department, Universiti Teknologi, PETRONAS, Bandar Seri Iskandar, 31750 Tronoh, Perak, Malaysia
e-mail: hasnamz@petronas.com.my

N. Yahya (ed.), *Carbon and Oxide Nanostructures*, Adv Struct Mater 5,
DOI 10.1007/8611_2010_13, © Springer-Verlag Berlin Heidelberg 2010,
Published online: 25 June 2010

maintaining their high resolution [6, 7]. In this chapter, the concepts of lithography and lithography techniques which include electron beam lithography technology systems are described. Current electron beam resists and their characteristics are discussed. A review of the application of fullerene and its derivatives as electron beam resists is presented. Finally, concepts of chemical amplification and current chemically amplified resists are discussed.

1 Lithography

Lithography is the process of transferring patterns to a substrate and can be used to fabricate integrated circuits. A beam of radiation, such as photons or ions is projected onto a suitable resist material coated on a wafer, causing chemical changes in the resist. For instance, the solubility of the resist in a certain solvent may change [8]. In photolithography, a photopolymer is exposed to visible light or UV radiation through a mask. The solubility of the exposed area of the polymer is increased (positive tone resist) or decreased (negative tone resist) and the exposed, or unexposed, resist is removed respectively, using a developing chemical, producing the desired pattern on the substrate for further processing. The principle of operation of lithography is shown in Fig. 1. A radiation sensitive material, known as a resist, is coated onto the substrate as shown in Fig. 1a. The resist is then exposed to a beam of radiation through a mask which shadows certain areas of the resist, as in Fig. 1b. The exposed areas of the resist undergo chemical alteration. In the case of a positive resist, its solubility is increased relative to the unexposed resist allowing it to be removed by a developer solvent, while the exposed areas of a negative resist become less soluble and are left behind upon development as shown in Fig. 1c [9].

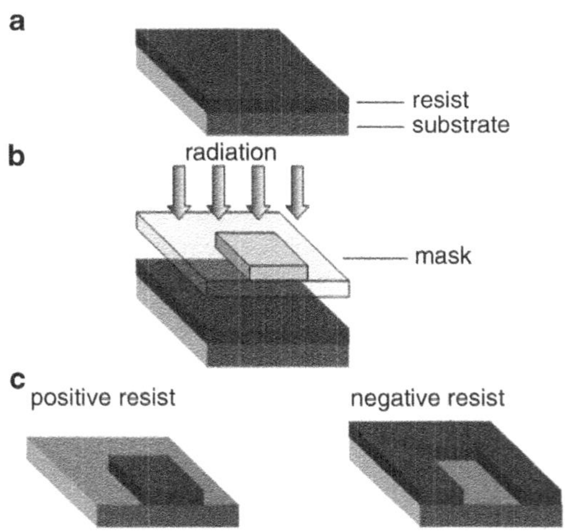

Fig. 1 The principles of lithography. (**a**) The substrate is coated with a radiation sensitive resist. (**b**) The resist is irradiated through a mask causing chemical modifications in the exposed areas. (**c**) In a positive tone resist, the exposed areas are removed upon development while in a negative tone resist, the exposed areas are retained

The final objective of the lithographic process is the accurate replication of the pattern originally specified by the device designer onto the substrate. Its success depends on the physics and chemistry of resist exposure and development, and the ensuing pattern transfer.

1.1 Photolithography

Photolithography (also known as optical lithography) has been the technique of choice for the fabrication of microdevices for more than 50 years, and still is the workhorse for volume manufacturing of integrated circuits [10]. Light (visible or near-UV) is projected through a mask, defining the desired pattern, and then focused onto a photoresist on the wafer. The pattern defined in the polymer film on the substrate is then used as a mask for further processing of the substrate. There are three primary exposure methods: contact, proximity, and projection.

In contact exposure, the wafer image is formed by placing the photoresist-coated wafer in contact with the mask and exposing it to light through the mask [11]. The mask used is a transparent glass plate with light-blocking patterns formed, for instance, by a metal coating. The pattern produced on the wafer is the same size as that on the mask. Due to the contact between resist and mask, very high resolution is possible since, as shown in Fig. 2, it is not affected by diffraction.

However, this method causes contamination of the mask by the resist and tends to trap debris between the resist and mask, causing damage to the mask and defects in the resist. Furthermore, uniform contact between mask and wafer is difficult due to surface structures and warping of wafer. Mask fabrication is also difficult since

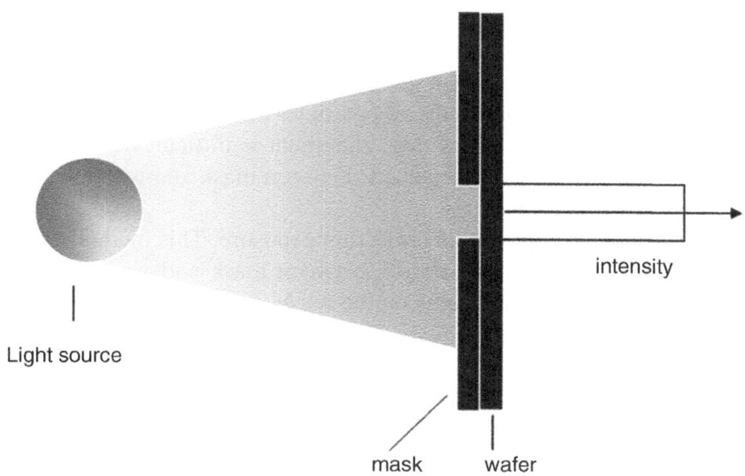

Fig. 2 Contact exposure is not affected by diffraction. The light intensity distribution projected through the mask is shown to be uniform

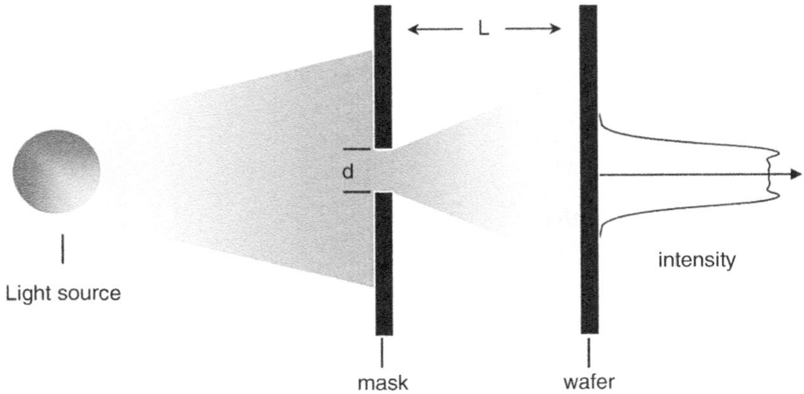

Fig. 3 Near field (Fresnel) diffraction of light through an opening, d, when $L < d^2/\lambda$ in proximity exposure. The light intensity distribution shows the diffraction effect

the patterns on the mask are of the same size as the desired final features. Due to these problems, contact exposures are rarely used.

In proximity exposure, a small gap, 10–25 μm wide, is maintained between the wafer and the mask during exposure. As in contact exposure, this method also has a 1:1 magnification ratio, (i.e. the size of the pattern imaged on the wafer is the same as that on the mask). The gap helps to reduce wear and tear to the mask due to contact. However the resolution of the method is not as good as contact printing due to near field (Fresnel) diffraction. Diffraction is an inevitable consequence of the wave nature of light, and is responsible for the spreading of light as it passes through an opening with a dimension close to (or less than) the order of its wavelength. Some of the exposing light propagates at divergent angles causing a larger area of the resist to be exposed, as shown in Fig. 3.

Other than diffraction, there are a few other disadvantages of proximity printing. Although problems with trapped dust and particles are reduced due to the gap, they are not totally eliminated. Besides that, alignment is difficult due to wafer warp, which causes a variation in the distance L between mask and wafer. Furthermore, 1:1 mask fabrication is difficult.

The third method of exposure is projection exposure. This method avoids mask damage entirely with a total separation between mask and wafer. In projection lithography, an image of the patterns on the mask, usually demagnified by four or five times, is projected onto the resist-coated wafer, which is many centimetres away, using a system of lenses [11] as shown in Fig. 4. A stepper system with a high speed stage that moves the wafer is used. Instead of a full field exposure, where the entire wafer is exposed at once, as in contact and proximity exposure, a stepper exposes only part of the wafer at a time and repeats the process until the entire wafer is exposed [8]. Furthermore, the demagnification of the pattern allows less constraint on the mask pattern accuracy, since the final patterns are four or five times smaller than those on the mask, and therefore any defect or variation on the

Fig. 4 Schematic diagram of projection photolithography

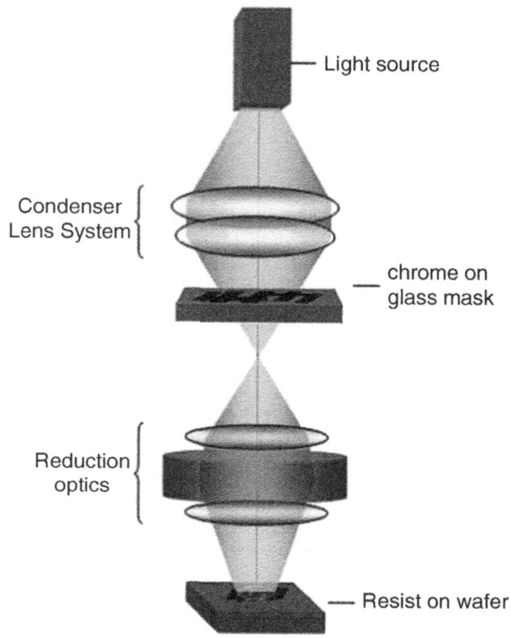

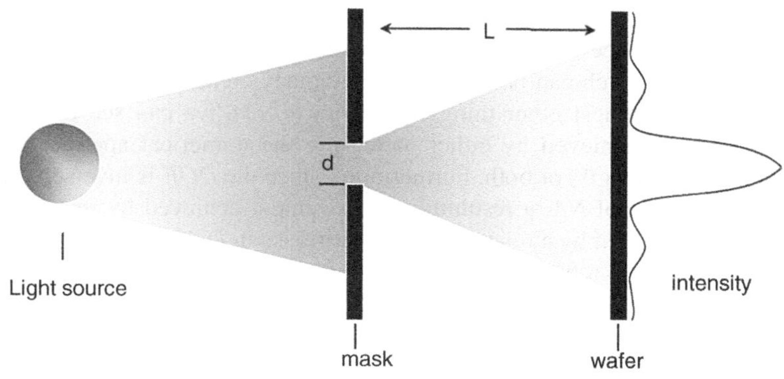

Fig. 5 Far field (Fraunhoffer) diffraction of light through an opening d when $L > d^2/\lambda$ in projection photolithography. The light intensity distribution shows the diffraction effect

patterns would be significantly reduced. The problem with projection photolithography is that as sizes of exposed details approach the wavelength of the exposing radiation, diffraction effects at the edges of the patterns become prominent. The large separation between mask and wafer, L, causes far field (Fraunhoffer) diffraction, as shown in Fig. 5.

The limits and performance of projection photolithography can be characterized by two equations. The minimum resolved feature size, $d_{\min}$, or resolution, is given by the equation

$$d_{\min} = k_1 \frac{\lambda}{NA} \quad (1)$$

where λ is the wavelength of the exposing light and k_1 is a characteristic constant of the specific lithographic process, which typically has a value in the range of 0.5–1.0 [8]. The value of k_1 generally depends on the lithographic equipment, resist, process parameters, the type of mask and the pattern being imaged. *NA* is the numerical aperture of the optical system, which is equal to the sine of the angle subtended by the objective lens of the system multiplied by the refractive index of the surrounding medium (~1 for air) [12]. The *NA* of optical lithography systems today ranges between 0.5 and 0.6 [13]. Based on (1) and the values of k_1 and *NA*, the minimum feature would be approximately equal to the wavelength of the light used. The depth of focus, *DOF*, or the range over which the image is adequately sharp, is given by the equation

$$DOF = k_2 \frac{\lambda}{NA^2} \quad (2)$$

Here, k_2 is another characteristic constant of the specific lithographic process, and is usually about 1.0. Depth-of-focus is important in defining whether an optical projection system is physically realizable. A large value of *DOF* is desirable since it increases the tolerance of the process to deviation of the substrate surface from perfect planarity, which can be caused by previously created surface structures or wafer warping, amongst other things. Referring to (1), we can see that a better resolution can be achieved by either increasing the numerical aperture (*NA*) or reducing the wavelength, or both. Furthermore, since the *DOF* is inversely proportional to the square of *NA*, a resolution improvement achieved by increasing *NA* would be accompanied by a relatively larger decrease in *DOF*. Therefore, reducing wavelength is usually a better option.

In the past, resolution improvement in optical lithography has typically been accomplished by decreasing the wavelength of light used. The progression from using g-line (436 nm) to the i-line (365 nm) light source, to using deep ultraviolet of wavelengths 248 and 193 nm, has improved the imaging resolution from 1 μm to sub-100 nm [14]. Deep Ultraviolet Lithography (DUV) can generate devices as small as 65 nm. However, reducing wavelength also decreases the depth of focus although to a lesser extent than increasing *NA*. To counteract this, the proportionality constants k_1 and k_2 have been improved by applying better resists, resist processes and resolution enhancement technologies, which have been used to improve lithographic performance at or below the diffraction limit.

Gordon Moore predicted in 1965 that the number of transistors in an integrated circuit would double every 18 months [1]. The famous Moore's Law has proven very accurate at projecting future processing power [15], as the device density of modern computer components (i.e. the number of transistors per unit area) continues to grow exponentially [16]. A higher resolution is always desired, as it means more complex integrated circuits containing larger numbers of components per unit area of substrate material can be produced. Smaller devices mean a reduced interconnect length, reducing the distance electrons have to travel and thus increasing switching speed and improving efficiency. For example in microprocessors, the minimum gate length of the transistors determines the speed of information processing of the chips [17]. Power dissipation due to heat is also reduced with reduced size. Smaller features also mean lighter and more portable devices. In sensors, the sensitivities are increased with reductions in size due to the corresponding increase in surface area [18, 19]. In terms of economy, the reduction in size means a reduction in material used and ultimately a reduction in cost, since, if the cost per wafer is maintained, there will be a significant reduction in cost per component. In general, miniaturization leads to systems with improved capability and lower prices.

So far, photolithography has shown remarkable progress in improving resolution to answer the demands of Moore's law and the semiconductor roadmaps via the reduction of radiation wavelength to 193 nm and the application of wavefront engineering. However, each shift in wavelength had to be accompanied by expensive efforts to develop appropriate light sources and imaging optics. The wavefront engineering used to improve resolution requires an increase in production cost due to mask complexity. Even then, further reduction of feature sizes would require further wavelength reduction. 157-nm lithography used to be the likely answer. Unfortunately, the development of lithography using this wavelength faces several issues. The lens and mask material used at other wavelengths are unsuitable for use with 157 nm radiation. At this wavelength, the only suitable lens material is CaF_2 (calcium fluoride), the supply of which is believed to be inadequate for volume production [20]. Moreover, CaF_2 lenses have poor longevity [21], and the effect of birefringence, a phenomenon of double refraction of light as it passes through an optically anisotropic medium such as CaF_2, is more prominent at this low wavelength [20]. The other problem is to develop a cost-effective mask pellicle, which is the soft material used to protect the mask from contamination; since the organic material used for other wavelengths do not provide the necessary transmittance and longevity. Furthermore, there is a lack of resists suitable for this wavelength. The limits of photolithography and issues of cost and throughput have motivated the development of alternative techniques of lithography.

1.2 Other Lithography Techniques

Due to the limits of photolithography, several alternative lithography techniques have been studied. However, most of these techniques have problems with cost, throughput, and/or practicality. Some notable techniques include dip-pen lithography, imprint, ion projection lithography (IPL) and X-ray lithography.

1.2.1 Dip-Pen Nanolithography

Dip pen nanolithography (DPN) is a scanning probe nanopatterning technique in which an atomic force microscope (AFM) tip is used to deliver molecules to a surface via a solvent meniscus. In DPN, the AFM tip acts as a "nib" of a pen, a solid-state substrate as "paper" and molecules with a chemical affinity for the solid-state substrate as "ink". Capillary transport of molecules from the AFM tip to the solid substrate is used to directly "write" patterns consisting of a small collection of molecules onto the substrate [22]. This direct-write technique offers high-resolution patterning capabilities for a number of molecular and biomolecular 'inks' on a variety of substrates, such as semiconductors and metals. The main advantage of DPN is its simplicity, and ability to achieve resolution comparable to more expensive and sophisticated competitive lithographic methods. However, DPN has major drawbacks which include its very slow speed writing, inherent to the AFM motion. Moreover, the 'ink' needs to be replenished periodically, which involves dismounting the AFM probe, interrupting the writing process.

1.2.2 Nanoimprint Lithography

Nanoimprint Lithography (NIL) is a method of transferring patterns by mechanical means. NIL uses a hard mould, which contains nanoscale features defined on its surface, to emboss a resist cast under controlled temperature and pressure conditions, creating a thickness contrast in the resist [23]. There are several variants of nanoimprint lithography, with slight process differences with one another. In Thermoplastic Nanoimprint Lithography, a thin layer of imprint resist, a thermal plastic polymer, is spin-coated onto the substrate. Then the mould, or template, is pressed against the resist and the assembly is heated until the polymer film melts and conforms into the patterns on the template. After being cooled down, the template is separated from the substrate, leaving behind a patterned resist. In Photocurable Nanoimprint Lithography, a transparent template is pressed into a low viscosity photocurable resist liquid such that it conforms to the topology of the template. Then, instead of heating as in the Thermoplastic Nanoimprint, the resist is irradiated with UV light to cure it, producing a relatively rigid polymer network [24].

Although NIL has proved to be successful in nanopatterning, and is capable of 14 nm pitch lines [24], it has several limitations as a flexible lithographic technique. First of all, the pattern can be degraded during removal of the mould from the resist. Furthermore, NIL requires perfect planary of both substrate and template for precise pattern transfer. Fabrication of a good quality 1× template is also a challenge. The direct contact between template and resist results in fast wear and contamination. Since the features of the mould physically deform and displace the polymer, larger features on the mould would displace more polymer material over larger distances. Therefore, larger features would be more difficult to imprint.

Besides that, the high viscosity of the fluid and the complexity of the mould pattern can result in an incomplete pattern transfer [23].

1.2.3 Ion Projection Lithography

Ion Projection Lithography (IPL) uses lightweight ions, such as H+, H_2+, H_3+, or He+, to expose the photoresist. The ions are accelerated by an electric field and the beam usually operates between 50 and 150 keV [25]. The ions pass through a patterned stencil mask and are focused and projected on to the wafer by electrostatic lenses. The wavelength of ions is extremely short; H+ with energy 150 keV has a wavelength of less than 10^{-4} nm; and thus diffraction can be ignored. As the ions penetrate the resist, they lose a large proportion of their energy to the electrons in the molecules of the resist and substrate, so their path is almost straight line enabling good resolution. Feature sizes of 65 nm have been achieved with IPL and a throughput of 50–120 wafers per hour of 200 mm wafers has been estimated. However, IPL has some major disadvantages. The only viable mask is a stencil mask, with holes in the mask to allow ions to pass through. This severely circumscribes the available patterns, not allowing for instance, annular structures (donut shapes) in a single mask. To form closed paths on the wafer, two masks are required with extremely tight alignment. The high energy ion beams can induce erosion and damage to both mask and wafers. Absorbed ions cause mask heating that leads to distortion to the patterns in the mask. Furthermore, ion beams require vacuum operation which limits access to the lithography machine.

1.2.4 X-Ray Lithography

The X-ray exposure system is similar, in principle, to photolithographic contact printing, with an X-ray beam, of wavelength between 0.4 and 4 nm, being used to expose the sensitive material instead of a light beam. The X-ray beam is usually generated by a synchrotron radiation source. A mask, made of a thin membrane which allows X-ray to pass through, such as silicon nitride, with patterns made of X-ray absorbing material, such as gold, is used [26]. In X-ray proximity lithography, the mask is held within a few microns of the resist-coated substrate, allowing large areas to be exposed, and is thus suitable for mass production of circuits. X-ray lithography has produces features as small as 20 nm [27]. One problem with X-ray lithography is that, since the mask is placed very close to the substrate, its pattern must be the same size as the final desired features on the chips, making mask fabrication more difficult and expensive. The mask must also be very thin to be transparent to the radiation, and requires extremely tight alignment. Mask damage would also occur quickly due to the high energy of X-ray radiation. In addition to that, the cost of X-ray lithographic tools, which include the synchrotron X-ray source and mask, is very high [28].

2 Electron Beam Lithography

In electron beam lithography (EBL), electrons are used to write a pattern in an electron sensitive resist coated on the substrate. The technique consists of scanning a beam of electrons across a surface covered with a resist film, thus depositing energy in the desired pattern into the film [29]. The main advantage of EBL over other methods of lithography is its very high resolution which is due to the very small spot size of the electron beams, and the ability to create arbitrary patterns rather than requiring a mask. The wavelength of the electron beam is so small that diffraction is negligible. The de Broglie wavelength of an electron is given by the equation

$$\lambda = \frac{hc}{eV} \tag{3}$$

where λ is in meters and the accelerating voltage V is in Volts. For example, electrons accelerated at 10–100 keV would have a wavelength of 0.0123–0.0039 nm. Therefore, the theoretical resolution is not limited by diffraction. Arbitrary patterns are possible because the electron beam can be steered using electrostatic and electromagnetic fields which can be generated from a virtual (computer) mask.

The most important current use of EBL is in photomask production. Masks are made by coating a chrome clad glass plate with e-beam sensitive resist layer, which is subsequently exposed and developed to generate the required pattern on the mask. In contrast to optical lithography systems, electron-beam lithography systems are not limited by diffraction, but instead, their ultimate attainable resolution is limited by beam–solid interactions, beam diameter which in turn is affected space charge, and the resist used. There are two basic types of electron beam exposure equipment – “direct write” and “projection systems” [30].

2.1 *Direct Write*

Direct write EBL systems are the most common. Most direct write systems use an electron beam with small diameter, that is moved with respect to the wafer to expose the wafer one ‘pixel’ at a time. The electron beam is focused to a fine spot at the surface of the resist, and is scanned electronically to trace out the desired pattern in the resist film [29], blanking the beam to move from one structure to the next. Direct write systems can be classified as raster scan or vector scan. In raster scan, the beam travels over the entire substrate, turning on and off depending whether the area is to be exposed or not. With the vector scan, the beam scans selected areas only. After a certain area is completely scanned, the beam is turned off and moved to another area that needs to be exposed. The raster scan is more common since it is simpler and cheaper. However it is comparatively slow. The vector scan, on the other hand, is faster, but since it requires more complicated hardware and software, is more expensive. Direct write is a method that can be used to generate submicron patterns, but, since the process takes a long time, it is not suitable for industrial mass production of circuits.

2.2 Projection Systems

Electron projection lithography (EPL) is basically similar to optical lithography but, instead of a quartz mask with chromium patterns, uses a solid membrane with holes (stencil mask) to pattern the electron beam [13]. Electrons are absorbed in the solid parts of the mask. The short wavelength of the electron beam allows EPL systems to achieve extremely small feature sizes, significantly less than 50 nm. The projection system can expose a large area in a short time, but patterns can only be reproduced from an appropriate mask fabricated by some other method. As mentioned earlier, stencil masks limit the possible patterns, since they do not allow, for instance, annular structures in a single mask, so at least two masks are required to form closed paths on the wafer. Another problem with EPL is that the electrons absorbed in the mask cause it to heat up and distort. One of the first EPL tools, Scattering with Angular Limitation Projection Electron beam Lithography (SCALPEL), used masks with scattering contrast to overcome the mask heating problem. The SCALPEL mask was made up of thin membrane of low atomic number material which allows electron to pass through, and patterns made up of a high atomic-number material which is transparent but scattered electrons at large angles, to be subsequently stopped by a separate aperture plate [31]. However, there were some problems with the SCALPEL technology. The mutual repulsion of electrons, or space charge effect, led to the defocusing of the beam. When low beam current was used to reduce space charge effects, it resulted in a very low throughput ($\sim$1.5 wafers per hour per μA of current) [32]. Furthermore, although the mask membrane was transparent to the electrons, some small energy loss did occur which led to image blur, affecting the resolution. When the beam current was increased for higher throughput, the space charge effects added to the image blur. If the beam size was enlarged to increase the effective electron field at the wafer (another way to increase throughput) it compromised the support of the mask design and reduced pattern placement performance. Hence, generally, the major problem for SCALPEL was throughput.

3 Electron Beam Resists

Electron beam resists are the recording and transfer media for electron beam lithography, with a function similar to film in photography. They are usually polymeric materials that are modified by exposure to electrons. Following exposure, the resist is developed by immersion in a suitable solvent which would remove either the exposed or unexposed areas of the resist. After development, the resist pattern is most often then used either as a mask for subsequent deposition of a material, such as dopants or a metal layer, or as an etch mask during etching of the underlying material to alter its topology. Therefore, as well as sensitivity to electrons, the resist must be able to protect the underlying substrate during subsequent processing, that is, it should be mechanically stable to allow high quality deposition and durable enough to resist etchants.

For pattern recording, the resist must be capable of easy application to give uniform and reproducible film thickness, sufficiently sensitive to the exposing radiation to be economically useful and have adequate resolution. For subsequent process steps, it must have strong adhesion to the substrate, be durable, and have a sufficiently high melting point so as to allow deposition and etching, but also be easily and completely removable after the subsequent processing [30].

3.1 Exposure of Electron Beam Resists

There are two types of resist, positive tone and negative tone. In the former, the exposed areas are removed by a suitable developer, whereas the unexposed areas are removed for a negative resist. Typically a positive tone resist would be used when the retained area is more than 50% of the overall area, and negative tone for less than 50%. When an electron strikes the resist, it can cause a number of different reactions. For instance, two molecules may crosslink forming a larger less soluble molecule (negative tone).

Alternatively, a polymer chain may be broken into smaller fragments (chain scission) increasing solubility (positive tone). Both reactions can happen at the same time, but in most materials, one reaction dominates over the other. After irradiation, there is a net increase or decrease in the solubility of the resist [30] resulting in either positive or negative tone. Electron irradiation can also cause an extensive rupture of the main chain to form volatile fragments so that no development is required, as in self-developing resists, which can only be positive tones. Another type of reaction is a change in polarity, bringing about a dual tone resist. An example is the mixture of poly(4-*t*-butoxycarbonyl oxystyrene) and an 'onium' salt. Exposure leads to a polar compound which would produce a positive tone image if developed in polar solvents, and a negative tone one if developed in non-polar solvents [30].

3.2 Electron Solid Interaction

Although the resolution of electron beam lithography is not limited by diffraction, it is very much limited by the scattering of electrons which occur when they enter the resist and penetrate further into the substrate. The two types of scattering are forward and back-scattering. Forward scattering occurs when electrons are deflected by small angles as they enter a resist. These small angle deflections lead to an overall increase of the beam diameter which results in the exposure of a larger area in the resist, reducing resolution. The extent of the diameter increase depends on both the beam energy and resist thickness. Since forward scattering causes the beam diameter to increase as the electron penetrates further into the resist, its effect can be reduced by using thinner films. Forward scattering can also be reduced by increasing the beam energy.

Backscattering is a process whereby electrons are deflected at large angles and occurs mostly in the substrate, which typically has a higher atomic number than the resist. Backscattering can be reduced by using electrons with lower energy, or using substrates with lower atomic weight. Backscattered electrons sometimes return into the resist at very large distances from the point where they first entered it, causing the proximity effect, a phenomenon where certain areas of a resist receive a larger dose than intended, due to backscattered electrons from neighbouring areas, as shown in Fig. 6. The proximity effect is one of the most serious problems experienced in EBL.

When the electrons in the electron beam enter, or are backscattered into, the resist they can deposit energy in the film. These electrons can collide inelastically with the electrons in the molecules of the resist or substrate, dissipating much of their energy in the form of secondary electrons. These secondary electrons have energies in the 2–50 eV range and are responsible for the majority of exposure of the resist [29]. They lead to an effective widening of the beam diameter, affecting the resolution but since their range in resist is only a few nanometers, they contribute little to the proximity effect.

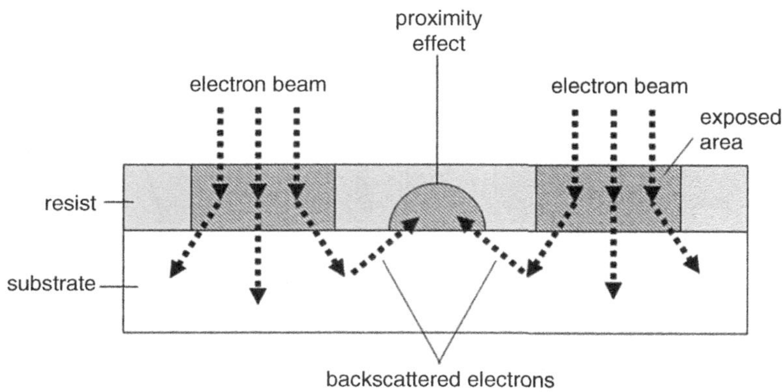

Fig. 6 Exposure of resist by backscattered electrons from neighbouring areas results in proximity effect

3.3 *Characteristics of Resists*

There are a number of characteristics that determines whether a resist is useful, the main ones being sensitivity, resolution, contrast, and etch resistance.

3.3.1 Sensitivity

The sensitivity of a resist is a measure of the dose of electrons required to expose it. A high sensitivity resist is generally desirable because less time is required to expose patterns. The sensitivity can be determined by plotting the thickness of resist after development against the exposure dose. This is also called the response curve, and is shown in Fig. 7. When a resist is exposed to radiation, its solubility gradually changes. For a positive resist, it becomes more soluble, so the thickness after development would decrease with dose. For a negative resist, the solubility decreases, so the retained film thickness increases with dose. To plot the response curve, areas on the resist are exposed to a range of electron doses, and then developed. The dose, D, of electrons received by an area of the film is given by equation:

$$\mathrm{D} = \frac{It}{A} \tag{4}$$

where I is the sample current of electrons in amps, t, the exposure time in seconds and A, the area of the exposure site in square centimeters. The retained film thickness in the exposed areas is measured using a surface profiler and is plotted against the dose received by that area, on a log-linear scale. The sensitivity of a positive resist is defined as the dose necessary to clear all of the film from the substrate (D_2 in Fig. 7a), while for a negative resist it is defined as the dose at which 50% of the resist thickness is retained (D_3 in Fig. 7b [33]. The sensitivity of a resist

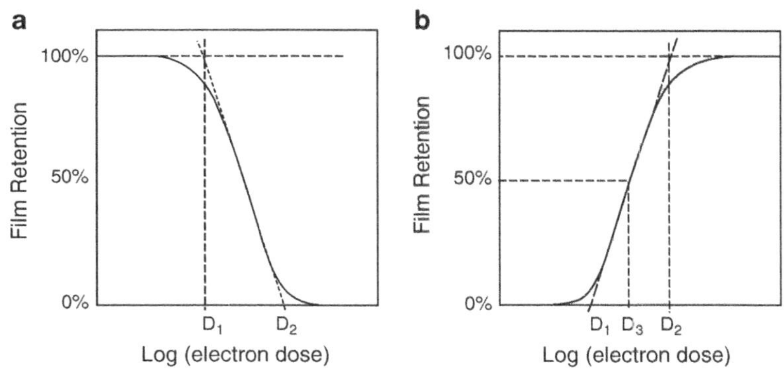

Fig. 7 Response curves of a resist upon electron irradiation and development for (**a**) positive resist and (**b**) negative resist

depends on several factors. In most positive resists, the sensitivity increases with the molecular weight of the resist polymer [30]. This is due to the fact that only a few chain scissions are necessary to substantially alter the solubility of a high molecular weight polymer. A larger molecule would also have a larger cross section and hence a higher probability of electron interactions.

The sensitivity is also affected by the electron energy. An electron with lower energy is more likely to interact with the resist layer rather than reaching the underlying substrate. Therefore, the required exposure dose is reduced. Hence, as long as the electron has enough energy to expose even the bottom of the resist film (thereby ensuring that the pattern is not undercut, and hence removed, on development), the sensitivity is higher at lower electron energy. Another factor that affects the sensitivity of the resist is the atomic weight of the substrate. The heavier the atoms of the substrate, the higher the electron cross section and therefore the greater the level of backscattering, returning the electrons back into the resist, which also leads to an apparent increase in the sensitivity of the resist [30].

3.3.2 Resolution

The resolution of a resist is usually indicated by the minimum feature size that can be resolved by the resist. However, it can also be defined as the minimum half pitch of dense lines and spaces, which is a more useful measure of lithographic use [13]. In practice, the smallest possible linewidth achievable would depend on not only the resist, but the lithographic system and processing conditions. The beam size and current of the system can determine minimum feature size. Resolution also depends on the electron energy. As the electron energy is increased, the effect of forward scattering become less pronounced, producing wall profiles that are more nearly vertical, thus better resolution may be achieved [29]. However, greater electron energy produces more back scattering leading to proximity effects which would affect the resolution of dense features. The use of thinner resist reduces the effects of forward scattering, and thus increases the resolution [3]. Although lower energy electrons undergo more forward scattering compared to high energy ones, the range of scattering depends on the electron energy. For example, at electron energy 5 keV and below, both forward and backscattering are significantly reduced. Therefore, using sufficiently low energy electrons can also improve resolution. As the energy is increased from 5 to 20 keV, the effect of both forward and backscattering significantly increases the exposed resist area, while at 20 keV and above, the almost straight path of the electrons leads to the improvement of resolution.

3.3.3 Contrast

The contrast of a resist is a measure of how fast its solubility changes when it is exposed to radiation. It is defined as the slope of the response curve shown in Fig. 6. In the curve, D_1 is the largest dose at which a positive resist retains its original thickness, and D_2 the smallest dose for it to be completely removed, so the contrast, γ, is defined as

$$\gamma = \frac{1}{\log_{10}\left(D_2/D_1\right)} \tag{5}$$

which is the slope of the response curve [30]. For a negative resist, the contrast is still similarly defined, with D_1 being the largest dose at which the film is completely removed, and D_2 the smallest dose where it retains its original thickness. A resist with high contrast would transition from unexposed to exposed over a small range of doses, indicated by a more vertical slope on the graph. A low contrast resist would have a wide range of doses over which the resist is only partially exposed. A high contrast resist is usually desirable since it tends to have more vertical feature sidewalls upon development compared to lower contrast resist. This is due to the lower current at the edge of the electron beam, causing a lower dose received at the edge of the pattern. For a high contrast resist, the effect is negligible, however, for a lower contrast one, the partially exposed resist at the edge of a pattern results in a sloping wall.

3.3.4 Etch Resistance

To achieve its function, a resist must be able to protect the underlying substrate during subsequent processing. One of the processes used to alter the substrate is etching, a technique that removes substrate material from the uncovered areas. A good resist should be highly resistant to the etching process to allow high aspect ratio to be realised. Besides, poor etch resistance causes an uneven film surface during etching that can be transferred on to the substrate. An important parameter in etching is the etch selectivity. This is the ratio between the etch rate of the material to be etched (the substrate) and the etch rate of the mask material (the resist). An ideal etch would be one where the substrate is removed whilst the resist is unaffected. A sample can be etched using a corrosive liquid (wet etch), gas or plasma (dry etch), or a beam of ions. Etching by wet chemicals is isotropic and can result in undercutting. Dry etching, such as reactive ion etching, can be highly anisotropic, resulting in a more faithful transfer of pattern.

3.4 Current Resists

Resists are grouped into two main categories, positive and negative tone, according to whether the exposed areas are removed, or retained, after exposure. Initially, electron beam resists were composed of electron sensitive polymers such as poly (methacrylates) or poly(sulphones) which readily form smooth, amorphous films by spin-coating. Examples of positive-tone electron beam resists include PMMA (Poly(methyl methacrylate)), PBS (Polybutene-1-sulphone) [30] EBR-9, and

ZEP [29], while negative-tone resists include polystyrene [30], HSQ (Hydrogen SilsesQuioxane) [34, 35] and SU-8 [36].

3.4.1 PMMA (Poly(Methyl Methacrylate))

PMMA was one of the first materials used for electron beam lithography [29]. It has been shown to provide resolution that is among the highest in any lithographic application. It is a commonly used positive resist and has demonstrated features down to <5 nm [37]. Exposure to electron irradiation causes a chain-scission producing low-molecular weight polymer fragments which are soluble in a developer [38] as shown in Fig. 8.

Its sensitivity is ~10 μC/cm^2 at 2 keV, falling to 90 μC/cm^2 at 20 keV. Structural modification by the incorporation of highly electron-withdrawing groups had produced PMMA derivatives with sensitivities as low as 1 μC/cm^2 at 10 keV [39]. When exposed to more than ten times the optimal positive dose, PMMA will crosslink, forming a negative resist. Its negative-tone resolution has been shown to be about 10 nm [40]. Despite excellent sensitivity and resolution, PMMA suffers from poor resistance to corrosive etching conditions [38].

3.4.2 Hydrogen SilsesQuioxane

Hydrogen SilsesQuioxane (HSQ) can be described as a siloxane-based polymer wherein every silicon atom is bound to three oxygen atoms and one hydrogen atom with the exception of the terminal silicon atom, where one of the oxygen atoms is replaced by an H group [34, 35]. Alternatively, HSQ can be described as caged oligomer structures with the general formula $(HSiO_{3/2})2n$ as shown in Fig. 9. HSQ is a high-resolution negative-tone inorganic resist for electron beam lithography. Isolated 6-nm-wide lines and 27 nm period gratings using 50 keV electrons have been demonstrated [41]. HSQ also shows negative-tone resist behaviour for X-ray, and photolithography with wavelengths from 157 nm and below [42]. In extreme ultraviolet (EUV) lithography, HSQ has been shown to have a sensitivity of 11.5 mJ/cm^2. Although HSQ is capable of high resolution and has demonstrated

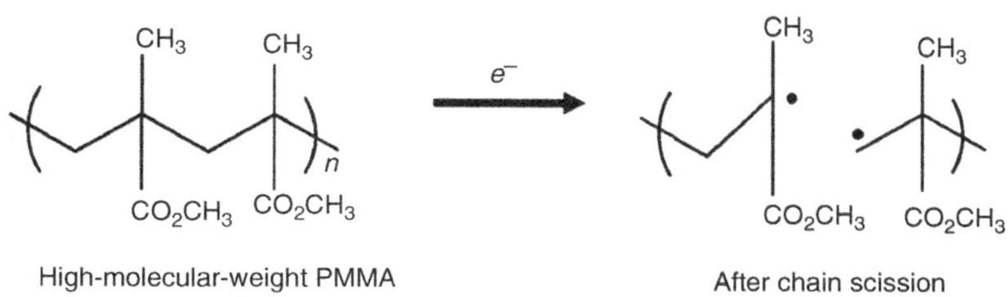

Fig. 8 Chain scission of poly(methyl methacrylate). After [38]

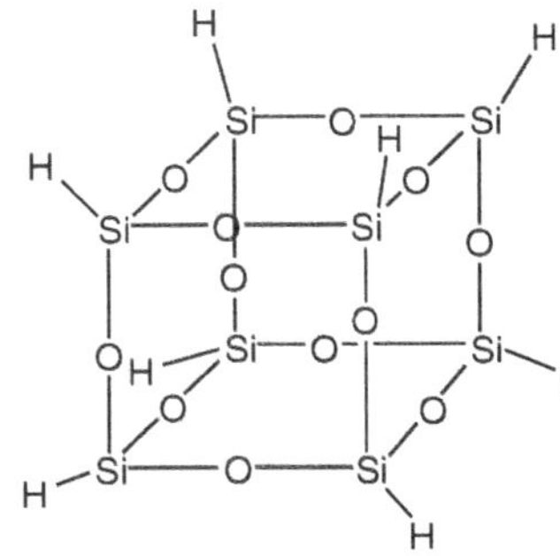

Fig. 9 Chemical structure of Hydrogen SilsesQuioxane

etch durability comparable to novolac resist [43], its sensitivity in EBL is only ~200 μC/cm^2 at 50 keV [34] making it unsuitable for high throughput applications.

3.4.3 Other Polymer Resists

Another example of positive-tone resist is PBS, poly(butane-1-sulfone), which was developed by Bell Laboratories, RCA and IBM [38]. It has been widely used in photomask production due to its high-sensitivity of 1–2 μC/cm^2, and has a resolution down to ~500 nm. However, as the industry progressed to smaller feature sizes, this resolution was not good enough, and since its performance with regards to etch resistance and CD linearity was poor, PBS was no longer acceptable. ZEP, another chain-scission positive-tone resist, developed by Nippon Zeon, consists of a copolymer of α-chloromethacrylate and α-methylstyrene. It has a sensitivity of 15–30 μC/cm^2 at 25 keV. Despite its resolution being almost as high as PMMA, enabling lines of widths 10 nm with pitch 50 nm to be fabricated in the resist [44], ZEP does not fully satisfy the industry's requirements in terms of etch resistance and contrast. A new nanocomposite resist incorporating a fullerene derivative into ZEP520 increased its resolution and etch resistance with a slight decrease of sensitivity [45].

In general, most polymers crosslink upon sufficient electron irradiation [30], so they can be used as negative-tone resists. However, not all of them have the sensitivity, resolution or other properties required for commercial application.

3.4.4 Molecular Resists

Although most conventional resists are polymer-based, as in PMMA and HSQ, there are some problems with them, such as line edge roughness, which is the deviation of a feature edge from a smooth, ideal shape and usually measured in terms of the root-mean square of the edge deviation from a best fit straight line [46], and pin-holes, which are minute holes or defects with diameters as small as a few nm. Furthermore, the resist resolution is limited by the size of the polymer molecule, usually defined in terms of the radius of gyration, which is the average distance of the chain segments from the center of the gravity, at least several nanometers for most polymers. This constraint is especially significant in negative-tone resists, where irradiation causes polymerization or crosslinking of molecules, resulting in larger insoluble molecules so that the resolution does not even approach the molecular size [30]. These problems might be avoided by using low molecular weight resists. These relatively new resists include calixarene derivatives [47–49], catechol [50], fullerene and its derivatives [2–5], and resists based on liquid crystals, such as triphenylene derivatives [51, 52]. These mainly negative tone resists are able to provide high resolution due to their low molecular size and weight. However, not all molecular compounds can be used as resists since most of them tend to crystallise upon spin coating, forming rough surfaces, making them unsuitable for resist application.

3.4.5 Calixarene Derivatives

Calixarenes are macrocyclic molecules comprising of phenolic units linked by methylene bridges to phenolic hydroxyl groups as shown in Fig. 10. They have small molecular size (~1 nm) and sensitivity in the ~mC/cm^2 range [53]. Calixarenes have high melting point (~300°C) and are stable in air. The phenol derivative in calixarene causes it to have high durability to plasma etching. Several derivatives of calixarene have demonstrated negative tone e-beam resist properties [47–49].

Catechol

Another family of low molecular weight resists, which also have cyclic ring structures, are catechol derivatives. The general chemical structure of a catechol is shown in Fig. 11. It was found that this derivative, with three catechol groups

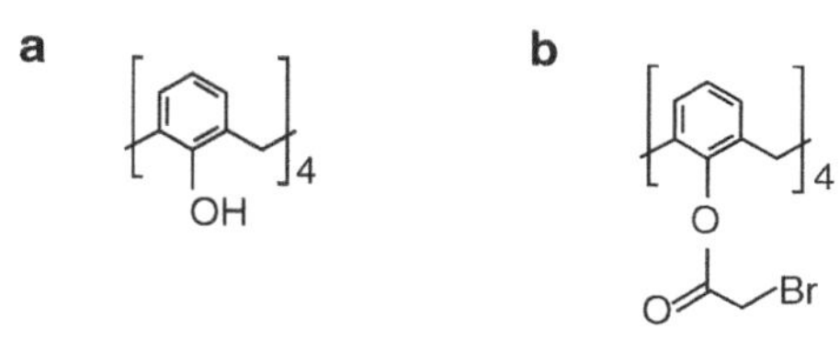

Fig. 10 Structures of two calixarene derivatives, (**a**) calix[4]arene and (**b**) tetra (bromoacetoxy)calix[4]arene

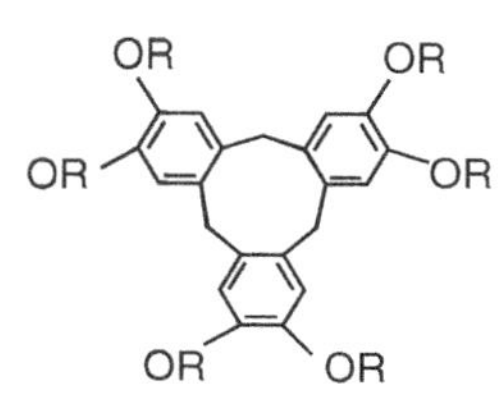

Fig. 11 Chemical structure of a catechol derivative. R can be COOtBu, $CH_2COOtBu$, or CH_2COOH

linked by three CH_2 units, acts as a positive-tone photoresist material. [50] This derivative is an amorphous compound and can form smooth films. Another derivative, also with three catechol groups, hexahydroxy cyclotribenzylene, protected with p-*t*-butoxy carbonylmethyl groups, works as a positive tone electron beam resist with a high sensitivity and contrast, and high durability to CF_4 plasma etching. [54] It has a poor resolution, however, allowing a 90-nm space pattern at 10 $\mu C/cm^2$ under 50-keV electron beam.

Triphenylene Derivatives

Triphenylene is a disc-shaped molecule with a planar structure and threefold rotation. Its main structure consists of a polycyclic aromatic hydrocarbon made up of four fused benzene rings. Low molecular weight resists based on the triphenylene molecule have been studied by Robinson et al. [51, 52] and it was shown that these resists can be used as negative tone electron beam resists, with sensitivity between 1.5 and 6.5 mC/cm^2 at 20 keV with monochlorobenzene as developer. These polysubstituted triphenylene derivatives, shown in Fig. 12, which are liquid crystalline, readily form smooth films when spin coated using chloroform as a solvent without a need for a post application bake, a heating process applied to most resist after spin coating to remove the casting solvent. One of the derivatives, the molecule 2,3,6,7,10,11-hexapentyloxytriphenylene demonstrated high resolution capability, allowing 14 nm patterns to be realized in the resist at 30 keV beam energy [52].

The derivative has a very small molecular size (744 AMU) [51]. The derivative C5/C5, also demonstrated positive tone behavior at doses lower than 250 $\mu C/cm^2$ and negative tone behaviour at higher doses when pentanol was used as a developer [55]. Furthermore, the resist has a high etch-durability.

Fullerene (C60) Derivatives

Ever since their discovery by Kroto et al. in 1985 [56] fullerenes have attracted a lot of attention. Fullerene (C60) is a closed hollow spherical cage comprising of 60 carbon atoms arranged in interlocking hexagons and pentagons, as shown in Fig. 13 [56]. The application of fullerene as a negative resist was first studied by Tada and Kanayama [2] who verified that electron beam irradiation on fullerene C60 films reduced its solubility in organic solvents such as monochlorobenzene, showing that this material could be used as a negative electron beam resist. Its small

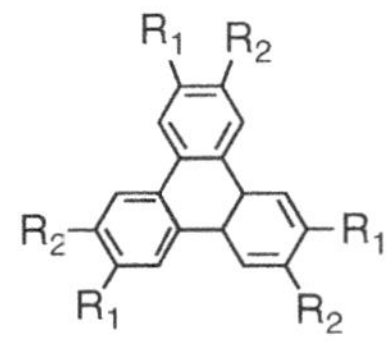

Fig. 12 Chemical structure of polysubstituted triphenylene derivatives.
R1 = OCnH2n + 1, and
R2 = OCnH2n + 1,where
$n = 0, 1, 2, 3\ldots$

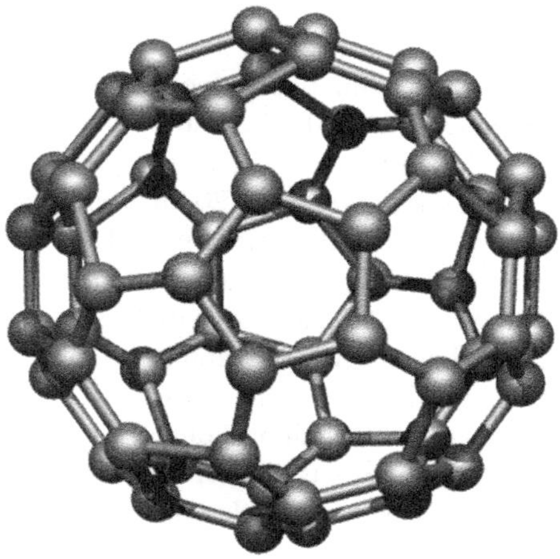

Fig. 13 Chemical structure of fullerene (C60)

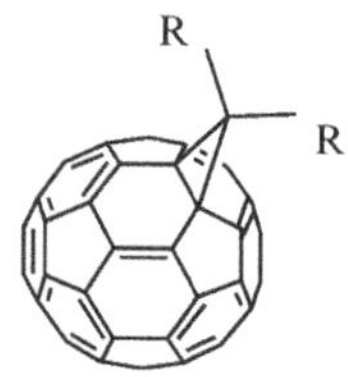

Fig. 14 A fullerene derivative, with a functional group, R attached to the fullerene cage via a methano bridge

molecule (<1 nm diameter) enables the resist to have a resolution of at least 20 nm. The resist has a sensitivity of ~10 mC/cm^2 and very high dry-etch durability [2].

However, the C60 film is unsuitable for spin coating due to poor solubility in all solvents, and must be deposited on the substrate by vacuum sublimation, making it difficult for industrial application. Furthermore, its sensitivity is two orders of magnitude lower than that of PMMA. Robinson et al. [3, 4, 5] demonstrated that chemical modification of C60 by adding functional groups to the C60 cage can significantly enhance the resist properties. Methano derivatives of the C60 (methanofullerenes) were formed by attaching a functional group to the fullerene cage via a three-carbon ring called a methano bridge as shown in Fig. 14. These derivatives dissolve readily in chloroform and smooth films of resists can be easily prepared by spin coating from their solutions. Films of nine different C60 methanofullerene monoadducts, exhibited negative tone resist behaviour with sensitivities between 8.2 and 0.85 mC/cm^2 when irradiated with a 20 keV electron beam, more than an order of magnitude higher than that of pure C60.

It is believed that the addition of functional groups cause a considerable charge localization, and consequently bond weakening, making them more susceptible to electron beam damage, thus increasing their sensitivity. Raman spectra of exposed and unexposed films implied that, instead of polymerization, the e-beam irradiation of the fullerene derivatives leads to cage fragmentation and the formation of a disordered network of graphite-like particles which are mechanically resistant to being washed away by the solvent and highly etch resistant [5]. In the case of derivatives with two polyether chains, the sensitivity was found to be linearly dependent upon the derivative mass. In addition, this derivatives has high dry-etch durability with etch ratios twice those of a standard novolac-based resist. Features with widths of 20 nm were produced using these compounds.

The study of the derivatives of fullerene was further pursued by Tada et al. [58] who also showed that the sensitivities of the methanofullerene resists were dependent on the methano bridge in the side chains. A derivative with two diels-alder side chains has a sensitivity about the same as that of C60, while derivatives with one methano side chain showed sensitivities in the order of mC/cm^2. By synthesizing a multi-adduct methanofullerene, which had four to six side chains, the sensitivity of the derivative was further improved to 0.38 mC/cm^2 at 20 keV, two orders of magnitude better than that of C60.

The primary objectives of e-beam lithography are high resolution and high speed (high sensitivity). Higher-voltage e-beam exposures could be the best solution to get the resolution and accuracy required in advanced masks due to their finer beam profile. However, the higher accelerating voltages lead to lower resist sensitivity because fewer electrons interact in the resist layer, but travel through to the underlying substrate [38]. Therefore, an enhancement of resist sensitivity is necessary to achieve an economically feasible throughput while maintaining the high resolution of electron beam lithography.

4 Chemically Amplified Resist

Whilst electron beam lithography is able to achieve resolutions that are simply not possible using conventional optical lithography, its low throughput rate has always been a disadvantage. A dramatic increase in resist sensitivity is one way to achieve an economically feasible throughput. Chemically amplified resists have become technologically important to increase throughput not only for photolithography but for other lithography techniques [15]. The concept of chemical amplification of photoresists to boost their sensitivity was introduced in 1982 and is now well accepted by the lithographic community [59]. The concept of chemical amplification aimed to dramatically increase the sensitivity of a resist and was invented at IBM Research in 1980 [60]. It is now well accepted by the lithography community. Chemically amplified resists (CAR) have higher sensitivity, thus increasing wafer throughput. In chemically amplified resist systems, a catalytic species generated by irradiation induces a cascade of subsequent chemical reactions, providing a gain mechanism [59, 60]. The original chemical amplification scheme included cross-linking through ringopening polymerization of pendant epoxide groups for negative resist, shown in Fig. 15a, depolymerization for self-developing positive resist, and deprotection of pendant groups to induce polarity change for dual-tone imaging as shown in Fig. 15b.

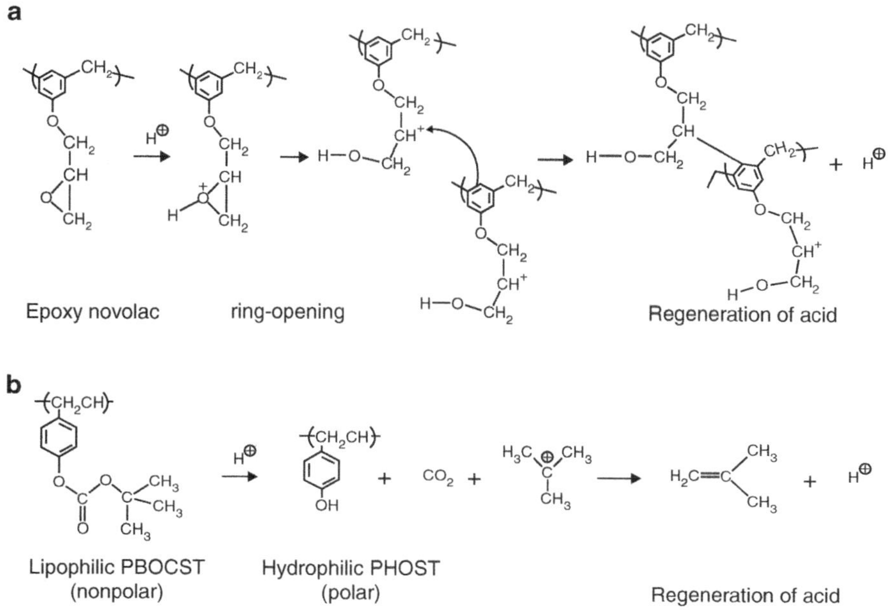

Fig. 15 Chemical amplification schemes, (**a**) cross-linking through ring-opening polymerization of pendant epoxy group for negative-resist systems, and (**b**) deprotection of pendant groups to induce polarity change

All three of these reactions are acid-catalyzed [61]. As already mentioned in Sect. 3.1, self-developing systems require no subsequent development due to the volatile products, however, they are not widely used since these products cause contamination to the lithography tools. Therefore some of these systems were later modified from self-development to thermal development, as in the depolymerization of polyphthalaldehyde [61]. Acid-catalyzed reactions form the basis of many chemically amplified resist systems for microlithography applications today, either via cross-linking to reduce solubility for negative tone resists or deprotection reactions (removal of a solubility inhibitor) for positive tone resists applications [62]. Cross-linked resists generally offer more resistance to subsequent processing than non-cross-linked resists [15].

CARs generally contain more than one component; for instance, a base polymer, a photo-acid generator (PAG), and sometimes a separate cross-linker. During exposure the photoacid generator is decomposed, releasing acid. The acid then diffuses during a postexposure bake (PEB) and catalyzes cross-linking reactions to render the resist insoluble for negative resist or deprotection/chain-scission to render it soluble for positive resist. Many different chemical amplification schemes have been investigated, and various process conditions have been studied, to attempt to achieve resists with increased sensitivity, optimized resolution, and acceptable etch resistance, suitable for nano-scale lithography.

4.1 Chemically Amplified Electron Beam Resists

The most widely used and studied chemically amplified resist family are the Shipley Advanced Lithography (SAL) resists produced by Rohm and Haas Electronics Materials. The negative-tone SAL601 resist for instance, has three components; a base polymer, an acid generator, and a crosslinking agent. It has been shown to have a sensitivity of 4 $\mu C/cm^2$ and displays high-dry etching resistance similar to that of a conventional diazoquinone based positive resist [63]. A minimum feature size of 20 nm isolated line was obtained in the resist, but required stringently controlled PEB conditions, and showed significant line edge roughness [64]. SAL601-ER7 consists of a novolak polymer contributing to high dry-etching resistance, a melamine resin as the crosslinker and a bromic triazine compound as the acid generator [65]. SAL605, the successor of SAL601, has the same basic chemistry but is about three times more sensitive [66].

Another negative-tone chemically amplified resist is the epoxy novolak SU-8, which is composed of a Bisphenol A Novolak epoxy oligomer and a triarylsulfonium hexafluroantimonate salt as photoacid generator [36]. Its sensitivity ranges from 1 to 4 $\mu C/cm^2$ when irradiated with 20 keV electrons, depending on the PEB conditions [67]. The resist has been shown to resolve features as small as 30 nm using 40 keV electrons at minimum line dose of 0.03 nC/cm [36] and 75-nm wide lines using 50 keV electrons at a dose of $\sim$3.6 $\mu C/cm^2$ [68]. However, the resist has poor dry etch resistance [36]. A number of chemically amplified DUV resists, have been shown to work effectively for electron beam lithography [69–71]. UVIII, a positive tone photoresist from Rohm and Haas has a resolution of 60 nm at a dose of 40 $\mu C/cm^2$ and beam energy of 50 keV [70], and sub-50 nm at 60 $\mu C/cm^2$ at 100 keV [71]. UVN-2 and UV-5, also from Rohm and Haas, also have high resolution capabilities in electron beam lithography. The negative-tone UVN-2 is a successor of UVN, a negative resist with a sensitivity of 20 $\mu C/cm^2$ at 50 kV which has an ultimate resolution of 70 nm isolated lines or 140 nm lines and spaces. UVN-2 allows 50 nm resolution of single line widths at 30 $\mu C/cm^2$ nominal dose [69]. UV-5 shows positive tone behaviour with a sensitivity of 12 $\mu C/cm^2$ and 50 nm resolution. The resist switches to negative-tone behaviour at higher doses, and shows negative-tone resolution of 90 nm at a dose of 500 $\mu C/cm^2$ [69].

Two chemically amplified resists from Clariant are the positive-tone AZPF514 and negative-tone AZPN114. AZPF514 is a sensitive resist, about 25 $\mu C/cm^2$ but unable to achieve resolution better than 150 nm [71]. AZPN114 also has a high sensitivity, about 30% better than SAL601, and a resolution of 30 nm [72].

4.2 Chemically Amplified Fullerene Resists

The sensitivities of several fullerene derivatives have been greatly improved by chemical amplification via the incorporation of crosslinkers and photoacid generators [6, 7]. Chemical amplification of the derivative, MF03-01 shown in Fig. 16a

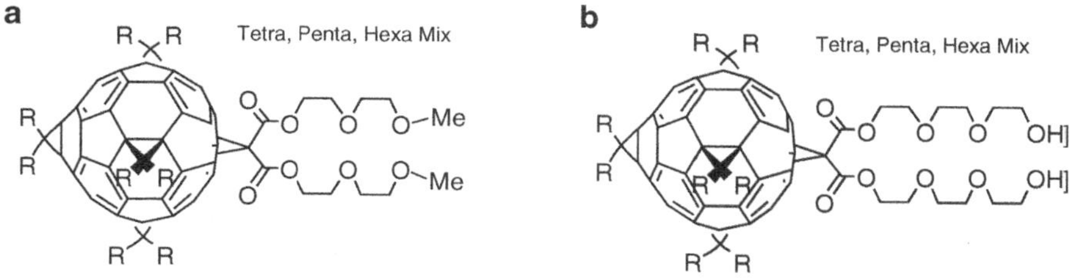

Fig. 16 Chemical structures of the fullerene derivatives (**a**) MF03-01 and (**b**) MF03-04

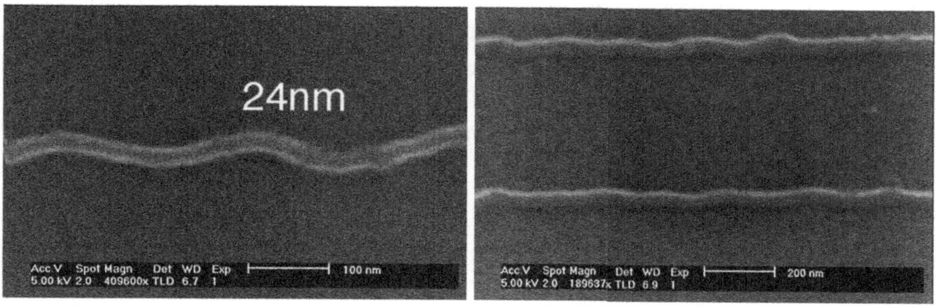

Fig. 17 SEM micrograph of lines defined in CA MF03-04 (**a**) at dose 200 pC/cm and (**b**) 800 pC/cm viewed at 30°C angle

with triphenylsulfonium triflate as photoacid generator and hexamethoxymethylmelamine as crosslinker demonstrates encouraging sensitivities as high as 12 μC/cm^2. This high sensitivity could be due to the high number of addends in MF03-01 which provides many crosslinking sites. Unfortunately, fine patterning of the film showed rounding of feature edges, which could be due to acid diffusion [57].

The methanofullerene MF03-04, shown in Fig. 16b, is similar to MF03-01 but with hydroxyl termination of the addends. Pure MF03-04 has a sensitivity of 550 μC/cm^2. Chemical amplification of the derivative MF03-04, demonstrated a significant sensitivity enhancement when photoinitiator bis[4-di(phenylsulfonio) phenyl]sulfide bis(hexafluorophoshate) and an epoxy novolac crosslinker were added. It was shown that increasing the photoinitiator and crosslinker concentration increased the sensitivity, reaching a maximum of 8 μC/cm^2. This CA resist also showed a high resolution, enabling linewidths of 24 nm to be fabricated in the resist, albeit with some swelling as shown in Fig. 17. Electron Cyclotron Resonance Microwave Plasma Etching (ECR) etching of the CA MF03-04 demonstrated a normalised etch durability of 2.55 that of the silicon substrate, approaching that of SAL601 under the same etch conditions [57].

The superior sensitivity of the chemically amplified MF03-04 compared to that of the other derivatives could be due to its addends. Comparing MF03-04 with MF03-01 which is also a hexaadduct but with methyl terminated suggests that the OH termination in MF03-04 is also responsible for its crosslinking ability.

MF03-04 with its high number of OH groups can be thought of as analogous to a polyol molecule commonly seen in epoxy crosslinker chemistry. The high number of epoxy groups in the crosslinker CL05-03 enabled it to work better in enhancing the sensitivities of the resists as compared to the other crosslinkers used.

5 Conclusion

The application of fullerene and its derivatives as resists for electron beam lithography has been discussed. The small molecular size of the fullerene enables high resolution patterning to be achieved. Addition of functional groups to the fullerenes improves the resist performance and increases the sensitivity of the resist. Chemical amplification of the resists via addition of photoacid generator or crosslinker further improves their sensitivities, thus making its use practical for an economically feasible throughput.

References

1. Moore, G.E.: Electronics **38**, 114 (1965)
2. Tada, T., Kanayama, T.: Jpn J. Appl. Phys. **35**, L63 (1998)
3. Robinson, A.P.G.: PhD thesis, University of Birmingham, UK (1999)
4. Robinson, A.P.G., Palmer, R.E., Tada, T., Kanayama, T., Preece, J.A., Philp, D., Jonas, U., Deiderich, F.: Chem. Phys. Lett. **289**, 586 (1998)
5. Robinson, A.P.G., Palmer, R.E., Tada, T., Kanayama, T., Shelley, E.J., Philp, D., Preece, J.A.: Chem. Phys. Lett. **312**, 469 (1999)
6. Robinson, A.P.G., Zaid, H.M., Gibbons, F.P., Palmer, R.E., Manickam, M., Preece, J.A., Brainard, R., Zampini, T., O'Connell, K.: Microelectron. Eng. **83**(4–9), 1115–1118 (2006)
7. Gibbons, F.P., Zaid, H.M., Manickam, M., Preece, J.A., Palmer, R.E., Robinson, A.P.G.: Small **3**(12), 2076–2080 (2007)
8. Levinson, H.J., Arnold, W.H.: In: Rai Chaudhury, P. (ed.) Handbook of Microlithography, Micromachining, and Microfabrication, vol. 1. IEE, London (1997)
9. Ledwith, A.: In: Moss, S.J., Ledwith, A. (eds.) The Chemistry of the Semiconductor Industry. Blackie and Son Ltd, London (1987)
10. Ronse, K.: Microelectron. Eng. **67/68**, 300 (2003)
11. Lin, B.J., Rai-Choudhury, P.: In: Rai Chaudhury, P. (ed.) Handbook of Microlithography, Micromachining, and Microfabrication, vol. 1. IEE, London (1997)
12. Wallraff, G.M., Hinsberg, W.D.: Chem. Rev. **99**, 1801 (1999)
13. Harriott, L.R.: Proc. IEEE **89**, 366 (2001)
14. Brainard, R.L., Barclay, G.G., Anderson, E.H., Ocola, L.E.: Microelectron. Eng. **61/62**, 707 (2002)
15. Peckerar, M.C., Perkins, F.K., Dobisz, E.A., Glembocki, O.J.: In: Rai Chaudhury, P. (ed.) Handbook of Microlithography, Micromachining and Microfabrication, vol. 1. IEE, London (1997)
16. Brennan, K.F., Brown, A.S.: Theory of Modern Electronic Semiconductor Devices. Wiley, New York (2002)
17. Ito, T., Okazaki, S.: Nature **406**, 1027 (2000)

18. Chan, S., Li, Y., Rothberg, L.J., Miller, B.L., Fauchet, P.M.: Mat. Sci. Eng. C Biomim. **15**, 277 (2001)
19. Takhistov, P.: Biosens. Bioelectron. **19**, 1445 (2004)
20. Mulkens, J., McClay, J., Tirri, B., Brunotte, M., Mecking, B., Jasper, H.: SPIE Proc. **5040**, 753 (2003)
21. Rothschild, M., Forte, A.R., Kunz, R.R., Palmateer, S.C., Sedlacek, J.H.C.: IBM J. Res. Dev. **41**, 49 (1997)
22. Piner, R.D., Zhu, J., Xu, F., Hong, S., Mirkin, C.A.: Science **283**, 661 (1999)
23. Guo, L.J.: J. Phys. D Appl. Phys. **37**, R123 (2004)
24. Austin, M.D., Ge, H., Wu, W., Li, M., Yu, Z., Wasserman, D., Lyon, S.A., Chou, S.Y.: Appl. Phys. Lett. **84**, 5299 (2004)
25. Melngailis, J., Mondelli, A.A., Berry III, I.L., Mohondro, R.: J. Vac. Sci. Technol. B **16**, 927 (1998)
26. Cerrina, F.: In: Rai Chaudhury, P. (ed.) Handbook of Microlithography, Micromachining and Microfabrication, vol. 1. IEE, London (1997)
27. Simon, G., Haghiri-Gosnet, A.M., Bourneix, J., Decanini, D., Chen, Y., Rousseaux, F., Launois, H., Vidal, B.: J. Vac. Sci. Technol. B **15**, 2489 (1997)
28. Silverman, J.P.: J. Vac. Sci. Technol. B **15**, 2117 (1997)
29. McCord, M.A., Rooks, M.J.: In: Rai Chaudhury, P. (ed.) Handbook of Microlithography, Micromachining, and Microfabrication, vol. 1. IEE, London (1997)
30. Roberts, E.D.: In: Moss, S.J., Ledwith, A. (eds.) The Chemistry of the Semiconductor Industry, p. 197. Blackie and Son Ltd, London (1987)
31. Berger, S.D., Gibson, J.M., Camarda, R.M., Farrow, R.C., Huggins, H.A., Kraus, J.S., Liddle, J.A.: J. Vac. Sci. Technol. B **9**, 2996 (1991)
32. Harriott, L.R.: J. Vac. Sci. Technol. B **15**, 2130 (1997)
33. Manako, S., Fujita, J., Ochiai, Y., Nomura, E., Matsui, S.: Jpn. J. Appl. Phys. **36**, 7773 (1999)
34. Henschel, W., Georgiev, Y.M., Kurz, H.: J. Vac. Sci. Technol. B **21**, 2018 (2003)
35. Hacker, N.P.: MRS Bull. **22**, 33 (1997)
36. Aktary, M., Jensen, M.O., Westra, K.L., Brett, M.J., Freeman, M.R.: J. Vac. Sci. Technol. B **21**, L5 (2003)
37. Yasin, S., Hasko, D.G., Ahmed, H.: Appl. Phys. Lett. **78**, 2760 (2001)
38. Medeiros, D.R., Aviram, A., Guarnieri, C.R., Huang, W.-S., Kwong, R., Magg, C.K., Mahorowala, A.P., Moreau, W.M., Petrillo, K.E., Angelopoulos, M.: IBM J. Res. Dev. **45**, 639 (2001)
39. Matsuda, S.: Polym. Eng. Sci. **17**, 410 (1977)
40. Tada, T., Kanayama, T.: J. Vac. Sci. Technol. B **13**, 2801 (1995)
41. Word, M.J., Adesida, H., Berger, P.R.: J. Vac. Sci. Technol B **21**, L12 (2003)
42. Peuker, M., Lim, M.H., Smith, H.I., Morton, R., van Langen-Suurling, A.K., Romijn, J., van der Drift, E.W.J.M., van Delft, F.C.M.J.M.: Microelectron. Eng. **61/62**, 803 (2002)
43. van Delft, F.C.M.J.M., Weterings, J.P., van Langen-Suurling, A.K., Romijn, H.: J. Vac. Sci. Technol. B **18**, 3419 (2000)
44. Ishii, T., Tamamura, T., Shigehara, K.: Jpn. J. Appl. Phys. **39**, L1068 (2000)
45. Ishii, T., Murata, Y., Shigehara, K.: Jpn. J. Appl. Phys. **40**, L478 (2001)
46. Patsis, G.P., Gogolides, E., Van Werden, K.: Jpn. J. Appl. Phys. **44**, 6341 (2005)
47. Fujita, J., Ohnishi, Y., Ochiai, Y., Nomura, E., Matsui, S.: J. Vac. Sci. Technol. B **14**, 4272 (1996)
48. Yasin, S., Hasko, D.G.: J. Vac. Sci. Technol. B **19**, 311 (2001)
49. Manako, S., Ochiai, Y., Yamamoto, H., Teshima, T., Fujita, J., Nomura, E.: J. Vac. Sci. Technol. B **18**, 3424 (2000)
50. Kihara, N., Saito, S., Ushirogouchi, T., Nakase, M.: J. Photopolym. Sci. Technol. **11**, 553 (1998)
51. Tada, T., Kanayama, T., Robinson, A.P.G., Palmer, R.E., Allen, M.T., Preece, J.A., Harris, K.D.M.: Microelectron. Eng. **53**, 425 (2000)

52. Robinson, A.P.G., Palmer, R.E., Tada, T., Kanayama, T., Allen, M.T., Preece, J.A., Harris, K.D.M.: J. Vac. Sci. Technol. B **18**, 2730 (2000)
53. Sailer, H., Ruderisch, A., Kern, D.P., Schurig, V.: J. Vac. Sci. Technol. B **20**, 2958 (2002)
54. Saito, S., Kihara, N., Ushirogouchi, T.: Microelectron. Eng. **61/62**, 777 (2002)
55. Robinson, A.P.G., Palmer, R.E., Tada, T., Kanayama, T., Allen, M.T., Preece, J.A., Harris, K.D.M.: J. Phys. D Appl. Phys. **32**, L75 (1999)
56. Kroto, H.W., Heath, J.R., O'Brien, S.C., Curl, R.F., Smalley, R.E.: Nature **318**, 162 (1985)
57. Zaid, H.M.: PhD thesis, University of Birmingham, UK (2006)
58. Tada, T., Uekusa, K., Kanayama, T., Nakayama, T., Chapman, R., Cheung, W.Y., Eden, L., Hussain, I., Jenning, M., Perkins, J., Phillips, M., Preece, J.A., Shelley, E.: Microelectron. Eng. **61–62**, 737 (2002)
59. Ito, H.: Jpn. J. Appl. Phys. **31**, 4273 (1992)
60. Ito, H.: J. Polym. Sci. **41**, 3863 (2003)
61. Ito, H.: IBM J. Res. Dev. **41**, 69 (1997)
62. Hinsberg, W.D., Wallraff G.M., Allen, R.D., et al.: Kirk-Othmer Encyclopedia of Chemical Technology, vol. 15. Wiley, New York (1998)
63. Liu, H., de Grandpre, M.P., Feely, W.E.: J. Vac. Sci. Technol. B **6**, 379 (1988)
64. Yoshimura, T., Nakayama, Y., Okazaki, S.: J. Vac. Sci. Technol. B **10**, 2615 (1992)
65. Azuma, T., Masui, K., Takigami, Y., Sasaki, H., Sakai, K., Nomaki, T., Kato, Y., Mori, I.: Jpn. J. Appl. Phys. **30**, 3138 (1991)
66. Dentinger, P.M., Taylor, J.W.: J. Vac. Sci. Technol. B **15**, 2632 (1997)
67. Kudryashov, V., Yuan, X.-C., Cheong, W.-C., Radhakrishnan, K.: Microelectron. Eng. **67–68**, 306 (2003)
68. Pépin, A., Studer, V., Decanini, D., Chen, Y.: Microelectron. Eng. **73–74**, 233 (2004)
69. van Delft, F.C.M.J.M., Holthuysen, F.G.: Microelectron. Eng. **46**, 383 (1999)
70. Macintyre, D., Thoms, S.: Microelectro. Eng. **35**, 213 (1997)
71. Cui, Z., Prewett. P.: Microelectron. Eng. **46**, 255 (1999)
72. Cui, Z., Gerardino, A., Gentili, M., DiFabrizio, E.: J. Vac. Sci. Technol. B **16**, 3284 (1998)

Hydrogenated Amorphous Carbon Films

Suriani Abu Bakar, Azira Abdul Aziz, Putut Marwoto, Samsudi Sakrani, Roslan Md Nor, and Mohamad Rusop

Abstract Hydrogenated amorphous carbon (a-C:H) thin film is one of the most studied materials due to its unique features. The a-C:H thin film is a remarkable material because of its novel optical, mechanical and electrical properties and its similarities to diamond. In this chapter we reviewed the structural and optical properties of hydrogenated amorphous carbon (a-C:H) thin films prepared in a DC-PECVD reactor. Both power and ion bombardment energy were continuously changed during the deposition, as a results of varying deposition parameters such as

S.A. Bakar
Department of Physics, Faculty of Science and Technology, Universiti Pendidikan Sultan Idris, 35900 Tanjung Malim, Perak, Malaysia
e-mail: absuriani@yahoo.com

A.A. Aziz
NANO-SciTech Centre, Institute of Science, Universiti Teknologi MARA, 40450 Shah Alam, Selangor, Malaysia
e-mail: azira.aziz@gmail.com

P. Marwoto
Physics Department, Faculty of Mathematics and Science, Semarang State University, 50229, Semarang, Indonesia
e-mail: pmarwoto@yahoo.com

S. Sakrani
Physics Department, Faculty of Science, Universiti Teknologi Malaysia, 81310, Skudai, Johor, Malaysia
e-mail: samsudi@difz2.fs.utm.my

R. Md. Nor
Department of Physics, Faculty of Science, University of Malaya, 50603, Kuala Lumpur, Malaysia
e-mail: rmdnor@um.edu.my

M. Rusop
NANO-SciTech Centre, Institute of Science, Universiti Teknologi MARA, 40450 Shah Alam, Selangor, Malaysia
e-mail: rusop8@gmail.com

N. Yahya (ed.), *Carbon and Oxide Nanostructures,* Adv Struct Mater 5,
DOI 10.1007/8611_2010_15, © Springer-Verlag Berlin Heidelberg 2010,
Published online: 27 June 2010

chamber pressure, electrode distance, CH_4 flow rate, and substrate temperature. The films properties ranged from polymer-like to graphite-like a-C:H films, as the power and ion energy increased. The structure and the optical properties of a-C:H films were analyzed by infrared and Raman spectroscopy, UV–Vis Spectrophotometer and photoluminescence. This is to extract the information on sp^3/sp^2 and hydrogen contents, optical gap, E_0 and photoluminescence properties of a-C:H films. The films were found to consist of sp^2 clusters of which the size increases with increasing power and ion bombardment energy during the deposition, resulting in lower hydrogen, sp^3 content, optical gap and photoluminescence response. The increased in hydrogen termination from the films at higher ion energies results in bigger cluster size and produced graphitic films.

1 Introduction

Thin films are solid materials of either metal, semiconductor or insulator, deposited onto substrates at film thickness in the range of 10–1,000 nm [1]. Beyond this range they are referred to as thick films. The term ultra thin film is used for films with thickness smaller than 100 nm. Its thickness creates characteristics such as physical, chemical and mechanical properties which is different compared to the original bulk material and this contribute to a new phenomena. Substrate is any kind of solid that can support the formation of thin film on it. It is usually a material that does not interact with the film. A clean and smooth surface of substrate is required in order to obtain good quality and homogeneous films. Nowadays, we can see thin films widely used in electronics, optoelectronic devices, optical application and surface engineering applications [2].

Thin films can be produced by utilizing several methods. Generally, the preparation of thin films can be classified into two methods; physical and chemical techniques [3]. Some examples of physical vapor deposition are vacuum evaporation and sputtering. Meanwhile there are a variety of the chemical vapor deposition processes such as plasma enhanced (assisted) chemical vapor deposition (PECVD, PACVD), low-pressure chemical vapor deposition (LPCVD) and etc. Thin film production can be either in the form of single crystal, polycrystalline or amorphous, this depends on several factors such as temperature, pressure and etc during the deposition process.

Hydrogenated amorphous carbon (a-C:H) thin films is one of the most studied material due to its unique features. The a-C:H thin films is remarkable material because of its novel optical, mechanical and electrical properties and its similarities to diamond [4–6]. The main properties of diamond material are low optical absorption in UV, visible and IR regions, high electrical resistivity and thermal conductivity, extremely hard, low coefficient of friction and etc. [7–9]. The a-C:H material is not as excellent in optical, mechanical and electrical material as a single crystal diamond, but it is adequate for numerous applications such as protective coating in areas namely optical window, scratch resistant, magnetic storage disk, biomedical coating and low friction wear resistant coatings for moving part in tools [6, 10].

The a-C:H thin films, is one of the diamond-like carbon (DLC) form. Aisernberg and Chabot [11] were the first to produce diamond-like carbon by ion beam deposition techniques. Angus et al. [12] defined pure DLC composed of carbon and hydrogen into two categories; hydrogenated amorphous carbon (a-C:H) and amorphous carbon (a-C). The a-C:H films contains less than 10–60% hydrogen (incorporations of hydrogen in the film are important to obtain diamond-like properties). The a-C films contain less than 1% hydrogen.

There are several deposition methods for a-C:H thin films production. These deposition methods can be categorized into two; physical vapor deposition (PVD) and plasma enhanced chemical vapor deposition (PECVD). PVD involves the sputtering of carbon atom from a solid target by energetic gas species, normally argon (Ar) ions [1]. PECVD method is widely used to deposit a-C:H films with better quality films as compared to chemical vapor deposition (CVD) method. PECVD involves a chemical process which takes place in the vapor phase at the substrate surface. As a result the films were deposited onto the substrate. The deposition process was conducted in low pressure ambient in the vapor phase.

In the PECVD method, there are two types of plasma species that contributes to the film growth; the radicals, (chemically active neutral species) and ions that diffuse from the plasma and drift toward the substrate surface [10]. The deposition parameters strongly affect the plasma species and the ionic energies as well as densities at the substrate [13].The deposition of a-C:H films by PECVD occur at low substrate temperatures, in contrast to CVD. In the PECVD method the high substrate temperature is overcome by applying electric field in the reactant gases to produce a significant number of free radicals, ion and etc. Due to the relatively low temperature, PECVD is a low-cost process as well. The power supply used to create the discharge in PECVD include RF [13–15], microwave (MW) – RF PECVD [16], RF – pulse DC mode PECVD [17], DC saddle field glow discharge [10, 18], DC – RF PECVD [4–19], electron cyclotron resonance (ECR) – MW Plasma Chemical Vapor Deposition (ECR-MPCVD) [20], microwave electron cyclotron resonance (ECR) – RF discharge PECVD [21, 22] etc.

Other than that, the study of a-C:H is very important since the structure of a-C:H is hardly understood as it consists of both sp^3 and sp^2 hybridized carbon. Both σ and π bond in a-C:H is definitely the difficulties faced in the analysis of the structure of a-C:H. Although, large amount of research has been done on a-C:H and as a-C:H films are already used in many applications, much about its properties have not been clearly comprehended. Until now, the Robertson model [6] is the most referred and successfully model. According to this model, amorphous carbon consists of sp^2 clusters, which are embedded in sp^3 bonded matrix. Hence it is the sp^2 sites that forms band edges and controls the optical properties while sp^3 sites controlled mechanical properties.

In this chapter, we review the structural and optical properties of a-C:H thin films obtained by direct current-plasma enhanced chemical vapor deposition (DC-PECVD) method. The a-C:H properties, are highly affected by experimental parameters such as chamber pressure, electrode distance, methane (CH_4) flow rate, substrate temperature and etc. By changing the deposition parameters of a-C:H films, one can expect to obtain different sp^3 and sp^2 bond distributions and different structures in the films because it involves the change in plasma power and ion bombardment energy. Hence, the change in the properties of these films can be suited to specific applications.

2 Method of Preparing a-C:H Thin Film

2.1 Direct Current-Plasma Enhanced Chemical Vapor Deposition System

The PECVD method a is well-known method for the deposition of a-C:H thin film. Among of them is DC-PECVD method other than of radio frequency (RF)-PECVD. The DC-PECVD system consists of a deposition chamber (plasma reactor), power supply and gas system.

2.1.1 Plasma Reactor

The plasma reactor is constructed from stainless steel which is cylindrical in shape, having diameter of 15.24 cm and 20 cm in height. It is equipped with access door for putting in the sample and taking it out and a view port to observe the process, which is taking place inside the reactor. There are also channels for gas supply, vacuuming, electrical feed through (for DC power supply and heater) and linear motion feed through. Both anode and cathode are constructed from steel plate. Anode having diameter of 6.8 cm and thickness of 0.5 cm is connected to the DC power supply as shown in Fig. 1. Cathode which has the same diameter and thickness with anode is placed on the heater. It is then connected to the direct current and grounded together with the reactor's chamber wall. The cathode's distance to the anode can be changed in order to produce optimum plasma (1.5–4.0 cm). The cathode also acts as the substrate holder, where as the a-C:H films that were deposited on a corning glass substrates was placed on the cathode. This system is also shielded to avoid discharge from occurring at areas outside both electrode and to confine the plasma so that it stay in area between the two electrodes.

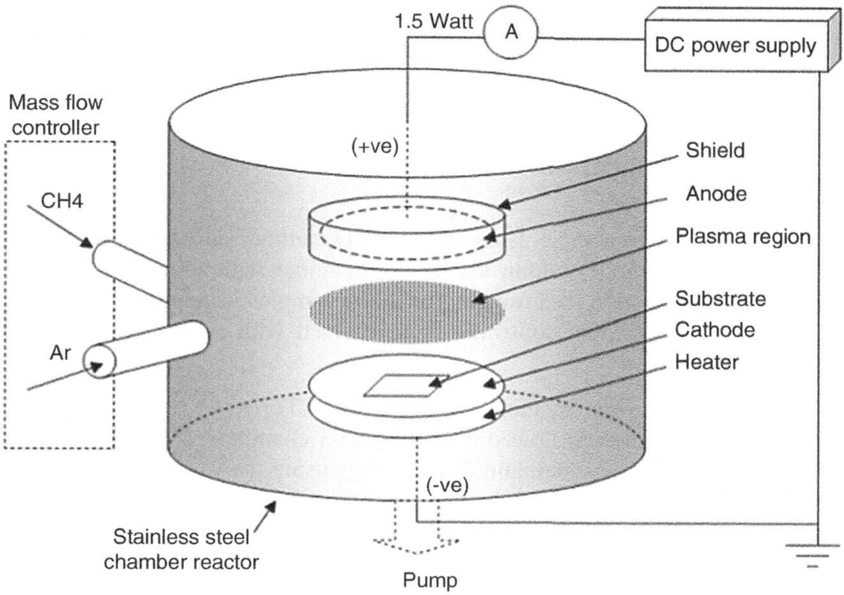

Fig. 1 DC-PECVD reactor schematic

2.1.2 Power Supply

Power supplies consist of two types:

1. Power supply for plasma formation
   The PECVD system used direct current power supply (DC). It provides 1.5 kV of voltage to produce maximum power of 1.5 W.

2. Power supply for heater
   The heater is capable of heating up the substrate at temperatures between 100 and 250°C.

2.1.3 Gas System

The gas systems consist:

1. The gases used: methane (CH_4), argon (Ar), and nitrogen (N_2).
2. Mass flow controller (MFC) used to measure and control the mass of gases that flow into the reactor chamber.
3. Vacuum system consists of vacuum pump (rotary mechanical pump), pressure gauge (Pirani gauge) and pressure controller. A vacuum of 10^{-2} Torr was achieved in the deposition chamber prior to the deposition.

This PECVD system is also equipped with an exhaust system in which all the unwanted gases from the reaction in the reactor will be ejected out using a pump via chimney.

3 Sample Preparation

In order to prepare the sample using DC-PECVD method, substrate was put on the substrate holder (cathode) within the reactor chamber with adjustable anode distance using linear motion feed trough. The entire samples were prepared in 5 hours (h). Before the deposition, substrates were cleaned with argon plasma with the following steps:

1. Reactor chamber were vacuumed to $\sim 2 \times 10^{-2}$ Torr.
2. Argon gas (3 sccm) was flowed into the reactor chamber.
3. Chamber pressure was maintained at 0.1 Torr using the variable valve.
4. DC power supply was turned on at 1.5 kV until Ar plasma was generated.
5. After 15 min, DC power supply was turned off.
6. The flow of Ar gas was stopped and the reactor chamber was vacuumed again.

After the cleaning process with Ar plasma was completed, the films deposition process was started. The deposition steps were as follows:

1. Reactor chamber were vacuumed to $\sim 2 \times 10^{-2}$ Torr.
2. Source gas (CH_4) with the desired rate was flowed into the reactor chamber (the operating deposition parameters were explained in the next section).
3. Chamber pressure was maintained at certain pressure.
4. DC power supply was turned on at 1.5 kV until plasma was generated.
5. After 5 h, DC power supply was turned off.
6. The flow of source gas was then stopped and the reactor chamber was vacuumed again to clean up all the gas residues within the chamber.
7. After the chamber pressure reaches $\sim 2 \times 10^{-2}$ Torr, the vacuumed pump was turn off.
8. Nitrogen gas was flowed into the reactor chamber until chamber pressure is equal to atmospheric pressure.
9. Take out the sample for structural and optical characterization.

4 Substrate Preparation

Prior to the growth process, the substrates were ultrasonically cleaned in order to achieve a good and uniform thin films deposition on the substrate surface.

The following procedure is followed:

1. A 30 min ultrasonic cleaned substrate in trichloroethylene.
2. Before continuing with acetone, methanol, the substrates were ultrasonic rinsed with deionized water (DI water).

3. A 30 min ultrasonic cleaned in acetone.
4. A 30 min ultrasonic cleaned in methanol.
5. And blow-dry the substrates using a hair dryer.
6. Put the substrates in a desiccator to avoid any contamination.

5 Deposition Parameters

Deposition parameters play an important role in the production of a-C:H films. Each parameter gives different effect to the a-C:H properties. Therefore optimization is required for excellent quality a-C:H. For a-C:H films which were deposited by DC-PECVD method the following parameter were experimented: chamber pressure, electrode distance, CH_4 flow rate and substrate temperature. Each set have five different samples.

The first set was deposited at chamber pressure between 0.1 and 0.8 Torr with all other parameter remain constant (Set 1) as shown in Table 1. The second set was deposited at electrode distances from 1.5 to 3.5 cm with all other parameter fixed (refer to Table 2). The next set was deposited by varying the CH_4 flow rate between 1 and 9 sccm (Set 3) as tabulated in Table 3. Finally, Set 4 was deposited by increasing the substrate temperature between 30 and 230°C while other parameters were fixed (Table 4). Through this various parameter done a set of optimum deposition parameters was obtained; consist of optimum chamber pressure, electrode distance, CH_4 flow rate and substrate temperature.

Table 1 Deposition parameters for different chamber pressure (Set 1)

Sample	Chamber pressure (Torr)	Power (W)
A01	0.1	0.3976
A02	0.2	0.2989
A03	0.4	0.2340
A04	0.6	0.1968
A05	0.8	0.1634

With other parameters fixed at: electrode distance, 2.5 cm, CH_4 flow rate, 5 sccm, substrate temperature, 30°C and deposition time, 5 h

Table 2 Deposition parameters for different electrode distance (Set 2)

Sample	Electrode distance (cm)	Power (W)
A06	1.5	0.3740
A07	2.0	0.3162
A08	2.5	0.2989
A09	3.0	0.1530
A10	3.5	0.1290

With other parameters fixed at: chamber pressure, 0.2 Torr, CH_4 flow rate, 5 sccm, substrate temperature, 30°C and deposition time, 5 h

Table 3 Deposition parameters for different CH_4 flow rate (Set 3)

Sample	CH_4 flow rate (sccm)	Power (W)
A11	1	0.3380
A12	3	0.3150
A13	5	0.2989
A14	7	0.2288
A15	9	0.2016

With other parameters fixed at: chamber pressure, 0.2 Torr, electrode distance, 2.5 cm, substrate temperature, 30°C and deposition time, 5 h

Table 4 Deposition parameters for different substrate temperature (Set 4)

Sample	Substrate temperature (°C)	Power (W)
A16	30	0.3150
A17	80	0.3213
A18	130	0.3328
A19	180	0.3445
A20	230	0.4218

With other parameters fixed at: chamber pressure, 0.2 Torr, electrode distance, 2.5 cm, CH_4 flow rate, 3 sccm and deposition time, 5 h

6 Sample Characterizations

6.1 Structural and Optical Analysis

Infrared and Raman spectroscopy, UV–Visible Spectrophotometer and photoluminescence are normally used to investigate the structural and optical properties of a-C:H thin films.

6.1.1 Infrared Spectroscopy

The infrared spectroscopy characterization of the a-C:H samples was done within the region from 400 to 4,000 cm^{-1}. The basic principle of infrared spectroscopy is the vibrations of the atomic molecule. The sub-molecular groups in the complex molecules structure possess their own natural vibration frequencies, which is generally in the infrared region. During infrared radiation (with broad range of the frequencies) on the sub-molecule, the sub-molecular groups in the molecule absorb the radiation frequencies, which meet their natural frequencies. Accordingly, the specific frequencies or wave numbers at which a molecule absorbs infrared radiation with the structure of the molecule can be feasibly correlated. The hydrogen incorporation into a-C:H films is possibly in the form of CH, CH_2, or CH_3 groups and the carbon to hydrogen bonding maybe sp^3, sp^2, or sp hybridized [10].

During infrared measurement, a bare corning glass substrate was used as a reference to cancel the substrate effects. The resolution of the measurement was 4 cm^{-1} and 100 scans to ensure a good signal to noise ratio. The final infrared spectrum should be free of all the environmental contributions (water vapor and CO_2) the background scanned should be done first.

6.1.2 Laser Raman Spectroscopy

Raman spectroscopy is often said to be complementary to infrared spectroscopy [23]. In Raman measurement, the sample is irradiated by intense laser beam of frequency, ν_o, (typically argon (Ar^+) laser wavelength of 514.5 nm) and the scattered light is detected. The scattered lights consist of two types; Rayleigh and Raman scattering. The Rayleigh scattering is an elastic collision between the incident photon and the molecule; therefore the frequency of the scattered photon is the same as that of the incident photon. This is the strongest component of the scattered radiation. Raman scattering is an inelastic collision between the incident photon and the molecule. The signal is very weak and is $\sim 10^{-5}$ of the incident beam which occurs in the range of ultraviolet to infrared region. The Raman spectrum line depends on the emitted photon frequency, the photon that shifts to a lower frequency is known as *Stokes line*. For a higher frequency of emitted photon this is known as an *anti-Stokes line*. This is due to less energetic final vibrational state as compared to the initial state. With the existence of frequency lines, which represents the vibration by certain material bonding, therefore the presence of amorphous structure could be known [23].

During Raman spectra measurement, the spectra of a-C:H films were scanned from 0 to 2,000 cm^{-1}. The spectra were curve-fitted using a program called GRAMS/32. The disorder (D) and graphite (G) peaks were fitted with Gaussian lineshapes on a quadratic baseline, and the substrate peak (if present at 1,095 cm^{-1}) was fitted with a Gaussian lineshape on a linear baseline. Values for the position, intensity, and full width half maximum (FWHM) of the G and D peaks respectively were recorded from the curve fitting.

6.1.3 UV/Visible Spectrophotometer

The main purpose of using UV–Vis spectrophotometer is to estimate the a-C:H thin film optical band gap (E_0). The optical band gap was obtained by the linear extrapolation of the $\sqrt{\alpha h\nu}$ versus photon energy, $h\nu$ graph using the Tauc plot [10]. The transmittance, reflectance, and absorption measurement were performed in a working range of 300–3,000 nm. Two radiation sources were used which is halogen/tungsten lamp for VIS/NIR region (340–2,500 nm) and deuterium lamp for UV region (190–350 nm). The scanning procedure starts by placing bare corning glass substrate in the reference holder and performing base line correction. The base line operation is to perform background correction over a certain selected wavelength range. Then placed the sample in the sample holder and the transmission scanning was obtained over the selected wavelength.

6.1.4 Photoluminescence Spectroscopy

One of the interesting features of a-C:H is its capability to exhibit strong photoluminescence (PL) emission at room temperature. PL spectroscopy is a technique that uses the photo excitation and recombination of electron-hole pairs to determine the structural and optical properties of the material. The recombination mechanism of a-C:H is affected by the disorder which cause tail state within the gap and by the defect state at mid-gap (dangling bond). This would influence the luminescence intensity. The PL mechanism is an effective tool to extract the information about the nature and the energy state of disordered a-C:H material.

The PL recombination process consists of two consecutive steps as shown in Fig. 2. The electron and hole initially lose their energy by number of transitions as they thermalize into the tail states followed by recombination which can be either radiative or non-radiative. The recombination which involves an emission of photon is designated as radiative recombination process and for non-radiative recombination process it involves emission of phonon which provides other recombination pathway. This does not contribute to the luminescences spectrum [10, 24]. The PL excitations can be achieved using special xenon flash tube, which produces an intense, short duration pulse of radiation over the spectral range of instrument (200–800 nm) to create electron-hole pairs. Pre-scan were done on the sample before an excitation scan.

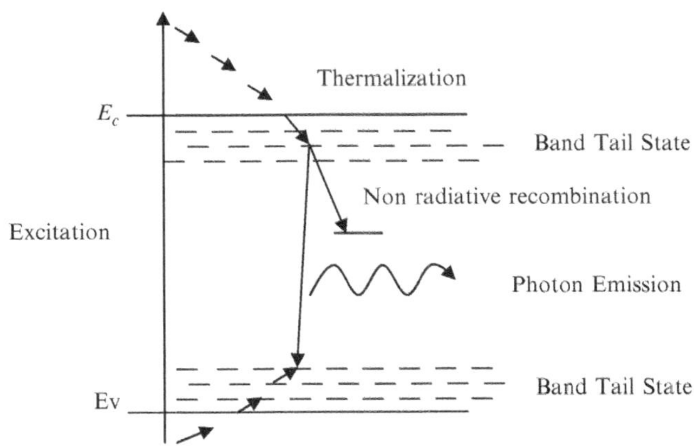

Fig. 2 Thermalization, radiatiave and non-radiative recombinations mechanism occurs in a-C:H structure [10]

6.2 Measurement of Film Thicknesses

Generally, there are several ways of measuring film thicknesses one of them is ellipsometry technique.

6.2.1 Ellipsometer

The basic principle of ellipsometry technique is by radiating the surface of the sample with monochromatic light (with known wavelength and polarity). Then, the polarity of the reflected light was analyzed. Multiple incident angle ϕ was used. Helium-neon laser with wavelength of 632 nm was used as light source. The equipment calibration was done based on standard silicon material by adjusting the analyzer meter and the pole on the ellipsometer equipment. Parameters obtained from the ellipsometer measurement were $A_{1'}$ and $A_{2'}$. Both were reading from the first and second analyzer respectively. Other than that, P_1 and P_2 were the reading from the first and second polarizer respectively. All the data were then entered into the BBC Ellips software in order to determine the thickness of the thin films.

7 Structural and Optical Properties of a-C:H Films

The results and discussions on the structural and optical properties of a-C:H are usually based on analysis from infrared and Raman spectroscopy, UV–Vis Spectrophotometer and photoluminescence.

7.1 Infrared Spectra

The structural characterization of a-C:H films is complicated due to their amorphous nature. The typical infrared spectra of a-C:H is a wide absorption band centered at about 2,900 cm^{-1}. Most of the researchers pay great attention to the C–H stretching band which is observed in the range of 2,800–3,200 cm^{-1}. In this region the complicated absorption bands originated from C–H vibrations with carbon atoms in various hybrid electronic states, such as sp^1, sp^2, sp^3, cyclic and aromatic states of carbon. The sp^1 hybrid is an unusual state in most types of a-C:H films [10]. The interpretation of the absorption peaks are tabulated in Table 5.

The example of infrared spectrum for a-C:H sample prepared by DC-PECVD method at plasma power, w from 0.2989 to 0.4218 W is illustrated in Fig. 3. The strongest and clearest absorption peaks were detected at 2,958, 2,932, 2,872 cm^{-1} which correspond to sp^3 CH_3 (asymmetric), sp^3 CH_2 (asymmetric) or sp^3 CH, and sp^3 CH_3 (symmetric) stretches respectively. From the spectrum, it can be concluded

Table 5 C–H stretch absorption bands for a-C:H

Wave number	Configuration	Olefinic or aromatic	Symmetric (S) or antisymmetric (A)
3,300–3,305	sp^1CH		
3,085	sp^2 CH_2	Olefenic	A
3,035–3,060	sp^2 CH	Aromatic	
3,020–3,025	sp^2 CH_2	Olefenic	
2,990–3,000	sp^2 CH	Olefenic	S
2,960–2,970	Sp^3 CH_3		A
2,975	sp^2 CH_2	Olefenic	S
2,955–2,962	sp^3 CH_3		A
2,920–2,925	sp^3 CH_2		A
2,920–2,925	sp^3 CH		
2,868–2,885	sp^3 CH_3		S
2,850–2,855	sp^3 CH_2		S

From Manage [10], Mutsukura et al. [13], Couderc et al. [25], Thomsen and Reich [26] and etc.

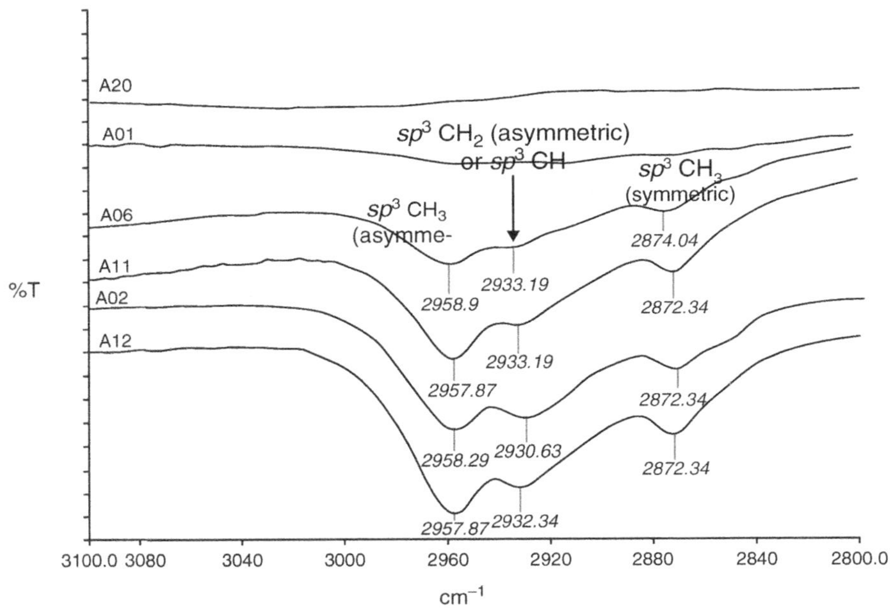

Fig. 3 Typical infrared absorption peak of C–H stretching vibrations of a-C:H thin films deposited at power; 0.2989–0.4218 W by DC-PECVD method

that higher plasma power causes an increase in ion energy, *E*, which easily breaks C–H bond. This led to the reduction of hydrogen content in the films, increased the sp^2 graphitic behaviour as detected at samples A20 and A01 [21, 27, 28]. As a result of higher plasma power consumption during deposition the films move from polymer to graphite like structure.

7.2 Raman Spectra

The Raman spectra of a-C:H films were dominated by the graphitic carbon features, the graphite (G) and disorder graphite (D) peaks at ~1,580 cm^{-1} and ~1,350 cm^{-1} respectively. Table 6 presents the interpretation of typical Raman features of a-C:H films according to Schwan et al. [29] Tamor and Vassell [30]. The G mode is a bond-stretching vibration of sp^2 sites that is arranged in olefinic chains or aromatic rings. The D mode is the vibrations of a six-membered aromatic ring, which is disorder active. It is only activated when sp^2 sites are in aromatic rings [29, 30].

For an entirely sp^3 bonded a-C material and pure crystalline graphite there is no D peak detected in Raman spectrum. In between these properties there is a ratio of G & D intensity peak (I_D/ I_G) [31], that is dependent on the size of graphitic clusters [32]. The I_D/I_G ratio can be correlated with qualitative information of sp^3/sp^2 ratio in the films. The increase of the I_D/I_G ratio indicates a higher disorder of a-C:H films and the decrease of the ratio of sp^3 content to sp^2-bonded carbon configuration [26, 33].

According to Ferrari and Robertson [34] the Raman spectrum of a film depends on four factors:

1. The clustering of the sp^2 phase
2. The extent of bond disorder
3. The presence of sp^2 rings or chains
4. The sp^2/sp^3 ratio

The positions, widths, and relative intensities of the D and G peaks are found to vary systematically with deposition parameters. The example of Raman spectrum of a-C:H films deposited by DC-PECVD method at power, w between 0.2989 and 0.4218 W is shown in Fig. 4. The details of the spectral were tabulated in Table 7. From the spectrum, it can be seen that plasma power affected the G and D peaks position, width, I_D/I_G ratio and curve background slope. The higher the power, the greater the G and D peaks shifted and I_D/I_G ratio increases as well. The higher power also causes the G peak to become narrower and background slope to reduce. This suggested that the films deposited at high power have high sp^2 content and larger cluster size compared with films deposited at low power.

Table 6 Positions and interpretation of Raman peaks [29, 30]

Peak	Wave number (cm^{-1})	Interpretation
D	~1,350	Microcrystalline mode of graphite sp^2 stretch vibration in aromatic ring which is activated by disorder
G	~1,580	Center vibration mode of graphite sp^2 stretch vibration of olefinic chains sp^2 stretch vibration in aromatic or condensed aromatic rings

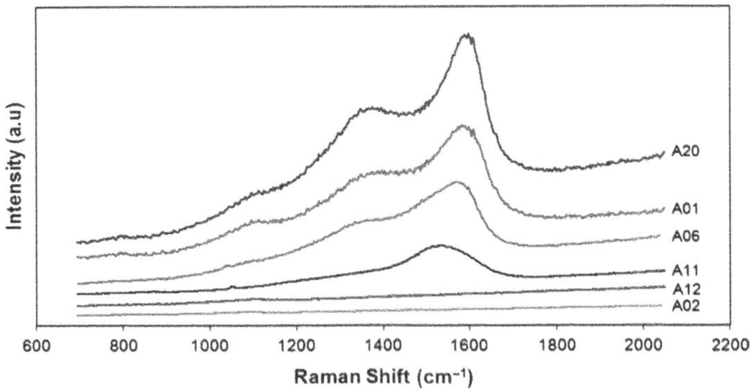

Fig. 4 Raman spectrum of a-C:H thin films deposited at various DC powers 0.2989–0.4218 W

Table 7 The position and width of G and D peaks, the slope value and I_D/I_G ratio of a-C:H in the 1,300–1,600 cm^{-1} range

Samples	D.C power, *w*	G-peak (cm^{-1})	G-width	Slope	D-peak (cm^{-1})	D-width	I_D/I_G Ratio
A20	0.4218	1,598	92.53	1.45	1,361	343.90	0.88
A01	0.3976	1,588	111.21	1.73	1,356	386.25	0.76
A06	0.3740	1,576	128.43	2.56	1,340	425.10	0.66
A11	0.3380	1,540	157.73	3.12	1,335	448.04	0.42
A12	0.3150	0	0	6.73	0	0	0
A02	0.2989	0	0	6.84	0	0	0

7.3 Optical Band Gap

The optical properties of a-C:H thin films have been extensively studied by many researchers with various forms of deposition technique. Table 8 show the correlation of film properties and deposition parameter on optical band gap, E_o. The E_o value can be tuned by adjusting the deposition parameter, as for higher E_o value the trend of deposition parameter such as substrate temperature, chamber pressure and etc. must be reduced to an optimum value. The reverse result; low E_o can be obtained by reversing deposition parameter to an increase value.

In order to determine the optical transition in a-C:H thin films, graphs of $(\alpha h\upsilon)^2$, $(\alpha h\upsilon)^{2/3}$, $(\alpha h\upsilon)^{1/2}$, $(\alpha h\upsilon)^{1/3}$ have been plotted versus photon energy, $h\upsilon$. Graphs of $(\alpha h\upsilon)^2$ and $(\alpha h\upsilon)^{2/3}$ versus photon energy determines the allowable and forbidden direct transition, respectively. Meanwhile, graphs of $(\alpha h\upsilon)^{1/2}$ and $(\alpha h\upsilon)^{1/3}$ versus photon energy shows the allowable and forbidden indirect transition. Optical band gap, E_0 can be gained by extrapolating the linear section of the curves until it crosses the x-axis (photon energy). The value at the intersection point on the x-axis is the optical band gap, E_0. Figure 5 shows the example of allowable indirect

Table 8 The correlation of film properties and deposition parameter on optical band gap, *E*

Film-property	Deposition parameter	E_os value	Reasons	Deposition technique
H content increase (polymer like a-C:H, soft film, in some study show low refractive index, *n*, value)	Power, *w*, substrate temperature, negative bias voltage, chamber pressure, – were all reduced	Increase	Increases in sp^3 content/ reduction of sp^2 content	DC saddle field glow discharge [10, 18] RF-PECVD [35–37] DC-PECVD [38] ECR-RF PECVD [39]
sp^2 content increase (graphite like structure, higher hardness, and *n* value)	Power, *w*, substrate temperature, negative bias voltage, chamber pressure, – were all increased	Decrease	Increment in graphitic cluster size and chain length	DC saddle field glow discharge [10, 18] ECR-microwave CVD [20] RF-PECVD [35–37] DC-PECVD [38]

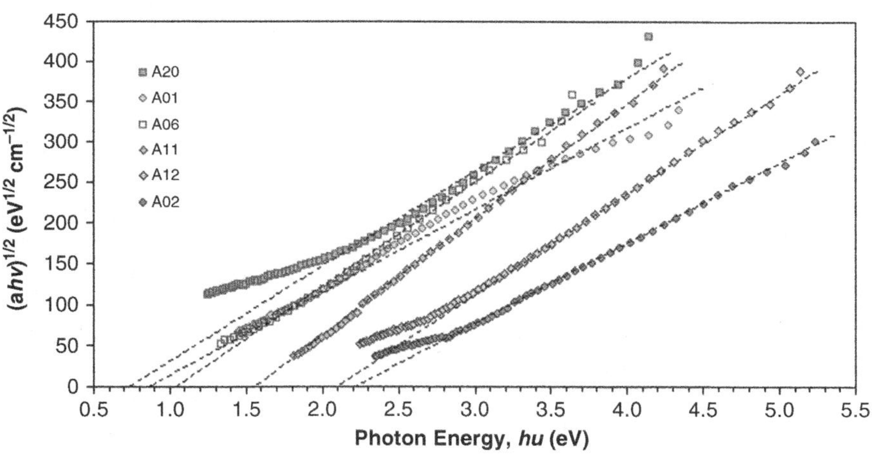

Fig. 5 Graph of $(\alpha h\upsilon)^{1/2}$ vs photon energy, $h\upsilon$ of a-C:H thin films at various DC powers 0.2989–0.4218 W

transition, $(\alpha h\upsilon)^{1/2}$ as a function of photon energy, $h\upsilon$ graph. The values of optical band gap of a-C:H thin films from DC-PECVD method for all type of optical transition are listed in Table 9. Generally, the table shows that the optical band gap values decreased with the increase of power from 0.2989 to 0.4218 W. It can be said that the decreased in sp^3 CH_3 bonding occurred in a-C:H films upon the increase of power during the deposition, led the decreased in the optical band gap. It was found that the E_0 value decreases when the sp^2 fraction in the films increased with increasing power and this was vice-versa for lower power value. This is also

Table 9 Values of optical band gap of a-C:H thin film at the respective DC power

Sample	DC power (W)	Optical band gap, E_0 (eV)			
		Allowable direct transition	Forbidden direct transition	Allowable indirect transition	Forbidden indirect transition
A02	0.2989	3.51	2.41	2.22	1.45
A12	0.3150	3.37	2.22	2.11	1.19
A11	0.3380	2.83	1.98	1.58	0.88
A06	0.3740	2.65	1.29	1.05	0.03
A01	0.3976	2.20	1.48	0.88	0.31
A20	0.4218	2.05	1.11	0.75	0.00

consistent with Robertson model [6]. According to this model, amorphous carbon consists of sp^2 clusters which were embedded in the sp^3 bonded matrix. Since sp^2 clusters form band edges, therefore they control the band gap. Bigger sp^2 clusters have lower E_0, while smaller sp^2 clusters have larger E_0 value.

7.4 *Photoluminescence*

A very broad photoluminescence (PL) spectra which cover entire visible range, centered at 2.1–2.3 eV have been reported by numerous study for a-C:H films. According to Rusli et al. [40], the PL intensity spectra could be associated as the number of dangling bonds in the films. The decreased in the PL intensity indicates high dangling bond concentration as less hydrogen attached to carbon atoms to terminate the bonds. For the relation between PL peak energy and optical gap shows that as optical gap decreases, the red shift of PL peaks occurs, however there was no clear relationship between them observed [41]. The photoluminescence (PL) data which was obtained from the a-C:H sample prepared at different power, *w* from 0.2989 to 0.4218 W using DC-PECVD is shown in Fig. 6. Here one can find that there were significant changes in the PL intensity and a slight peak shifting between them. As the structure in a-C:H films moves from a polymer-like (low power) to a graphite-like structure (high power), the PL intensity decreases. This was accompanied by a red shift of the PL peak energy, as the peak shifted from 1.93 to 1.42 eV. From the analysis, sample A02 showed highest PL peak energy value at 1.93 eV, containing smaller sp^2 cluster size compared with other samples.

The films which exhibit strong photoluminescence were the soft a-C:H films which have higher hydrogen concentration and greater optical band gaps. The red shift can also be explained by the increased in cluster size with the increasing power hence ion energy during deposition [10, 14, 40]. As the ion energy increases, the sp^2 cluster size in a-C:H increases, resulting in smaller optical band gaps, and finally, shift the photoluminescence peak energy towards lower energy.

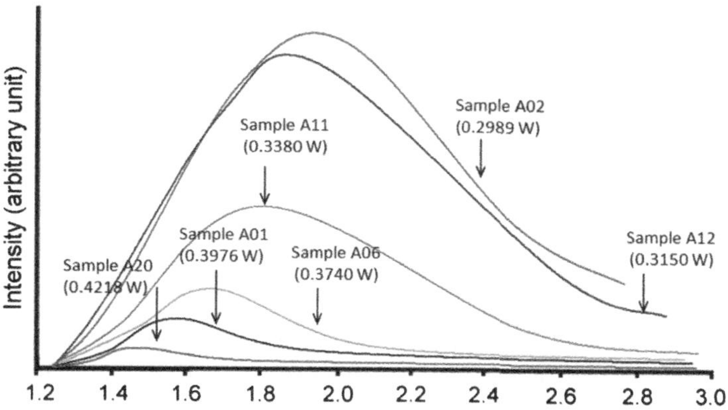

Fig. 6 The photoluminescence spectrum for a-C:H films deposited at various DC powers 0.2989–0.4218 W

8 Ions Bombardment Calculation

Ions play a crucial role in the growth mechanism of a-C:H thin film. Ions can create dangling bond sites on the film surface which are more likely causes the radicals to bond on the film surface through the dangling bond sites [10]. Therefore, the adsorption rate of the radicals on the surface is strongly dependent on the dangling bond generation rate [10, 13]. The generation of dangling bond depends on the energy and the flux of ions reaching the substrate. The major ionic species in the CH_4 plasma are the CH_3^+ ions [10]. The energy of CH_3^+ ions reaching the substrate increases with increasing power, *w*. The number of radicals which participate in the growth increases as well, with increasing power.

Hence the calculation of the ion bombardment was done as follow:
The ion bombardment, *E* is described as [6]:

$$E \sim V\left(\frac{\lambda_{mfp}}{L}\right) \tag{1}$$

where λ_{mfp} is the mean free path (mfp) of the particles within the ion sheath,

$$\lambda = \frac{kT}{\sqrt{2}p\sigma} \tag{2}$$

and *k* is the Boltzmann constant (1.38066×10^{-23} J K^{-1}). *T* is the plasma temperature in Kelvin (K). We assume that plasma temperature is somewhat above ambient since the electron and ion temperature is very high ~23,200 K and 500 K, respectively, while the neutral temperature is low ~293 K so the estimated value of plasma temperature is ~400 K. The plasma temperature will be higher than that if the substrate temperature is higher, ~500 K [42]. p is the pressure in Pascal and σ is the collision cross-section of CH_3^+ ions (~100 Å^2). We assume that the cross-section of CH_3^+ ions with the CH_4 molecules should be higher than CH_4 molecules with CH_4 molecules (since the cross-section of CH_4 is ~46 Å^2). Where L is the thickness of the ion sheath within DC glow discharge given by May [43] from the modified version of Morgan's equation [44]

$$L_{\max} = 2.12 \times 10^{-5}\{{}^1\!/_2 e/(3.5 \times 10^{19} p m_i \sigma)\}^{1/5}(V_{\max})^{3/5} \quad (3)$$

and p is the pressure, m_i is the mass of the major ionic species, σ and V_{max} are the collision cross-section of CH_3^+ and sheath potential. Here we assume that the voltage, V which is being measured is equal to sheath potential with plasma potential, V_p neglected since it is usually only around 10–30 V in the DC glow discharge [42]. For $p = 0.1$–0.2 Torr, $V = 490$–570 V we find that, $\lambda =$ 0.148–0.294 mm and $L = 4.45$–5.54 mm. Figure 7 shows an example plot of the sheath thickness against pressure and the line shows the value of sheath thickness at $p = 200$ mTorr for the sample A20. The sheath thicknesses which were obtained are consistent with visual observations during the experiment, which is about a couple of millimeter.

Figure 8 shows a plot of ion bombardment energy, E against power, w. Generally, it can be seen that E increases from 16.26 to 29.67 eV with increasing power, w from 0.2989 to 0.4218 W. It was found that sample A01 showed the highest E value which was 29.67 eV and sample A02 showed the lowest E value which was 16.26 eV. Rather than ion bombardment energy which contributes to the removal of hydrogen from the surface, the hot substrate (230°C) is also capable of removing the hydrogen by thermal evaporation. Since the hydrogen is low mass and the C–H bonding energy, E_B is low ~4 eV, compared to –C–C– (15.6 eV) the hydrogen will be preferentially evaporated [45].

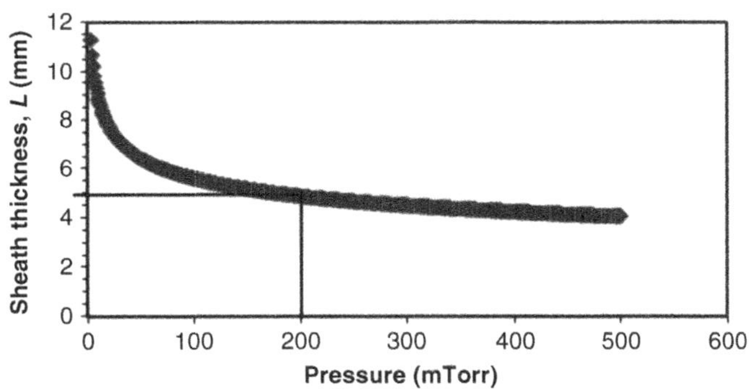

Fig. 7 A plot of the sheath thickness against pressure and the dashed line shows the value of sheath thickness at $p = 200$ mTorr for sample A20

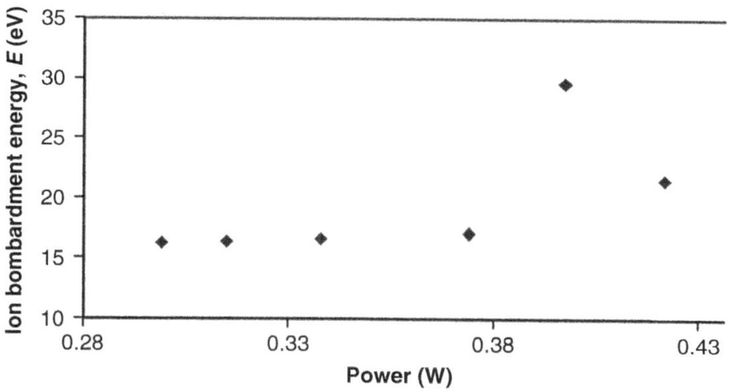

Fig. 8 A plot of ion bombardment energy, *E* of a-C:H thin films at different powers, *w*

The ion bombardment can causes the hydrogen to sputter from the surface [10, 36]. High energy ions can penetrate deeper into the growing film, compared to the low energy during the deposition. More hydrogen atoms were removed both from the surface and the subsurface atomics layers of the growing film. As a result, the H/C ratio in the film decreases with increasing ion energy. More sp^2 bonds and bigger clusters were formed as the ions energy increases. As for low energy ion, the removal of hydrogen from the growing films and creation of sp^2 sites were not efficient. The resulting films were soft a-C:H films with a high hydrogen concentration, a high sp^3 fraction (a large number of the sp^3 carbons were in the form of the sp^3 CH_3 configuration), and a low sp^2 fraction with small cluster. Due to the small cluster size, the soft a-C:H films have larger optical band gaps, as *E* decreased.

9 Conclusion

In this chapter the synthesis of a-C:H films using the DC-PECVD method was discussed in detailed. Experimental results on the effect of deposition parameters such as chamber pressure, electrode distance, CH_4 flow rate, and substrate temperature on the a-C:H structural and optical properties were presented and discussed. Structural and optical properties of a-C:H thin films based on infrared and Raman spectroscopy, UV–Vis spectrophotometry and photoluminescence spectroscopy analyses were presented. The sp^3/sp^2 and hydrogen contents, optical gap, E_0 and photoluminescence properties of a-C:H films were strongly dependent on the plasma power and ion bombardment energy which resulted from varying the deposition parameters. The most interesting feature of a-C:H is its various properties (either diamond or graphite-like structure) which can be done by manipulating the deposition parameter. Hence, each property can be customized for specific applications.

Acknowledgements I would like to express my gratitude to UPSI, MARA, UTM and UiTM for the financial and technical support on this project.

References

1. Chopra, K.L.: Thin Film Device Application. Plenum, New York (1969)
2. Kazmerski, L.L.: Polycrystalline and Amorphous Thin Films and Devices. Academic, New York (1980)
3. Ibrahim Abu Talib, Mustaffa Hj. Abdullah, Sahrim Hj. Ahmad: Sains Bahan, 2nd edn. Dewan Bahasa dan Pustaka, Kuala Lumpur (1993)
4. Dachuan, Y., Niankan, X., Zhengtang, L., Yong, H., Xiulin, Z.: Surf. Coatings Technol. **78**(1–3), 31–36 (1996)
5. Grill, A.: IBM. J. Res. Develop **43**(1/2), 147–160 (1999)
6. Robertson, J.: Mater. Sci. Eng. R **37**, 129–281 (2002)
7. May, P.W.: CVD diamond: a new technology for the future? Elsevier Science Ltd., 10–106 (1995)
8. May, P.W.: Phil. Trans. R. Soc. Lond. A **358**, 473–495 (2000)
9. Smith, J.A.: PhD thesis, University of Bristol, UK (2001)
10. Manage, D.P.: PhD thesis, University of Toronto, Canada (1998)
11. Aisenberg, S., Chabot, R.: J. Appl. Phys **42**(7), 2953–2958 (1971)
12. Angus, J.C.: Thin Solid Films **142**(1), 145–151 (1986)
13. Mutsukura, N., Inoue, S., Machi, Y.: J. Appl. Phys **72**(1), 43–53 (1992)
14. Rusli, Amaratunga, G.A.J., Silva, S.R.P.: Opt. Mater. **6**, 93–98 (1996)
15. Jing, Q.C.: PhD thesis, University of Texas at Dallas, US (1999)
16. Bouree, J.E., Godet, C., Etemadi, R., Drevillon, B.: Synth Met **76**, 191–194 (1996)
17. Taube, K.: Surf. Coatings Technol. **98**(1–3), 976–984 (1998)
18. Sagnes, E.: PhD thesis, University of Toronto, Canada (1998)
19. Cheng, Y.H., Wu, Y.P., Chen, J.G., Qiao, X.L., Xie, C.S., Tay, B.K., Lau, S.P., Shi, X.: Surf. Coatings Technol. **135**(1), 27–33 (2000)
20. Zhou, X.T., Lee, S.T., Bello, I., Cheung, A.C., Chiu, D.S., Lam, Y.W., Lee, C.S., Leung, K. M., He, X.M.: Mater. Sci. Eng. B **77**, 229–234 (2000)
21. Hong, J., Goullet, A., Turban, G.: Thin Solid Films **364**(1–2), 144–149 (2000)
22. Hong, J., Lee, S., Cardinaud, C., Turban, G.: J. Noncrystalline Solids **265**(1–2), 125–132 (2000)
23. Long, D.A.: Raman Spectroscopy. McGraw-Hill, London (1977)
24. Morigaki, K.: Physics of Amorphous Semiconductor. Imperial College Press, London (1999)
25. Couderc, P., Catherine, Y.: Thin Solid Films **146**(1), 93–107 (1987)
26. Thomsen, C., Reich, S.: Phys. Rev. Lett. **85**, 5214 (2000)
27. Cho, Y.O., Cho, J.M., Jun, W.J., Cha, O.H., Suh, E.K., Yu, K.H., Yu, S.C., Lee, J.K.: Diam. Relat. Mater. **11**(11), 1848–185 (2002)
28. Lacerda, R.G., Tojolan, V.S., Cox, D.C., Silva, S.R.P., Marques, F.C.: Diam. Relat. Mater. **11** (3–6), 990–984 (2002)
29. Schwan, J., Ulrich, S., Ehrhardt, H., Silva, S.R.P.: J. Appl. Phys. **80**(1), 440–447 (1996)
30. Tamor, M.A., Vassell, W.C.: J. Appl. Phys. **76**(6), 3823–3830 (1994)
31. Kleinsorge, B., Rodil, S.E., Adamopoulos, G., Robertson, J., Grambole, D., Fukarek, W.: Diam. Relat. Mater. **10**(3–7), 965–969 (2001)
32. Tuinstra, .F, Koenig, J.: J. Chem. Phys. **53**, 1126 (1970) in Schwan, J., Ulrich, S., Ehrhardt, H., Silva, S.R.P.: J. Appl. Phys. **80**(1), 440–447 (1996)
33. Lejeune, M., Durand-Drouhin, O., Henocque, J., Bouzerar, R., Zeinert, A., Benlahsen, M.: Thin Solid Films **389**(1–2), 233–238 (2001)

34. Ferrari, A.C., Robertson, J.: Interpretation of Raman spectra of disordered and amorphous carbon. Phys. Rev. B. **61**(20), 14095–14107 (2000)
35. Filik, J., May, P.W., Pearce, S.R.J., Wild, R.K., Hallam, K.R.: Diam. Relat. Mater. **12**(3–7), 974–978 (2003)
36. Budenzer, A., Dischler, B., Brandt, G., Koidl, P.: J. Appl. Phys **54**(8), 4590–4595 (1983)
37. Serra, C., Pascual, E., Maass, F., Esteve, J.: Surf. Coatings Technol. **47**(1–3), 89–97 (1991)
38. Suriani, A.B.: MSc thesis, Universiti Teknologi Malaysia, Malaysia (2005)
39. Durand-Drouhin, O., Zeinert, A., Benlahsen, M., Zellama, K., Kre, R., Turban, G., Grosman, A.: Diam. Relat. Mater **9**(3–6), 752–755 (2000)
40. Rusli, Amaratunga, G.A.J., Silva, S.R.P.: Thin Solid Films **270**(1–2), 160–164 (1995)
41. Demichelis, F., Tagliaferro, A., Gupta, D.D.: Surf. Coatings Technol. **47**(1–3), 218–223 (1991)
42. Chapman, B.: Glow Discharge Processes. Wiley-Interscience, New York (1980)
43. May, P.W.: PhD thesis, University of Bristol, UK (1991)
44. Morgan, R.A., PhD thesis, University of Sussex, UK (1985)
45. Craig, S., Harding, G.L.: Thin Solid Films **97**(4), 345–361 (1982)

Carbon Nanotubes Towards Polymer Solar Cell

Ishwor Khatri and Tetsuo Soga

Abstract The semiconducting polymer thin film has gained substantial interest in the research community because of the possibility to produce polymer based photovoltaic devices by roll to roll type manufacture, which is impossible by conventional technologies. Hole transferring semiconducting polymer and electron accepting fullerene (C_{60}) derivative are of special interest because of their stability and high power conversion efficiencies. With the discovery of new carbon "carbon nanotubes" (CNTs), researchers have started blending them with polymer for improving the solar cell efficiency. CNT-incorporated solar cell shows better power conversion efficiency than pristine solar cell without CNTs. This is because of the wonderful properties of CNTs. CNTs have outstanding properties like ballistic conductive, high aspect ratio, high surface area, flexible, strong, rigid, environmental stable, capability of charge dissociation, transportation and so on, which are believed to be an ideal material for fabricating high performance solar cell. In this chapter, some of the works done on polymer–CNTs based solar cells are summarized. A variety of device architecture and band diagram proposed by different authors have been included and described.

1 Introduction

A solar cell converts sunlight into electricity directly through the photovoltaic (PV) effect. In semiconducting material electrons live in range of define energy level called band. The conduction band is partially filled with electrons, creating negative charge. The valance band has areas where electrons are missing-known as hole (positive charge). In the absence of light, the positive and negative charges balance each other in the case of intrinsic semiconductors. When light energy

I. Khatri and T. Soga
Department of Frontier Materials, Nagoya Institute of Technology, Gokiso-cho, Showa-ku, Nagoya 466-8555, Japan
e-mail: ishwor_nep2000@yahoo.com; soga@nitech.ac.jp

N. Yahya (ed.), *Carbon and Oxide Nanostructures*, Adv Struct Mater 5,
DOI 10.1007/8611_2010_16, © Springer-Verlag Berlin Heidelberg 2010,
Published online: 27 June 2010

strikes on semiconductor pn junction, the electrons are dislodged causing electrons to move down an external circuit in the form of light-generating electricity. This phenomenon is called PV effect. Detail theoretical description is given in [1]. If light with photon energy is greater than optical band gap, free electrons and free holes are formed by optical excitation in the semiconductor. Different materials used for solar cells are: Silicon (Si), Copper indium gallium selenide [Cu(In,Ga)Se2], Cadmium Sulfide (CdS), Cadmium Telluride (CdTe), Titanium dioxide (TiO_2) etc. Silicon and compound semiconductor-based devices are dominating solar technology. However, the cost of these solar cells is much high to reach for daily life. So, low cost and high efficiency solar cells are yet to be realized for their commercialization.

In the search for alternative material, carbon is highly attractive because it is expected to have similar properties to silicon and it is highly stable. Carbon is a remarkable element that exist in a variety of forms ranging from insulator/semiconducting diamond (or diamond-like amorphous film [2]) to metallic/semi-metallic graphite (or graphene [3]), conducting/semiconducting fullerenes (e.g. C_{60} [4]) and carbon nanotubes (CNTs) [5].

On the other hand, the organic polymer-based materials are promising too because of their high manufacture using roll to roll or spray deposition. Attractive additional features are the possibilities to fabricate flexible devices, which can be integrated into curved surfaces, existing building structure. Furthermore, the polymer–CNTs composite materials have potential macroscopic device application, such as light-emitting diodes (LED), field-effect transitions (FET) and PV devices [6–9]. This is because organic materials having delocalized pi(π) electron system can absorb sunlight, create photogenerated charge carriers and transport these charge carriers towards the electrodes. Here, we discuss on the PV devices developed from combination of conjugated polymer and CNTs.

1.1 Device Fabrication

The general device design for organic-CNTs solar cell is sandwich structure (Fig. 1) consisting transparent conducting electrodes, (TCE) basically indium tin oxide (ITO) (marked with ‘a’ in Fig. 1), or fluorine tin oxide (FTO) coated on glass (or polyethylene terephthalate (PET) substrate, poly (3,4-ethylenedioxythiophene) poly(styrenesulfonate) (PEDOT:PSS) (marked with ‘b’ in Fig. 1), active layer (CNTs–polymer) (marked with ‘c’ in Fig. 1) and a thermally evaporated metal electrode on the top of the composite layer typically aluminium ‘Al’ (marked with

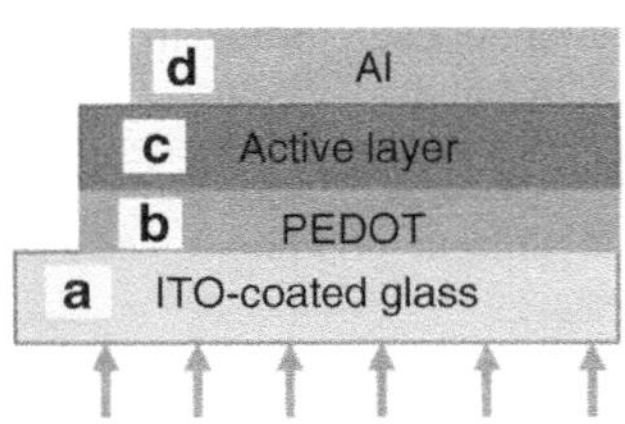

Fig. 1 Common device structure for organic-CNTs solar cell

'd' in Fig. 1). Somewhere, 'Al' is also used with an ultrathin lithium fluoride (LiF) underlayer. Layer of PEDOT:PSS and LiF are generally called buffer layer. Ultrasonication method is used to disperse CNTs in the conjugated polymer, followed by centrifugation to remove large aggregates. The composite solution is then deposited on the transparent electrode either by drop casting or spin coating.

Advantages of Buffer Layers

1. They minimize the band bending that arise from the Schottky barrier between the semiconducting active layer and the metallic electrodes
2. Photoelectron spectroscopy studies show that the work function of metal can be reduced by evaporating of LiF layer [10]
3. PEDOT:PSS improves the surface quality of the electrodes as well as facilitates the hole injection/extraction [11]

2 Organic Semiconducting Materials and Carbon Nanotubes (CNTs)

Organic semiconducting materials exist either in small size of small molecules with molecular weight of less than a few thousand atomic mass units (amu), or polymer with molecular with greater than 10,000 (amu). Dendrimes are another recent class of materials that have molecular weight range between small molecules and polymers. Dendrimes are also studied for photovoltaic devices. Semiconducting donor like organic polymer such as poly(3-hexylthiophene) (P3HT), poly[2-methoxy-5-(2′-ethyl-hexyloxy)-1,4-phenylene vinylene] (MEH-PPV), poly(3-octylthiophene) (P3OT) are derivative of phenylene vinylene, thiophene chain. They can be doped chemically, photochemically or electrochemically methods. Likewise, C_{60} or CNTs are electron acceptor [12]. The photoinduced charge transfer between P3OT and single wall carbon nanotubes (SWCNTs) is shown in Fig. 2.

2.1 Merits of Organic Semiconductor Materials

1. They show high absorption coefficients exceeding 10^5 cm^{-1} [13]
2. Electronic band gap can be engineered by chemical synthesis [14]

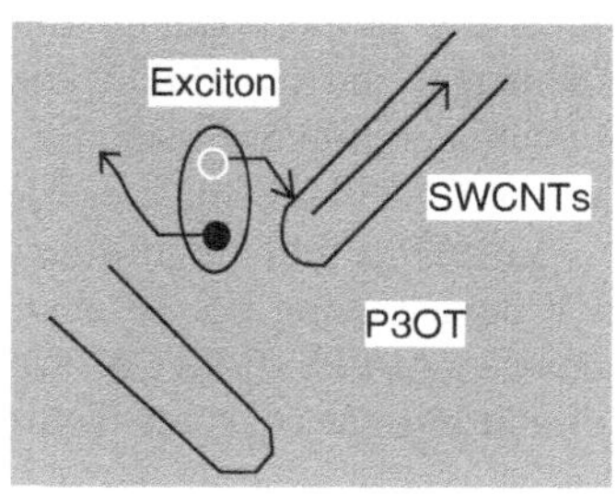

Fig. 2 Dissociation process of photogenerated exciton on polymer-SWCNTs blend. Electron is transferred to SWCNTs

3. Block polymer composition can be designed in such a way that one block of polymer can interacts with third compounds (like quantum dots, quantum wires) while the other block(s) will provide the solubility [15]
4. Charge carrier mobilities of organic polymers (about 10 cm^2/V s) are competitive to amorphous silicon [16]

2.2 Demerits of Organic Semiconducting Materials

1. Band gap of the organic materials is large. Majority of semiconducting polymer have band gap higher than 2 eV (620 nm). So, only a small portion of the incident solar light is absorbed, which limits the possible harvesting of solar photons [17].
2. In organic materials photoexcitation produces coulombically bound electron–hole pairs, called excitons. It is estimated that only 10% of the photoexcitations lead to free charge carriers in conjugated polymer [18].
3. Exciton diffusion length is very small about 10 nm.
4. Carrier mobilities of organic materials are low compare to single crystal Si.

3 Carbon Nanotubes (CNTs)

CNTs are promising materials that act as exciton dissociating centers and ballistically conductive agent with high carrier mobility. In addition they are strong, rigid, chemically inert and environmentally resistant. CNTs are the cylindrical roll up graphene sheet whose length is in the micro scale and diameter in nano scale. Thus, it signifies large aspect ratios ($>10^3$). CNTs are classified by the number of rolled up graphene sheets. A single walled carbon nanotubes (SWCNTs) have only one rolled up graphene sheet in cylindrical structure. Nanotubes with many rolled up graphene layers are called multi walled carbon nanotubes (MWCNTs). There is also a special class of MWCNTs called double walled carbon nanotubes (DWCNTs) which have two graphene layers rolled up with concentric center, that have different properties to SWCNTs and MWCNTs.

Van Hove Singularities exist in the local density of states (DOS) of SWCNTs due to their 1D nature of the conduction electron states. Band gap of SWCNTs is inversely proportional to the tube diameter. By combining CNTs of different diameters and chiralities, correspondingly, different band gaps, it is possible to obtain a continuous response over a spectral range [19]. A characteristic comparison of CNTs and single-crystal silicon is shown in Table 1 [20].

In 1999, a thin film of functionalized MWCNTs (f-MWCNTs) was used to make PV device using poly(p-phenylene vinylene) (PPV)[21]. Under illumination, the device (Al/PPV-fMWCNTs/ITO) produced an open circuit voltage (V_{oc}), short circuit current (I_{sc}), fill factor (FF) and efficiency (η) were 0.9 V, 0.56 $\mu A/cm^2$,

Table 1 Characteristics comparison of CNTs and single-crystal silicon (Source [20])

Properties	CNT	Silicon
Density (g/cm^3)	0.8–1.2	2.33
Band gap (eV at 300 K)	0.3–2.0	1.12
Resistivity (Ωcm at 300 K)	0.1	~1.0
Electron mobility (cm^2/Vs at 300 K)	10^8	<1,400
Hole mobility (cm^2/Vs at 300 K)	10^3	<500

0.23 and 0.081% respectively. The external quantum efficiency (EQE) of the device was twice than that of the ITO/PPV/Al devices. From observed EQE, it was estimated that holes are accepted by f-MWCNTs at the PPV-fMWCNTs interface. Later in 2002, it was also reported that SWCNTs transfer electron in the active layer of P3OT-SWCNTs composite [12]. In this device, arc generated SWCNTs were mixed with P3OT polymer. Diodes (Al/polymer-SWCNTs/ITO) show increase in the short circuit current by two order of magnitude compare to pristine polymer diode (Al/P3OT/ITO). It was proposed that the main reason for this increment is the photoinduced electron transfer at polymer/nanotube interface.

After these major breakthrough, most of the devices that are reported either claim in the increase of efficiency by incorporation of MWCNTs (enhancing hole conductivity) or SWCNTs (enhancing electron conductivity). In some literatures, thin films of SWCNTs have been used for hole conductivity, which are reported below.

3.1 Charge Separation

In organic materials, photon absorption generates bound electron–hole pairs, so-called "excitons". These excitons have to dissociate into free charges in order to get photovoltaic response. Materials of higher electron affinity (like C_{60}, PCBM, CNTs) can dissociate excitons. This allows the preferential transfer of the electrons into electron acceptor molecule (like C_{60}, PCBM, SWCNTs) while leaving holes to be transported through the polymer. This process of charge separation is known as photoinduced charge transfer.

Figure 3 shows schematic energy-band diagram of a simple device consisting of a single organic layer between two electrodes of different work functions, i.e. Au and ITO. The difference in the work function of these electrodes generates electric field which is sufficient to break up the photogenerated exciton. Since, exciton diffusion length is short, (about 1–10 nm) in single layer and the mobility is low, there are a lot of chances of recombination or break up to supply separate charges. Thus, quantum efficiency (QE) of this type of solar cell is less than 1%.

But in heterojunction (suppose P3HT-C_{60}), electrostatic forces result from the differences in electron affinity and ionization potential. If both electron affinity and ionization potential are greater in one material (the electron acceptor, say C_{60}) than the other (the electron donor, say P3HT) then the interfacial electric field drives

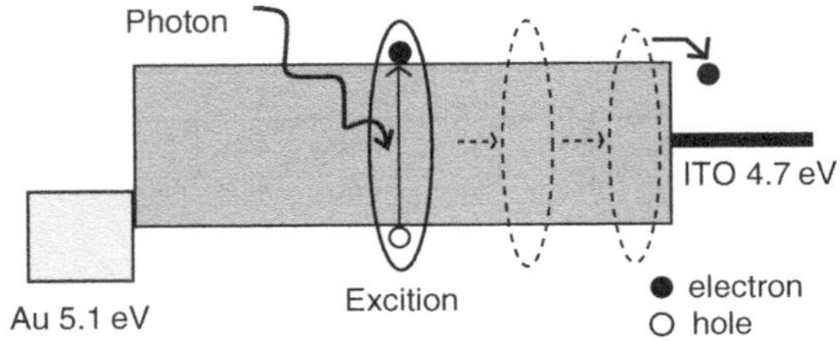

Fig. 3 Schematic energy-band diagram of a simple device consisting of single organic layer between two metal contacts. An electric field results from the difference in work functions of the contacts. Absorbed photons generate excitons which diffuse towards one or other contact (here it is diffusing towards ITO contact) where they may dissociate to yield charge pairs. Only the layer of organic material, which lies within an exciton diffusion length of a contact can contribute to the photocurrent

charge separation [22]. The local electric field is strong and may break up photogenerated excitons.

However, the better understanding of the charge separation in polymer–CNTs interface is still in progress. The large band gap of organic materials to small band gap semiconducting CNTs may generate 'build in voltage'. CNTs can be metallic or semiconductor depending in their chirality. The high electric field (or build in voltage) in these junctions can split up the excitons. The different band diagrams proposed in different literatures considering SWCNTs as metallic, semiconductor are discussed below.

3.2 *Improvement of Polymer Solar Cell by CNTs Incorporation*

The advantages of CNT incorporation in polymer solar cell are summarized as follows:

1. In SWCNTs-epoxy composites, the electrical conductivity has been claimed to rise by nearly 10^5 S/cm between 0.1 and 0.2% weight loading of SWCNTs [23]. Hence, CNTs improve low carrier mobilities of organic material.
2. CNTs provide high surface area ($\sim$1,600 m^2/g), which offers a tremendous opportunity for exciton dissociation [24].
3. Since SWCNTs have diameters in nanoscale and lengths on the order of microns, so at low doping of SWCNTs, percoration pathway are established providing the means for a high carrier mobility and efficient charge transfer. However, such pathways are not established in polymer solar cell. Recently, there are some reports of fabrication of polymer nanofibers for creating pathway in enhance hole mobility of polymer [25]. Thus, CNTs provide large interfacial area for low exciton diffusion length of polymer.

4. CNTs have low energy gap compared to large energy gap of organic polymer. The lower band gap of CNTs and higher band gap of polymer materials may generate a large build in voltage. Charges may separate at this built in field.
5. CNTs make percolation pathway for charge transformation. However, in polymer–C_{60} composition, electrons transfer by hopping.

Other beneficial properties of SWCNTs relevant to polymeric photovoltaic development include composite reinforcement and thermal management. Tensile strengths of SWCNTs have been estimated to equal ~20 GPa, while the Young's modulus measured by atomic force microscopy is ~1 TPa. This high Young's modulus and strength to weight ratio could provide much needed mechanical stability to large area thin film arrays [26 and therein].

4 CNT Incorporated Polymer Solar Cell

There are two ways to incorporated CNTs with polymer (1) layer assembly of CNTs at desired locations and (2) blending CNTs with the organic species.

4.1 Layer Assembly at Desire Location

Layer assembly is a way of using CNTs films at different layers of device. Chaudhary et al. [27] studied the effect of thin layer of SWCNTs by placing it at different layers in the device. They noted that SWCNTs on the hole collecting side of the active layer leads to an increase in the power conversion efficiency (PCE) of the PV devices from 4 to 4.9%. The success of the device performance depends on appropriate purification processes [28–31], chemical functionalization [31–35], individualization [36, 37], length shortening [37] and crystallinity [38, 39]. CNTs films are generally produced after functionalization. This is because f-CNTs get dissolved in polar and non polar solvent. Here, we describe one of the simple and popular methods of functionalization.

4.1.1 Functionalization of CNTs

Non-functionalized CNTs are insolubility. A number of methods have been reported to functionalize CNTs [30–34]. Edge sites of CNTs are much more reactive than the atoms in the interior of the graphene cylinder and chemisorbed foreign elements. Surface oxides are produced by liquid oxidants e.g. aqueous solutions of H_2O_2, NaOCl, $(NH_4)_2S_2O_8$, $AgNO_3$, H_2PtCl_6 etc. The surface oxides decompose to CO_2 or CO on heating. After cooling to room temperature, they may again react with oxygen (air) or water vapor. It is shown that SWCNTs whose lengths have been reduced by sonication (in range of 100–300 nm) can be soluble in common organic solvents by covalent functionalization [36]. Functionalization of CNTs starts from simple ultrasonication [40, 41] to complex chemical treatment. Titration, infrared spectroscopy (IR), X-ray photoelectron spectroscopy, thermal desorption spectroscopy, electrokinetic measurement are the most frequently used methods for the characterization of surface oxides [42]. Oxidation with HNO_3 is considered as mild oxidation method and often used because its oxidizing properties can be controlled by concentration and temperature. It is suggested that HNO_3 introduces carboxylic acid groups (COOH) only at those initial defects that already exist (like edges of CNTs). In contrast, sonication of CNTs in mixture of H_2SO_4 and HNO_3 increases the incidence of carboxylic acid groups not only at initial defect sites but also created new defect sites along the walls of CNTs. The oxidation processes begin at initial defects at first and followed throughout the CNTs length. Strong oxidation like cutting of CNTs [36] (begins from newly developed defect sites in CNTs) can be divided into two steps: (1) the defect generating step and (2) the defect consuming step. During defect generating step, the oxidants attack the graphene structure by electrophilic reactions and generate active sites like OH, C=O. During defect consuming step, the graphene structure of the tube was destroyed [31]. For chemistry of CNTs see reference [43]. Once oxidized, the nanotubes spontaneously dispersed in water and remain stable for many months with only a very small amount of aggregation materials precipitation over time. Precipitation could be accelerated by ionic additions, particularly acidic ones. The oxidation of nanotubes induces a negatively charged surface, particularly through the ionization of acidic surface groups [44, 45]. The resulting electrostatic repulsion leads to a dramatic increase in the stability of the colloidal dispersion. Figure 4 shows the FTIR measurements of acid treated CNTs samples at 120°C in H_2SO_4/HNO_3 (3:1). FTIR is a useful tool to characterize f-CNTs [46]. Theoretical investigation reveals that monocrystalling graphite belongs to the $D_{6h}{}^4$ space-group symmetry. Thus, an out-of-plane A_{2u} mode at around 868 cm^{-1} and an in-plane E_{1u} mode at around 1,588 cm^{-1} are infrared active [47, 48]. Fourier transform infrared (FTIR) spectra of purified CNTs (Fig. 4 green line) have –C=C– bond at around 1,636 cm^{-1}. During purification, CNTs were heated in deionized water in order to remove defects like CH and CH_2 groups. Hydrothermally initiated dynamic extraction (HIDE) method [29] at around 373 k disintegrate the CNTs soot. In this process, hydroxyle group may form. Water can attack a terminal –C=C– bond resulting in the formation of a hydroxyl group on one of the carbon atoms and a free hydrogen ion. The hydrogen ion will attach to the other carbon of the double bond pair,

$$\text{i.e.,} -C = C- \; + H_2 + H_2O \rightarrow -CH - CH_2(OH) \qquad (1)$$

resulting in a terminal hydroxyl group ($-CH_2OH$) [49]. Refluxing of CNTs in acid mixture (H_2SO_4/HNO_3) for about 2 h did not make any significant change in FTIR spectrums. However, symmetric and asymmetric C–H stretching mode appears (Fig. 4 blue line). In asymmetric mode one bond stretch while other is compressed.

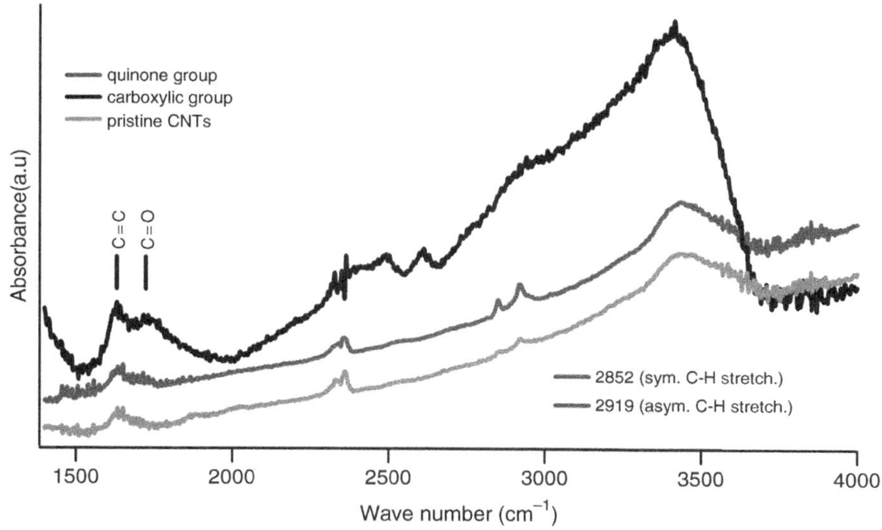

Fig. 4 IR spectra of pristine CNTs and that after 2 h liquid-phase treatment. Peaks at 2,852 cm^{-1} and 2,919 cm^{-1} assign to symmetric and asymmetric C–H stretch, respectively. Spectrum after 5 h liquid-phase treatment, peak at around 1,720 cm^{-1} assign to C=O vibration mode of carboxylic group is observed

This stage can be considered as defect generating stage. Under strong oxidation conditions, the graphene structure of CNTs is broken and the tube cut off. Zhang et al. [31] suggest that $KMnO_4$ and dilute HNO_3 are all inefficient in the defect consuming step whereas the acid mixture (H_2SO_4, HNO_3) is strong enough to cut the graphite structure. After 5 h of treatment, a sign of carboxylic group (COOH) was observed (Fig. 4, black line). Formation of COOH and OH group increases with increasing treatment time. Carboxylic group (COOH) helps to dissolve CNTs in solvent that makes possible of preparing CNTs films by spin coating or drop casting. T. V. Sreekumar et al. [50] observed conductivity in the plain of functionalize CNTs film is 1×10^5 S/m. This film reduces conductivity after annealing. However, in microwave functionalized SWCNTs film, conductivity was regain after annealing in inert atmosphere [51].

4.1.2 Chemical Treatment on Energy Level of CNT Films

The electronic structure of the SWCNTs is related to a 2-D graphene sheet, but because of 1-D nature of SWCNT the continuous electronic density of state (DOS) in graphite divides into a series of spikes that are referred to as Van Hove singularities. In visible-near infrared (vis-NIR) spectra of microwave functionalized SWCNTs (Fig. 5), Van Hove singularities in the electronic density of states

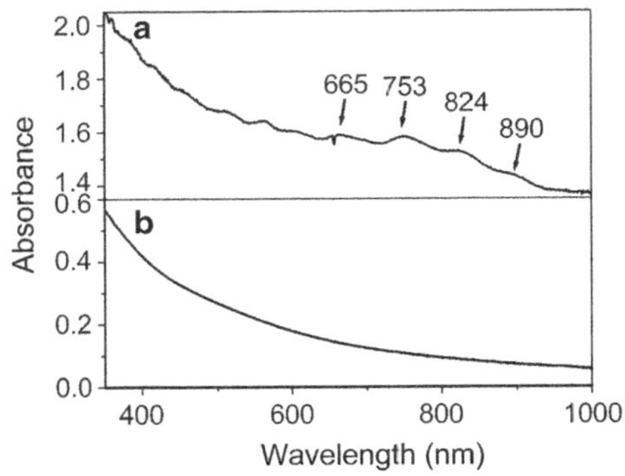

Fig. 5 Visible near infrared (vis-NIR) spectra of (**a**) pristine SWCNTs suspended in dimethylformamide (0.05 mg/ml) and (**b**) aqueous dispersion of microwave-reacted SWCNTs (0.5 mg/ml). Reuse with permission from author [51]. Copyright 2006, ACS publications

Table 2 Work function of MWCNTs and graphite determined from UPS (Source [52])

Sample	Work function (eV)
Purified MWCNTs	4.3
Air-oxidized MWCNTs	4.4
Plasma-oxidized MWCNTs	4.8
Acid-oxidized MWCNTs	5.1
Highly oriented graphite (HOPG)	4.4

(DOS) of the SWCNTs were observed at 665, 753, 824, and 890 nm whereas, they disappear in f-CNTs.

So, it can be said that chemical treatments affect DOS of valence band. H. Ago et al. [52] studied the change in the DOS of acid treated MWCNTs. They showed that oxidized MWCNTs affect the work functions. The calculated work function of MWCNTs using ultraviolet photoelectron spectroscopy (UPS) are listed in Table 2.

UPS is a powerful technique for the investigation of both the valence band DOS and the work function. Authors have proposed the following reasons for changing the work functions

1. The lower work function of MWCNTs to HOPG is due to destabilization of the π-electron due to the curvature of the graphene sheets.
2. The higher work function of functionalized MWCNTs is due to the reduction of the π conjugation of the MWCNTs, which reduces the pπ derived DOS. Such reduction is due to the transformation of the graphene layers to amorphous carbon phase with an sp^3 networks and due to the enhancement of surface dipoles because of oxygen-containing functional groups.

Thus, work function of CNTs may varies to the chemical treatments. During the time of purification, the work function of the CNTs may vary too, as air and oxygen atoms expect to make network with carbon.

Shortly, in the functionalization world of CNTs, addition of any impurities (such as oxygen, nitrogen) to CNTs is called dopant. This oxygen dopant in SWCNTs depletes electrons from valence band van Hove singularities, and introduce holes into the valence band that lead to an increase in the far-IR absorption as demonstrated by Itkis et al. [53]. For each functional group that is added, an equivalent number of π electrons are removed from the conjugated π system. One of the immediate results of functionalized CNTs is the disappearance of the characteristics van Hove singularities in the near-infrared region of the spectrum, which corresponds to the band gap transitions in semiconducting SWCNTs.

4.2 CNTs Films: An Alternative to ITO

Commercially available transparent conducting oxides (TCO) materials such as ZnO/Al, ITO work well in thin film PV applications. However, the current cost of PV power (thin film and wafer based) is still not competitive with electricity generated from fossil fuels. CNTs films can be an alternative TCO. Transparency and conductivity of CNTs films are the most important factors that affect the performance of PV device. They (Transparency and conductivity of CNTs films) depend on the parameter optimization like purification, deposition and types of CNTs used (arc discharge, laser ablation, chemical vapor deposition, Hipco). The growing technologies have made it possible for developing transparent, conductive CNTs films on plastic substrate [54]. The printing method produces relatively smooth, homogeneous films with a high transmittance and moderate resistance. The polymer layer on CNTs film produces EQE higher than polymer single layer, it is probable that the charges are preferentially accepted by CNTs at the polymer–CNTs interface [21]. CNTs films are important to ITO from following point of views:

1. High quality ITO is expensive.
2. ITO contains indium that might be too limited in supply for widespread use in solar energy application.
3. The ease of chemical charge transfer doping in CNTs help to change dopant concentration on the film to obtain transparency-versus-conductivity optimization.
4. CNT films are simple to deposit, flexible, reliable and robust.
5. Transfer printing methods have been developed for producing CNTs transparent electrodes on flexible substrates.
6. Commercially available CNTs were dissolved in solution with surfactants. The well dispersed and stable solutions were vacuum filtered over a porous membrane. Following drying, the CNTs films were lifted off with a stamp (like poly(dimethylsiloxane)(PDMS)) and transferred to a flexible poly-(ethylene terephthalate) (PET) substrate by printing [55].
7. The work function of CNTs can be varied by attaching different functional groups.
8. The resistivity and porosity of CNTs film can be reduced by adding suitable solution (like PEDOT:PSS [54]).

However, due to the irreproducibility of the drop-casting process and the differences in interaction with the depositing solution between the SWCNTs layer and PEDOT:PSS layers, these devices may not have the same layer thicknesses. It is also possible that the PEDOT layer completely isolates the SWCNT layer from the top contact, decreasing shunting of the device by errant nanotubes penetrating the entire active layer. Organic solar cell with carbon nanotube transparent electrode without using PEDOT:PSS has been reported [56]. However, PCE of PV device is still low in nanotube electrodes than ITO. In such devices, the FF appears to be series resistance limited. Series resistance includes poor conductivity of the CNTs film, conductivity of nonoptimized active layer and high contact resistance. For example, the series resistance observed with ITO and CNTs films in device fabrication was 416 and 700 respectively [56]. The sheet resistance of as-prepared transparent CNTs is higher than that of ITO.

4.3 CNTs Films as 3-D Charge Collector

Figure 6 shows large porosity of densified transparent CNTs (t-CNTs), which remains, even after deposition of thin layer of PEDOT:PSS. Instead of replacing ITO with CNTs film, R. Ulbricht et al. [57] proposed to use them as 3D charge collector. Hybrid anode: "t-CNT-ITO" shows twice higher short circuit photocurrent, compared to OPVs with only ITO or only t-CNTs anodes. Collection of an additional large number of photo-generated carriers can explain the increase in Jsc and η. In hybrid anode, due to the 3D architecture of CNTs film, spin coating of PEDOT; PSS on top of the CNT sheets improves densification and, probably, the intertube contacts and leave enough space for filling up the photoactive layer of P3HT-PCBM. It is proposed that if the t-CNTs would be planer, continuous film (like ITO) it would be like collecting the same amount of photocurrent of holes arriving to it from donor polymeric chains. Since, these planar t-CNTs would stay on the top of ITO, blocking it from direct contact with photoactive P3HT.

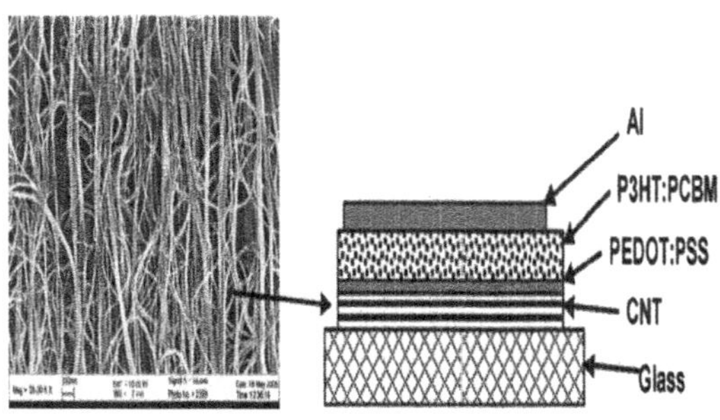

Fig. 6 Schematic of OPV device having MWCNTs transparent layer on glass as anode. SEM image of densified t-CNTs shows large porosity, which remains, even after deposition of thin layer of PEDOT:PSS (shown at inset). Reuse with permission from the publisher [57]. Copyright Elsevier 2007

4.4 CNTs Blend with Organic Polymer

4.4.1 CNTs in Organic Polymer

SWCNTs doped polymeric film for photovoltaic devices was reported in 2002, utilizing arc-generated SWCNTs-poly(3-octythiophene)-(P3OT) composites [12]. The results showed a diode response for devices constructed in the sandwich formation, containing the composite film between an ITO front contact and aluminum back contact. There was a photoresponse under AM1.5 illumination for both the pristine P3OT devices and a 1 wt% doped SWCNTs-P3OT. The photoresponse of 1 wt% doped SWCNTs-P3OT composite was reported to have an V_{oc} of 0.75 V and I_{sc} of 0.12 mA/cm^2, in comparison to a V_{oc} = 0.35 V and I_{sc} = 0.7 uA/cm^2 for the pristine P3OT device. It was suggested that V_{oc} obtained in polymer–CNTs composite cannot be explained by metal-insulator-metal (MIM) model proposed by Parker [58]. It was explained that in a short circuit condition, the negative electrode will form ohmic contact to the nanotube percolations paths and the positive electrode to the polymer, this implies that the Fermi energy is constant throughout the system. In this way, the built-in potential of the device is the built-in potential of the polymer-nanotube junction. Therefore, the upper limit of the V_{oc} must be equal to the difference in the P3OT highest occupied molecular orbital (HOMO) and the SWCNTs work function, which is 0.75 eV. Under illumination, electrons and holes travel to opposite contacts due to an internal electric field. Thereby, a collection of a short circuit photocurrent was guaranteed [59]. Potential energy band diagram is given in Fig. 7. This device had an increase in power conversion efficiency from 2.5×10^{-5} to 0.1% compared to P3OT alone. By varying the percentage of SWCNTs in the composite, the maximum efficiency was found at 1 wt% SWCNTs.

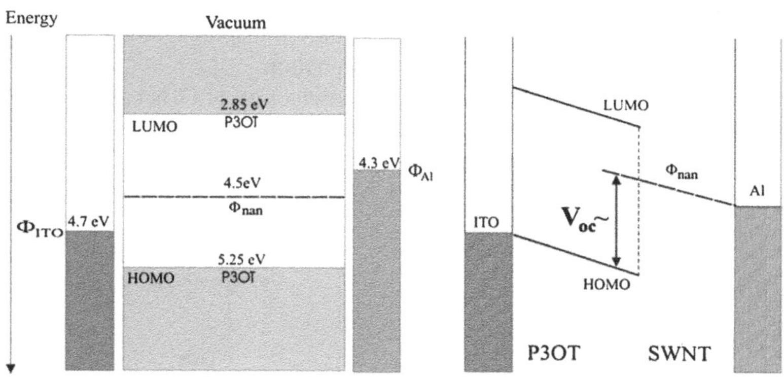

Fig. 7 Potential energy diagrams relative to vacuum level of a P3OT-SWCNTs bulk heterojunction under (**a**) flat band conditions and (**b**) under short circuit conditions, assuming no interfacial layer at the metal contacts and pinning of the Al and ITO to the energy states of the polymer and SWCNTs, respectively. Reuse with permission from the author [59]. Copyright Advanced Study Center Co. Ltd 2005

Higher CNTs loading limited the photocurrent due to a lower photogeneration rate from the polymer. V_{oc} of 0.95 V has obtained in similar device structure using highly purified SWCNTs [25]. Recently, other has also reported that in CNTs–polymer model, V_{oc} does not depend on the difference of work functions between two electrodes. Patyk et al. [60] demonstrated V_{oc} of 1.81 V in CNTs–polybithiophene photovoltaic devices.

Addition of dye on the CNTs–polymer [61], decoration of CNTs with quantum dots [62], functionalization of CNTs are some of the notable reports to increase PCE of polymeric solar cell. Merits of each additives are shortly described as below.

Merits of additions of dye:

1. Photocurrent can be improved by increasing light absorption in the active layer.
2. The mismatch of the optical absorption of the polymer to the solar spectrum can be overcome by the incorporation of dye with high absorption coefficient at the polymer/nanotube junction [63].
3. The dye can enhance the efficiency of the system by reducing the recombination of charge carriers. Charge separation and transfer from the dye to the polymer and nanotubes are shown in Fig. 8.
4. Dye functionalized CNTs make possible of its attachment to the CNTs surface so that exciton transfer from CNTs. For example. *N*-(1-pyrenyl)maleimide (PM) molecules irreversibly adsorb onto the inherently hydrophobic surfaces of SWCNTs in the organic solvent [61].

Merits on addition of quantum dots

1. The interactions between II and IV QDs (such as CdS, CdSe and CdTe) and CNTs are of special interest because they would have sufficient electron affinity to dissociate the electron–hole pair.
2. CdSe nanorods and tetrapods, CdSe quantum dots (QDs) are promising material for polymeric solar cell [65–67] because they transport charges with minimal coulombic interactions, preventing recombination.
3. Landi et al. [67] suggest that the hopping conduction of QDs can be changed by attaching QDs in ballistic conductor like SWCNTs.

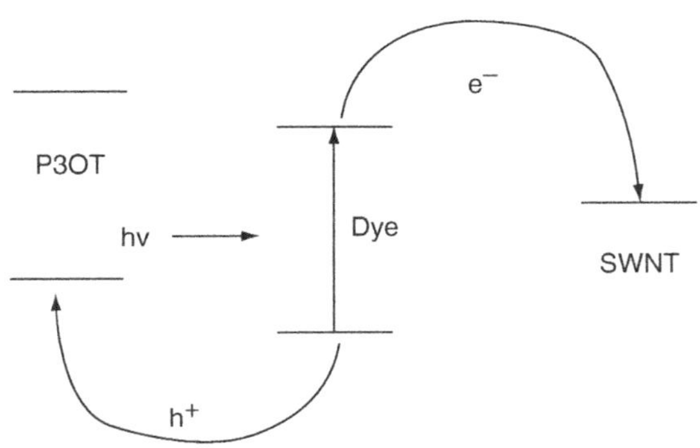

Fig. 8 Sensitisation of the polymer-nanotube solar cell by separate charge transfer from the dye to the polymer (holes) and the nanotubes (electron). Reuse from permission from author [64]. Copyright Elsevier 2003

4. QDs can be covalently attached to CNTs by simply acid treatment [68].
5. Quantum dots have been fabricated from a number of materials including TiO_2 [69], CdSe [70], CdS [71], carbon fullerenes [72].

On the other hand, metal particle decorated on CNTs is also an efficient way to increase efficiency. The acid sites for metal decoration are composed of functional groups (such as COOH, OH, NH), which act as nucleation centers for metal ions. For example, carboxylic group on the nanotube surface for ion exchange, replace the photon with metal ion as

$$\mathrm{CNTs - COOH + M^{+}X- \rightarrow CNTs - COO - M^{+} + HX[73]} \qquad (2)$$

Decoration of nanotubes by Au, Pt and Ag has been demonstrated by Satishkumar et al. [74]. It is expected that such metal particles help to capture a wider range of wavelength of sunlight and enhance splitting the excitons [75]. In C_{60} decorated SWCNTs, photoinduced charge separate at the polymer–C_{60} interface that is followed by electron transfer from C_{60} onto bounded SWCNTs. The SWCNTs network provides a direct path for faster electron transfer towards the electrode [62]. The C_{60}–SWCNTs complex is shown in Fig. 9.

4.4.2 Water Soluble Organic Solar Cell

Water soluble polymer and CNTs PV devices have also been reported [76, 77]. These types of devices increase the environmental compatibility. Water soluble polymer and oxidized MWCNTs in bilayer heterojunction configuration showed significant photoresponse. A homogeneous blend of acid oxidized MWCNTs and

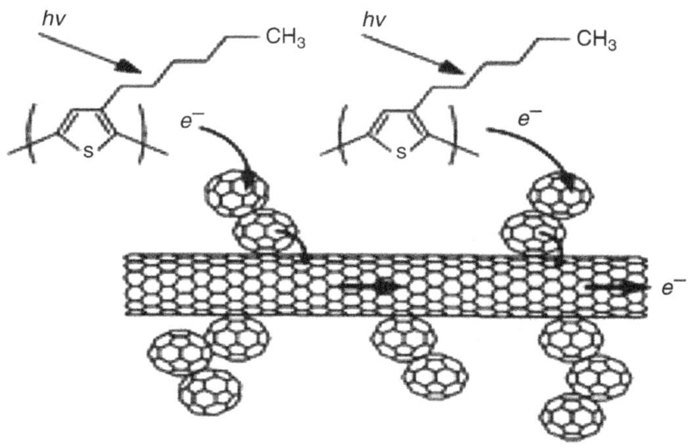

Fig. 9 Under light irradiation, photoinduced charge separation at the polymer–C_{60} interface. The SWCNTs network provides a direct path for faster electron transport towards the electrode. Reuse with permission from author [62]. Copyright 2007 The Royal Society of Chemistry

the water soluble polythiophene, sodium poly[2-(3-thienyl)-ethoxy-4-butylsulfonate] (PTEBS) have shown PCE of 0.56% [77]. Interestingly, the absorption spectra of PTEBS can be modified by changing the pH of the solution [78]. This type of solar cell is processed from an aqueous solution making it environmentally compatible with low cost fabrication for large area devices.

4.4.3 Semiconductor SWCNTs in Organic Polymer

Semiconductor SWCNTs can be an ideal material for photovoltaic effect [79–81]. Impurities and defects in the bulk semiconductor form defect states in the energy band gap. However, individual SWCNTs can form ideal p-n junction diodes, providing direct evidence for their structural purity [79]. Additionally, SWCNTs offer a wide range of band gaps to match the solar spectrum, enhanced optical absorption and reduce carrier scattering for hot carrier transport. The full band-gap of the SWCNTs can be utilized to make efficient PV devices. Figure 10a shows the device fabricated by Lee where SWCNTs are suspended as shown in the insert. Under illumination, SWCNT diode show significant PCE owing to enhance properties of an ideal diode. Figure 10b shows I–V characteristic under increased light intensity showing a progressive shift into the fourth quadrant where the diode generates power. The insert shows the linear increase in the current measurement.

Considering photoexcited effect of semi-conductor SWCNTs in polymer, Kazaoui et al. [19] suggested that in the near infrared range, semi-SWCNTs (transition $v_1^s \rightarrow c_1^s$) bounded electron (e) hole (h) pairs. They eventually dissociate into free e and h under electric field (either externally applied or built in due to the difference in the work function of Al and ITO). At short wave length, both semiconducting and metallic SWCNTs as well as MEHPPV are expected to be photoexcited but only semi-SWCNTs and MEHPPV may contribute to protocarrier generation photoexcited state as shown in Fig. 11. The energy levels labeled $(v_1^s,c_1^s),(v_2^s,cs2),(v_1^m,c_1^m)$ refer to the first and second pairs of van Hove singularities of semiconducting and metallic SWCNTs, respectively, while π and π^* are related to MEHPPV.

5 Organic/CNTs-Si Heterojunction Solar Cell

Since the discovery of photoinduced charge transfer between organic conjugated polymer and nanotubes, both MWCNTs and SWCNTs have been used to fabricate photovoltaic devices. f-MWCNTs [82], f-SWCNTs [62, 83], C_{60}-decorated SWCNTs [62], C_{60} decorated MWCNT [84] have shown better power conversion efficiency than pristine samples without CNTs. In such solar cells, it is suggested that MWCNTs enhance hole transport, whereas SWCNTs enhance electron transport. On the other hand, functional group of the CNTs make homogeneous blend of CNTs-polymer composite. This may be because CNTs are soluble in organic

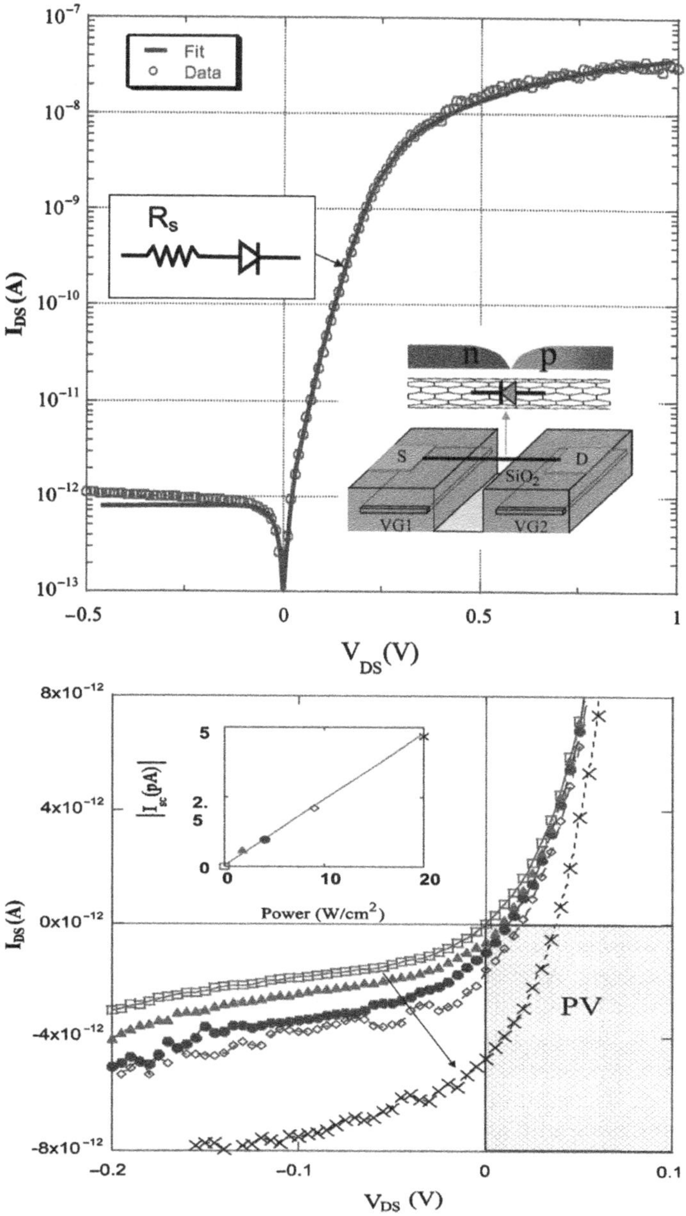

Fig. 10 (**a**) Fitting curve and data curve well matching in ideal SWCNTs diode. SWCNTs are suspended as shown in the insert. (**b**) I–V characteristics under increased light intensity showing a progressive shift into the fourth quadrant where the diode generates power. Reuse from permission from author [79]. Copyright American Institute of Physics 2005

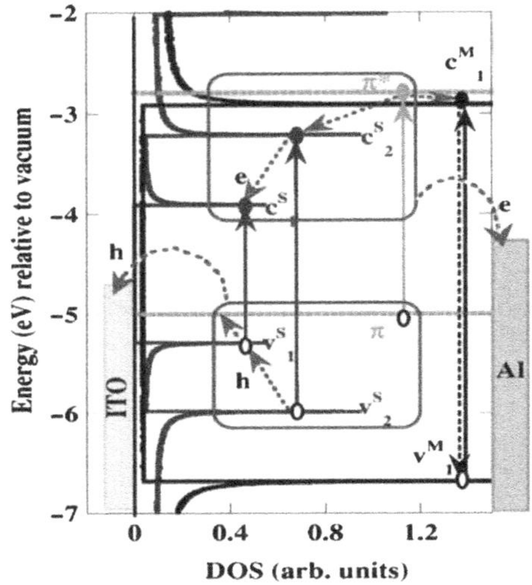

Fig. 11 Schematic energy diagram of an Al/SWCNT-MEHPPV/ITO device (before electric contact) assuming that MEHPPV (*green lines*) is doped with semiconducting (*blue lines*) and metallic (*black line*) nanotubes. Holes (h) and electrons (e) are collected at the Al and ITO electrodes, respectively. Reuse from permission from author [19]. Copyright American Institute of Physics 2005

solvent after functionalization. Furthermore, CNTs films have been extensively used for making PV devices.

Recently, it was reported that both f-MWCNTs and SWCNTs with conducting polymer play significant role to increase PCE of heterojunction PV device [85]. Functional group on the surface of MWCNTs was obtained through acid treatment. Poly(3-octylthiophene) (P3OT)/n-Si heterojunction photovoltaic cells were fabricated with and without CNTs(f-MWCNTs and SWCNTs) by keeping all other device parameters same. It was observed that CNTs embedded devices showed better performance than pristing without CNTs. CNTs are synthesized by ultrasonic spray pyrolysis method. Detail description about the system is given by Khatri et al. [86]. It is a simple, convenient method to synthesis SWCNTs, MWCNTs [85, 86]. MWCNTs were oxidized in the mixture of 3:1 concentration of H_2SO_4:HNO_3 and purified by washing in deionized water. Finally it was mixed with SWCNTs in chloroform solvent. After acid treatment, the tubes are not only cut into short pipes but also purified because of intercalation and exfoliation of graphite. This method is more effective than gas phase oxidation method because CNTs are believed to be functionalized throughout the tube length. It has been realized that acid-oxidized MWCNTs are dispersed in polar and nonpolar solvent, which in turn can be very useful for further processing.

Figure 12a shows the schematic of the P3OT/n-Si (organic–inorganic heterojunction) solar cell incorporating CNTs (f-MWCNTs and SWCNTs). Figure 12b shows I–V characteristics in dark and under AM 1.5 simulated solar radiation for *n*-Si/*f*-MWCNTs + SWCNTs/P3OT heterojunction solar cell with partially transparent gold upper electrode 20–30 nm. Direct contact was made to the silicon by conducting stainless steel stage, and the cell was illuminated from gold electrode side. Under illumination, I_{sc} and V_{oc} are about 6.16 mA/cm^2 and 0.44 V, respectively. The FF and white light conversion efficiency are about 0.36 and 0. 98%

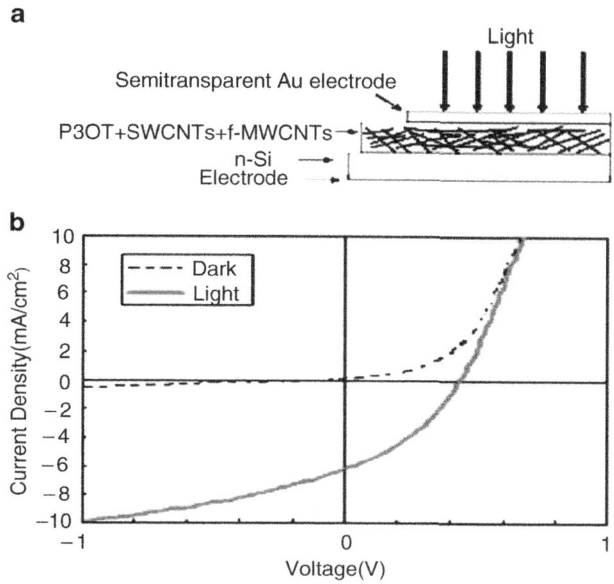

Fig. 12 (**a**) Schematic of the P3OT/n-Si heterojunction solar cell incorporating f-MWCNTs and SWCNTs in the polymeric layer (**b**) Current voltage (I–V) characteristics of the n-Si/f-MWCNTs + SWCNTs + P3OT/Au heterojunction solar cell in the dark and under AM 1.5 simulated solar radiation. Cell is illuminated from the semitransparent Au electrode side. Reuse with permission from author [85]. Copyright 2009 American Institute of Physics

Table 3 Photovoltaic characteristics of solar cells under AM 1.5 simulated solar radiation obtained from different references [85]

No.	Composite	Jsc (mA/cm^2)	FF	η (%)	Ref.
1.	C_{60}-O-MWCNTs/P3OT	1.68	0.27	0.11	[84]
2.	DWCNTs/P3HT	0.3398	0.17	0.026	[87]
3.	MWCNTs/P3OT	2.915	0.27	0.175	[88]
4.	C_{60} nanorod/P3OT	0.0098	0.1485	0.0002	[89]
5.	Pt-MWCNTs/P3OT	5.88	0.3876	0.775	[90]
6.	Cut-MWCNTs/P3OT	7.65	0.31	0.54	[91]

respectively. Table 3 lists photovoltaic properties of solar cells fabricated by similar conditions using various kinds of composites. It is obvious that *f*-MWCNTs + SWCNTs-P3OT composite showed better power conversion efficiency among these samples. A twin reference cell fabricated in the same manner with only P3OT film without CNTs with identical device parameters shows no good photovoltaic effect. With the introduction of *f*-MWCNTs and SWCNTs in the photoactive layer, there is large improvement in power conversion efficiency. It is believed that such enhancement in power conversion is due to the efficient electron transportation through SWCNTs and efficient hole transportation through *f*-MWCNTs. The work

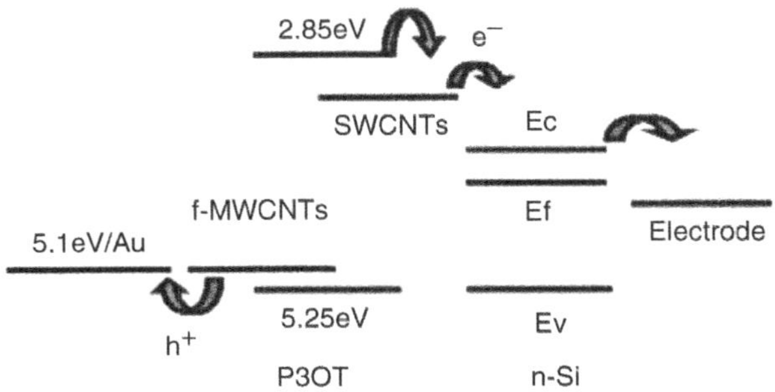

Fig. 13 Energy band diagram of the fabricated device showing band alignment for f-MWCNTs and SWCNTs. Reuse with permission from author [85]. Copyright 2009 American Institute of Physics

function of SWCNTs ranges from 3.4 to 4 eV, while for MWCNTs the range is from 4.6 to 5.1 eV. Acid oxidation introduces carboxylic acid groups on the surface of MWCNTs and produced higher work function 5.1 eV. It suggests that work functions of SWCNTs and *f*-MWCNTs are closer to the conduction band and valance band of P3OT, which signifies possible electron and hole transportation, respectively. Increased work function of *f*-MWCNTs brings MWCNTs in same level of Au (5.1 eV), which enhances hole transportation toward anode. Energetically favorable charge transportation and band diagram is shown in Fig. 13.

In the PV device with SWCNTs and f-MWCNTs, P3OT act as the photoexcited electron donor and SWCNTs as the electron acceptor and provide percolation paths. Similarly, The mobility of MWCNTs is several order higher in magnitude compared to that in the polymer, which may result the enhancement of hole transport.

6 Conclusion

In this chapter, we showed the growing research of CNTs towards organic solar cell. Devices containing CNTs in organic polymer have shown a photoresponse with high PCE. Attempts to improve the exciton dissociation and carrier transport in these devices by adding appropriate materials and quantum dots are the recent research topics.

References

1. Soga, T.: Nanostructured Materials for Solar Energy Conversion, Chapter 1. Elsevier, Amsterdam (2006)
2. Robertson, J.: Mater. Sci. Eng. R **37**, 129 (2002)
3. Geim, A.K., Novoselov, K.S.: Nat. Mater. **6**, 183 (2007)
4. Kroto, H.W., Health, J.R., O'Brien, S.C., Curl, R.F., Smalley, R.E.: Nature **318**, 162 (1985)

5. Iijima, S.: Nature **354**, 56 (1991)
6. Wang, X., Zhi, L., Tsao, N., Tomovic, Z., Li, J., Mullen, K.: Angew. Chem. Int. Ed. **47**, 2990 (2008)
7. Ma, W., Song, L., Yang, R., Zhang, T., Zhao, Y., Sun, L., Ren, Y., Liu, D., Liu, L., Shen, J., Zhang, Z., Xiang, Y., Zhou, W., Xie, Z.: Nano Lett. **7**, 2307 (2007)
8. Peng, H.: J. Am. Chem. Soc. **130**, 42 (2008)
9. Koezuka, H., Tsumara, A., Ando, T.: Synthetic Met. **18**, 699 (1987)
10. Jong, D., Friedlein, M.P., Osikowicz, R.W., Salaneck, W.R., Fahlman, M.: Mol. Cryst. Liq. Cryst. **455**, 193 (2006)
11. Frohne, H., Shaheen, S., Brabec, C., Muller, D., Sariciftci, N.S., Meerholz, K.: Chem. Phys. Chem. **9**, 795 (2002)
12. Kymakis, E., Amaratunga, G.A.J.: Appl. Phys. Lett. **80**, 112 (2002)
13. Hoppe, H., Sariciftci, N.S., Meissner, D.: Mol. Cryst. Liq. Cryst. **385**, 113 (2002)
14. Roncali, J.: Chem. Rev. **97**, 173 (1997)
15. Zou, J., Liu, L., Chen, H., Khondaker, S.I., McCullough, R.D., Huo, Q., Zhai, L.: Adv. Mater. **20**, 2055 (2008)
16. Dimitrakopoulos, C.D., Malenfant, P.R.L.: Adv. Mater. **14**, 99 (2002)
17. Nunzi, J.M.C.R.: Physique **3**, 523 (2002)
18. Miranda, P.B., Moses, D., Heeger, A.: J. Phys. Rev. B **64**, 81201 (2001)
19. Kazaoui, S., Minami, N., Nalini, B., Kim, Y., Hara, K.: J. Appl. Phys. **98**, 084314 (2005)
20. Zhu, H., Wei, J., Wang, K., Wu, D.: Sol. Energy Mater. Sol. Cell. **93**, 1461 (2009)
21. Ago, H., Petritsch, K., Shaffer, M.S.P., Windle, A.H., Friend, R.H.: Adv. Mater. **11**, 1281 (1999)
22. Halls, J.J.M., Pichler, K., Friend, R.H., Moratti, S.C., Holmes, A.B.: Appl. Phys. Lett. **68**, 3120 (1996)
23. Biercuk, M.J., Liaguno, M.C., Radosavljevic, M., Hyun, J.K., Johnson, A.T., Fischer, J.E.: Appl. Phys. Lett. **80**, 2767 (2002)
24. Cinke, M., Li, J., Chen, B., Cassell, A., Delzeit, L., Han, J., Meyyappan, M.: Chem. Phys. Lett. **365**, 69 (2002)
25. Berson, S., Bettignies, R.D., Bailly, S., Guillerez, S.: Adv. Funct. Mater. **17**, 1377 (2007)
26. Landi, B.J., Raffaelle, R.P., Castro, S.L., Bailey, S.G.: Prog. Photovoltaics **13**, 165 (2005)
27. Chaudhary, S., Lu, H., Muller, A.M., Bardeen, C.J., Ozkan, M.: Nano Lett. **71**, 973 (2007)
28. Harutyunyan, A.R., Pradhan, B.K., Chang, J., Chen, G., Eklund, P.C.: J. Phys. Chem. B **106**, 8671 (2002)
29. Tohji, K., Takahashi, H., Shinoda, Y., Shimizu, N., Jeyadevan, B., Matsuoka, I., Saito, Y., Kasuya, A., Ito, S., Nishina, Y.: J. Phys. Chem. B **101**, 1974 (1997)
30. Lian, Y., Maeda, Y., Wakahara, T., Akasaka, T., Kazaoui, S., Minami, N., Shimizu, T., Choi, N., Tokumoto, H.: J. Phys. Chem. B **108**, 8848 (2004)
31. Zhang, J., Zou, H., Qing, Q., Yang, Y., Li, Q., Liu, Z., Guo, X., Du, Z.: J. Phys. Chem. B **107**, 3712 (2003)
32. Liu, J., Rodriguez, M., Zubiri, I., Dossot, M., Vigolo, B., Hauge, R.H., Dossot, M., Fort, Y., Ehrhardt, J.J., McRae, E.: Carbon **45**, 885 (2007)
33. Chen, J., Hamon, M.A., Hu, H., Chen, Y.S., Rao, A.M., Eklund, P.C., Haddon, R.C.: Science **282**, 95 (1998)
34. Qin, Y., Shi, J., Wu, W., Li, X., Guo, Z.X., Zhu, D.: J. Phys. Chem. B **107**, 12899 (2003)
35. Kovtyukhova, N.I., Mallouk, T.E., Pan, L., Dickey, E.C.: J. Am. Chem. Soc. **125**, 9761 (2003)
36. Liu, J., Rinzler, A.G., Dai, H., Hafner, J.H., Bradley, R.K., Boul, P.J., Lu, A., Iverson, T., Shelimov, K., Huffman, C.B., Macias, F.R., Shon, Y.S., Lee, T.R., Colbert, D.T., Smalley, R.E.: Science **280**, 1253 (1998)
37. Pénicaud, A., Poulin, P., Derré, A., Anglaret, E., Petit, P.: J. Am. Chem. Soc. **127**, 8 (2005)
38. Georgakilas, V., Kordatos, K., Prato, M., Guldi, D.M., Holzinger, M., Hirsch, A.: J. Am. Chem. Soc. **124**, 760 (2002)
39. Zhao, W., Song, C., Pehrsson, P.E.: J. Am. Chem. Soc. **124**, 12418 (2002)

40. Koshio, A., Yudasaka, M., Zhang, M., Iijima, S.: Nano Lett. **1**, 361 (2001)
41. Niyogi, S., Hamon, M.A., Perea, D.E., Kang, C.B., Zhao, B., Pal, S.K., Wyant, A.E., Itkis, M.E., Haddon, R.C.: J. Phys. Chem. B **107**, 8799 (2003)
42. Boehm, H.P.: Carbon **40**, 145 (2002)
43. Niyogi, S., Hamon, M.A., Hu, H., Zhao, B., Bhowmick, P., Sen, R., Itkis, M.E., Haddon, R.C.: Acc. Chem. Res. **35**, 1105 (2002)
44. Esumi, K., Ishigami, A., Nakajima, A., Sawada, K., Honda, H.: Carbon **34**, 279 (1996)
45. Shaffer, M.S.P., Fan, X., Wandle, A.H.: Carbon **36**, 1603 (1998)
46. Chen, J., Rao, A.M., Lyuksyutov, S., Itkis, M.E., Hamon, M.A., Hu, H., Cohn, R.W., Eklund, P.C., Colbert, D.T., Smalley, R.E., Haddon, R.C.: J. Phys. Chem. B **105**, 2525 (2001)
47. Knight, D.S., White, W.B.: J. Mater. Res. **4**, 385 (1989)
48. Kastner, J., Pichler, T., Kuzmany, H., Curran, S., Blau, W., Weldon, D.N., Delamesiere, M., Draper, S., Zandbergen, H.: Chem. Phys. Lett. **221**, 5 (1994)
49. Scharf, T.W., Singer, I.L.: Third Bodies and Tribochemistry of DLC Coatings. SpringerLink, New York (2007)
50. Sreekumar, T.V., Liu, T., Kumar, S., Ericson, L.M., Hauge, R.H., Smalley, R.E.: Chem. Mater. **15**, 175 (2003)
51. Wang, Y., Iqbal, Z., Mitra, S.: J. Am. Chem. Soc. **128**, 95 (2006)
52. Ago, H., Kugler, T., Cacialli, F., Salaneck, W.R., Shaffer, M.S.P., Windle, A.H., Friend, R.H.: J. Phys. Chem. B **103**, 8116 (1999)
53. Itkis, M.E., Niyogi, S., Meng, M.E., Hamon, M.A., Hu, H., Haddon, R.C.: Nano Lett. **2**, 155 (2002)
54. Rowell, M.W., Topinka, M.A., McGehee, M.D., Prall, H.J., Dennler, G., Sariciftci, N.S., Hu, L.B., Grüner, G.: Appl. Phys. Lett. **88**, 233506 (2006)
55. Zhou, Y.X., Hu, L.B., Gruner, G.: Appl. Phys. Lett. **88**, 123109 (2006)
56. van de Lagemaat, J., Barnes, T.M., Rumbles, G., Shaheen, S.E., Coutts, T.J., Weeks, C., Levitsky, I., Peltola, J., Glatkowski, P.: Appl. Phys. Lett. **88**, 233503 (2006)
57. Ulbricht, R., Lee, S.B., Jiang, X., Lee, S.B., Zhang, M., Fang, S., Baughman, R.H., Zakhidov, A.: Sol. Energy Mater. Sol. Cell. **91**, 416 (2007)
58. Parker, I.D.: J. Appl. Phys. **75**, 1656 (1994)
59. Kymakis, E., Amaratunga, G.A.J.: Rev. Adv. Mater. Sci. **10**, 300 (2005)
60. Patyk, R.L., Lomba, B.S., Nogueira, A.F., Furtado, C.A., Santos, A.P., Mello, R.M.Q., Micaroni, L., Hummelgen, I.A.: Phys. Stat. Sol. **1**, R43 (2007)
61. Bhattacharyya, S., Kymakis, E., Amaratunga, G.A.J.: Chem. Mater. **16**, 4819 (2004)
62. Li, C., Chen, Y., Wang, Y., Iqbal, Z., Chhowalla, M., Mitra, S.: J. Mater. Chem. **17**, 2406 (2007)
63. Rahman, G.M.A., Guldi, D.M., Cagnoli, R., Mucci, A., Schenetti, L., Vaccari, L., Prato, M.: J. Am. Chem. Soc. **127**, 10051 (2005)
64. Kymakis, E., Amaratunga, G.A.J.: Sol. Energy Mater. Sol. Cell. **80**, 465 (2003)
65. Huynh, W.U., Dittmer, J.J., Alivisatos, A.P.: Science **295**, 2425 (2002)
66. Sun, B., Marx, E., Greenham, N.C.: Nano Lett. **3**, 961 (2003)
67. Landi, B.J., Castro, S.L., Ruf, H.J., Evans, C.M., Bailey, S.G., Raffaelle, R.P.: Sol. Energy Mater. Sol. Cell. **87**, 733 (2005)
68. Si, H.Y., Liu, C.H., Xu, H., Wang, T.M., Zhang, H.L.: Nanoscale Res. Lett. **4**, 1146 (2009)
69. Arango, A.C., Carter, S.A., Brock, P.J.: Appl. Phys. Lett. **74**, 1698 (1999)
70. Huynh, W.U., Peng, X., Alivisatos, A.P.: Adv. Mater. **11**, 923 (1999)
71. Greenham, N.C., Peng, X., Alivisatos, A.P.: Phys. Rev. B **54**, 17628 (1996)
72. Piok, T., Schroeder, R., Brands, C., Feflin, J.R., Leising, G., Graupner, W.: Synthetic Met. **121**, 1589 (2001)
73. Ebbesen, T.W., Hiura, H., Bisher, M.E., Treacy, M.M.J., Keyer, J.L.S., Haushalter, R.C.: Adv. Mater. **8**, 155 (1996)
74. Satishkumar, B.C., Vogl, E.M., Govindaraj, A., Rao, C.N.R.: J. Phys. D Appl. Phys. **29**, 3173 (1996)

75. Yoon, W.J., Jung, K.Y., Liu, J., Duraisamy, T., Revur, R., Teixeira, F.L., Sengupta, S., Berger, P.R.: Sol. Energy Mater. Sol. Cell. **94**, 128 (2010)
76. Rud, J.A., Lovell, L.S., Senn, J.W., Qiao, Q., Mcleskey, J.T.: J. Mater. Sci. **40**, 1455 (2005)
77. Miller, A.J., Hatton, R.A., Salva, S.R.P.: Appl. Phys. Lett. **89**, 123115 (2006)
78. Van, F.T., Carrier, M., Chevrot, C.: Synthetic Met. **142**, 251 (2004)
79. Lee, J.U.: Appl. Phys. Lett. **87**, 073101 (2005)
80. Gabor, N.M., Zhong, Z., Bosnick, K., Park, J., McEuen, P.L.: Science **325**, 1367 (2009)
81. Lee, J.U., Gipp, P., Heller, C.M.: Appl. Phys. Lett. **85**, 145 (2004)
82. Pradhan, B., Batabyal, S.K., Pal, A.J.: Appl. Phys. Lett. **88**, 093106 (2006)
83. Li, C., Mitra, S.: Appl. Phys. Lett. **91**, 253112 (2007)
84. Kalita, G., Adhikari, S., Aryal, H.R., Umeno, M., Afre, R., Soga, T., Sharon, M.: Appl. Phys. Lett. **92**, 063508 (2008)
85. Khatri, I., Adhikari, S., Aryal, H.R., Soga, T., Jimbo, T., Umneo, M.: Appl. Phys. Lett. **94**, 093509 (2009)
86. Khatri, I., Soga, T., Jimbo, T., Adhikari, S., Aryal, H.R., Umeno, M.: Diam. Relat. Mater. **18**, 319 (2009)
87. Somani, S.P., Somani, P.R., Umeno, M., Flahaut, E.: Appl. Phys. Lett. **89**, 223505 (2006)
88. Somani, S.P., Somani, P.R., Umeno, M.: Diam. Relat. Mater. **17**, 585 (2008)
89. Somani, P.R., Somani, S.P., Umeno, M.: Appl. Phys. Lett. **91**, 173503 (2007)
90. Somani, P.R., Somani, S.P., Umeno, M.: Appl. Phys. Lett. **93**, 033315 (2008)
91. Kalita, G., Adhikari, S., Aryal, H.R., Umeno, M., Afre, R., Soga, T., Sharon, M.: Appl. Phys. Lett. **92**, 123508 (2008)

Irregular Configurations of Carbon Nanofibers

Suriati Sufian

Abstract Carbon nanofiber (CNF) has been extensively produced as it shows outstanding physical properties which can be exploited in many applications. In addition to a long and straight nanofiber, several irregular configurations of CNF could be observed in a certain particle catalyst, reaction conditions and experimental techniques. One of the ways is by synthesizing carbon nanofiber using catalytic chemical vapor decomposition (C-CVD) method. This can be achieved by using Fe_2O_3 and NiO as the catalysts with ethylene and hydrogen as the carbon feedstock and carrier gas, respectively. The results from TEM analysis show that irregular CNF configurations are formed such as coiled, regular helical and twisted coil as well as corrugated CNF.

1 Introduction

Carbon nanofiber (CNF), also known as graphitic nanofiber, is a type of carbon nanomaterials that has a size in a range of 0.4–500 nm [1, 2]. CNF was initially known as herringbone multiwall nanotubes (*h*-MWNT). This is because in MWNT structure the concentric tubes of graphite are oriented parallel to the carbon filament axis with a hollow core in the middle but when the graphene make an angle with respect to the nanotube axis the texture is known as herringbone or fishbone structure [3]. Thus, they are referred as 'nanofibres' when the inner cavity is no longer opened all along the filament as it is for a genuine tube. Besides having graphene planes at a certain angle to the filament axis, some graphene planes are perpendicular to the axis and stacked as piled-up plates. Therefore, CNFs can be categorized into platelet and herringbone or fishbone structures as in Fig. 1.

S. Sufian
Department of Chemical Engineering, Universiti Teknologi PETRONAS, Bandar Seri Iskandar, 31750 Tronoh, Perak, Malaysia
e-mail: suriati@petronas.com.my

N. Yahya (ed.), *Carbon and Oxide Nanostructures*, Adv Struct Mater 5,
DOI 10.1007/8611_2010_26, © Springer-Verlag Berlin Heidelberg 2010,
Published online: 17 August 2010

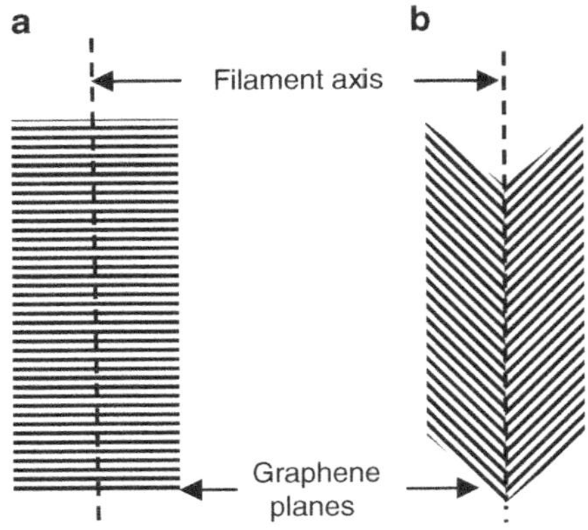

Fig. 1 Basic CNF structures consist of (**a**) platelet and (**b**) herringbone structures

The former structure is when the graphene layers are perpendicular to the filament axis while the latter structure is when the graphene layers are at a certain angle to the axis. The angle of conicity varies from 15° to 85° [4]. The herringbone structure can be a hollow core or solid structure [5].

However, some study have categorized carbon nanofibers into three unique conformations that are platelet, spiral and ribbon structures [6]. As described earlier, platelet consists of graphite platelet with carbon oriented perpendicular to the fiber axis. But spiral form is when platelet is in helical nanofibers parallel to the fiber axis while ribbon structure is when graphite platelets are aligned parallel to the fiber axis. Thus, CNFs are not just in the form in long, straight structure but in certain conditions, they could also exist in a non-straight structure such as spiral-like, coiled, regular helical, spring-like, double helical and branch-like structures. The formation of the bent or spiral-like is due to the incorporation of pentagon and hexagon pairs into hexagonal sheets, which then creates positively and negatively curved surfaces [7]. Another CNF structure is known as stacked cup where the nanofiber exhibits a continuous layer of rolled or spiral graphene along the fiber axis which yields a truncated cone arrangement along the axis with wide internal hollow space [5].

2 Growth Mechanism

The growth mechanism of CNF can be referred in Fig. 2, where ①, ② and ③ symbols represent the three steps that are step 1, 2 and 3, respectively in the growth mechanism process. At the beginning (step 1) absorption and decomposition of hydrocarbon molecules such as C_2H_4 and CH_4 occur on the surface of metal crystallographic faces [9]. Then at step 2, the dissolution and diffusion of carbon species into the nanoparticles interior occur to form a metal-carbon solid state solution. At step 3, nanofibers growth occurs when a supersaturation leads to carbon

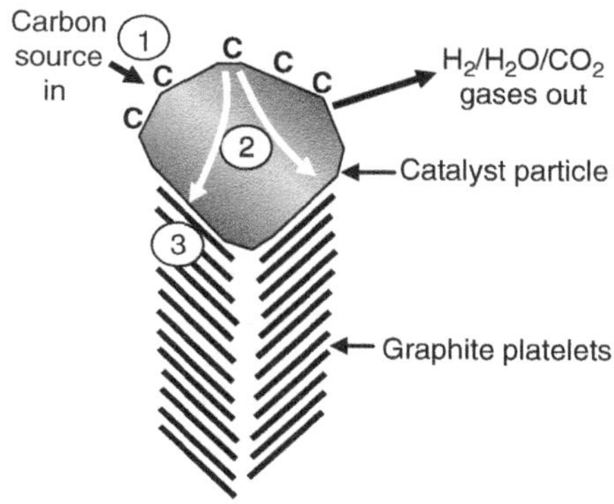

Fig. 2 Growth process of carbon nanofiber [8]

precipitation into a crystalline tubular form. The size of the metal catalyst nanoparticle generally dictates the diameter of the synthesized nanofiber [10].

The crystalline perfection degree of the deposited fiber is determined by the chemical nature of the catalyst particle, the reactant gas composition and the temperature. Surface science studies revealed that only certain faces of metal particle favor precipitation of carbon in the form of graphite while other faces could deposit less ordered carbon [11]. As long as the heat balance is maintained in the reaction system, the carbon structure will continue to grow. However, any perturbations in the chemistry of the leading face of the metal catalyst will cause a buildup of solid carbon at this interface. As a result, deactivation and termination in nanofiber growth will occur [9].

In general, nanostructures are formed by a bidirectional mode where precipitation of carbon occurs simultaneously from opposite faces of the catalyst particle. Thus, the catalyst particle remains within the structure throughout the growth process. Conversely, for herringbone case, the carbon deposition process occurs at the sides of the particle that is carried at the tip of the growing fiber. For platelet structure, the graphene planes are stacked in a direction parallel to the base of the particle and perpendicular to the fiber or filament axis. As for herringbone structure, the particles adopt a rhombic morphology where the nanofibers are precipitated from a pair of opposite faces to generate a structure that is aligned at an angle to the fiber axis [12].

3 Reaction Parameters

There are several parameters involved in the growth mechanism of CNF. As for a catalyst, controlling the size of the catalyst nanoparticles is crucial since it determines the final diameter of nanofiber and the type of [1]. Moreover, particles with diameters smaller than 25 nm will tend to produce nanotubes, whereas larger particles usually generate solid nanofibers. Apart from metal powder, catalyst can come in different configurations such as gauzes, wires, foils and supported metal particles like alumina and silica [9]. The catalysts usually come from transition metal oxides such as Cobalt (Co), Nickel (Ni), Copper (Cu), Molybdenum (Mo), Iron (Fe) as well as their alloys for heterogeneous process and metal-organic compound or a metallocene like ferrocene, nickelocene or cobaltocene for a homogeneous process.

Majority researchers used Ni for CNFs production which can be in the form of metals, films, alloys and different types of catalyst supported such as alumina and silica. However, most of the catalysts consist of bimetallic such as nickel (Ni) metal combined with other transition metals like Cu and Fe [13–15] or any other combination of bimetallic compound apart from Ni [16, 17]. In addition, other studies have done in using Ni along with other support such as zeolite or different types of carbon [18–20] while others used Ni alone without any support or other metal compound [21, 22].

Ni is the common catalyst used in CNF synthesis because it gives high yield as compared with other metal such as Fe [23]. This could probably due to that Ni has more active catalyst site that allows more carbon to be absorbed and saturated and finally precipitate on the surface of the catalyst particle to form graphite layers. In the CVD reaction, there are several types of hydrocarbon gases like ethylene (C_2H_4), acetylene (C_2H_2) and methane (CH_4) as well as carbon monoxide (CO) has been successfully utilized to produce both CNTs and CNFs [1, 3, 9].

In addition benzene and xylene are used as carbon feedstock, especially in a homogeneous process. Unsaturated gases like C_2H_4, C_2H_2 and benzene are used to synthesize MWNTs since they have high carbon content, whereas saturated gases like CO and CH_4 are used to synthesize SWNTs since they tend to produce highly graphitized filaments and fewer walls [1]. Along with these carbon rich gases and vapors, hydrogen is used to reduce the metal particles (that is oxides) during the heating as well as to decrease the formation of carbon deposits from pyrolysis of the carbon feedstock. Moreover, carrier gas such as argon, helium or nitrogen is used to flush the reactor initially to create an inert atmosphere.

Studies show that majority of the researchers use ethylene (C_2H_4) as the carbon feedstock [14–16, 18, 20, 24]. This is due to that unsaturated gas like C_2H_4 has high carbon content which is suitable in developing carbon nanomaterials like MWNTs and CNFs. Some researchers had a gas mixture of ethylene with other saturated carbon gas such as carbon monoxide that gives C_2H_4–CO [13] and methane that gives C_2H_4–CH_4 [19]. Other researchers used saturated gases as the carbon feedstock like CO [22] and CH_4 [17]. The reaction temperature in CVD is much less as compared with electric arc discharge, which can reach up to 4,000 K. For CNFs, they can be synthesized between 400 and 1,000°C [9]. The variation of the temperature depends on the catalyst activity, experimental configurations and carbon feedstock. Any temperature higher than 1,000°C could result in more formation of amorphous carbon [1].

The reaction time varies from minutes to hours depending on the gas flow rates and the length of the reactor. It is thought that prolong the reaction time resulting in longer CNFs produced. However, longer reaction hour could result in more carbon

Table 1 Developed carbon nanofibers from different carbon source and catalyst

Carbon source	Catalyst	Reference
C_2H_4	NiO	[21]
	Bimetallic Fe–Cu	[16]
	Ni/graphite microfibers	[18]
	Ni/Y zeolite based	[20]
	Metal powder	[24]
	Cu:Fe, Cu:Ni	[29]
	98 mg Ni, 2 mg Cu	[15]
C_2H_4, CO	Ni–Fe 100 mg	[13]
CH_4/C_2H_4	Ni/carbon	[19]
CO	0.2 g Ni	[22]
CH_4	Co on MgO, Al_2O_3, SiO_2	[17]

deposits. CNF will be continuously formed as a long straight nanofiber as long as the rate of precipitation of carbon from all four faces is identical, the. However, any perturbation in this pattern of behavior will give rise to abnormalities in the conformation of the fibers. As a result, coiled or helical configurations of nanofibers are formed instead [12].

The growth rate and the shape of nanofibers are controlled by the particle catalyst geometry and rate of precipitation of carbon on the surface. For example it is thought that due to various catalyst positions, a mixture of carbon nanostructures at different degree of crystallinity such as helical regularly coiled nanofibers and ribbon-like coiled nanofibers are observed [25]. Other researcher reported that nanofiber growth in helical morphology is due to the anisotropy catalyst particle surface which depends on the exposed crystal facet of the particle [26]. In addition, the diameter of the coil is increasing with temperature in which the diameter is varied from 100 nm to 10 μm for temperature variation between 250 and 400°C [27, 28].The study of CNF formation using different carbon feedstocks and catalysts can be summarized in Table 1 as above.

4 Characterization of CNF

For microscopic study analysis, CNF can be characterized using electron microscopes such as scanning electron microscope (SEM) and transmission electron microscope (TEM). The instruments use a beam of highly energetic electrons to image the samples. The electrons are used since the wavelengths are much shorter than that of lights, which result in higher resolution of image [30]. The examination of image can yield to the following information of the surface and morphology of the material. SEM and TEM analyses can be studied not just to observe the typical CNF structures that are long and straight but also to observe different configurations of nanofiber. These configurations include hollow nanofibers [5, 31], coiled and helical nanofiber [27, 32, 33], triple helical nanofiber [34] and ribbon-like nanofiber [25, 35].

Apart from electron microscopy analysis, CNF can be characterized based from spectroscopic study analysis that is by using Raman spectroscopy. This advanced and powerful technique is used to analyze materials that contain carbon element without any contact of microprobe and can perform under atmospheric pressure or in solution. This is because the resonance phenomena and sensitivity of the tube structure provides a strong excitation wavelength dependence of the spectra resulting from the electronic band structure [36]. One of the features in Raman spectroscopy analysis is *D* mode or disorder band which is located between 1,330 and 1,360 cm^{-1}. This excitation is expected to be observed in carbon nanomaterials like MWNTs and CNFs. When it is observed in SWNTs, it is assumed that the feature observed is due to defects in the tubes. The next feature known as *G* mode or Tangential Mode (TM). It corresponds to the stretching mode in the graphite plane. This mode is located around 1,580 cm^{-1}. The shoulder at higher frequency (1,618 cm^{-1}) or D′ band is typical of defective graphitelike materials and can be smaller in better quality samples [37].For CNF samples analysis, the samples could have higher graphitization if *G* band is higher than *D* band [31, 38] or have more carbon structure defects if *D* band is higher than *G* band [22, 24].

Another significant CNF characterization can be achieved using X-ray diffractometer (XRD). This technique is used to reveal detailed information about the chemical composition and crystallographic structure of natural and manufactured materials [39]. It can determine the crystal structure of an unknown material as well as measure the size, shape and internal stress of small crystalline regions [40]. A typical XRD pattern exhibits the most distinguished carbon peak at $2\theta = 26.24°$ with intensity of graphite (0 0 2) which signifies high crystallization carbon structure with the d_{002} spacing is greater than 0.34 nm. This shows that it is a turbostratic carbon which is higher than for ideal graphite (0.335 nm) [41]. Other carbon peaks are observed as (1 0 0), (0 0 4) and (1 1 0) [19, 20, 22, 31].

5 Synthesis of Carbon Nanofibers

CNF was synthesized using catalytic chemical vapor deposition (C-CVD) method. The experimental setup as seen in Fig. 3 consists of horizontal tube furnace with iron (III) oxide (Fe_2O_3) catalyst was spread along the ceramic boat which was then placed inside the tube reactor. There were three gas lines that are Argon (Ar), hydrogen (H_2) and ethylene (C_2H_4) which were controlled by two-way valves. The three-way valve was used to release an excess pressure as indicated by the pressure indicator whereas the dreschel bottles were used to detect any leakage in the system.

Initially, the air in the tube reactor was vacuumed before purging it with Argon (Ar) up to 1 atm. Then, the furnace was set at 600°C of reaction temperature with an increment of 10°C min^{-1}. When the temperature reached equilibrium, a gas mixture of ethylene and hydrogen (C_2H_4/H_2) was flowed into the reactor. The formation of CNFs took place when the gas mixture was in contact with the catalyst in the

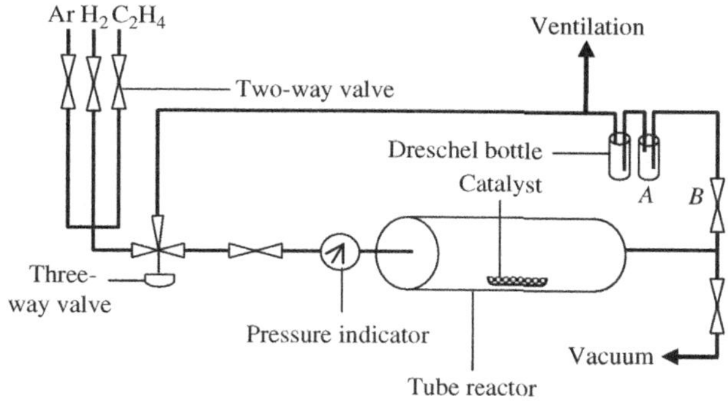

Fig. 3 Schematic diagram of GNF experimental setup

Fig. 4 Process flow of CNF synthesis

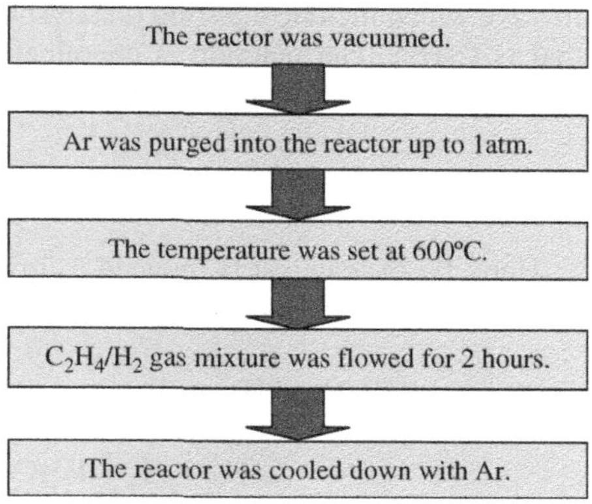

ceramic boat. The reaction took place at atmospheric pressure and the process flow of GNF synthesis is summarized in Fig. 4. Similar procedure was done using Nickel oxide (NiO) as the catalyst particles.

For sample characterization, X-ray diffraction (XRD) studies were carried out using Bruker AXS D8-Advance diffractometer with Cu Kα radiation of wavelength, $\lambda = 0.15406$ nm. The operating voltage and current were at 40 kV and 40 mA, respectively. Each sample was scanned with 2θ of 20–80° and the data was collected in step mode with an interval of 0.02°. In order to determine the composition of the samples, direct comparison method was done by using powder diffraction databases, ICDD (International Centre for Diffraction Data) or formerly known as JCPDS (Joint Committee on Powder Diffraction Standards).

For spectroscopy study, the Raman spectroscopy analysis was carried out using a Labram HR800 Horiba Jobin Yvon at room temperature with Argon-ion laser operated at 514.5 nm. The Raman spectra were obtained with a fixed grating over a range of 100–1,800 cm^{-1}.

The surface and morphology of CNFs were studied using field emission scanning electron microscopy (FESEM) known as SUPRA 55VP is used at accelerating voltage of 0.02–30 kV. The inner structure of CNF samples were analyzed using transmission electron microscopy (TEM) observation that was performed by Jeol JEM 2000FX II at accelerating voltage of 200 kV and Phillips TEM 400 at accelerating voltage of 100 kV. Samples for TEM were prepared by ultrasonic dispersion of the GNF samples with ethanol [42]. Then, the suspension was dropped onto carbon grid and dried in the air before the TEM analysis was carried out.

6 Reaction Mechanism

The CNF synthesis is known as a heterogeneous process since there are two states involve which include the solid (catalyst that is Fe_2O_3) and vapor (carbon feedstock that is C_2H_4). The reaction is chemically defined as catalysis-enhanced thermal cracking are represented in the following equation:

$$C_2H_4 + H_2 \xrightarrow{\text{Fe or Ni}} 2C + 3H_2 \tag{1}$$

Here, Fe_2O_3 catalyst is reduced to iron carbide (Fe_3C) and later to iron (Fe) in the following reactions:

$$3C_2H_4 + H_2 + 3Fe_2O_3 \rightarrow 6H_2 + H_2O + 4CO_2 + 2Fe_3C \tag{2}$$

$$C_2H_4 + H_2 + Fe_3C \rightarrow 3H_2 + 3C + 3Fe \tag{3}$$

For NiO, the catalyst is reduced to nickel carbide (Ni_3C) and later to nickel (Ni) in the following reactions:

$$2C_2H_4 + H_2 + 6NiO \rightarrow 3H_2 + 2H_2O + 2CO_2 + 2Ni_3C \tag{4}$$

$$C_2H_4 + H_2 + Ni_3C \rightarrow 3H_2 + 3C + 3Ni \tag{5}$$

As for carbon (C), there are three possible structural carbons that could be formed such as amorphous carbon layers, filaments of amorphous carbon and graphite layers covering metal particles that are CNFs.

The growth mechanism of the developed CNF is based on the tip growth. Prior to the growth, the formation of small domain of catalyst metal particle that acts as nucleation seeds occur. The size of the nucleation seeds correspond to the CNF diameter [43]. At the beginning of the growth, hydrocarbon molecules from ethylene (C_2H_4) absorb and decompose onto the surface of metal oxide (Fe_2O_3 or NiO) crystallographic faces as described in other research study [9]. Later, the dissolution and diffusion of carbon species into the nanoparticles interior occur to form a

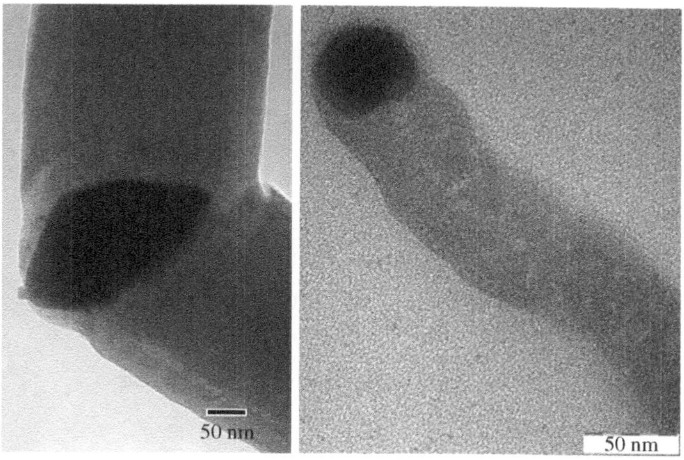

Fig. 5 TEM image shows the tip growth of CNF on iron catalyst (the *dark spot*) in (**a**) two directions direction and (**b**) one direction

metal-carbon solid state solution that is metal carbide (Fe_3C or Ni_3C). Soon the growth of nanofibers occurs since the supersaturation leads to carbon precipitation at the 'cold-end' of the catalyst particle. These carbon precipitates into a form of crystalline tubular.

A herringbone is formed if carbon deposition process occurs at the sides of the particle that is carried at the tip of the growing fiber while a platelet is formed when the graphene planes are stacked in a direction parallel to the base of the particle and perpendicular to the fibre or filament axis [12]. Both diameters of these two structures are determined from the size of the metal catalyst nanoparticle [10].

Generally, nanofibers are formed on the metal surface in two opposite directions. This can be seen in TEM image in Fig. 5. The Fe catalyst particle underwent reconstruction during the catalyst reduction and became a rhombohedral-shaped particle (dark area) and the growth of nanofibers is observed in two directions. From the TEM observation, majority of iron-based catalyst CNFs are in the form of platelet and herringbone structures. As for the latter structure, the hexagonal shape of the catalyst that leads to graphene plane at the angle of 45° to the fiber axis. Some tip growths occurred in one direction.

7 X-Ray Diffraction and Raman Spectroscopy Analyses

A typical XRD pattern for iron-based catalyst CNF sample is shown in Fig. 6a. The pattern exhibits the most distinguished carbon peak at $2\theta = 26.24°$ with intensity of graphite (0 0 2) which signifies high crystallization carbon structure. The d_{002} spacing is greater than 0.34 nm indicates that it is a turbostratic carbon which is

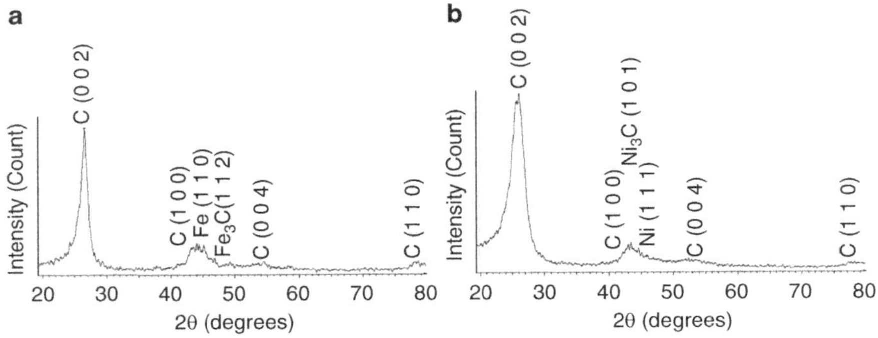

Fig. 6 XRD pattern of CNF with (**a**) Fe_2O_3 and (**b**) NiO as the catalysts

higher than for ideal graphite (0.335 nm) [41]. Other carbon peaks are observed as (1 0 0), (0 0 4) and (1 1 0). Moreover, several peaks due to catalyst remnants such as iron (Fe) peaks (1 1 0) and iron carbide (Fe_3C) at (1 1 2) are detected as in Fig. 6a. Figure 6b displays XRD patterns of CNFs using NiO as the catalyst. Several carbon diffraction lines such as (0 0 2) at high intensity and others at low intensity at (1 0 0), (0 0 4) and (1 1 0) can be seen in the figure. Ni metal at strongest peak can be seen at the strongest peak (1 1 1) and other peaks at (2 0 0) and (2 2 0). In addition, there are several low peaks are detected as nickel carbide (Ni_3C) such as (1 0 1) and Ni (1 1 1).

Basically, Raman modes in CNF consist of *G* band (tangential) at 1,580–1,600 cm^{-1} and *D* band (disorder induced) at 1,300–1,400 cm^{-1} [44] which originates from double resonant Raman scattering. In the experimental results of the Raman spectra can be seen in Fig. 7 where the band at 1,582 cm^{-1} is assigned to the C=C stretching vibrations with sp^2 bonding which is referred to the *G*-mode (first-order Raman). This is due to the in-plane displacement of carbon atoms in hexagonal sheets and it correlates with the XRD diffraction line (0 0 2) [45, 46]. The band at 1,350 cm^{-1} symbolizes the defective structure that is *D* band with sp^3 bonding. The *D* band is due to highly sensitive to structural disorder within graphite planes [41]. It is suggested that the disorder could be due to the defects of curved graphene sheets, tube ends and surviving impurities [47]. In addition, it is believed that the defect could be due to irregular configuration and corrugated CNFs which would be discussed further in this chapter.

8 Surface and Morphology

The bright field TEM images reveal the internal carbon structure of GNF as herringbone (or fish bone) and platelet. The herringbone structure consists of graphene planes at the angle of 45° to the carbon filament axis (Fig. 8a) while the

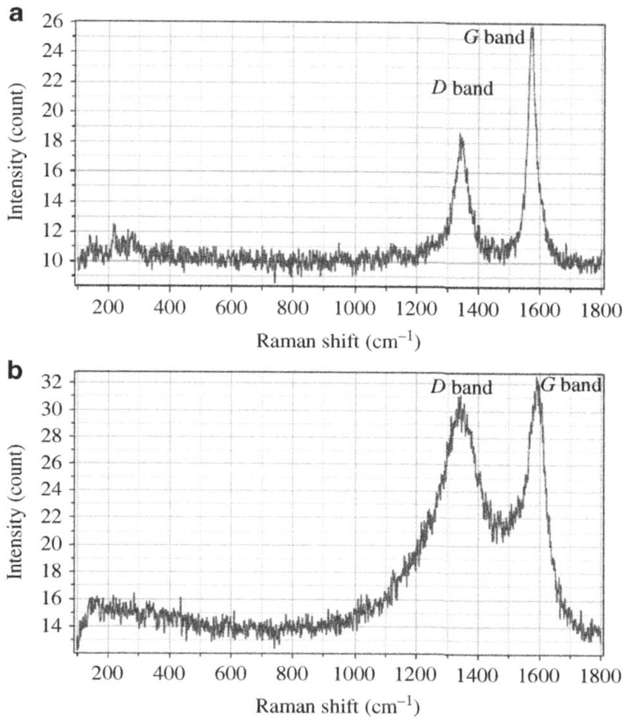

Fig. 7 XRD pattern of CNF with (**a**) Fe_2O_3 and (**b**) NiO as the catalysts

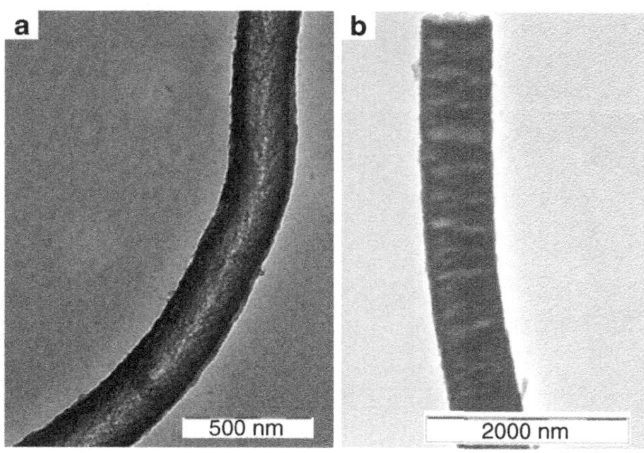

Fig. 8 TEM images showing sample GNFs with (**a**) herringbone and (**b**) platelet structures

platelet structure consists of graphene planes is perpendicular to the carbon filament axis (Fig. 8b). In addition, the developed GNF could exist as hollow GNF platelet structure. This is due to the increase in reaction temperature which will be discussed in the next paragraph.

GNFs would be continuously formed a long straight nanofibers if the precipitation of carbon from all faces is identical [12]. Unfortunately, any perturbation in this behavior will results in abnormalities of fiber conformation. Thus, coiled and spiral perturbations are likely to form. According to [48], the spiral structure is due to an equal diffusion of carbon through the catalyst particle which results in anisotropic growth. Different diffusional pathways taken by carbon atoms through the carbide phase lead to different rates of carbon growth and therefore twisted or spiral growth occurred.

A few examples of irregular configurations can be seen in Fig. 9. The coil diameter is estimated to be more than 1.5 μm. This is in good agreement with other findings [27, 28]. For regular helical as observed in Fig. 9b, the bidirectional growth from a single catalyst particle is in a symmetric growth mode which is resulted from helical reversals [32, 33].

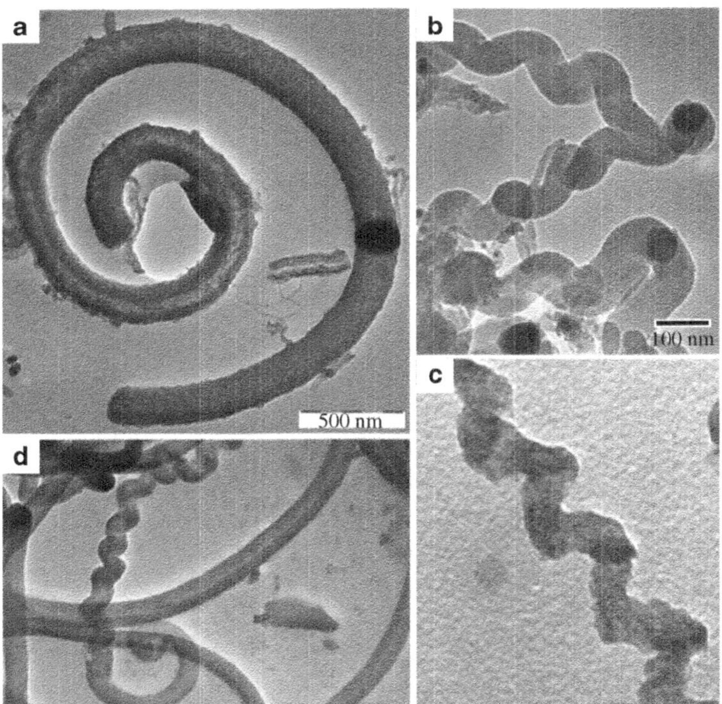

Fig. 9 Various spiral CNF structures that is (**a**) coiled (**b**) regular helical (**c**) twisted coil (**d**) partly helical and partly coiled

The irregular configurations of CNFs could also be observed in FESEM analysis where the coiled and double-helical structures of CNFs were found in iron-based and nickel-based catalyst samples, respectively. The later structure was developed from two nanofibers based from two different catalyst particles which are grown in double-helical configurations (refer to Fig. 10).

The growth of CNFs could also result in a formation of corrugated on the surface that is known as corrugated carbon nanofibers (C-CNFs) [49]. The growth mechanism of C-CNF has been described in several literatures [49, 50]. Basically, when the surfaces of catalyst particles absorb carbon source gas, they maintain the state of absorption until the particles become saturated with the gases. As a result, the particles burst out all the gases uncontrollably and produce high density carbon disc (refer to Fig. 11). Later, since all the carbons have been discharged, the absorption takes place again. At this time while waiting for the catalyst to become saturated again, low density of carbon discs are grown before the catalyst particles burst out the carbon to form high density carbon discs. These processes are repeated for several times.

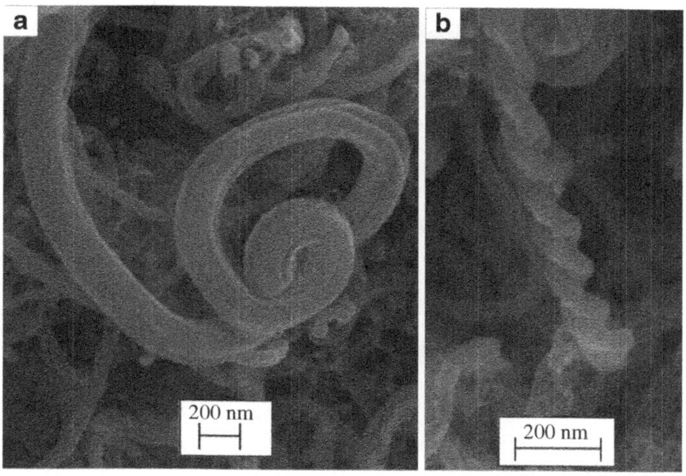

Fig. 10 FESEM images of (**a**) coiled and (**b**) regular helical of CNF structures

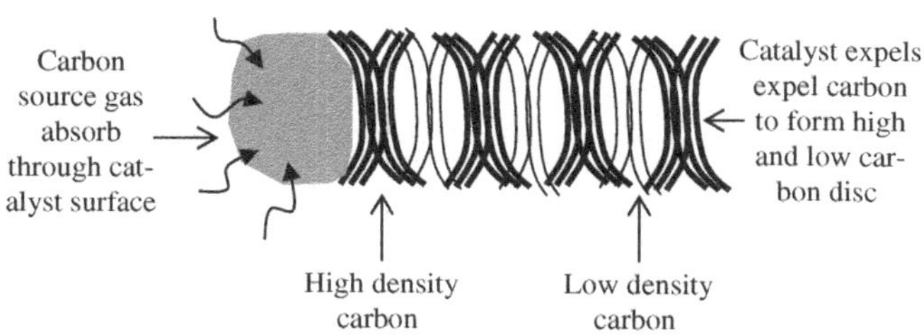

Fig. 11 Formation of corrugated shape of CNF

Based on TEM morphology observation it is found that there is a formation of C-CNFs in all samples during the experiment. Most C-CNFs are found in thicker nanofibers (more than 200 nm) and a few of them can be observed in thickness less than 100 nm. Some good examples of formation of C-CNF and GNF in a particular sample can be seen in Fig. 12. Thus, the nanofiber formed a corrugated shape as described in Fig. 11.

The thicker nanofiber (from a bigger catalyst particle) shows corrugated shape is labeled as 'X' while the thinner nanofiber (from a smaller catalyst particle) shows a normal CNF platelet structure is labeled as 'Y'. The reason is that for smaller catalyst particle, the absorption time for carbon gases is short and the gas expelling effect is so fast until it appears almost continuously. Thus, the corrugated shape pattern is not observed [49]. C-CNFs can be found in different shape such as coil shape as in Fig. 13. Most of C-CNF structures are not as smooth as CNFs. In some situations, certain CNFs exhibit corrugated and compartmentalized morphologies until the surface becomes uneven and ridged as seen in Fig. 13b. Such phenomenon

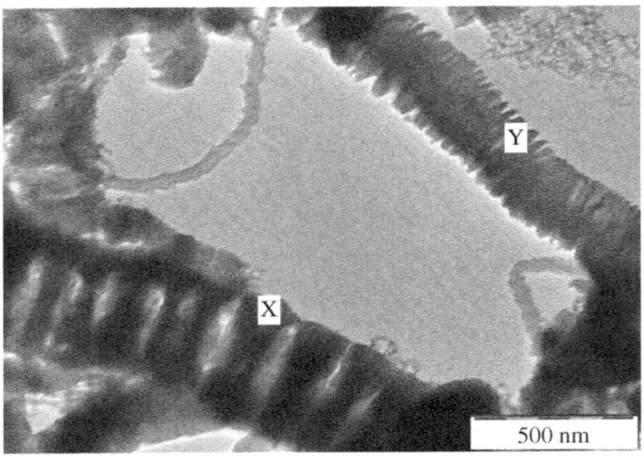

Fig. 12 Both CNFs (labeled as 'Y') and C-CNFs (labeled as 'X') are developed

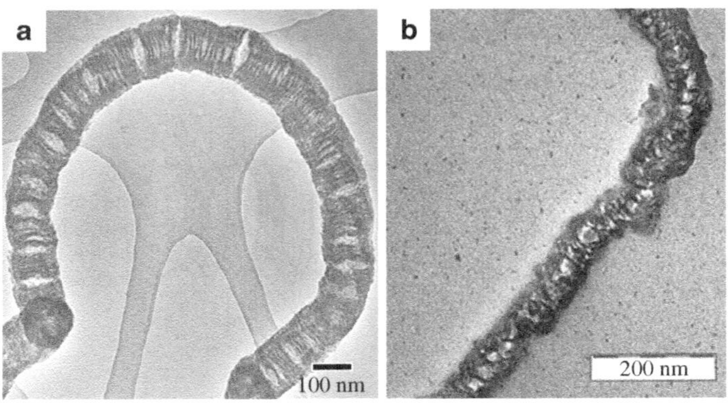

Fig. 13 C-CNFs at different configurations that are (**a**) coiled and (**b**) straight

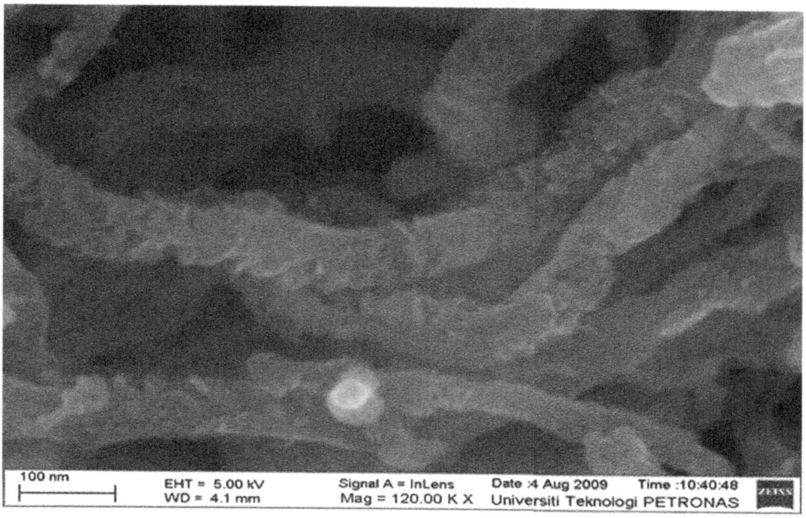

Fig. 14 FESEM image of long straight C-CNFs with uneven surface structures

has also been observed by other researcher [51]. Similar observation can be seen in FESEM image in Fig. 14 where the long C-CNFs have rough uneven graphitic plane structures.

9 Conclusion

Synthesis of carbon nanofibers using Fe_2O_3 and NiO as the catalysts at 600°C with ethylene as the carbon feedstock and hydrogen as the carrier gas was achieved using C-CVD method. Apart from observing typical long, straight CNF structure of platelet and herringbone, the synthesis has resulted in several irregular configurations such as coiled, regular helical and twisted coil. This is due to the particle catalyst geometry and the different diffusional pathways taken of carbon atoms through the carbide phase which then leads to different growth rate. In addition, corrugated carbon nanofibers were also observed in the samples where high and low densities of carbon discs were formed alternately. Such phenomena occurred when the absorbed carbon in a catalyst was saturated and burst out uncontrollably to form high density of carbon disc. More studies would be carried out to investigate other effects that contribute to the configurations and the properties of in each configuration for future applications.

Acknowledgement I would like to express my gratitude to Universiti Teknologi PETRONAS for the financial and technical support on this project.

References

1. Teo, K.B.K., Singh, C., Chhowalla, M., Milne, W.I.: Encyclopedia of Nanoscience and Nanotechnology, pp. 1–22 (2003)
2. Serp, P., Corrias, M., Kalck, P.: Appl. Catal. A Gen. **253**, 337–358 (2003)
3. Bhushan, B.: Springer Handbook of Nanotechnology, p. 1347. Springer, Berlin (2004)
4. Ko, F.K.: Nanotubes and Nanofibers, pp. 37–108. CRC, Boca Raton (2006)
5. Martin-Gullon, I., Vera, J., Conesa, J.A., González, J.L., Merino, C.: Carbon **44**, 1572–1580 (2006)
6. Park, C., Baker, R.T.K.: J. Phys. Chem. B **102**, 5168–5177 (1998)
7. Huang, S., Dai, L.: J. Nanopart. Res. **4**, 145–155 (2002)
8. De Jong, K.P., Geus, J.W.: Catal. Rev. Sci. Eng. **42**, 481–510 (2000)
9. Baker, R.T.K.: Encyclopedia of Materials – Science and Technology, pp. 932–941. Elsevier, Amsterdam (2001)
10. Dresselhaus, M.S.: Carbon-based nanostructures. http://www.wtec.org/loyola/nano/US.Review/09_02.htm, accessed date: 8 February 2006, last updated: 2006 (2006)
11. Baker, R.T.K.: Properties and applications of graphite nanofibers, Northeastern University (1998)
12. Rodriguez, N.M., Chambers, A., Baker, R.T.K.: Langmuir **11**, 3862–3866 (1995)
13. Yu, Z., Chen, D., Tøtdal, B., Holmen, A.: Mater. Chem. Phys. **92**, 71–81 (2005)
14. Lueking, A.D., Pan, L., Narayanan, D.L., Clifford, C.E.B.: J. Phys. Chem. B **109**, 12710–12717 (2005)
15. Luxembourg, D., Flamant, G., Guillot, A., Laplaze, D.: In: 1st European Hydrogen Energy Conference 2003, France, 2–5 September 2003 (2003)
16. Marella, M., Tomaselli, M.: Carbon **44**, 1404–1413 (2006)
17. Takenaka, S., Ishida, M., Serizawa, M., Tanabe, E., Otsuka, K.: J. Phys. Chem. B **108**, 11464–11472 (2004)
18. Pham-Huu, C., Vieira, R., Louis, B., Carvalho, A., Amadou, J., Dintzer, T., et al.: J. Catal. **240**, 194–202 (2006)
19. Takehira, K., Ohi, T., Shishido, T., Kawabata, T., Takaki, K.: Appl. Catal. A Gen. **283**, 137–145 (2005)
20. De Lucas, A., Garrido, A., Sánchez, P., Romero, A., Valverde, J.L.: Ind. Eng. Chem. Res. **44**, 8225–8236 (2005)
21. Bououdina, M., Grant, D., Walker, G.: Carbon **43**, 1286–1292 (2005)
22. Kayiran, S.B., Darkrim, F.L., Gicquel, A., Bernier, P., Gadelle, P., Levesque, D.: In: 1st European Hydrogen, Energy Conference 2003, France, 2–5 September 2003 (2003)
23. Sufian, S.: PhD thesis, Universiti Teknologi PETRONAS, Malaysia (2010)
24. Rzepka, M., Bauer, E., Reichenauer, G., Schliermann, T., Bernhardt, B., Bohmhammel, K., et al.: J. Phys. Chem. B **109**, 14979–14989 (2005)
25. Saveliev, A.V., Merchan-Merchan, W., Kennedy, L.A.: Combust. Flame **135**, 27–33 (2003)
26. Qin, Y., Zhang, Z., Cui, Z.: Carbon **42**, 1917–1922 (2004)
27. Qin, Y., Zhang, Z., Cui, Z.: Carbon **41**, 3072–3074 (2003)
28. Yang, S., Chen, X., Kusunoki, M., Yamamoto, K., Iwanaga, H., Motojima, S.: Carbon **43**, 916–922 (2005)
29. Lueking, A.D., Yang, R.T., Rodriguez, N.M., Baker, R.T.K.: Langmuir **20**, 714–721 (2004)
30. Davis, M.: 2009. http://faculty.evansville.edu/md7/bact02/microscope%20copy/microscope_files/microscope.ppt, accessed date: 1 March 2009, last updated: 2003 (2003)
31. Endo, M., Kim, Y.A., Hayashi, T., Yanagisawa, T., Muramatsu, H., Ezaka, M., et al.: Carbon **41**, 1941–1947 (2003)
32. Qin, Y., Yu, L., Wang, Y., Li, G., Cui, Z.: Solid State Commun. **138**, 5–8 (2006)
33. Qin, Y., Zhang, Q., Cui, Z.: J. Catal. **223**, 389–394 (2004)
34. Cheng, J., Zhang, X., Tu, J., Tao, X., Ye, Y., Liu, F.: Mater. Chem. Phys. **95**, 12–15 (2006)

35. Vera-Agullo, J., Varela-Rizo, H., Conesa, J.A., Almansa, C., Merino, C., Martin-Gullon, I.: Carbon **45**, 2751–2758 (2007)
36. Horiba, J.Y.: Carbon Nanotubes – What Information Does Raman Microscopy Bring? The Application Notebook (2005)
37. Jorio, A., Pimenta, M.A., Souza Filho, A.G., Saito, R., Dresselhaus, G., Dresselhaus, M.S.: New J. Phys. **5**, 139.1–139.17 (2003)
38. Endo, M., Kim, Y., Hayashi, T., Takeuchi, K., Terrones, M., Dresselhaus, M.S.: Encyclopedia of Chemical Processing, pp. 333–335. CRC, Boca Raton (2005)
39. Panalytical, X-ray diffraction. http://www.panalytical.com/index.cfmpid=135, accessed date: April 2009, last updated: 2009 (2009)
40. Moeck, P.: X-ray diffraction (XRD) **47**. http://web.pdx.edu/~pmoeck/phy381/Topic5a-XRD.pdf, accessed date: last updated: 31 March 2004 (2004)
41. Reshetenko, T.V., Avdeeva, L.B., Ismagilov, Z.R., Pushkarev, V.V., Cherepanova, S.V., Chuvilin, A.L., et al.: Carbon **41**, 1605–1615 (2003)
42. Zhang, M., Xu, C.L., Wu, D.H., Cao, L.M., Wang, W.K.: J. Mater. Sci. Lett. **19**, 511–514 (2000)
43. Boskovic, B.O., Stolojan, V., Zeze, D.A., Forrest, R.D., Silva, S.R.P., Haq, S.: J. Appl. Phys. **96**, 3443–3446 (2004)
44. Hartman, A.Z., Jouzi, M., Barnett, R.L., Xu, J.M.: Phys. Rev. Lett. **92**, 236804-1–236804-4 (2004)
45. Jawhari, T., Roid, A., Casado, J.: Carbon **33**, 1561–1565 (1995)
46. Reshetenko, T.V., Avdeeva, L.B., Ismagilov, Z.R., Chuvilin, A.L.: Carbon **42**, 143–148 (2004)
47. Ebbesen, T.W., Nanotubes, C.: Preparation and Properties. CRC, Boca Raton, FL (1997)
48. Park, C., Keane, M.A.: J. Catal. **221**, 386–399 (2004)
49. Koizumi, N., MIinorikawa, Y., Yakabe, H., Kimura, H., Kurosu, T., Ida, M.: Jpn. J. Appl. Phys. **45**, 6517–6523 (2006)
50. Wang, H., Wu, Y., Choong, C.K.S., Zhang, J., Teo, K.L., Ni, Z., et al.: IEEE Nano **1**, 219–222 (2006)
51. Kordatos, K., Vlasopoulos, A.D., Strikos, S., Ntziouni, A., Gavela, S., Trasobares, S.: Electrochim. Acta **54**, 2466–2472 (2009)

Molecular Simulation to Rationalize Structure-Property Correlation of Carbon Nanotube

Abhijit Chatterjee

Abstract The aim of this chapter is to discuss the role of computer simulation in the domain of nano materials with a special emphasis on carbon nanotube. In recent years, nanotubes have been a major focus of nanoscience and nanotechnology. It is a self-growing field of study attracting tremendous interest, insight and effort in research and development around the world for its multi domain applications. Nanotube has been discovered a decade ago. With lots of its potentiality it is still hard to rationalize the structure property correlation, which is otherwise impossible without using computer modeling. Molecular modeling is a method, which combines computational chemistry techniques with graphics visualization for simulating and predicting three-dimensional structures, chemical processes, and physico-chemical properties of molecules and solids. The current chapter while covering the issues of electronic structural issues of nanotubes will especially focuses on the sensing issue. It will cover the role of reactivity index to design new carbon nanotubes efficient for sensing or storage capability at par with the global concern of environmental safety. We wish to show the capability of molecular modeling as a state of art to design new futuristic materials of interest to satisfy industrial needs.

1 Introduction

Carbon materials are found in a variety of different forms, these include graphite, diamond, carbon fibers, fullerenes and carbon nanotubes. A fullerene by definition is a closed convex cage molecule containing only hexagonal and pentagonal faces; the best known of these is C-60 or the buckyball. A carbon nanotube is considered

A. Chatterjee
Accelrys K.K., Material Science, 3-3-1 Nishishinbashi, Tokyo 105-0003, Japan
e-mail: achatterjee@accelrys.com

N. Yahya (ed.), *Carbon and Oxide Nanostructures*, Adv Struct Mater 5,
DOI 10.1007/8611_2010_11, © Springer-Verlag Berlin Heidelberg 2010,
Published online: 25 June 2010

to be a type of giant linear fullerene. It was the discovery of fullerenes by Harold Kroto of Sussex University and Richard Smalley at Rice University, which led to the realization that nanotubes might exist with a diameter similar to those of the fullerene molecules. Carbon nanotubes were first discovered in 1991 by Sumio Iijima of the NEC Corporation in Tsukuba, Japan [1]. The samples were created by a direct-current arc discharge between carbon electrodes in a noble gas and the structures were examined by transmission electron microscopy (TEM). Single-wall nanotubes were then subsequently discovered 2 years later by Iijima [2] group at NEC and Donald Bethune [3] group at Almalden Research Center in California. Whereas the multi-wall nanotubes had diameters of the order tens of nanometers across, the typical diameter of a single-wall nanotube was just 1 or 2 nm with the lowest diameter being approximately.

1.1 Hybridization in a Carbon Atom

Carbon is the sixth element in the periodic table, it has six electrons and is listed at the top of column IV. These electrons occupy the 1s2, 2s2 and 2p2 atomic orbital. The 1s2 contains two strongly bound core electrons and the 2s22p2 contain the more weakly bound valence electrons. These valence electrons form 2s, 2px, 2py and 2pz orbital and it is these that are important in forming covalent bonds in carbon materials. Now because the energy difference between the upper 2p energy levels and lower 2s level is small compared to the binding energy of the chemical bonds, these four electrons can mix with each other leading to an increase in the binding energy of the Carbon atom and its neighbors. This mixing of 2s and 2p orbital is called Hybridization and the mixing of a single 2s level with $n = 1, 2, 3$ 2p electrons is called spin hybridization. When dealing with carbon nanotubes the sp2 hybridization is important, here three σ bonds per carbon atom are formed and these bonds make a skeleton for a planar structure in two dimensions. This two-dimensional honeycomb structure can be seen in Figure 1 where the atoms would be located on the six points of the hexagons. A single wall carbon nanotube (SWCNT) can be described as a rolled up sheet of graphene that forms a cylinder with a diameter between 0.7 and 10 nm, though most have diameters are less than 2 nm. The two-dimensional graphene structure is simply a single atomic layer of graphite, which is the three-dimensional form of carbon. The structure of the single wall carbon nanotube (SWCNT) is determined by the orientation of the hexagonal ring of the carbon atoms in the honeycomb lattice relative to the axis of the nanotube. Figure 2 shows some examples of the tubes and it can be seen from here that the direction of the hexagonal ring can be taken as almost arbitrary without any distortion to the ring. However, there will be distortion due to the curvature of the tube. This means there are many possible structures for a carbon nanotube even though the tube is always cylindrical in shape.

In the above diagram Figure 1, the vector OA is the chiral vector, with a chiral angle of q, for the $(N,M) = (5,2)$ nanotube. The rectangle $OBB'A$ is the repeat unit

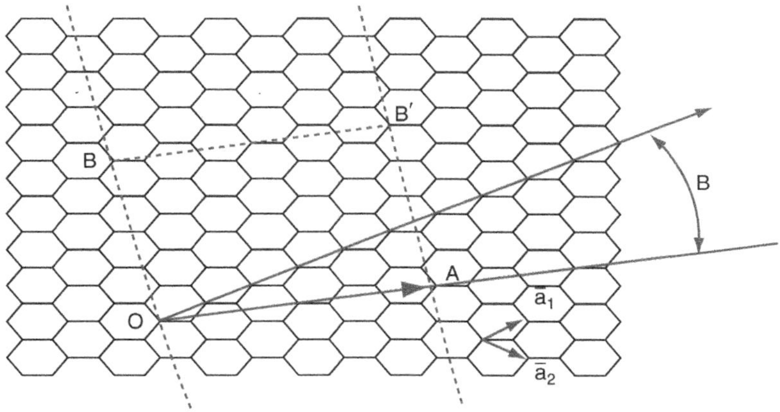

Fig. 1 Model of the two dimesional graphene sheet

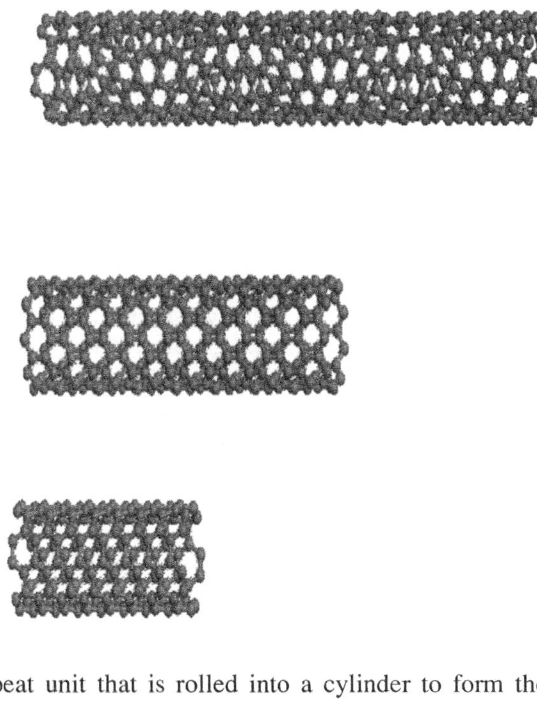

Fig. 2 The model for 9,0 zigzag; 5,5 arm-chair and 6,4 chiral nanotube

for the nanotube. It is this repeat unit that is rolled into a cylinder to form the nanotube. The basic structure of a SWCN is specified by a single vector called the chiral vector Ch. This vector is defined as OA and this is the section of the nanotube perpendicular to the nanotube axis, which when rolled up is the circumference of the tube. Figure 1 shows the unrolled lattice of the nanotube, and in this case, the direction of the nanotube axis is given by the vector OB. The lattice can then be rolled to form a cylinder by lining up the points so that O is on top of A and B is on top of B. The chiral vector Ch is then determined by the real space lattice vectors a1 and a2, which are also defined in Fig. 1 and this leads to the following equation:

$$C_h = na_1 + ma_2 = (n, m) \tag{1}$$

The main symmetry classification of a tube is defined as being either achiral or chiral. An achiral tube is defined as one where the mirror image of the tube is identical to the original one. A chiral tube is then defined as a tube where the mirror image cannot be superposed onto the original tube and they show spiral symmetry. There are two types of achiral tubes, which are named armchair and zigzag nanotubes. These two types can be seen in Fig. 2 where (5, 5) is an armchair tube and (9, 0) is a zigzag tube. These two tubes get their names from the shape of the cross-sectional ring at the edge of the nanotubes. An armchair nanotube is then defined as the case where n = m in (1), a zigzag corresponds to the case where m = 0. All other chiral vectors corresponds to the production of a chiral nanotube but this is subject to the constraint that m is less than or equal to n.

There has been a high level of interest in understanding the electronic properties of single walled carbon nanotubes (SWNT) due to potential applications in microelectronics as a new class of electronic materials and energy storage as a new type of electrode material. It is well known that the chemical and physical properties of SWNT differ substantially with changes in nanotube diameter and chirality [4]. It is proposed theoretically that a very small diameter zigzag SWNT can undergo spontaneous coalescence, or cross-linking, forming a two-dimensional layered structures [5]. These structures of ordered sheets of cross-linked, deformed nanotubes are highly unique and we anticipate that these hypothetical materials should exhibit novel chemical and physical properties. The layered structure also makes it feasible to intercalate a variety of chemical species (dopants) into the material, analogous to what has been observed for graphite. The intercalation of molecular and atomic species in graphite has resulted in a large number of novel compounds with unique properties, including high electrical conductivity and, in some cases, novel magnetic and thermal transport properties [6]. Among the graphite intercalation compounds, those intercalation compounds of the alkali metals have been studied most extensively [7]. A significant level of interest in lithium intercalation has resulted from the use of graphitic carbons as anode materials in lithium ion batteries. Certain examples, such as C24K, can also reversibly absorb gases, including hydrogen, and may undergo phase transitions under certain conditions [8]. Since the discovery of single- and multi-walled carbon nanotubes, the intercalation chemistry and doping of carbon nanotubes have been an area of considerable activity [9, 10].

1.2 Electronic Structure

One of the main features of the carbon nanotube is that theory predicts that some will act as semiconductors while others will act as nearly perfect metallic conductors. The electronic structure of a single wall carbon nanotube is obtained from that of the graphene sheet. In rolling up the graphene sheet, periodic boundary conditions are imposed in the direction of the circumference denoted by the chiral vector Ch. The wave vector associated with Ch becomes quantized, while the wave vector associated with the direction along the axis remains continuous for a tube of infinite length. These periodic boundary conditions are used to obtain the energy eigenvalues of a nanotube and define the small number of allowed wave vectors k, in the direction of the circumference.

$$C.k = 2\pi j \tag{2}$$

This means that the energy bands of a nanotube consist of a set of one dimensional energy dispersion relations which are cross sections of the one found in graphene. The energy of the electronic states in graphene as a function of the wave vector, k, is plotted near the Fermi energy. When the energy dispersion relations of two-dimensional graphite are folded so that the wave vectors are parallel to k2, as shown in Fig. 3a. Thus N pairs of 1-D energy dispersion relations are obtained. N is the number of hexagons in the cut out sheet of graphene, which is defined from the chiral vector and the length of the tube. These energy dispersion relations correspond to the energy dispersion relations of a single wall carbon nanotube. These energy dispersion relations are the cross-sections of the energy dispersion surface of 2-D graphene, where the cuts have been made on the lines of intersections

$$kK_2/|K_2| + \pi K_1 \tag{3}$$

If for a particular (n, m) nanotube, the cutting line passes through a K point where the energy bands of 2-D graphite are degenerate then the one dimensional

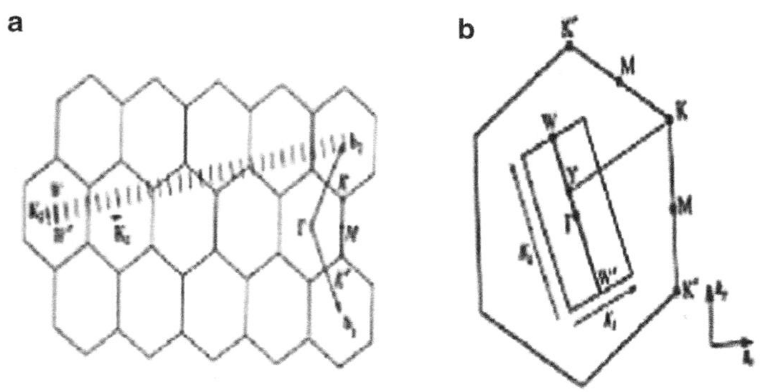

Fig. 3 (**a**) Shows the Brillion zone for a carbon nanotube, (**b**) shows the condition for metallic energy bands

energy bands have a zero energy gap. Also the density of states at the Fermi level has a finite value for these tubes and they are therefore metallic. However, if the cutting line does not pass through one of these K points, then the carbon nanotube is predicted to show semiconducting behavior with a band gap between the valence and conduction bands.

The condition for obtaining a metallic energy band is that the ratio of the length of vector YK to that of K_1 in Fig. 3b is an integer. Since the vector YK is given by

$$Y\vec{K} = \frac{2n+m}{3}K \tag{4}$$

the condition for metallic nanotubes is that (2n + m) or equivalently (n − m) is a multiple of 3. This leads to the cases that all armchair nanotubes are metallic, and zigzag nanotubes are only metallic if n is a multiple of 3.

1.3 Applications of Carbon Nanotubes

The numerous properties that carbon nanotubes possess could lead to many diverse applications for their use. One of the primary uses could be in electronic devices where their size could be an important fact in the drive towards smaller and smaller devices, which is reaching the nanometer scale. The nanotube may eventually be used as the interconnections between electronic components and as logic and storage devices. Another application is to use tubes as devices to store energy, where they could be used as batteries or fuel cells. One of the most promising applications in this area is the hydrogen fuel cell, which may be used as a source of portable energy in the future. The hydrogen fuel cell works by converting the hydrogen to water and releasing electronic energy in the process. The main problem with hydrogen is the fact that it is a flammable gas, which means it is difficult to store. There are some materials that absorb hydrogen and then release it again, Palladium for example. However, it was discovered in 1993 [2] that it was possible to store various gases in single wall carbon nanotubes, hydrogen being one of these. This marks them out as one of the prime candidates for use as a fuel cell. Carbon nanotubes have a structure which is almost perfect, the absence of virtually any defects leads to an enhancement in the strength of the material and these tubes rank among the strongest known materials. Nanotubes are exceptionally rigid with a Young's modulus in the terapascal range, this makes them almost an order of magnitude more rigid than steel. In fact single wall carbon nanotubes are so strong and light that it would be possible to construct a wire that extended into space without it breaking under its own weight. These features indicate that single wall carbon nanotubes may well replace synthetic graphite fibers and revolutionize the materials industry. There are also many other applications where the carbon nanotube may be useful, from electronic transistors to field emission lamps for use in displays and their unique properties will mean that they will become important in many applications as research into them carries on. There are also many other applications where the carbon nanotube may be useful, from electronic transistors to field emission lamps for use in displays and their unique properties will mean that they will become important in many applications as research into them carries on. An exemplary application of reactivity in Gas sensor with single-wall carbon nanotube: In this part, we wish to explore interatomic interaction as well intermolecular interaction through the center of activity. Since the discovery of the structure of carbon nanotubes (CNTs) or SWNT, much effort has been devoted to finding uses of these structures in applications ranging from filed-emission devices to other nanodevices [11, 12]. Kong et al. [13] proposed for the first time the use of CNTs as gas sensors. Experimental data have shown that transport properties of SWNT change dramatically upon exposure to gas molecules at ambient temperature [14]. Main advantage of the open SWNT bundles is that they provide a larger number of adsorption sites. As a result, the adsorption capacity is significantly increased and several new structures and phase transitions were observed [15]. A recent study of Andzelm et al. [16] indicate that the semiconducting SWNTs can serve as gas sensors for several gases like CO, NH_3, H_2, etc. However, NH_3 shows an intriguing behavior compared to other gasses. NH_3 molecule can binds weakly with CNTs, yet can change the conductance significantly. This discrepancy was explained by assuming that the NH_3 binds at defects. For a semiconducting SWNT exposed to 200 ppm of NO_2, it was found that the electrical conductance can increase by three orders of magnitude in a few seconds. On the other hand, exposure to 2% NH_3 caused the conductance to decrease up to two orders of magnitude [17]. Sensors made from SWNT have high sensitivity and a fast response time at room temperature, which are important advantages for sensing applications.

It is very difficult to obtain conductance by quantum mechanical calculation as it will be very much CPU intensive, but measurement of conductance is an utmost important parameter to prove the efficiency of the nanotubes as gas sensors, which is the experimental way of measuring sensors. Thus, a method was developed by calculating the change in the reactivity index before and after the application of the reaction field. The reactivity index provides information about the activity of the gas molecules over SWNT, if the activity changes then the sensing behavior will change. This is a simplistic approach, which is cost-effective to new material design for the sensor industry.

2 Model

The lattice parameter is a = b = 11.483001 and c = 24.595121; alpha = beta = 90.0 and gamma = 120.0. We consider here a arm-chair CNT (10, 10) with diameter of 13.56 Å, as this SWNT is the most frequently observed experimentally. The armchair SWNT is comprised of cylindrical sheet of hexagons with long axes

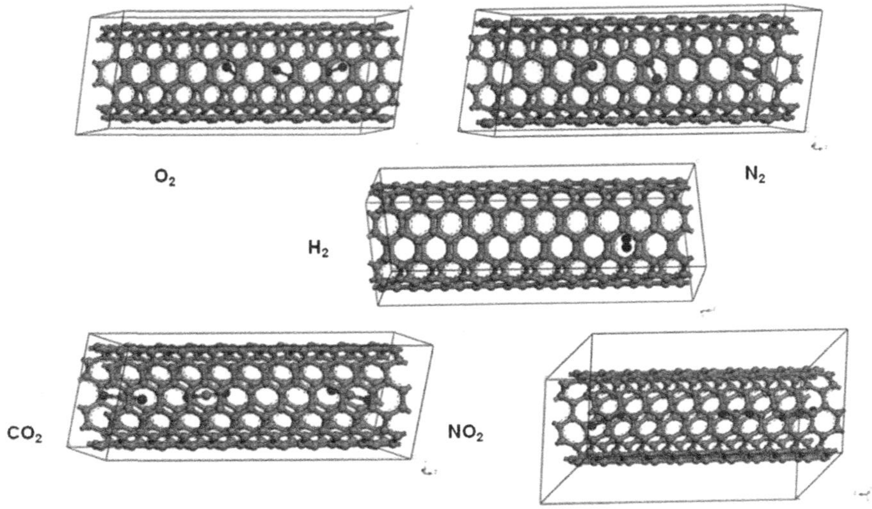

Fig. 4 Model of CNT with adsoprbed molecules within

perpendicular to the tube axis, while they are parallel in the zigzag SWNT. A nanotube is formed by rolling a graphene sheet into a cylinder. Its structure can be uniquely characterized by defining the indices N and M of the chiral vector:

$$Ch = N^{*}a1 + M^{*}a2 \tag{5}$$

The vectors a1 and a2 are the principle lattice vectors of a graphene sheet. The choice of such a tube is based on the following arguments:

- It displays metallic properties
- Its diameter ranges within the experimental distribution of tube sizes
- Its length corresponds to a reasonable compromise which allows us to neglect the side effects for the molecular adsorption and to preserve relatively short computational times
- The adsorption of a single molecule on this tube (Fig. 4) yields a coverage equal to 5×10^{-3} if we consider that one carbon hexagon on the tube corresponds to one adsorption site

3 Method

Absolute adsorption isotherms were computed using a Grand Canonical Monte Carlo (GCMC) calculation algorithm, which allows displacements (translations and rotations), creations and destructions. The GCMC method was convenient to simulate adsorption [18] and carried out by the Sorption module in Materials Studio software of Accelrys Inc, which can execute a series of fixed pressure simulations at a set temperature in a single step. At the start of the Sorption run, a fixed pressure simulation is performed for a specified number of steps between the start and end fugacity set. In the default setup, the start fugacity of each component is increased linearly to the end fugacity with an equidistant step. However, it is also possible to regulate those steps to space logarithmically; in that case, the logarithm of the start fugacity of each component is increased linearly to the logarithm of the end fugacity in equidistant steps (the fugacity thus increases exponentially). Sorption result reports the sum of component fugacities as the total fugacity. This is equal to the pressure if the reservoir is an ideal gas. The equilibration steps and the production steps were set to 1,000,000 and 1,000,000,000, respectively Monte Carlo moves. We have saved the frames for every 1,000 steps. Materials Studio database provides various types of forcefields. The GCMC workflow (Fig. 5) looks like the following Figure 5.

We have performed the GCMC calculation on the selected CNT with all the diatomic molecule. The adsoprtuon isotherm is calculated for all the respective molecules. The configurations are sampled from a grand canonical ensemble. In the grand canonical ensemble, the fugacities of all components, as well as the temperature, are fixed as if the framework was in open contact with an infinite sorbate reservoir with a fixed temperature. The reservoir is completely described by the temperature plus the fugacities of all components and does not have to be simulated explicitly. So once can find the probability of the loading of the molecule in the vacant space of the periodic structure. During the course of the simulation, sorbate molecules within the framework are randomly rotated and translated, and, in

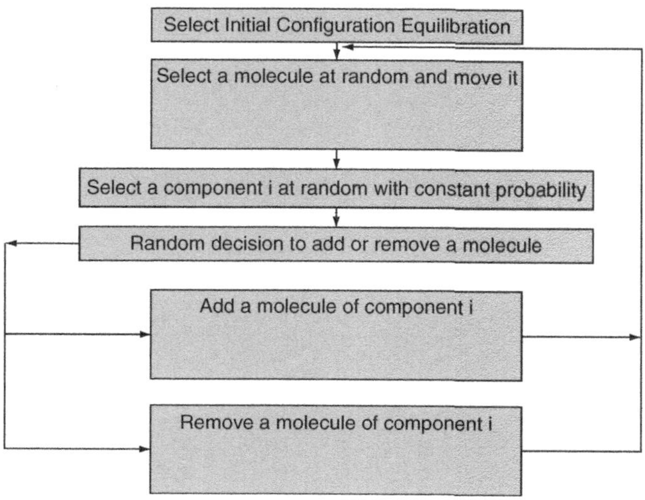

Fig. 5 Grand Canonical Monte Carlo (GCMC) workflow

addition, sorbates are randomly created in and deleted from the framework. The configuration that results from one of these steps is accepted or rejected according to the selection rules of the Monte Carlo method being used for the simulation.

4 Grand Canonical Monte Carlo (GCMC) Results

The results are shown below in Fig. 6, the results show the loading of different gases in the nanotube. This trend will be then compared to identify the capability of the gases to adsorb in CNT and therefore may act as a selective adsorption. The results were compared with experimental adsorption to see if they match.

The results were further compared with experimental observation [19]. It is seen that the adsorption isotherm type looks same as that of experiment, to justify the reliability of our model. The results are shown in Fig. 7. Similar trend is as well observed for the experimental numbers.

Nitrogen adsorption on two types of single walled CNT bundles at both subcritical and supercritical temperature is studied to understand the role of external surface in the type of isotherm [20]. It is observed that on an infinite periodic hexagonal bundle without an external surface a subcritical adsorption of type 1 with two steps is formed. At higher coverage multilayer adsorption and wetting occur on the external surface as the bulk phase approaches saturation. The experimental scenario is shown below in Fig. 8, which can be compared with the isotherm of

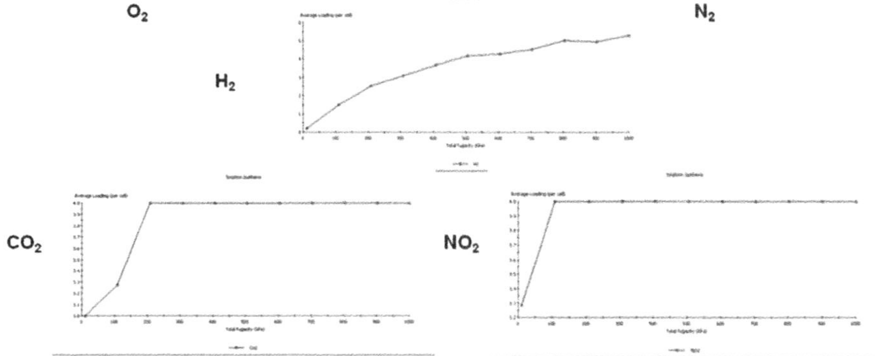

Fig. 6 Adsorption isotherm for individual gas molecules with CNT

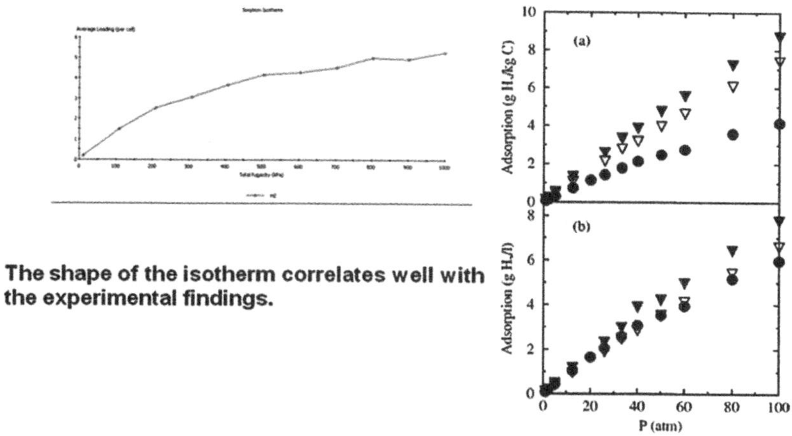

Fig. 7 Comparison of adsorption isotherm for adsorption of H_2 molecules with CNT

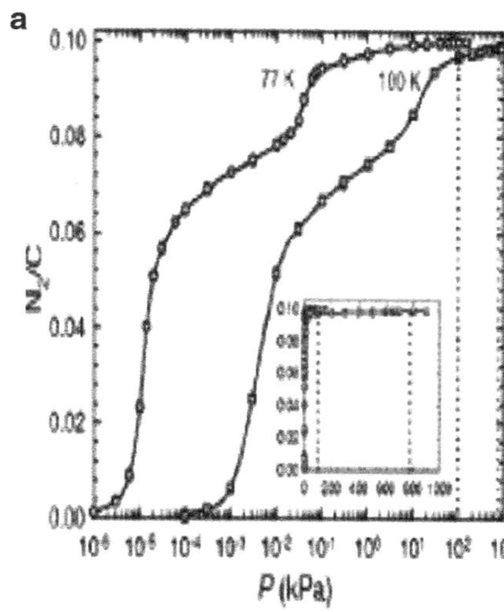

Fig. 8 Nitrogen adsorption isotherms on a semilog arithmic scale of bulk pressure on the periodic bundle at 77 and 100 K

Fig. 6 with a consideration that the one step is due to the single CNT used in simulation.

The GCMC results shows that the trend matches with experiment. The loading steadily increases with hydrogen probably due to size. There is a variation of order in gases except hydrogen; there is no drastic difference. Carbon dioxide and nitrogen dioxide show a sort of similar trend. The adsorbed molecules are at a average distance of roughly around 2.0 Å with a cut off distance used of 5 Å.

5 Density Functional Theory (DFT) Calculation Method

All the calculations related to the transition state were performed with Density Functional theory (DFT) [21] using DMol^3 [22–24] code of Accelrys Inc. BLYP exchange correlation functional [25, 26] and DNP basis set [27]. The transition state calculations were performed using the synchronous transit methods as included in the DMol^3 module of Accelrys Inc.

6 Reactivity Concept and Theory

The Hard and Soft Acid and Bases (HSAB) principle classifies the interaction between acids and bases in terms of global softness. Pearson proposed the global HSAB principle [28, 29]. The HSAB principles classify the interaction between acids and bases in terms of global softness and this has been validated further [30–39]. The global hardness was defined as the second derivative of energy with respect to the number of electrons at constant temperature and external potential, which includes the nuclear field. The global softness is the inverse of global hardness. Pearson also suggested a principle of maximum hardness (PMH) [40, 41], which states that, for a constant external potential, the system with the maximum global hardness is most stable. PMH has also been studied extensively to further probe into both inter- and intra-molecular interactions [42–47]. In recent days, DFT has gained widespread use in quantum chemistry [48–51]. For example, some DFT-based local properties, e.g., Fukui functions and local softness have already been successfully used for the reliable predictions in various types of electrophilic and nucleophilic reactions. On the other hand, the reactivity index finds its application in material designing. We proposed a reactivity index scale for heteroatomic interaction with zeolite frame work [52]. Moreover, the role of water during nucleation process can also be monitored efficiently in terms of solvation energy to rationalize the fundamental mechanism of crystal growth [53]. A range of reactivity index along with DFT was fruitful to determine the activity of nitrogen heterocyclics present in biomacromolecules and its suitable sorbent from the dioctahedral smectite family [54]. In another approach, a novel function λ was introduced for quantitative description of weak adsorption cases, which was so far qualitative inside the domain of DFT. In addition to that, several calculations were performed to derive group softness [55] for inter and intra-molecular reactivity for nitro aromatics and its adsorption over clay matrices and contributed to the development of the methodology and its application in various systems [56–58].

In density functional theory, hardness (η) is defined as [41]

$$\eta = {}^1\!/_2(\partial^2 E/\partial N^2)v(r) = {}^1\!/_2(\partial\mu/\partial N)_v \tag{6}$$

Where E is the total energy, N is the number of electrons of the chemical species and the chemical potential.

The global softness, S, is defined as the inverse of the global hardness, η.

$$S = {}^{1}\!/_{2\eta} = ({}^{\partial N}\!/_{\partial \mu})_v \tag{7}$$

Using the finite difference approximation, S can be approximated as

$$S = {}^{1}\!/_{(IE - EA)} \tag{8}$$

where IE and EA are the first ionization energy and electron affinity of the molecule, respectively.

The Fukui function f(r) is defined by [42]

$$f(r) = [{}^{\partial \mu}\!/_{\partial v(r)}]_N = [{}^{\partial \rho(r)}\!/_{\partial N}]_v \tag{9}$$

The function 'f' is thus a local quantity, which has different values at different points in the species, N is the total number of electrons, μ is the chemical potential and v is the potential acting on an electron due to all nuclei present. Since $\rho(r)$ as a function of N has slope discontinuities, (6) provides the following three reaction indices [42]:

$$f^-(r) = [{}^{\partial \rho(r)}\!/_{\partial N}]_v^- \text{ (governing electrophilic attack)}$$

$$f^+(r) = [{}^{\partial \rho(r)}\!/_{\partial N}]_v^+ \text{ (governing nucleophilic attack)}$$

$$f^0(r) = 1/2[f^+(r) + f^-(r)] \text{ (for radial attack)}$$

In a finite difference approximation, the condensed Fukui function [42] of an atom, say x, in a molecule with N electrons are defined as:

$$f^0(r) = 1/2[f^+(r) + f^-(r)] \text{ (for nucleophilic attack)} \tag{10}$$

$$f_x^- = [q_x(N) - q_x(N-1)] \text{ (for electrophilic attack)}$$

$$f_x^0 = [{}^{q_x(N+1) - q_x(N-1)}\!/_{2}] \text{ (for radical attack)}$$

where q_x is the electronic population of atom x in a molecule.

The local softness s(r) can be defined as

$$s(r) = ({}^{\partial \rho(r)}\!/_{\partial \mu})_v \tag{11}$$

Equation (3) can also be written as

$$s(r) = [{}^{\partial \rho(r)}\!/_{\partial N}]_v [{}^{\partial N}\!/_{\partial \mu}]_v = f(r) S \tag{12}$$

Thus, local softness contains the same information as the Fukui function f(r) plus additional information about the total molecular softness, which is related to the global reactivity with respect to a reaction partner, as stated in HSAB principle. Thus the Fukui function may be therefore is thought of as a normalized local softness. Atomic softness values can easily be calculated by using (12), namely:

$$s_x{}^+ = [q_x(N+1) - q_x(N)]S \tag{13}$$

$$s_x{}^- = [q_x(N) - q_x(N-1)]S$$

$$s_x{}^0 = [q_x(N+1) - q_x(N-1)/_2]S$$

We have further explained the interaction energy scheme as follows. Let us consider a process where AB is the final product formed at equilibrium after combination of isolated A and B species present. With the existing knowledge, this is recognized that A and B interacts in two steps: (1) interaction will take place through the equalization of chemical potential at constant external potential and (2) A and B approach the equilibrium state through changes in the electron density of global system generated by making changes in the external potential at constant chemical potential. Thus, within DFT we can write

$$\Delta E_{inter} = E[\rho AB] - E[\rho A] - E[\rho B] \tag{14}$$

where ρAB, ρA, ρB are the electron densities of the systems AB at equilibrium and of the isolated systems A and B, respectively or in terms of the potentials we can write

$$\Delta Einter = \Delta Ev + \Delta E\mu \tag{15}$$

where, $\Delta Ev = -1/2\left[(\mu A - \mu B)^2/(SA + SB)\right]$

$$\Delta E\mu = -1/2NAB^2k\left[1/(SA + SB)\right]$$

NAB total number of electrons, *k* is the proportionality constant between SAB and SA + SB, product of N^2 and K is λ.

$$\Delta E\mu = (-1/2)\lambda/(SA + SB)$$

If the interaction is taking place through *j* site of *A*,

$$\lambda Aj = qAj^{eq} - qAj^0 \tag{16}$$

qAj^{eq} is density of *j*th atom of A in complex AB and qAj^0 is the density in isolated system.

7 Molecular Modeling

Here, we have first optimized all the molecules and half of the CNT using the same level of theory with DFT as mentioned earlier. We have then calculated the Fukui function for the individual molecules. The results for the global softness, Fukui function, and the local softness values were calculated for the nucleophilic and electrophilic behavior and shown in Table 1, Figs. 9a, b and 10a, b represents the

Table 1 Reactivity parameters for all the interacting molecules

Global Softness, Fukui Function and Local Softness of the Interacting Molecules					
System	Global softness	f_x^+	s_x^+	f_x^-	s_x^-
N_2 (free)	1.956	0.5	0.98	0.5	0.98
N_2 (adsorbed)	1.956	0.22	0.43	0.22	0.43
O_2 (free)	1.589	0.22	0.34	0.22	0.34
O_2 (adsorbed)	1.589	0.24	0.38	0.24	0.38
H_2 (free)	2.150	0.36	0.77	0.36	0.77
H_2 (adsorbed)	2.150	0.07	0.15	0.07	0.15
C of CO_2 (free)	2.666	0.22	0.59	0.45	1.20
C of CO_2 (adsorbed)	2.666	0.30	0.78	0.50	1.30
O of CO_2 (free)	2.666	0.27	0.72	0.39	1.04
O of CO_2 (adsorbed)	2.666	0.01	0.03	0.04	0.11
N of NO_2 (free)	2.680	0.21	0.56	0.21	0.56
N of NO_2 (adsorbed)	2.680	0.36	0.96	0.26	0.69
O of NO_2 (free)	2.680	0.14	0.37	0.14	0.37
O of NO_2 (adsorbed)	2.680	0.03	0.08	0.12	0.32

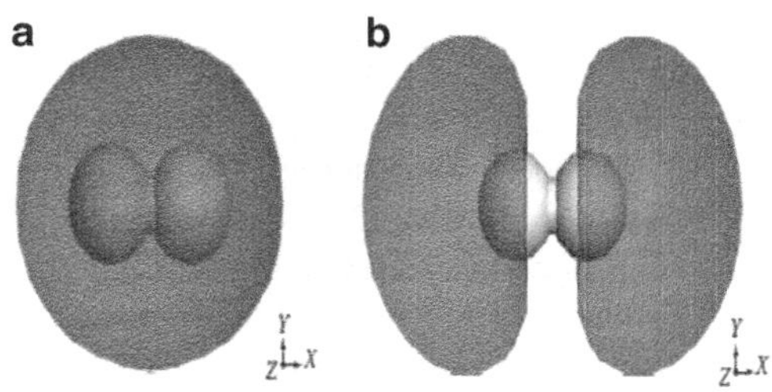

Fig. 9 (**a**) The electrophilic Fukui function of hydrogen is plotted as an isosurface with a grid of 0.2 Å. (**b**) The nucleophilic Fukui function of hydrogen is plotted as an isosurface with a grid of 0.2 Å

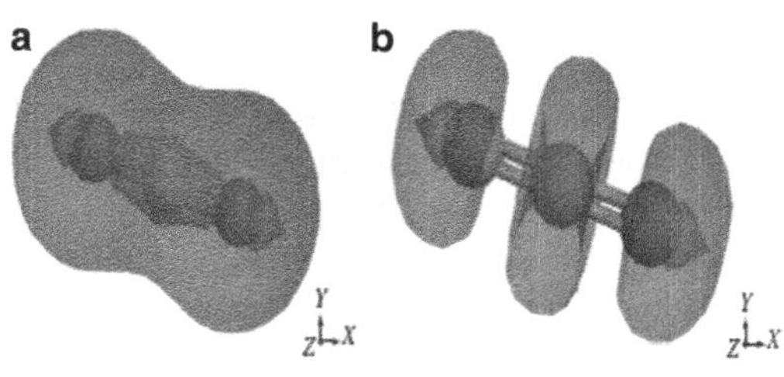

Fig. 10 (**a**) The electrophilic Fukui function of carbon dioxide is plotted as an isosurface with a grid of 0.2 Å. (**b**) The nucleophilic Fukui function of carbon dioxide is plotted as an isosurface with a grid of 0.2 Å

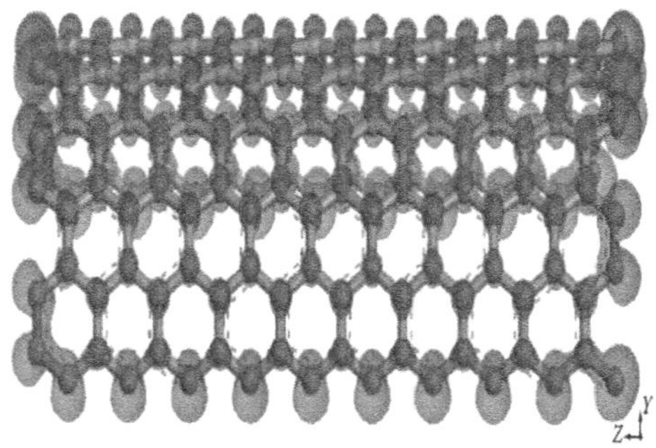

Fig. 11 The electrophilic Fukui function of single wall CNT is plotted as an isosurface with a grid of 0.2 Å. The localized minima as obtained after the GCMS simulation with carbon dioxide adsorption over single wall CNT with a fixed fugacity of 100 kPa

plot of electrophilic and nucleophilic Fukui function for the H_2 and CO_2 molecules, respectively. We have also calculated the same Fukui function and plotted the electrophilic function only for the CNT. A drastic change in the Fukui function values was observed for H_2 and CO_2. Hydrogen shows a lower softness values compared to that of CO_2. The order of activity for all the interacting molecules after adsorption over CNT is $CO_2 > NO_2 > N_2 > O_2 > H_2$. To validate this order we have optimized the half of the nanotube first and monitored the electrophilic Fukui function (Fig. 11). This is followed by the adsorption of H_2 and CO_2 through a grand canonical Monte Carlo (GCMC) simulation methodology using the Sorption tools of Accelrys [59]. The configurations are sampled from a grand canonical ensemble. In the grand canonical ensemble, the fugacity of all components, as well as the temperature, are fixed as if the framework was in open contact with an infinite sorbate reservoir with a fixed temperature. The reservoir is completely described by the temperature plus the fugacity of all components and does not have to be simulated explicitly. The adsorption isotherm for the gas molecule has also been calculated to compare with the experiment. A Langmuir-type isotherm for a fixed pressure of gas was observed. The minimum energy adsorption configuration is shown in Figs. 12 and 13, respectively, for carbon dioxide and hydrogen. We have as well plotted the electrophilic Fukui function for the adsorption complex of carbon dioxide and hydrogen as oriented over the CNT, as shown in Figs. 14 and 15, respectively. We have then taken that geometry of the local minima and optimized the geometry with DFT by using the same level of theory in which we have optimized the whole complex molecule. The next step is the calculation of binding energy in the presence and absence of the electric field. We have calculated the binding energy in presence of electric field of 0.05 a.u. in the same direction with the CNT length. There results of binding energy are shown in Table 2. There is a variation in the order of energy, but the trend remains same. The energy gap looks very narrow for the adsorption complex. Finally, we have performed only the interaction energy calculation using DFT. No change in the order of activity was found, but the numbers are more robust than the binding energy. The order obtained by binding energy follow the same order as obtained from Fukui function calculation. This shows that the method is robust and can be dependable for localized

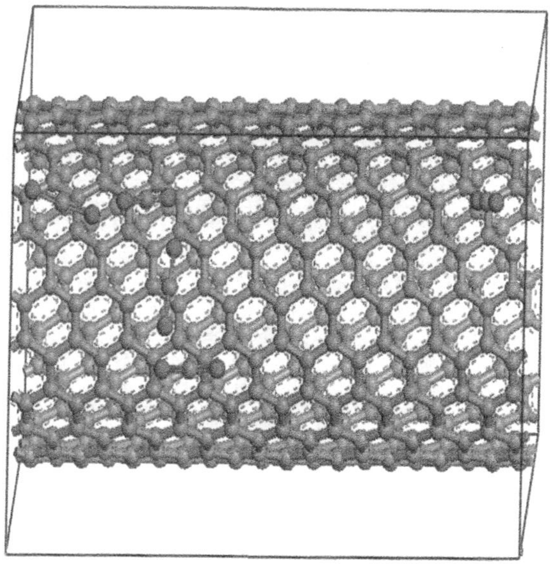

Fig. 12 The localized minima as obtained after the GCMS simulation with carbon dioxide adsorption over single wall CNT with a fixed fugacity of 100 kPa

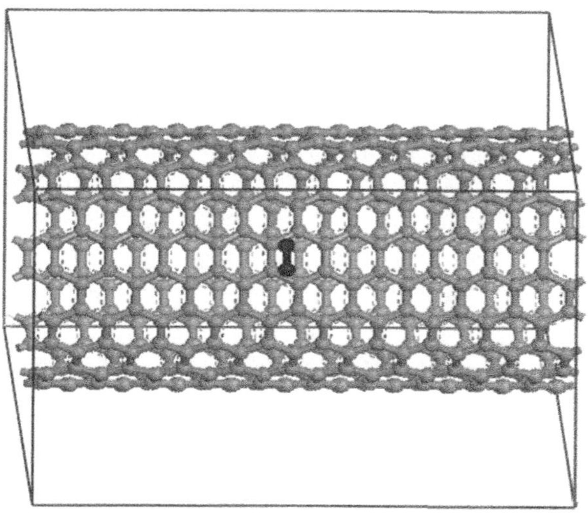

Fig. 13 The localized minima as obtained after the GCMS simulation with hydrogen adsorption over single wall CNT with a fixed fugacity of 100 kPa

interactions. This is an example to design new material for again an application with a recent need where experimentation is hard and designing is difficult.

In summary, we have performed first principles calculations on the electronic properties of a nanotube upon adsorption of gas molecules. We found that all molecules are weakly adsorbed on SWNT with small charge transfer, while they

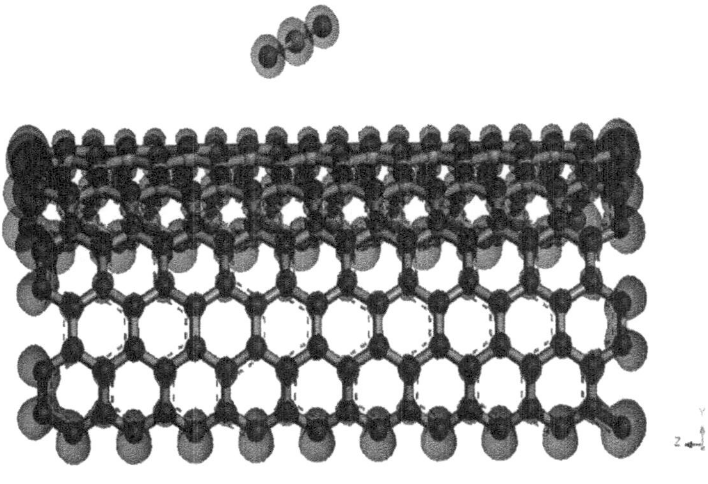

Fig. 14 The electrophilic Fukui function of carbon dioxide adsorbed over single wall CNT is plotted as an isosurface with a grid of 0.2 Å

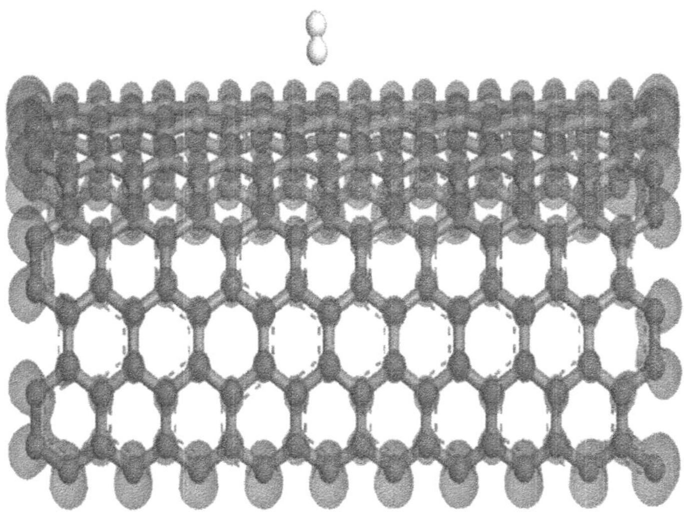

Fig. 15 The electrophilic Fukui function of hydrogen adsorbed over single wall CNT is plotted as an isosurface with a grid of 0.2 Å

can be either a charge donor or an accepter of the nanotube. The adsorption of some gas molecules on SWNTs can cause a significant change in electronic and transport properties of the nanotube due to the charge transfer and charge fluctuation. The molecule adsorption on the surface or inside of the nanotube bundle is stronger than that on an individual tube.

Table 2 Binding energy of the individual molecule over CNT

Binding Energy and the Interaction Energy for the Interacting Molecule with SWNT			
Molecule	Binding energy (eV)		λ (eV)
	Normal	Presence of electric field 0.05 a.u. in the tube direction	
N_2	0.42	0.46	1.56
O_2	0.37	0.54	1.07
H_2	0.15	0.65	0.86
CO_2	0.78	0.18	2.91
NO_2	0.56	0.32	1.81

There are different areas of CNT, but the simulation technology can help design the material in terms of structure property. Carbon nano tubes interacting with metal nanoparticles are gaining considerable interest as sensing materials, catalysts, in the synthesis of metallic nanowires, as well as in nanoelectronics applications as Field-Effect-Transistor (FET) devices. A systematic study of electron-beam-evaporation-coating of suspended SWNT with various metals reveals that the nature of the coating can vary significantly depending upon the metal. Thus, Ti (Titanium), Ni, (Nickel) and Pd (Palladium) form continuous and quasi-continuous coating, while Au (Gold), Al (Aluminum) and Fe (Iron) form only discrete particles on the SWNT surface. In fact, Pd is a unique metal in that it consistently yields good contacts (i.e. low contact resistance) to both metallic and semiconducting nanotubes. For p-doped semiconductors one expects the contact resistance to go down even further if a higher work function metal, e.g., Pt (Platinum) is used. Unfortunately, Pt appears to form poor contacts to both metallic and semiconducting SWNTs with lower p-channel conductance than Pd-contacted junctions. However, the above result is in apparent disagreement with computed interaction energy of a single metal atom on a SWNT, which follows the trend Eb(Pt) > Eb(Pd) > Eb (Au), where Eb denotes the binding energy of the metal atom to the SWNT. It does not explain why Pt consistently makes worse contacts than Pd, and why Ti, in spite of its good wetting of a CNT surface, yields good contacts only rarely [59]. Carbon nanotubes are a hot research area, fuelled by experimental breakthroughs that have led to realistic possibilities of using them in a host of commercial applications: field emission-based flat panel displays, novel semiconducting devices in microelectronics, hydrogen storage devices, chemical sensors, and most recently in ultra-sensitive electromechanical sensors. As a result they represent a real-life application of nanotechnology.

However, two major challenges remain an obstacle to the full commercialization of nanotube-based nanotechnologies and molecular electronic devices:

- The manipulation of individual tubes is difficult owing to their size, and
- The ability to manipulate nanotube properties to suit the application has to be achieved

In semiconducting nanotubes, introducing impurities, a process known as doping, is the main method of tuning properties to make electronic devices. Doping is also a way of creating chemically active impurity sites. Using CASTEP, the researchers [60] found that, at low concentrations of nitrogen impurity (less than 1 atom%), the impurity site becomes chemically and electronically active. In addition, the team found that an inter-tube covalent bond can form between neighboring nanotubes with impurity sites facing each other. With the advance of molecular modeling methodology it is now possible to calculate NMR chemical shifts for nanotubes. The ^{13}C NMR chemical shifts of fluorinated semiconducting single-walled carbon nanotubes (SWNTs) were computed using a gauge including projector augmented plane wave (GIPAW) density functional method. The chemical shifts of the fluorinated carbons (C_α) were rather insensitive to the degree and pattern of functionalization, as well as to the nanotube radius. The calculated shifts were typically between 82 and 84 ppm, which is in excellent agreement with a recent experimental value of 83.5 ppm. Because of the insensitivity of the shifts to details of the nanotube's electronic structure and diameter, the NMR signals of the C_α carbons are a useful indicator of the degree of functionalization in a heterogeneous bulk sample. At high degrees of functionalization, the shifts of atoms neighboring C_α might also be useful indicators of the functionalization pattern [61].

Another important area is in the usage of nanotube in electronic material. Using first-principles density functional theory calculations, two types of junction models constructed from armchair and zigzag carbon nanotube (CNT) insertion into a graphene matrix have been envisioned. It has been found that the insertion of the CNT into the graphene matrix leads to the formation of C–C covalent bonds between graphene and the CNT that distort the CNT geometry. However, the hydrogenation of the suspended carbon bonds on the graphene resumes the graphene-like structure of the pristine tube. The calculated band structure of armchair CNT insertion into graphene or hydrogenation graphene opens up a band gap and converts the metallic CNT into a semiconductor. For the zigzag CNT, the sp^3 hybridization between the graphene and nanotube alters the band structure of the tube significantly, whereas saturating the dangling bonds of terminal carbon atoms of graphene makes the CNT almost keep the same character of the bands as that in the pristine tube. The synthesis of our designed hybrid structures must be increasingly driven by an interest in molecules that not only have intriguing structures but also have special functions such as hydrogen storage [62].

8 Conclusion

In this chapter, we have presented an overview of nanotube with discussion of the material, the electronic property and their applications. We have chosen one specific application of CNT as gas sensors. We have then elaborated the reactivity index theory from concept to industrial application. We have demonstrated that a theory within the DFT domain based on the theory of electro negativity and explored in the realm of electron affinity and ionization potential is capable to deliver a simple correlation to predict the intermolecular and intra-molecular interaction. If one can predict the localized interaction between interacting species carefully, then it will be possible to rationalize many chemical phenomena. The main issue of industry is to reduce cost and to design novel material for a specific application, which is time-consuming due to the trial and error process involved in this and as well expensive. They need a reliable as well faster way to screen the reactants and propose the products, which can be handled well by computer simulation technology with current reactivity index. We here have tried to share with you its capability through the CNT application examples to show that reactivity is an emerging area for material designing from nanocluster through nanowire, nanotube to biomaterial applications. The effort will only be successful if one believes in this and tries to explore around to make it more robust and develop the way to apply this unique theory to all possible material of choice and interest. We have also shared some of the key features of the newest first principle calculation to design utility for CNT matrices.

References

1. Iijima, S.: Nature **354**, 56–58 (1991)
2. Iijima, S., Ichihashi, T.: Nature **363**, 603–605 (1993)
3. Bethune, D.S., Klang, C.H., de Vries, M.S., Gorman, G., Savoy, R., Vazquez, J., Beyers, R.: Nature **363**, 605–607 (1993)
4. Saito, R., Dresselhaus, G., Dresselhaus, M.S.: Physical Properties of Carbon Nanotubes. Imperial College Press, London (1998)
5. Cheng, H., Pez, G.P., Cooper, A.C.: Nano Lett. **3**, 585–587 (2003)
6. Inagaki, M.: J. Mater. Res. **4**, 1560 (1989)
7. Bartlett, N., McQuillan, B.W.: In: Whittingham, M.S., Jacobson, A.J. (eds.) Intercalation Chemistry. Academic, New York (1982)
8. Ajayan, P.M., Iijima, S.: Nature **361**, 333–334 (1993)
9. Chambers, A., Park, C., Baker, R.T.K., Nelly Rodriguez, M.: J. Phys. Chem. B **102**, 4253–4256 (1998)
10. Shimoda, H., Gao, B., Tang, X.P., Kleinhammes, A., Fleming, L., Wu, Y., Zhou, O.: Phys. Rev. Lett. **88**, 155021 (2002)
11. De Heer, W.A., Chatelain, A., Ugarte, D.: Science **270**, 1179–1180 (1995)
12. Kong, J., Franklin, N.R., Zhou, C.W., Chapline, M.G., Peng, S., Cho, K.J., Dai, H.J.: Science **287**, 622–625 (2000)
13. Collins, P.G., Bradley, K., Ishigami, M., Zettl, A.: Science **287**, 1801–1804 (2000)
14. Jakubek, J.Z., Simard, B.: Langmuir **20**, 5940–5945 (2004)
15. Andzelm, J., Govind, N., Maiti, A.: Chem. Phys. Lett. **421**, 58–62 (2006)
16. Tombler, T.W., Zhou, C., Kong, J., Dai, H.: Appl. Phys. Lett. **76**, 2412–2416 (2000)
17. Zhao, J., Buldum, A., Han, J., Lu, J.P.: Nanotechnology **13**, 195–200 (2002)
18. Frenkel, D., Smit, B.: Understanding Molecular Simulation: From Algorithms to Applications, 2nd edn. Academic, San Diego (2002)
19. Wang, Q., Johnson, J.K.: J. Chem. Phys. **110**, 577 (1999)
20. Jian, J., Sandler, S.I.: Phys. Rev. B **68**, 245412 (2003)
21. Kohn, W., Sham, L.J.: Phys. Rev. A **140**, A1133–A1138 (1965)
22. Delley, B.: J. Chem. Phys. **92**, 508–517 (1990)
23. Delley, B.: J. Chem. Phys. **94**, 7245–7250 (1991)
24. Delley, B.: J. Chem. Phys. **113**, 7756–7764 (2000)
25. Delley, B.: J. Phys. Chem. **100**, 6107–6110 (1996)
26. Becke, A.D.: J. Chem. Phys. **88**, 2547–2553 (1988)

27. Lee, C.T., Yang, W.T., Parr, R.G.: Phys. Rev. B **37**, 785–789 (1988)
28. Pearson, R.G.: J. Chem. Educ. **64**, 561–567 (1987)
29. Parr, R.G., Pearson, R.G.: J. Am. Chem. Soc. **105**, 7512–7516 (1983)
30. Geerlings, P., De Proft, F.: Int. J. Quant. Chem. **80**, 227–236 (2000)
31. Parr, R.G., Yang, W.: J. Am. Chem. Soc. **106**, 4049–4050 (1984)
32. Ayers, P.W., Levy, M.: Theor. Chem. Acc. **103**, 353–360 (2000)
33. Chattaraj, P., Lee, K.H., Parr, R.G.: J. Am. Chem. Soc. **113**, 1855–1856 (1991)
34. Baeten, A., Tafazoli, M., Kirsch-Volders, M., Geerlings, P.: Int. J. Quant. Chem. **74**, 351–355 (1999)
35. Mendez, F., Romero, MdL, De Proft, F., Geerlings, P.: J. Org. Chem. **63**, 5774–5778 (1998)
36. Mendez, F., Tamariz, J., Geerlings, P.: J. Phys. Chem. A **102**, 6292–6296 (1998)
37. Ayers, P.W., Parr, R.G., Pearson, R.G.: J. Chem. Phys. **124**, 194107 (2006)
38. Ayers, P.W.: J. Chem. Phys. **122**, 141102 (2005)
39. Ayers, P.W.: **135**, 161–190 (2007)
40. Parr, R.G., Chattaraj, P.K.: J. Am. Chem. Soc. **113**, 1854–1855 (1991)
41. Pearson, R.G.: Int. J. Quant. Chem. **56**, 211–215 (2004)
42. Chattaraj, P.K.: Indian Natl. Sci. Acad. A **62**, 513–519 (1996)
43. Chattaraj, P.K., Parr, R.G., Sen, K.D., Mingos, D.M.P.: Springer, Berlin, vol. 80, pp. 11–25 (1993)
44. Ayers, P.W., Parr, R.G.: J. Am. Chem. Soc. **122**, 2010–2018 (2000)
45. Torrent-Sucarrat, M., Luis, J.M., Duran, M., Sola, M.: J. Chem. Phys. **117**, 10561–10570 (2002)
46. Torrent-Sucarrat, M., Luis, J.M., Duran, M., Sola, M.: J. Am. Chem. Soc. **123**, 7951–7952 (2001)
47. Parr, R.G., Yang, W.T.: Annu. Rev. Phys. Chem. **46**, 701–728 (1995)
48. Parr, R.G., Yang, W.T.: Density-Functional Theory of Atoms and Molecules. Oxford University Press, New York (1989)
49. Geerlings, P., De Proft, F., Langenaeker, W.: Chem. Rev. **103**, 1793–1873 (2003)
50. Chermette, H.: J. Comp. Chem. **20**, 129–154 (1999)
51. Ayers, P.W., Anderson, J.S.M., Bartolotti, L.J.: Int. J. Quant. Chem. **101**, 520–534 (2005)
52. Chatterjee, A., Iwasaki, T., Ebina, T.: J. Phys. Chem. A **103**, 2489–2494 (1999)
53. Chatterjee, A., Ebina, T., Iwasaki, T.: J. Phys. Chem. A **105**, 10694–10701 (2001)
54. Chatterjee, A., Iwasaki, T.: J. Phys. Chem. A **105**, 6187–6196 (2001)
55. Chatterjee, A., Iwasaki, T., Ebina, T., Mizukami, F.: J. Chem. Phys. **118**, 10212–10220 (2003)
56. Chatterjee, A., Onodera, Y., Ebina, T., Mizukami, F.: J. Mol. Graph. Model. **22**, 93–104 (2003)
57. Chatterjee, A., Suzuki, T., Onodera, Y., Tanaka, D.A.P.: Chemistry **9**, 3920–3929 (2003)
58. Chatterjee, A., Onodera, Y., Ebina, T., Mizukami, F.: J. Chem. Phys. **120**, 3414–3424 (2004)
59. Maiti, A., Ricca, A.: Chem. Phys. Lett. **395**, 7–11 (2004)
60. Nevidomskyy, A.H., Csa´nyi, G., Payne, M.C.: Phys. Rev. Lett. **91**(10), 105502 (2003)
61. Zurek, E., Pickard, C.J., Autschbach, J.: J. Phys. Chem. A **113**(16), 4117–4124 (2009)
62. Mao, Y., Zhong, J.: New J. Phys. **11**, 093002–093012 (2009)

Carbon Nanostructured Materials

Azira Abdul Aziz, Suriani Abu Bakar, and Mohamad Rusop

Abstract In recent years, a lot of work has been focused on the synthesis of novel materials, clusters, and molecules which are unique in many ways. Numerous attempts to synthesize the theoretically predicted solids have been published. This chapter summarised the carbon materials in various forms; crystalline and non-crystalline. Carbon constitutes a class of new materials with a wide range of compositions, properties, and performance. Due to its unique optical and electrical properties, carbon has potential applications in vast fields especially in semiconductor devices. The structure and properties of the various crystalline carbon materials are reviewed. Related carbon based materials such as fullerenes, carbon fiber, glassy carbon, carbon black, amorphous carbon, diamond, graphite and buckminsterfullerene mentioned briefly as well as carbon nanotubes (CNTs). The CNTs preparation and characterization methods are presented and discussed in depth. However, it can be stated that a fascinating new field in the area of carbon has been discovered, which gives motivation for further studies dedicated to fundamental questions as well as the exploitation of the novel materials for industrial applications.

A.A. Azira (✉) and S.A. Bakar
NANO-SciTech Centre, Institute of Science, Universiti Teknologi MARA, 40450 Shah Alam, Selangor, Malaysia

Faculty of Applied Sciences, Universiti Teknologi MARA, 40450 Shah Alam, Selangor, Malaysia
e-mail: aziraaziz@yahoo.com; absuriani@yahoo.com

M. Rusop
NANO-SciTech Centre, Institute of Science, Universiti Teknologi MARA, 40450 Shah Alam, Selangor, Malaysia
Faculty of Electrical Engineering, Solar Cell Laboratory, Universiti Teknologi MARA, 40450 Shah Alam, Selangor, Malaysia
e-mail: rusop8@gmail.com

N. Yahya (ed.), *Carbon and Oxide Nanostructures*, Adv Struct Mater 5,
DOI 10.1007/8611_2010_14, © Springer-Verlag Berlin Heidelberg 2010,
Published online: 26 June 2010

1 Introduction

Carbon has been known since ancient times in the form of soot, charcoal, graphite and diamonds. Ancient cultures did not of course realize that these substances were different forms of the same element. 'Carbon' is derived from the Latin carbo, meaning charcoal. Carbon based materials, clusters and molecules are unique in many ways and allotropes of carbon are inter convertible to each other suitable temperature and pressure. Carbon is most commonly obtained from coal deposits, although it usually must be processed into a form suitable for commercial use. Three naturally occurring allotropes of carbon are known to exist: amorphous, graphite and diamond. Under ambient condition, the graphite phase with strong in-plane trigonal bonding is a stable phase. Under high pressure (60,000 atm) and temperature (2,000 K) graphite can be converted to diamond and when exposed to irradiation or heat, diamond will quickly transform back to the more stable graphite phase. Carbon has atomic number of 6 and is classified in group IV of the second period of the periodic table and has $1s^2 2s^2 2p_x{}^1 p_y{}^1$ electronic ground state configuration. In the graphite structure, strong in plane bonds are formed which is denoted by trihedral sp^2 and in diamond structure, they are tetrahedrally bonded sp^3 configuration.

2 Carbon Structures

Carbon is a fascinating and very unique element because it can assume various forms and structures. It is very abundant and is the basis of organic life. Carbon has two features which, taken together, make it quite unique: a carbon atom can bond with another carbon atom in several configurations (different hybridizations of the C-C bond), and can also bond with many other elements, among those hydrogen, nitrogen and oxygen. In order to understand the nature of the carbon bond it is necessary to examine the electronic structure of the carbon atom. Carbon contains six electrons, which are distributed over the lowest energy levels of the carbon atom. The structure is designated as follows ($1s^2$), (2s), ($2p_x$), ($2p_y$), ($2p_z$) when bonded to atoms in molecules electron. The configuration of ground state (lowest energy state) of carbon is shown in Fig. 1 below.

The lowest energy level 1s with the quantum number N = 1 contains two electrons with oppositely paired electron spins. The electron charge distribution in an s state is spherically symmetric about the nucleus. The 1s electrons do not participate in the chemical bonding. The next four electrons are in the N = 2 energy state, one in a spherically symmetric s orbital, and three in p_x, p_y, and p_z orbitals,

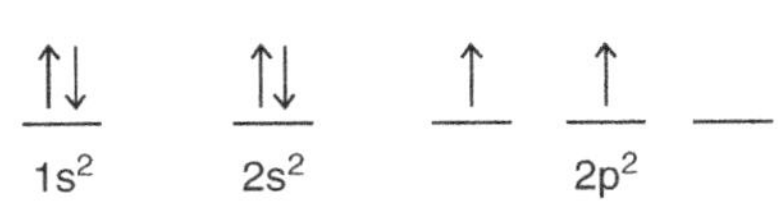

Fig. 1 The configuration of ground state (lowest energy state) of carbon

which have the very directed charge distributions shown in Fig. 1, oriented perpendicular to each other. The outer s orbital together with the three p orbitals form the chemical bonds of carbon with other atoms. The charge distribution associated with these orbitals mixes (or overlaps) with the charge distribution of each other atom being bonded to the carbon. In effect, the electron charge between the two carbon atoms of a bond can be viewed as the glue that holds the atoms together.

From the ground state electron configuration, one can see that carbon has four valence electrons, two in the 2s subshell and two in the 2p subshell. The 1s electrons are considered to be core electrons and are not available for bonding. There are two unpaired electrons in the 2p subshell, so if carbon were to hybridize from this ground state, it would be able to form at most two bonds. Recall that energy is released when bonds form, so it would be to carbon's benefit to try to maximize the number of bonds it can form. For this reason, carbon will form an excited state by promoting one of its 2s electrons into its empty 2p orbital and hybridize from the excited state. By forming this excited state, carbon will be able to form four bonds. The excited state configuration is shown in Fig. 2 below.

In order to determine the hybridization on a carbon atom, Lewis structure must be drawn. From the Lewis structure, the numbers of groups around the central carbon need to be counted. A group represents the regions of electron density around the carbon, and may be single, double or triple bonded. The number of groups represents how many hybrid orbitals have formed. The number of hybrid orbitals formed equals the number of atomic orbitals mixed. The description of the atomic orbitals mixed is equivalent to the hybridization of the carbon atom.

The Lewis structure shows four groups around the carbon atom. This means four hybrid orbitals have formed. In order to form four hybrid orbitals, four atomic orbitals have been mixed. The s orbital and all three p orbitals have been mixed, thus the hybridization is sp^3. By using the atomic orbitals of excited state carbon found in the valence shell. The four sp^3 hybrid orbitals will arrange themselves in three dimensional space to get as far apart as possible (to minimize repulsion). The geometry that achieves this is tetrahedral geometry, where any bond angle is 109.5° as described in Fig. 3.

Each hybrid orbital contains one electron. A hydrogen 1s orbital will come in and overlap with the hybrid orbital to form a sigma bond (head-on overlap). The Lewis structure shows three groups around each carbon atom. This means three hybrid orbitals have formed for each carbon. In order to form three hybrid orbitals, three atomic orbitals have been mixed. The s orbital and two of the p orbitals for each carbon have been mixed, thus the hybridization for each carbon is sp^2. By referring to Fig. 4, using the atomic orbitals of excited state carbon found in the valence shell.

The three sp^2 hybrid orbitals will arrange themselves in three dimensional space to get as far apart as possible. The geometry that achieves this is trigonal planar

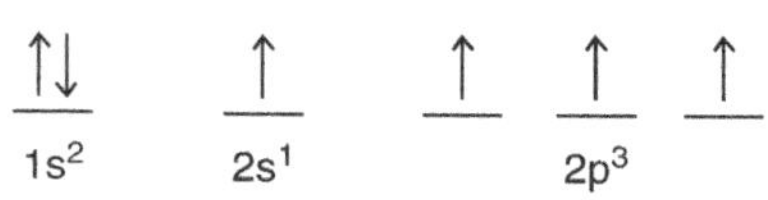

Fig. 2 The excited state configuration of carbon

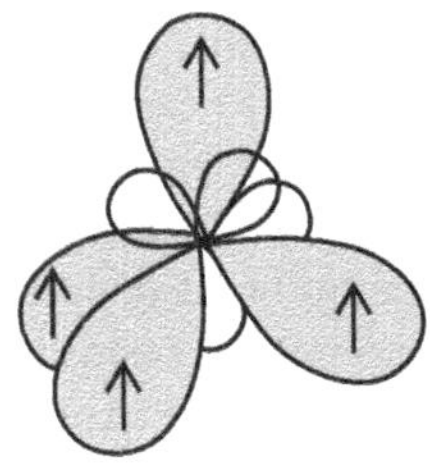

Fig. 3 The sp^3 hybridization structure (three-dimension)

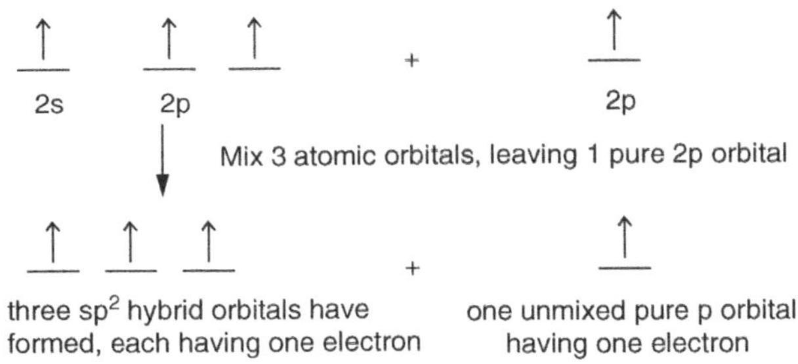

Fig. 4 The schematic diagram of sp^2 hybrid orbitals

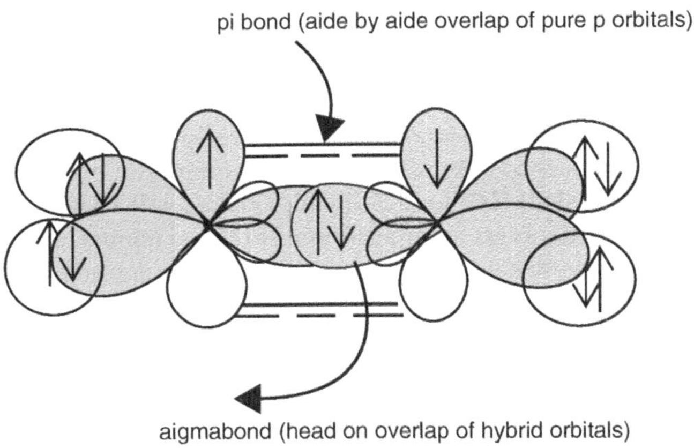

Fig. 5 The head-on overlap of sp^2 orbitals

geometry, where the bond angle between the hybrid orbitals is 120°. The unmixed pure p orbital will be perpendicular to this plane and each carbon atom is sp^2, and trigonal planar as shown in Fig. 5.

The head-on overlap of sp^2 orbitals forms a bond and the side by side overlap of pure p orbitals forms a pi bond between the carbon atoms. This accounts for the carbon-carbon double bond. Each carbon is trigonal planar with a bond angle of 120°. By referring to acetylene, C_2H_2 the Lewis structure, shows two groups around

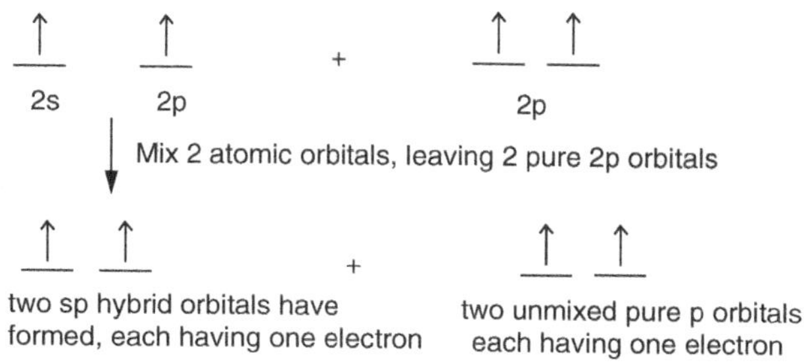

Fig. 6 The schematic diagram of two sp hybrid orbitals

each carbon atom. This means two hybrid orbitals have formed. In order to form two hybrid orbitals, two atomic orbitals have been mixed. By using the atomic orbitals of excited state carbon found in the valence shell as shown in Fig. 6.

The two sp hybrid orbitals arrange themselves in three dimensional space to get as far apart as possible. The geometry which achieves is linear geometry with a bond angle of 180°. The two pure p orbitals which were not mixed are perpendicular to each other. The triple bond consists of one sigma bond and two pi bonds. The geometry around each carbon is linear with a bond angle of 180°.

Solid carbon has two main structures called allotropic forms that are stable at room temperature: diamond and graphite. Diamond consists of carbon atoms that are tetrahedrally bonded to each other through sp^3 hybrid bonds that form a three dimensional network. Each carbon has four nearest-neighbour carbons. Graphite has a layered structure with each layer, called a graphite sheet, formed from hexagons of carbon atoms bound together by sp^2 hybrid bonds that make 120° angles with each other. Each carbon atom has three nearest-neighbour carbons in the planar layer. The hexagonal sheets are held together by weaker van der Waals forces. Van der Waals forces exists from the clusters of gases tend to be larger because of their atoms have closed shells that are held together by much weaker forces.

Crystalline carbon can be found in essentially two forms in nature, namely graphite and diamond. They correspond to two different ways of forming a bond between carbon atoms, namely the sp^2 (typical of graphite, with three nearest neighbours arranged in the same plane) and sp^3 bond (typical of diamond, with four nearest neighbours located at the tips of a regular tetrahedron). Although they are both formed of pure carbon, their chemical and physical properties are very different and in some aspects completely opposite. In graphite, the planar graphene sheets of sp^2-bonded carbon atoms glide easily in a direction parallel to the planes, resulting in a very soft material while diamond is among the hardest materials known. Graphite is a zero gap semiconductor while diamond is a high band gap semiconductor. Graphite is opaque, while diamond is transparent. It exists in different allotropic forms that give rise to its versatile behavior. In the amorphous form it is powdery in nature and black in color. It becomes the hardest substance known and has a shining appearance in the allotropic form of diamond. Catenation, the self-linking property of carbon atoms is responsible for much of organic

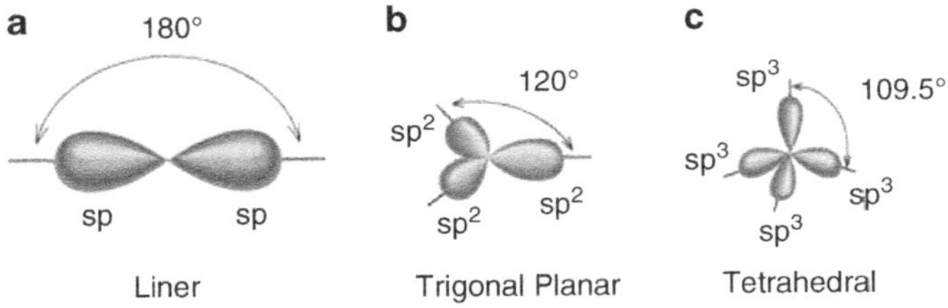

Fig. 7 The different hybridisations of carbon (**a**) sp^1, (**b**) sp^2, (**c**) sp^3

chemistry. The role of carbon clusters and carbon clouds in the interstellar region and in atmosphere remains to be understood [1]. In 1980, only three forms of carbon, namely diamond, graphite and amorphous (non-crystalline carbon).

The chemical element carbon can combine with itself and other elements in three types of hybridisations. This gives the rich diversity of structural forms of solid carbon and is the basis of organic chemistry and life. In the sp^3 hybridisation four equivalent $2sp^3$ hybrid orbitals are tetrahedrally oriented around the atom (Fig. 7) and can form four equivalent σ bonds by an overlap with orbitals of other atoms. An example is the ethane molecule (C_2H_6) where a Csp^3-Csp^3 σ bond (or C-C) is formed between two carbon atoms by the overlap of sp^3 orbitals, and three Csp^3-H1s σ bonds are formed on each C atom. In the sp^2 hybridisation three equivalent $2sp^2$ orbitals are formed and one unhybridised 2p orbital is left. They are coplanar and oriented at 120° to each other and form σ bonds by an overlap with orbitals of neighbouring atoms as e.g. in ethane (C_2H_4). The remaining p orbitals on each C atom form a π bond by the overlap with the π orbital from the neighbouring C atom. Such bonds formed between two C atoms are represented as $Csp^2{=}Csp^2$ (or C=C). Figure 7 shows the different hybridisations of carbon sp^1, sp^2 and sp^3.

In the sp^1 hybridisation two linear $2sp^1$ orbitals are formed and two 2p orbitals are left. Linear σ bonds are formed by the overlap of the $2sp^1$ hydride orbitals of neighbouring atoms as for example in the ethyne molecule (acetylene). Two π bonds are formed with the overlapping unhybridised π orbitals of the two C atoms. These bonds are represented as $Csp \equiv Csp$ (or $C \equiv C$). In the aromatic carbon-carbon bond exemplified by the aromatic molecule benzene (C_6H_6) the carbon atoms are bonded with sp^2 σ bonds in a regular hexagon. The ground state π orbitals are all bonding orbitals and are fully occupied; there is a large delocalisation energy that contributes to the stability of the molecule. The aromatic carbon-carbon bond is denoted as $Car \cong Car$.

3 New Carbon Structures

Until 1964 it was generally believed that no other carbon bond angles were possible in hydrocarbon, that is, compound containing only carbon and hydrogen atoms. In that year Phil Eaton of the University of Chicago synthesized a square carbon

Table 1 Types of sp^n hybridization, the resulting bond angles, and examples of molecules

Types of hybridization	Diagonal sp	Trigonal sp^2	Tetrahedral sp^3
Orbital used for bond	s, p_x	s, p_x, p_y	s, p_x, p_y, p_z
Example	Acetylene C_2H_2	Ethylene C_2H_4	Methane CH_4
Bond angle	109°28′	180°	120°

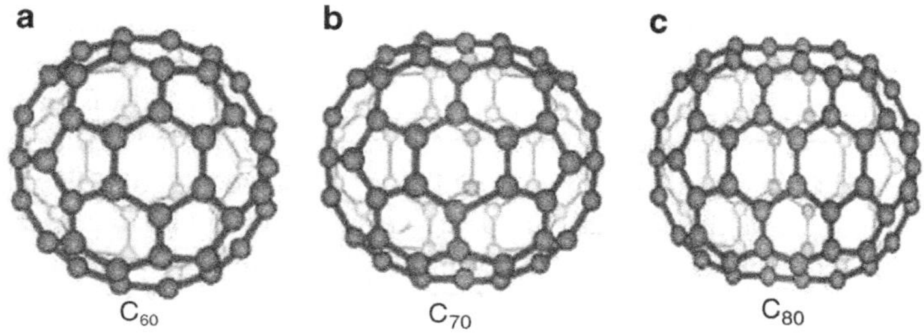

Fig. 8 Structure of (**a**) fullerene, C_{60}, (**b**) C_{70} and (**c**) C_{80} [2]

molecule, C_8H_8, called cubane. In 1983, L. Paquette of Ohio State University synthesized a $C_{20}H_{20}$ molecule having a dodecahedron shape, formed by joining carbon pentagons, and having C-C bond angles ranging from 108° to 110°. The synthesis of these hydrocarbon molecules with carbon bond angles different from standard hybridization values of Table 1 has important implication for the formation of carbon nanostructures, which would also require different bonding angles.

3.1 Fullerenes

Today there are whole families of other forms of carbon. Laser evaporation of a carbon substrate using the apparatus in a pulse of He gas can be used to make carbon clusters. The neutral cluster beam is photoionized by a UV laser and analyzed by mass spectrometer. The first of these to be discovered was buckminsterfullerene (also called buckyball and fullerene C_{60}). The discovery of fullerene [2], a new form of carbon, was perhaps a serendipity. But it led to a number of other fundamental discoveries. Fullerenes of various sizes and shapes have been reported subsequently. It would be fair to say that the discovery of carbon nanotubes in 1991 was a by-product of the fullerene production process [3].

The discovery of the existence of a soccer-ball-like molecule containing 60 carbon atoms was named fullerene. Fullerene C_{60}, (buckyball), is the first spherical carbon molecule with carbons arranged in a soccer ball shape (Fig. 8). In the structure there are 60 carbon atoms (hence C_{60}) and a number of five-membered rings isolated by six-membered rings [4]. It may well be that these objects can be used as ball bearings in some of Drexler's mechanical devices. The second spherical carbon molecule in the same group is the rugby ball, C_{70}, whose structure has extra six-membered carbon rings (Fig. 8), but there are also a large number of other potential structures containing the same number of carbon atoms (isomers) depending on whether five-membered rings are isolated or not, or whether seven-membered rings are present. Many other forms of fullerenes up to and beyond C_{120} have been characterized and it is possible to draw lots of structure with five-membered rings in different positions and sometimes together.

The important fact for nanotechnology is that the atom can be placed inside the fullerene ball. Atoms contained within the fullerene are aid to be endohedral and they can also be bound to fullerenes outside the ball as salts if the fullerene can gain electrons. The structure is then $M_x^{+}C_{60}^{n-}$, where M^+ is a cation and x is the number to balance the charge on the fullerene. In this case the cation is said to be exahedral [5–8].

A sketch of the molecule is shown in Fig. 8. It has 12 pentagonal (five sided) and 20 hexagonal (six sided) faces symmetrically arrayed to form a molecular ball. The ball-like molecules bind with each other in the solid state to form a crystal lattice having a face centered cubic structure. In the lattice each C_{60} is separated from its nearest neighbour by 1 nm (the distance between their centers is 1 nm), and they are held together by van der Waals forces. Larger fullerenes such as C_{70}, C_{76}, C_{80} and C_{84} have also been found. The interesting about the fullerene is, in practical terms, they many have a number of applications. For example, they have been used as lubricants because the tiny balls can roll between surfaces (it turns out that pure fullerenes are not good for this; they must be changed chemically first by having other atoms bonded around the ball). Also, they turn out to have strong optical effects (i.e. they change their properties upon irradiation with light, UV, in most cases), which could be useful in photolithography.

3.2 Carbon Fibers

Carbon fibers represent important class of graphite-related materials. Many precursors can be used to synthesize carbon fibers with high mechanical strength, each having different cross-sectional morphologies, when the as-prepared vapor-grown fibers were heat-treated to 3,000°C, forms facets, of all carbon fibers. These faceted are closest to crystalline graphite in both crystal structure and properties [9].

3.3 Glassy Carbon

Glassy carbon (GC) is another common material, which is manufactured as a commercial product by slow, controlled degradation of certain polymers at temperature typically on the order of 900–1,000°C [10]. The name glassy carbon is thus given to family of disordered carbon materials, which are glass-like and can be easily polished to attain a black, shiny appearance.

3.4 Carbon Black

Classical carbon blacks represent many types of finely divided carbon particles that are produced by hydrocarbon dehydrogenation [10, 11] and are widely used in industry as a filler to modify the mechanical, electrical, and optical properties of the materials in which they are dispersed [10, 12]. Various types of industrial carbon blacks are produced by various methods for example; thermal blacks are typically obtained by thermal decomposition of nature gas, channel blacks by partial combustion of natural gas, etc. One of the characteristic signatures associated with carbon blacks is a concentric organization of the graphite layers in each individual particle.

3.5 Amorphous Carbon

In addition to the crystal carbon, non-crystalline carbon constitutes a class of new materials with a wide range of composition, properties, and performance. This field of non-crystalline carbon is of interest both technologically to materials scientists and also at a more fundamental level to solid-state chemists and physicists. The physical properties of the various non-crystalline forms of carbon are compared with those of diamond, graphite and C_{60} in Table 2.

Non-crystalline carbon mainly include amorphous carbon (a-C) and amorphous hydrogenated carbon (a-C:H). a-C and A-C:H refer to highly disordered network of carbon atoms having no long-range order, but some short-range order, can be considered to be intermediate between diamond, graphite and hydrocarbon polymers, in that they can contain variable amounts of sp^2 and sp^3 sites and hydrogen. Since the nature of short-range order varies significantly from one preparation method to another, the properties of amorphous carbon likewise vary according to preparation method [13]. Two parameters, the carbon bonding and the hydrogen content, are most sensitive for characterizing the short-range order, which many exist on a length scale of ~10 A. Thus the sp^2-bonded carbons of a-C may cluster into tiny warped layered regions; the sp^3-bonded carbon may also cluster and segregate, as may the hydrogen impurities, which are very effective in passivating the dangling bonds.

Table 2 Properties of various forms of carbon

Materials	Density (g cm^{-3})	Hardness (GPa)	% sp^3	at. %H	Energy (Gap eV)
Diamond	3.515	100	100		5.5
Graphite	2.267		0		0.04
C_{60}			0	0	1.8
Glassy C	1.3–1.55	2–3	~0		0.01
a-C, evap	1.9–2.0	2–5	1		0.4–0.7
a-C, MSIB	3.0	30–130	90 ± 5	<9	0.5–1.5
Pda-C:H, hard	1.6–2.2	10–20	30–60	10–40	0.8–1.7
Pda-C:H, soft	0.9–1.6	<5	50–80	40–65	1.6–4
Polyethylene	0.92	0.01	100	67	6

The introduction of disorder and sp^3 defects creates a semiconductor with localized states near the Fermi level and an effective band gap between mobile filled valence band states and empty conduction states. The greater is the disorder, the greater the concentration of sp^3 bonds.

a-C and a-C:H have been prepared by variety of deposition method, such us chemical vapor deposition, ion beam sputtering, laser plasma etc. The properties of the amorphous carbon are strongly dependent on the deposition technique, e.g. amorphous carbon prepared by evaporation tends to have smaller band gap ($E_g \sim 0.4$–0.7 eV) compared with that of prepared by ion deposition for which the E_g is in the range of range of 0.4–0.3 eV [14, 15].

3.6 Diamond

Diamond exists in a cubic and hexagonal form (Lonsdaleite). In the most frequent cubic form each carbon atom is linked with four other carbon atoms by four sp^3 σ bonds in a tetrahedral array with a C-C bond length of 1.544 Å [16]. This is nearly 10% larger than in graphite. However the atomic density (1.77×10^{23} cm^{-3}) is 56% higher than in graphite. The crystal structure is zinc blend type (FCC) with a diatomic basis. The second carbon atom is at position (¼, ¼, ¼) in the unit cell and the lattice constant is $a_0 = 3.567$ Å (Fig. 9a). The physical properties of diamond are given by its structure.

Diamond is a wide-gap semiconductor (5.47 eV), the hardest material in nature (Most hardness 10) and has the highest atomic density. Diamond, as also graphite (in-plane) have the highest thermal conductivity (~25 W cm^{-1} K^{-1}) and the highest melting point (4,500 K). The hexagonal diamond (Lonsdaleite) has a wurtzite crystal structure (Fig. 9, right) and a C-C bond length of 1.52 Å [16]. The gravimetric density of both types of diamond is 3.52 g cm^{-3}.

3.7 Graphite

In graphite the atoms are arranged in layers of a honeycomb network in which the carbon atoms are bonded with sp^2 σ bonds and a delocalized π bond. In the most common hexagonal crystal form of graphite the layers are stacked in an ABAB... sequence (called Bernal stacking) (Fig. 10). The in-plane nearest neighbour distance aC-C is 1.421 Å [17] and the lattice constant is $a_0 = 2.461$ Å. The c-axis lattice constant is $c_0 = 6.708$ Å and the interplanar distance $c_0/2$. A minor component of well-crystallised graphite is the rhombohedral form of graphite in which the graphene (single layer of crystalline graphite) layers are stacked in the ABCABC... sequence. The lattice constant is also $a_0 = 2.456$ Å and $c_0 = 3$ $(3.438) = 10.044$ Å. The Bernal AB stacking of graphite is more stable than the ABC stacking. The density of both forms of graphite is 2.26 g cm^{-3} [18]. The weak interlayer bonding of graphite originates from the small overlap of the π-orbitals between atoms of adjacent layers and not to Van der Waals bonding [19].

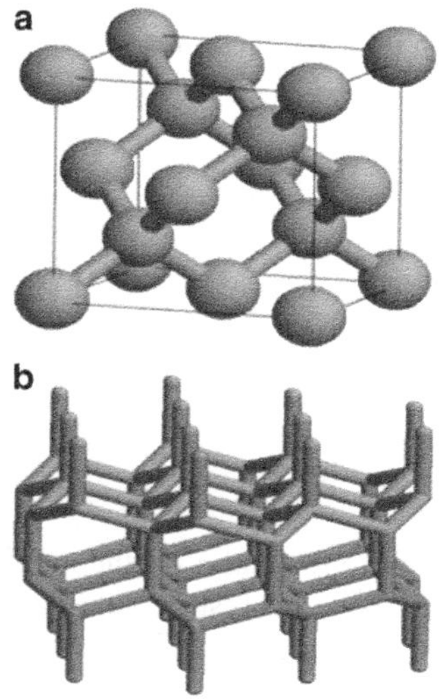

Fig. 9 Diamond in (**a**) cubic form and (**b**) hexagonal Lonsdaleite

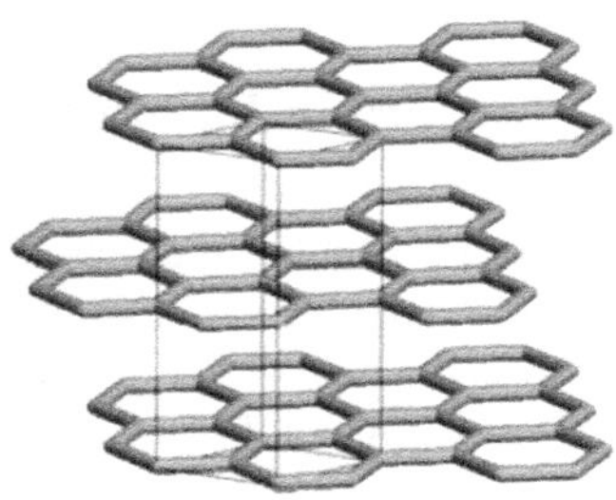

Fig. 10 Hexagonal graphite (ABAB stacking) with unit cell

3.8 Buckminsterfullerenes

Experimental and theoretical work has shown that the most stable form of carbon clusters form linear chains [20] for clusters of up to about 10 atoms. For clusters that have 10 to 30 carbon atoms the ring is the most stable form [21]. Carbon clusters between 30 and 40 carbon atoms are unlikely and clusters above 40 atoms form cage structures. Especially stable structures are the C_{60} (Fig. 11), whose structure was identified the first time by Kroto et al. in 1985 [8]. The carbon atoms are located at the 60 vertices of a truncated icosahedron that has 90 edges and 32 faces of which 12 are pentagons and 20 hexagons, consistent with Euler's theorem for polyhedra:

$$f + v = e + 2 \tag{1}$$

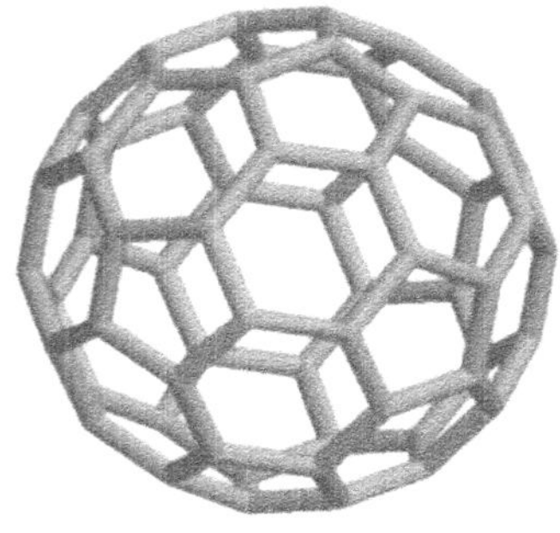

Fig. 11 C_{60} Buckminsterfullerene

where *f*, *v* and *e* are the number of faces, vertices, and edges of the polyhedra. The average nearest-neighbour C-C distance is with aC-C = 1.44 Å almost equal to that in graphite. Each carbon atom is trigonally bonded to three other carbon atoms in an sp^2-derived bonding configuration. The curvature of the trigonal bonds in C_{60} leads to some admixture of sp^3 bonding, characteristic for tetrahedrally bonded diamond, but absent in graphite [22]. Further stable fullerenes are C_{70}, C_{78} and C_{80}.

3.9 *Carbon Nanotubes*

Carbon nanotubes are cylindrical molecules ~1 nm in diameter and 1–100 μm in length. They are constituted of a hexagonal network of carbon atoms and can essentially be thought of as a layer of graphite rolled-up into a cylinder [3]. Properties of the nanotube researched include the different arrangements of the nanotube hexagonal structure such as the zigzag and the armchair. Carbon nanotubes are the strongest fibers known. A single perfect nanotube is about 10–100 times stronger than steel per unit weight. A carbon nanotube is synthesized through three methods: the carbon arc method where ordered nanotubes are produced from ionized carbon plasma, the laser vaporization of carbon targets and catalytic vapor deposition. A new method proposed by the authors for growing carbon nanotubes from renewable material is fluidized floating catalyst method [23–27]. Currently, carbon nanotubes can be applied as semiconductors for transistors, space transportation and data storage for computers. The carbon nanotube is still currently under heavy research and many possible applications are being conceived. Carbon nanotubes are constituted of a hexagonal network of carbon atoms and can essentially be thought of as a layer of graphite rolled-up into a cylinder (a commonly mentioned noncarbon variety is made of boron nitride) a distinction between multi-walled and single-walled nanotubes. Carbon nanotubes can also come in circular form. Carbon nanotubes surprisingly have very high tensile strength and they have varying electrical properties, which depending on the way the graphite sheet is being rolled up and other factors), and can be insulating, semiconducting or conducting (metallic).

Currently, the physical properties are still being discovered and disputed. What makes it so difficult is that nanotubes have a very broad range of electronic, thermal, and structural properties that change depending on the different kinds of nanotube (defined by its diameter, length, and chirality. Nanotubes can either act as electrical conductors, semiconductors or insulators. This makes them very good choices for nanoscale wires and electrical components. Nanotubes exhibit electrical conductivity as high as copper, thermal conductivity as high as diamond and strength 100 times greater than steel at one sixth the weight. Carbon nanotubes are quite popular now for their prospective electrical, thermal, mechanical and chemical applications.

Carbon nanotubes can be thought to be formed by the rolling of graphite layers. The rolling of a single layer of graphite results in a single-walled carbon nanotube (SWCNT), whereas the rolling of more than one layer around a central axis produces multi-walled carbon nanotubes (MWCNT). Depending on the way of rolling of graphene sheets (as shown in Fig. 12), nanotubes of different types, viz. armchair, zig-zag and chiral could be produced. They can be represented using the method given by Hamada et al. [28]. For example, to realize an (n, m) nanotube, one has to move n times a_1 from the selected origin and then m times a_2. On rolling the graphite sheet these points coincide to form the (n, m) nanotube. Thus armchair, zig-zag and chiral nanotubes can be represented as (n, n), $(n, 0)$ and (n, m) respectively. Like many solids, carbon nanotubes are not devoid of defects. Such defects in carbon nanotubes at the intramolecular junction, formed by interposing one or more topological pentagon/heptagon defects in the hexagonal structure between two nanotube segments of different helicity was predicted first theoretically [29] and observed later experimentally [30]. Depending on the orientation in the hexagonal network, the pentagon-heptagon pairs can create a small deformation in the nanotube. It is considered to be a local defect in a straight nanotube. Defects in nanotubes can be produced by controlled electron irradiation, for example. This causes atom removal and creates vacancies, which cluster into large holes in the structure. Molecular simulations have shown that such nanotubes undergo surface reconstruction and nonhexagonal rings such as squares, pentagons, heptagons and octagons so formed disappear finally through Stone-Wales mechanism and lead to five-, six-, and seven membered rings [31]. Coalescence and welding of carbon nanotubes have also been reported [32]. Welding of nanotubes creates molecular junctions of various geometries like X, Y and T [33, 34]. These defects in carbon nanotubes seem to be responsible for the changes in many of their properties. The mechanical properties of carbon nanotubes have been discussed extensively in the last 1 decade. The Young's modulus and tensile strengths of carbon nanotubes are very high compared to that of steel. Reinforcement of these carbon nanotubes can be done to fully exploit the above tensile strengths by the coiling of the tubes. Many models have been proposed for these coiled nanotubes. Recently, Szabó et al. [35] reported the inclusion of impurities like N and S atom and non-hexagonal rings causing the coiling of carbon nanotubes. The folding of haeckelite sheets, which are made up of pentagon, hexagon and heptagons, has also been predicted. All the proposed models explain either the local defect in the straight nanotube or the non-uniform coils of carbon nanotube.

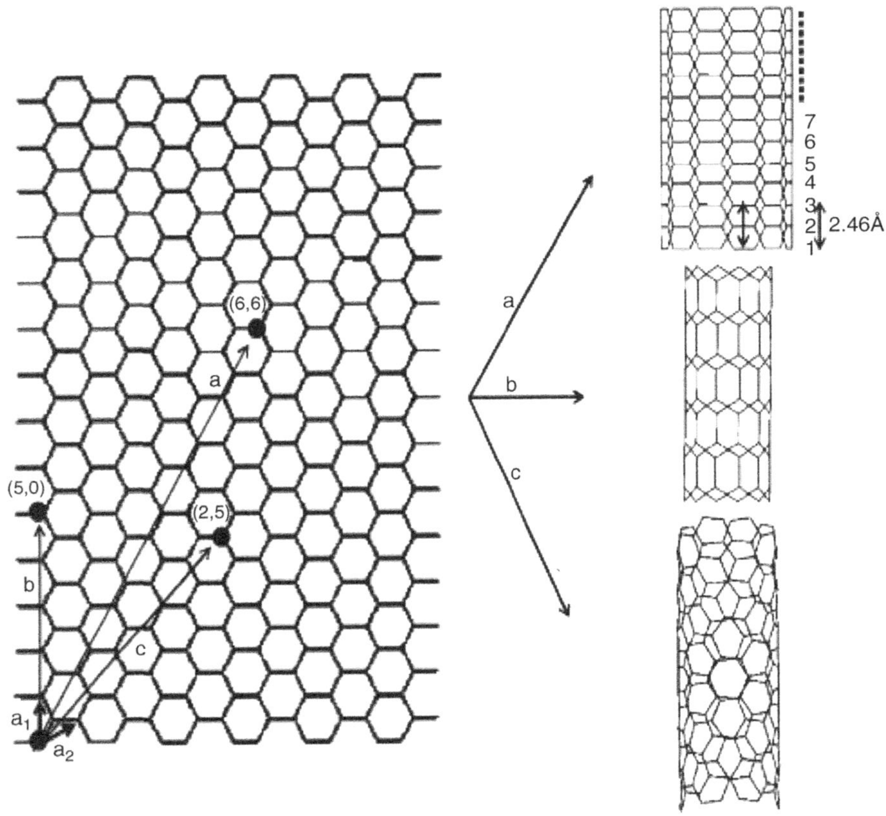

Fig. 12 Rolling of a graphite layer to form single-walled carbon nanotubes of (**a**) armchair, (**b**) zig-zag and (**c**) chiral type. The numbering in the case of the armchair nanotube shows the numbering of the layers running perpendicular to the tube axis as described in the text

3.10 *Catalyst Nanoparticles for Carbon Nanotubes*

Catalysts speed up chemical reactions but can be recovered unchanged at the end of the reaction. They can also direct the reaction towards a specific product and allow chemistry to be carried out at lower temperature and pressures with higher selectivity towards the desired product. As a result they are used very extensively in chemical industry. There are two kinds of catalysts. Heterogeneous catalysts are insoluble in the medium in which the reaction is taking place so that reactions of gaseous or liquid reagents occur at the surface, whilst homogeneous catalysts are dissolved in the reaction medium and hence all catalytic sites are available for reaction. Some of the properties of catalysts are collected in Table 3, where homogenous and heterogeneous catalysts are compared.

Table 3 Comparison of homogeneous and heterogeneous catalysts [36]

	Heterogeneous	Homogeneous
Catalyst from	Solid, often metal or metal oxide	Metal complexes
Mode of use medium	Fixed bed or slurry	Dissolved in reaction
Solvent	Usually not required	Usually required – can be product or byproduct
Selectivity	Usually poor	Can be tuned
Stability	Stable to high temperature	Often decompose <100°C
Recyclability	Easy	Can be very difficult
Special reactions	Haber process, exhaust clean up etc	Hydroformylation of alkenes, methanol carbonylation, asymmetric synthesis etc

The *d*-orbital transition metals known to have strong catalyst activity in general [37]. Transition metal carbides and nitrides are used in a wide range of applications due to their unique physical and chemical properties [38]. They also exhibit high catalytic activity in different reactions including hydrogenation [39–41], hydrodesulfurization and hydrodenitrogenation [42–46].

The catalytic properties of metals should be regarded as only one of the physical and chemical properties of metals requiring theoretical interpretation. The difficulties encountered in developing a satisfactory theory of the metallic state are well known. Two extreme types of theory have been proposed: in the Electron Band, electrons are assumed to be delocalized over the whole metallic crystal, but at least in its simplest form. This theory cannot account for the regular variation in crystal structure across the Transition Series. While for the Valence Bond Theory, in which the cohesive properties of metals are described in terms of localized bonds between atoms, has been extended to describe this phenomenon [47]. Both theories have been of some assistance in accounting for catalytic properties, but both have their limitations. It is therefore worth examining whether the description of metallic behavior in terms of molecular orbitals is fruitful when applied to surface phenomena. It is worth emphasizing that these new approach, due especially to Goodenough [48], is not a completely new theory [49] but rather an attempt to unify the two theories mentioned above. It applies to the molecular orbital theory: it is best described in relation to the face-centered cubic metals to which family belong most of the metals of catalytic interest. In these metals each atom is in contact with 12 others, four in each of three planes mutually at right angles, while six others octahedrally disposed lie at a slightly greater distance. According to the Crystal Field Theory the five d-orbitals are split in an octahedral environment into two groups.

The two higher energy orbitals are taken to lie along the Cartesian axes (Fig. 13a) and are designated $d_{x^2-y^2}$ and d_{z^2}, or collectively in the language of group theory as e_g. The three lower energy orbitals then lie between the Cartesian axes (Fig. 13b) and so are designated d_{xy}, d_{yz} and d_{xz} or collectively as t_{2g}. As each of these orbitals has four lobes, it is reasonable to say that the 12 near neighbours in the face-centered

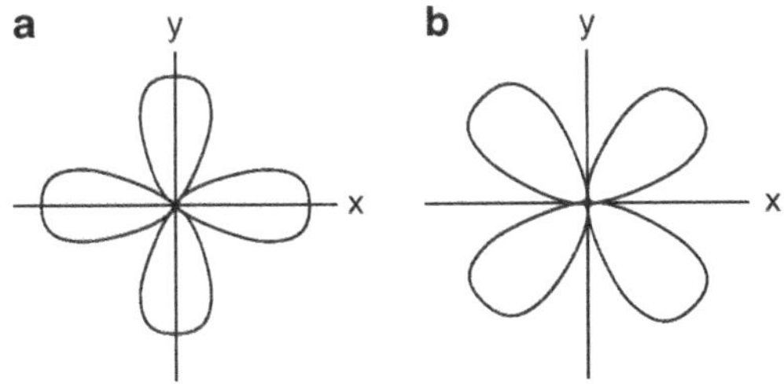

Fig. 13 Representation of (**a**) the $d_{x^2-y^2}$ orbital and of (**b**) the d_{xy} orbital, examples respectively of the e_g and t_{2g} families of orbitals in an octahedral complex

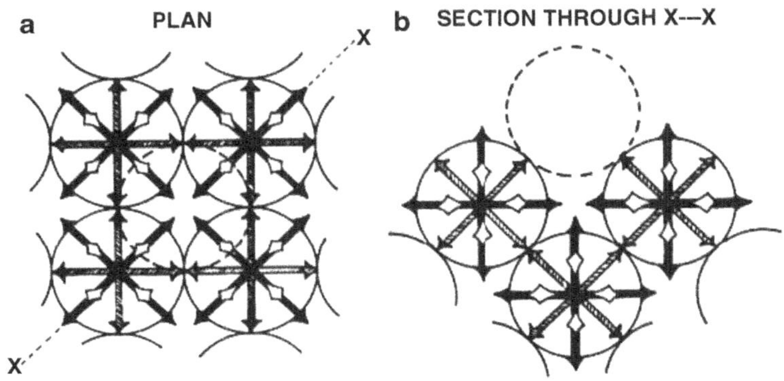

Fig. 14 The emergence of orbitals from the (100) face of a face-centered cubic metal: (**a**) plan; (**b**) section; *Filled arrows*: e_g orbitals in plane of paper, *Hatched arrows*: t_{2g} orbitals in plane of paper, *Open arrows*: t_{2g} orbitals at 45° to plane of paper

cubic crystal are bonded to the central atom by t_{2g} orbitals which overlap to form a broad collective t_{2g} band, while the *six* next nearest neighbours are bonded by e_g orbitals. In fact, however, these latter orbitals do not overlap significantly because the next nearest neighbours are too far apart, so that electrons in the e_g orbitals occupy localized states around each atom. It is now an easy matter to evaluate the manner in which the two groups of orbitals emerge from the three low index planes. This is illustrated in Figs. 14 and 15 for the (100) and (110) planes respectively.

The former consists of a square array of atoms (Fig. 14a): the non-bonding e_g orbitals in the plane of the surface are shown in black while the bonding t_{2g} orbitals are hatched. Furthermore, there are four t_{2g} orbitals emerging from each atom at 45° to the plane of the surface, and these are shown in white: one of these from each of four atoms is directed towards the position occupied by an atom in the next layer. Figure 14b shows a section normal to the surface. Figure 15a shows the plan of the (110) plane: this consists of parallel rows of next nearest neighbors bonded as usual by shaded t_{2g} orbitals. Other t_{2g} orbitals emerge, four from each atom (shown in white) at 30° to the plane of the surface and are directed towards the sites of atoms in the next layer. Figure 15b shows a section normal to the surface: this plane differs

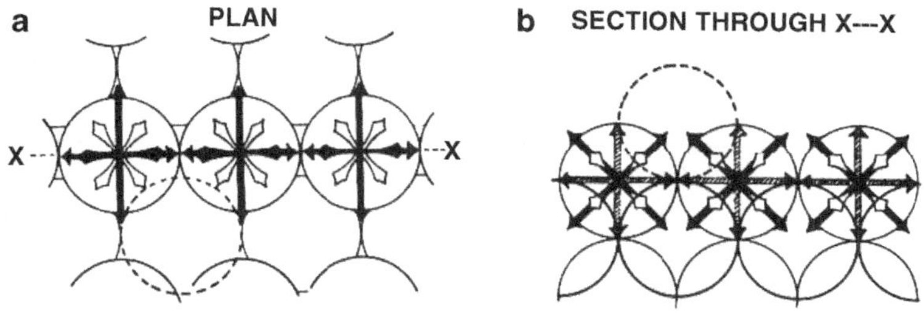

Fig. 15 The emergence of orbitals from the (110) face of a face-centered cubic metal: (**a**) plan; (**b**) section; *Filled arrows*: e_g orbitals in plane of paper or emerging at 45°, *Hatched arrows*: t_{2g} orbitals in plane of paper, *Open arrows*: t_{2g} orbitals at 30° to plane of paper

from the (100) in that it is a t_{2g} orbital which emerges normal to the surface from each atom. Having obtained the direction of emergence of the t_{2g} and e_g orbitals from the low index planes of the crystal, it is now necessary to try to decide on the extent: to which they are occupied by electrons. If in Group VIII, (nickel, palladium and platinum), according to Goodenough [48], the 0.55 d-band holes are distributed between the two bands as follows : t_{2g} band, 0.41 holes: e_g band, 0.14 holes. Both bands are therefore substantially filled, and supposed that both can participate in forming covalent bonds with adsorbed species. The atom in the centre foreground is deeply imbedded in the surface: it is bonded by overlap of its spherical 1s orbital with four e_g orbitals from the surrounding metal atoms and by a further e_g orbital from the metal atom below it. This must therefore be quite a strongly bound state. The atom on the left is bonded only by this latter e_g orbital and is likely to be much more weakly bound. The atoms on the left and right are bonded simply by the normal-emergent t_{2g} orbitals: the one on the right is, however, bonded to a metal atom in a layer one below the surface, but both should be weak states of binding. The position of the atom in the centre is more complicated: it is bonded by two t_{2g} orbitals emerging from the atoms below it and by two e_g orbitals emerging from the atoms below it and by two e_g orbitals from the atoms on either side of it. This should therefore be a strongly bound state. Space does not permit the elaboration of these ideas to the adsorption of other atoms and molecules.

Mostly 3d-metals are used to catalyze formation of CNTs because of their ability to decompose hydrocarbons. However, they are not all equivalent and, in particular, alloying them with each other or with other non-transition metals dramatically changes (generally improves) their catalytic properties. The role of the non-transition metal is to disperse and to stabilize the catalyst metal-so an effect on the geometric aspect (for example, support catalyst). On the other hand, effects of alloying two/three transition metals on e.g. the electronic structure or the carbon solubility and diffusion, observed improvement in the catalyst performances as studied here. Furthermore, overlap of catalyst 3d-orbitals with carbon orbitals after hydrocarbon dissociation could be a relevant parameter for the CNT growth process. The link between the size of the catalyst nanoparticles and the diameter of the CNTs and the consequently absolute necessity to prepare stable nanoparticles of controlled size has been emphasized. To meet this severe constraint high-technological efforts have to be made. Though, other parameters concerning the catalyst morphology have an influence on the growth process like the crystallographic orientation of the nanoparticles. The preparation method of the catalyst particles also plays its part in the complex problem of the factors influencing the catalytic properties of the metal nanoparticles. The choice of the preparation method determines the morphology of the obtained nanoparticles and the use of porous support to limit the size of the nanoparticles. But the influence of the preparation method is not limited to morphological effects and can directly affect catalytic properties. A given catalyst can yield completely different results upon CVD when supported on different materials. So not only the catalyst has to consider but the catalyst/support. The interactions between catalyst and support are found to be essential. First, they partly determine the morphology of the nanoparticles and, second, they also alter the electronic structure of the nanoparticles, which altering their catalytic properties. These interactions depend on both support and catalyst materials but also on their crystallographic orientations, on the surface roughness and porosity of the support. Finally, pretreatment can be essential for activating the catalyst by reducing it to its actual catalytic form (pure metal) or by forming nanoparticles from the annealing. Moreover, the parameters concerning the catalyst nanoparticles only cannot explain whether CNTs will grow or not and how they will grow. The catalytic process requires a hydrocarbon gas (or CO) coupled with a supported catalyst in charge of its dissociation and the subsequent growth of the CNT. The catalyst/gas or rather the trio support/catalyst/gas should be considered for a complete understanding of the growth process. Other important factors influence the kinetic and thermodynamic aspects of the growth process, e.g. temperature, pressure, flow rate and reaction time. Many authors explain their success in producing SWNTs by the very small size of their catalyst nanoparticles. However, the differentiation between SWNT and MWNT growth processes is not really clear and kinetic arguments are often invoked.

3.11 Preparation of Carbon Nanotubes

Here, we presented result of carbon nanotubes synthesized by a newly developed method; fluidized floating catalyst method [23] as shown in Fig. 16.

The growth of CNTs was carried out in a tubular furnace with a horizontal quartz tube at atmospheric pressure using fluidized floating catalyst method. Firstly, carrier gas (Argon) was flushed for about 10 min before the furnace was turned on to remove the surrounding air and to create an inert atmosphere. Then, the precursor (camphor oil) and catalyst (Fe/Co/Al) mixtures with ratio (1:1:1) were placed together in a quartz boat and pushed into the center of the quartz tube. Then the quartz tube was heated in Ar ambient with a flow rate of 50 sccm to ensure no oxygen. When the temperature of the furnace rapidly increased to 650°C, it was

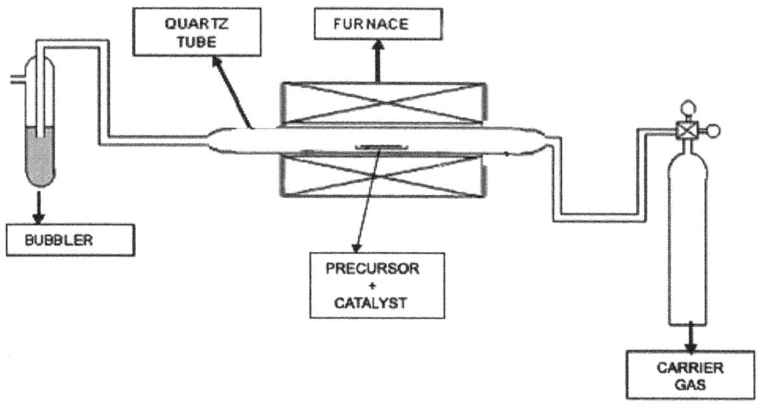

Fig. 16 A schematic sketch of modified fluidized floating catalyst method [23]

discovered that CNTs were deposited in the quartz boat. Carbon deposited materials formed were removed from the catalyst boat. The temperature was maintained for a further period of 30 min for annealing process after deposition. Finally, the quartz tube was cooled down to room temperature in Ar ambient with a flow rate of 50 sccm. The carbon deposited materials formed were removed from the quartz boat.

3.11.1 Surface Morphology

The FESEM micrograph shows the CNTs grown was randomly curled as shown in Fig. 17. When camphor oil decomposed into vapor state, CNTs was found to be deposited in the presence of catalyst. From here, we noticed the conversion of carbon to CNT was dependent on the amount of catalyst used. Different percentage of catalyst used to deposit CNTs which are 0.25% (FCM1), 0.50% (FCM2), 0.75% (FCM3), 1.00% (FCM4) and 1.25% (FCM5) [23]. The CNTs yield was estimated by weighing the amount of CNTs obtained and it was found to be the highest at 0.75% of catalyst. Therefore, 0.75% Fe/Co/Al can be considered as optimum catalyst concentration for CNTs synthesized by fluidized floating catalyst method.

The morphology of CNTs investigated by Samant et al. [50] on effects of ferrocene using naphthalene as precursor by Thermal-CVD method observed spider-like web structure CNTs. The SEM recorded on the product formed in the presence of Ar gas is shown in Fig. 18.

As seen, the substrate is occupied with densely packed, CNTs with average tube-diameter of about 70 nm. Not much variation in size or packing density was observed in the micrographs recorded at various other areas. Small amount of amorphous carbon in the form of globules (as marked with red arrow) is also seen in the picture. The bright, spherical structures at one of the ends of most of the tubes are attributed to the iron nanoparticles formed by decomposition of

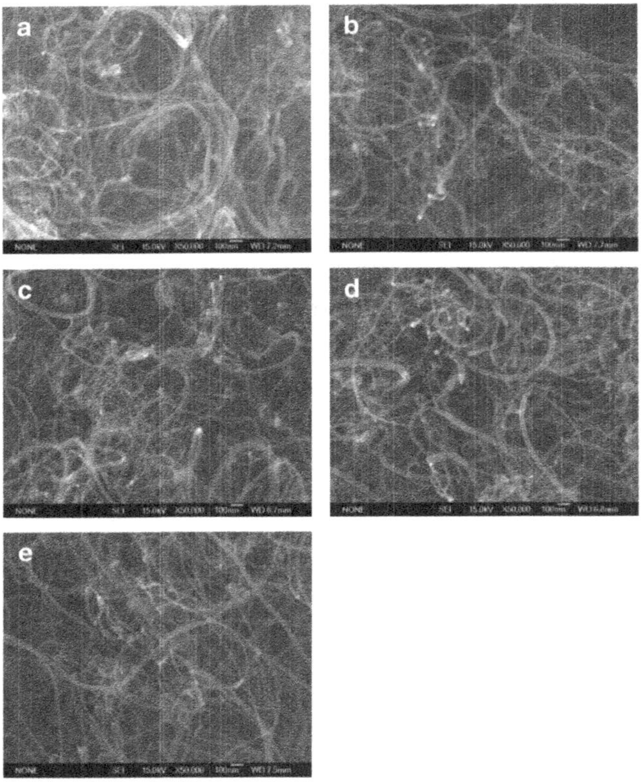

Fig. 17 FESEM images of CNTs using (**a**) 0.25% (FCM1), (**b**) 0.50% (FCM2), (**c**) 0.75% (FCM3), (**d**) 1.00% (FCM4) and (**e**) 1.25% (FCM5) of catalyst (Fe/Co/Al) by fluidized floating catalyst method [23]

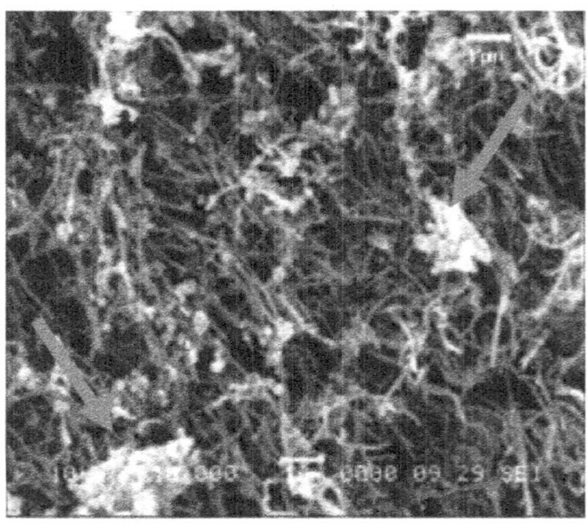

Fig. 18 SEM image of the CNTs deposited on fused silica plate in the presence of Ar [50]

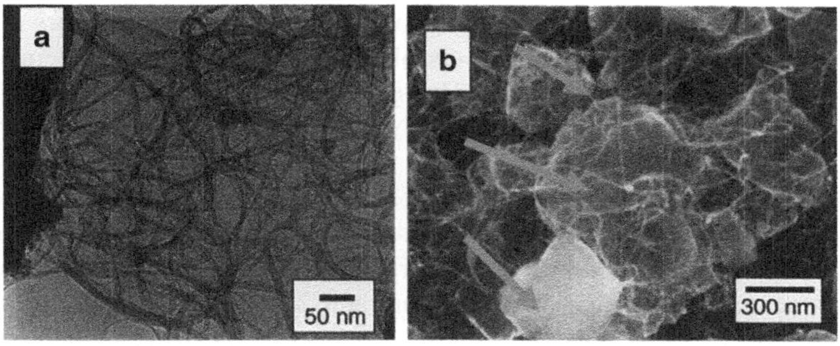

Fig. 19 Lower magnification TEM image (**a**) and SEM image (**b**) of 'as grown' SWNTs [51]

ferrocene [50]. By using alcohol as the carbon source, a new simple catalytic chemical vapor deposition technique to synthesize high purity single-walled carbon nanotubes at low temperature is demonstrated by Maruyama et al., [51] at 550°C. The morphology of the deposited CNTs is shown in Fig. 19.

In brief, iron and cobalt acetate were dissolved in ethanol and mixed with Y-type zeolite powder. The amounts of Fe and Co were 2.5 wt% each. From the SEM image in 19(b), a small amount of 'as grown' sample was placed with a conductive tape used for the observation. Here, web-like growth of bundles of SWNTs also observed from zeolite particles around 300 nm. However, no other structures such as amorphous carbon are observed from both (a) and (b) except the zeolite flakes (as marked with green arrow) which used as the support catalyst. The impurities such as amorphous carbon, multi-walled carbon nanotubes, metal particles and carbon nanoparticles are completely suppressed even at relatively low reaction temperature due to the etching effect of decomposed OH radical attacking carbon atoms with a dangling bond. The high-purity synthesis at low temperature promises large scale production at low cost and the direct growth of SWNTs on conventional semiconductor devices already patterned with aluminum [51].

The role of catalyst can yield completely different results upon CVD when supported on different materials. So not only the catalyst has to consider but the catalyst/support. The interactions between catalyst and support are found to be essential. First, they partly determine the morphology of the nanoparticles and, second, they also alter the electronic structure of the nanoparticles, which altering their catalytic properties. These interactions depend on both support and catalyst materials but also on their crystallographic orientations, on the surface roughness and porosity of the support. Finally, pretreatment can be essential for activating the catalyst by reducing it to its actual catalytic form (pure metal) or by forming nanoparticles from the annealing. Moreover, the parameters concerning the catalyst nanoparticles only cannot explain whether CNTs will grow or not and how they will grow. The catalytic process requires a hydrocarbon gas (precursor) with a supported catalyst in charge of its dissociation and the subsequent growth of the CNT. The catalyst/gas or rather the trio support/catalyst/gas should be considered for a complete understanding of the growth process. Other important factors influence the kinetic and thermodynamic aspects of the growth process, e.g. temperature, pressure, flow rate and reaction time. However, the differentiation between SWNT and MWNT growth processes is not really clear and kinetic arguments are often invoked.

Tao et al. [52] also employed ferrocene as catalyst but interestingly as precursor as well. CNTs have been synthesized via directly pyrolyzing only as the precursor in the autoclave. The nanotubes with several micrometers in length have outer and inner diameters in the range of 40–100 nm and 2–40 nm, respectively. Yield of ~70% of CNTs can be obtained without any accessorial solvents and catalysts. Experimental results showed that a temperature higher than 600°C in conjunction with proper pressure was favorable for achievement of the nanotubes [52]. Figure 20a, b show SEM and TEM images of carbon nanotubes, respectively. It can be seen that large quantity of nanotubes can be achieved by this method.

The yield of carbon nanotubes estimated through SEM and TEM observations of the as-prepared samples was 70%. Most of the carbon nanotubes with close ends and curly morphologies have outer diameters of 40–100 nm, inner diameters of 20–40 nm and length up to several micrometers, and no metal catalyst particles were detected from TEM image (Fig. 20b). The size of the catalyst nanoparticles seems to be the determining factor for the diameter of the CNT grown on it. Beyond this size correlation, only small nanoparticles are able to catalyze formation of CNT. This can be explained on the one hand by the fact that such very small nanoparticles can exhibit peculiar electronic properties and thus catalytic properties due to the unusual high ratio surface atom/bulk atom. On the other hand, the growth mechanism involving the formation of a carbon cap on the nanoparticle surface to reduce its unusual high surface energy. The crystallographic orientation of the catalyst nanoparticle too can be crucial for CNT growth.

Pawan et al., deposited CNTs by chemical vapor deposition [53]. Methane (CH_4) and hydrogen (H_2) were used as precursor gases for the CNT growth with Ni as the catalyst. Figure 21 shows the SEM micrograph of the as-grown CNTs deposited

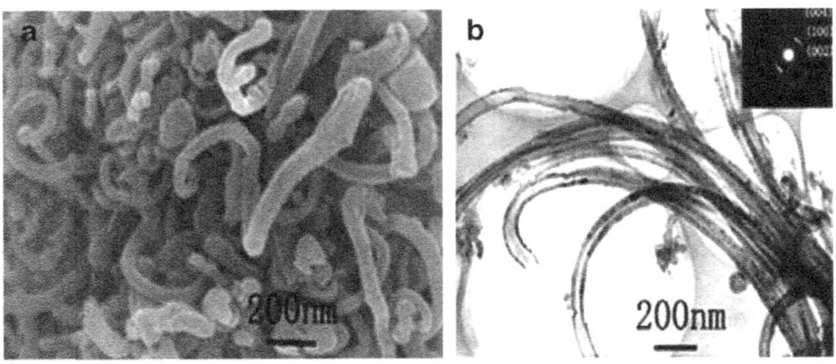

Fig. 20 (**a**) SEM image and (**b**) TEM image of the carbon nanotubes; (inset 3b) SAED pattern of individual carbon nanotube [52]

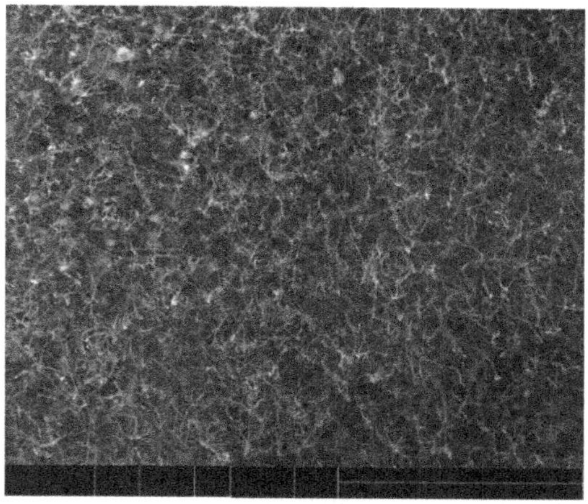

Fig. 21 SEM micrograph of the as-grown CNTs deposited using the CVD method [53]

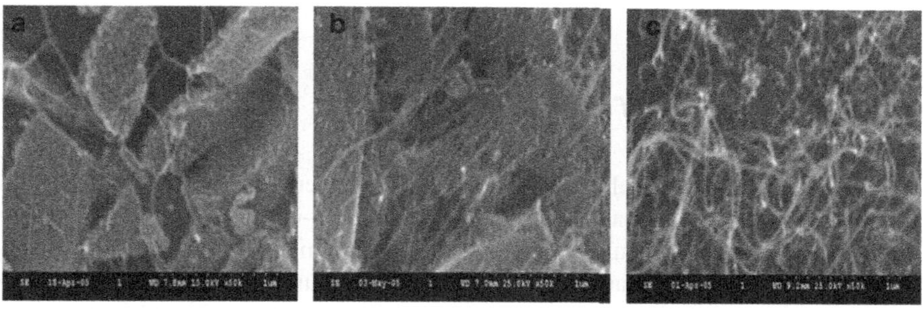

Fig. 22 SEM images of CNTs grown on catalysts with different Mo concentrations: (**a**) Fe:Mg:Mo = 0.1:12:0.1, (**b**) Fe:Mg:Mo = 0.1:12:0.5, and (**c**) Fe:Mg:Mo = 0.1:12:1 [54]

using CVD. The tubes were long, dense and showed a typical MWNTs structure. By using Ni as the catalyst, the area density of the deposited CNTs was high and the tubes appeared parallel to the substrate.

Mixing of two or more different metal catalysts yields improvement of the catalyst performance in terms of quality of obtained product or lowering the reaction temperature. Study on the effect of Fe catalyst with different ratio of Mo and Mg was investigated by Yangfang et al. [54]. High quality, high yield CNTs were synthesized on a composite catalyst using catalytic chemical vapor deposition. The composite catalysts Fe/MgO, Mo/MgO and (Fe, Mo)/ MgO, prepared via the sol gel method using citric acid as fuel, were investigated for the production of CNTs. Only the (Fe, Mo)/MgO catalyst could support CNTs growth with high yield in this study. The different mole ratios between Fe, Mo, and Mg resulted in changes in product structure, diameter size, and yield. Decreasing the Fe concentration reduces the structural defects, and by increasing the Mo concentration, the yields of CNTs clearly increase as observed in Fig. 22.

The influence of catalyst composition on the growth of CNTs was studied in this paper. Among the four catalysts, Fe-Mg-O, Fe-Mo-O, Mg-Mo-O, and Fe-Mg-Mo-O, made by the sol-gel method, only Fe-Mg-Mo-O supported the production of CNTs with high yield. In the catalyst, Fe was considered as the catalytic center to magnetize carbon atoms for the CNTs growth, but the crystallite size of Fe was very important in carbon nanofibers growth. The larger the Fe particle size, the more carbon nanofibers were formed as the size of the catalyst nanoparticle is important to determine the diameter of the nanotubes. By controlling the Fe concentration in the catalyst was necessary in preparing CNTs with few carbon nanofibers and in simultaneously controlling the diameter of the CNTs. Mo in the catalyst Fe-Mg-Mo-O was proved to accelerate carbon depositing, which favored the high yield of CNTs. Higher Mo concentration facilitated the structural transformation from carbon soot to carbon nanotubes. The ratio harmony of the metal element and crystallite size of the particles in the composite catalyst were very important. The catalyst composition Fe:Mg:Mo = 0.1:12:1 in our study was successful in CNTs growth with high yield, high purity, fewer walls, and smaller diameter [54].

3.12 Applications of Carbon Nanotubes

CNTs have extraordinary electrical conductivity, heat conductivity and mechanical properties. They are probably the best electron field-emitter possible. They are polymers of pure carbon and can be reacted and manipulated using the tremendously rich chemistry of carbon. This provides opportunity to modify the structure and to optimise solubility and dispersion. Very significantly, CNTs are molecularly perfect, which means that they are free of property-degrading flaws in the nanotube structure. Their material properties can therefore approach closely the very high levels intrinsic to them. These extraordinary characteristics give CNTs potential in numerous applications.Here, we discussed the application of CNTs in electrochemical system.

3.12.1 Lithium-Ion Battery

The outstanding mechanical properties and the high surface-to-volume ratio (due to their small diameter) make carbon nanotubes potentially useful as anode materials [55–60] or as additives [61] in lithium-ion battery systems. An electrode containing 10 wt% of carbon nanotubes as the additive displays a homogeneous distribution of nanotubes in synthetic graphite. The cyclic efficiency of a synthetic graphite anode as a function of the weight percent of carbon nanotube. With increasing weight percent of carbon nanotubes, the cyclic efficiency of the synthetic graphite battery anode increases continuously, and, in particular, when 10 wt% of the nanotubes was added, the cyclic efficiency was maintained at almost 100% up to 50 cycles. At higher concentrations, the nanotubes interconnect graphite powder particles together to form a continuous conductive network.

3.12.2 Additives to the Electrodes of Lead-Acid Batteries

In order to increase the conductivity of electrodes in lead-acid batteries, different weight percents of carbon nanotubes are added to the active anode material (with average diameters of ca. 2–5 mm) of the positive electrode. The resistivity of the electrode is lowered for the case of 1.5% nanotube addition. When this sample (0.5–1 wt%) is incorporated in the negative electrode, the cycle characteristics are greatly improved compared with those of an electrode without additive [61]. This is probably due to the ability of carbon nanotubes to act as a physical binder, resulting in electrodes that undergo less mechanical disintegration and shedding of their active material. Therefore, it is expected that the use of carbon nanotubes as an electrode's filler should produce an enhanced cyclic behavior for electrodes in lead-acid batteries compared with electrodes using conventional graphite powder, because the unusual morphology of the carbon nanotube, such as the concentric orientation of their graphite crystallites along the fibre cross-section, induces a high resistance towards oxidation, and furthermore the nanotube network embedded in the polymer would enhance the reactivity of the electrode.

3.12.3 The Electric Double-Layer Capacitor

The merit of the electric double-layer capacitor (EDLC) is considered to be a high discharge rate [62], which makes them applicable as a hybrid energy source for electric vehicles and portable electric devices [63]. EDLCs containing carbon nanotubes in the electrode exhibit relatively high capacitances resulting from the high surface area accessible to the electrolyte [60, 64, 65]. On the other hand, the most important factor in commercial EDLCs is considered to be the overall resistance. In this context, carbon nanotubes with strong electrical and mechanical properties can be used as an electrical conductive additive in the electrode of EDLC. It has been demonstrated that the addition of carbon nanotubes results in an enhanced capacity at higher current densities, when compared with electrodes containing carbon black [66].

3.12.4 Fuel Cells

Fuel cells have been considered as the next generation of energy devices because this type of system transforms the chemical reaction energy from hydrogen and oxygen into electric energy [67]. Carbon nanotubes decorated with metal nanoparticles as the electrode have doubled the performance of the fuel cell, owing to the increased catalytic activity of nanotube-based electrodes [68–70]. In this context, we have reported the efficient impregnation of Pt nanoparticles (outer diameter less than 3 nm) on cup-stacked type carbon nanotubes [71]. The method involves dispersion of the fibers in H_2PtC_{16}, followed by low-temperature annealing. The Pt particle deposition is always homogeneous, and can be controlled selectively on the outer or inner core using the hydrophobic nature of the material. Since the Pt particle activity on the fibers is high, this material could find application as an efficient catalyst or biological device. It may be that carbon nanotube technology will contribute to the development of fuel cells as a catalyst support and also as a main component of bipolar systems. However, additional basic and applied research is necessary.

3.12.5 Multifunctional Fillers in Polymer Composite

It has been shown that carbon nanotubes could behave as the ultimate one-dimensional material with remarkable mechanical properties [72, 73]. The density-based modulus and strength of highly crystalline SWNTs are 19 and −56 times that of steel. Based on a continuum shell model, the armchair tube exhibits larger stress-strain response than the zigzag tube under tensile loading. Strong mechanical properties of carbon nanotubes due to a strong carbon-carbon covalent bond are highly dependent upon the atomic structure of nanotubes and the number of shells. Moreover, carbon nanotubes exhibit strong electrical and thermal conducting properties. Therefore, carbon nanotubes (single and multi-walled) have been studied intensively as fillers in various matrices, especially polymers [74–78]. The best use of the intrinsic properties of these fibrous nanocarbons in polymers can be achieved by optimizing the interface interaction of the nanotube surface and the polymer. Therefore, surface treatments via oxidation could be used in order to improve adhesion properties between the filler and the matrix. This results in a good stress transfer from the polymer to the nanotube. There are various surface oxidative processes, such as electrochemical, chemical and plasma techniques. From the industrial point of view, ozone treatment is a very attractive technique. In addition, the dispersion of the nanotubes/nanofibres in the polymer should be uniform within the matrix.

The smallest working composite gear has been prepared by mixing nanotubes into molten nylon and then injecting into the tiny mould. As shown in Fig. 12, this gear is as small as the diameter of a human hair. This piece exhibits a high mechanical strength, high abrasion resistance and also good electrical and thermal conductivity. When cup-stacked-type carbon nanotubes are incorporated in polypropylene, the improvement of the tensile strength is very noticeable (up to 40%). This remarkable result can be explained by the particular morphology of cup-stacked-type carbon nanotubes. In other words, a large portion of edge sites on the outer surface of the nanotubes might act as nucleation sites, resulting in good adhesion between nanotubes and polymers (good stress transfer). Recently, various studies on the nucleation effect of nanotubes on the crystallization of semi-crystalline polymers have been reported [79–81].

4 Conclusion

Carbon materials have various forms, crystalline and non-crystalline carbon constitutes a class of new materials with a wide range of compositions, properties, and performance. Carbon based material can be prepared by a variety of techniques such as CVD, PLD etc. Due to its unique optical and electrical properties, carbon has potential applications in various fields especially in semiconductor devices. Many interesting crystalline phases with useful properties have been predicted which make carbon promising target for preparative materials scientists and chemists. The majority of the experimental approaches are based on vapor phase deposition routes, which furnished many interesting properties. This chapter has shown the effect of the catalyst on the morphology of the carbon nanotubes. The size, substrate and the composition of the catalyst seems to be the determining factors on the formation and diameter of the nanotubes. Interactions between catalyst and support are essential since they can greatly affect the electronic structure of the nanoparticles and their morphology and in turn their catalytic properties. Hence, selecting the best catalyst and substrate is crucial to synthesize carbon nanotubes. In summary it can be stated that a fascinating new field in the area of carbon has been discovered, which gives motivation for further studies dedicated to fundamental questions as well as the exploitation of the novel materials for future industrial applications.

Acknowledgment The author would like to thank NANO-SciTech Centre and Faculty of Applied Sciences, UiTM and RMI, UiTM for the facilities and financial support on this project.

References

1. Allamandola, L.J., Hudgins, D.M., Bauschlicher Jr., C.W., Langhoff, S.R.: Astron. Astrophys. **352**, 659–664 (1999)
2. Kroto, H.W., Heath, J.R., O'brien, S.C., Curl, R.F., Smalley, R.E.: Nature **318**, 162–163 (1982)
3. Iijima, S.: Nature **354**(6348), 56–58 (1991)
4. Yakobson, B.I., Smalley, R.E.: Am. Sci. **85**, 324–337 (1997)
5. Kroto, H.W., Allaf, A.W., Balm, S.P.: Chem. Rev. **91**, 1213–1235 (1991)
6. Kraetschmer, W., Lamb, L.D., Fostiropoulos, K., Huffman, D.R.: Nature **347**, 354–58 (1990)
7. Haufler, R.E., Coceicao, J., Chibante, L.P.F., Chai, Y., Byrne, N.E., et al.: J. Phys. Chem. **94**, 8634–8640 (1990)
8. Dresselhaus, M.S., Dresselhaus, G., Eklund, P.C.: Science of Fullerenes and Carbon Nanotubes. Academic, San Diego (1996)
9. Dresselhaus, M.S., Gresselhaus, G., Sugihara, K., Spain, I.L., Goldberg, H.A.: Graphite Fibers and Filaments, vol. 5. Springer Series in Materials Science, Springerverlag, Berlin (1988)
10. Lahaye, J., Prado, G.: In: Walker Jr, P.L. (ed.) Chemistry and Physics of Carbon, vol. 16. Marcel Dekker, New York (1978)
11. Donnet, J.B., Bansal, R.C., Wang, M.J.: Carbon Black. Marcel Dekker, New York (1993)
12. McClure, J.W.: In: Carter, D.L., Bate, R.T. (eds.) Proceeding of the International Cong. On Semimetals and Narrow Gap Semiconductors. Pergamon, New York (1971)

13. Charlier, J.J.C.: PhD thesis, Dept. of Phys. Mater., Catholic Univ. of Louvain (1994)
14. Hauser, J.J.: Solid State Commun. **17**, 1577–1580 (1975)
15. Bubenzer, A., Dischler, B., Brandt, G., Koidi, P.: J. Appl. Phys. **54**, 4590–4595 (1983)
16. Ebbesen, T.W.: Carbon Nanotubes. CRC, Boca Raton, New York, London, Tokyo (1997)
17. Ruffieux, P.: Thesis, University of Fribourg, Switzerland (2002)
18. Rohlfing, E.A., Cox, D.M., Kaldor, A.: J. Chem. Phys. **81**, 3322–3330 (1984)
19. Carman, H.S., Compton, R.N.: J. Chem. Phys. **98**, 2473–2476 (1993)
20. Kroto, H.W., Heath, J.R., O'Brien, S.C., Curl, R.F., Smalley, R.E.: Nature **318**, 162–163 (1985)
21. Dresselhaus, M.S., Dresselhaus, G., Saito, R.: Carbon **33**, 883–891 (1995)
22. McEnaney, B.: In: Burchell, T.D. (ed.) Carbon Materials for Advanced Technologies. Pergamon, Amsterdam, Lausanne, New York, Oxford, Shannon, Singapore, Tokyo (1999)
23. Azira, A.A., Zainal, N.F.A., Nik, S.F., Mohamad, F., Kudin, T.I.T., Soga, T., Abdullah, S., Rusop, M.: Mater. Res. Innovat. **13**(3), 182–184 (2009)
24. Azira, A.A., Zainal, N.F.A., Nik, S.F., Rusop, M.: AIP Conf. Proc. **1136**, 720–724 (2009)
25. Azira, A.A., Zainal, N.F.A., Nik, S.F., Rusop, M.: AIP Conf. Proc. **1136**, 735–739 (2009)
26. Azira, A.A., Zainal, N.F.A., Nik, S.F., Rusop, M.: AIP Conf. Proc. **1136**, 740–744 (2009)
27. Azira, A.A., Zainal, N.F.A., Nik, S.F., Soga, T., Abdullah, S., Rusop, M.: Int. J. Nanosci. **8**(4 & 5), 351–357 (2009)
28. Hamada, N., Sawada, S., Oshiyama, A.: Phys. Rev. Lett. **68**, 1579–1581 (1992)
29. Dunlap, B.I.: Phys. Rev. B **46**, 1933–1936 (1992)
30. Amelinckx, S., Zhang, X.B., Bernaerts, D., Zhang, X.F., Ivanov, V., Nagy, J.B.: Science **265**, 635–639 (1994)
31. Ajayan, P.M., Ravikumar, V., Charlier, J.C.: Phys. Rev. Lett. **81**, 1437–1440 (1998)
32. Charlier, J.C.: Acc. Chem. Res. **35**, 1063–1069 (2002)
33. Terrones, M., Banhart, F., Grobert, N., Charlier, J.C., Terrones, H., Ajayan, P.M.: Phys. Rev. Lett. **89**(075505), 1–4 (2002)
34. Deepak, F.L., Govindaraj, A., Rao, C.N.R.: J. Chem. Sci. **118**, 9–14 (2006)
35. Szabó, A., Fonseca, A., Nagy, J.B., Lambin, Ph, Biro, L.P.: Carbon **43**, 1628–1633 (2005)
36. Saito, R., Takeya, T., Kimura, T.: Phys. Rev. **B 57**, 4145–4153 (1998)
37. Cole Hamilton, D.J., Tooze, R.P.: Catalysis by Metal Complexes, vol. 30. Springer, Netherlands (2006)
38. Atkins, P.W.: Physical Chemistry, 6th edn. W.H. Freeman & Co., New York (1997)
39. Oyama, S.T.: The Chemistry of Transition Metal Carbides and Nitrides. Blackie Academic and Professional, London (1996)
40. Lee, J.S., Yeom, M.H., Park, K.Y., Nam, I.S., Chung, J.S., et al.: J. Catal. **128**, 126–136 (1991)
41. Oyama, S.T.: Catal. Today **15**, 179–200 (1992)
42. DaCosta, P., Lemberton, J.L., Potvin, C., Manoli, J.M., Perot, G., et al.: Catal. Today **65**, 195–200 (2001)
43. Liaw, S.J., Raje, A., Chary, K.V.R., Davis, B.H.: Appl. Catal. A Gen. **123**, 251–271 (1995)
44. Raje, A., Liaw, S.J., Chary, K.V.R., Davis, B.H.: Appl. Catal. A Gen. **123**, 229–250 (1995)
45. Ramanathan, S., Oyama, S.T.: J. Phys. Chem. **99**, 16365–16372 (1995)
46. Sajkowski, D. J., Oyama, S. T.: ACS Div. Fuel Chem. Preprints **35**(2), 233–236 (1990)
47. Melo-Banda, J.A., Dominguez, J.M., Sandoval-Robles, G.: Catal. Today **65**, 279–284 (2001)
48. Altmann, S.L., Coulson, C.A., Hume-Rothery, W.: Proc. R. Soc. A **240**, 145–159 (1957)
49. Goodenough, J.B.: Magnetism and the Chemical Bond. Interscience, New York (1963)
50. Samant, K.M., Haram, S.K., Kapoor, S.: J. Phys. **68**(1), 51–60 (2007)
51. Maruyama, S., Kojima, R., Miyauchi, Y., Chiashi, S., Kohno, M.: Chem. Phys. Lett. **360**, 229–234 (2002)
52. Tao, C., Zhiyong, F., Guifu, Z., Qixiu, H., Biao, H., Xiao, Z.Y., Youjin, Z.: Bull. Mater. Sci. **29**(7), 701–704 (2006)
53. Pawan, K.T., Singha, M.K., Abha, M., Umesh, P., Misraa, D.S., et al.: Thin Solid Films **469–470**, 127–130 (2004)
54. Yanfang, C., Doh, H.R., Yun, S.L.: Met. Mater. Int. **14**(3), 385–390 (2008)

55. Endo, M., Nakamura, J., Sasabe, Y., Takahashi, T., Inagaki, M.: Trans. IEE Jpn. **A115**, 349–356 (1995)
56. Tatsumi, K., Zaghib, K., Abe, H., Higuchi, S., Ohsaki, T., Sawada, Y.: J. Power Sources **54**, 425–427 (1999)
57. Frackowiak, E., Gautier, S., Gaucher, H., Bonnamy, S., Beguin, F.: Carbon **37**, 61–69 (1999)
58. Wu, G.T., Wang, C.S., Zhang, X.B., Yang, H.S., Qi, Z.F., He, P.M., Li, W.Z.: J. Electrochem. Soc. **146**, 1696–1701 (1999)
59. Gao, B., Kelinhammes, A., Tang, X.P., Bower, C., Wu, Y., Zhou, O.: Chem. Phys. Lett. **307**, 153–157 (1999)
60. Frackowiak, E., Beguin, F.: Carbon **40**, 1775–1787 (2002)
61. Endo, M., Kim, Y.A., Hayashi, T., Nishimura, K., Matushita, T., Miyashita, K., Dresselhaus, M.S.: Carbon **39**, 1287–1297 (2001)
62. Conway, B.E.: Electrochemical Supercapacitors – Scientific Fundamentals and Technological Applications. Kluwer Academic, Dordrecht (1999)
63. Miyadera, K.: Toyota Tech. Rev. **52**, 22–27 (2002)
64. Niu, C., Sickel, E.K., Hoch, R., Moy, D., Tennent, H.: Appl. Phys. Lett. **70**, 1480–1482 (1997)
65. An, K.H., Kim, W.S., Park, Y.S., Moon, J.M., Lim, S.C., Lee, Y.S., Lee, Y.H.: Adv. Mater. **11**, 387–392 (2001)
66. Takeda, T., Takahashi, R., Kim, Y.J., Koshiba, K., Ishi, K., Kasahi, T., Endo, M.: TANSO **196**, 14–18 (2001)
67. Williams, M.C.: Fuel Cells **2**, 87–91 (2001)
68. Britto, P.J., Santhanam, K.S.V., Rubio, A., Alonso, A., Ajayan, P.M.: Adv. Mater. **11**, 154–157 (1999)
69. Che, G., Lakshmi, B.B., Fisher, E.R., Martin, C.R.: Nature **393**, 346–349 (1999)
70. Yoshitake, T., Shimakawa, Y., Kuroshima, S., Kimura, H., Ichihashi, T., et al.: Physica B **323**, 124–126 (2002)
71. Endo, M., Kim, Y.A., Ezaka, M., Osada, K., Yanagisawa, T., Hayashi, T., Terrones, M., Dresselhaus, M.S.: Nano Lett. **3**, 723–726 (2003)
72. Lu, J.P.: J. Phys. Chem. Solids **58**, 1649–1652 (1997)
73. Li, F., Cheng, B.S., Su, G., Dresselhaus, M.S.: Appl. Phys. Lett. **77**, 3161–3163 (2000)
74. Ajayan, P.M., Stephan, O., Colliex, C., Trauth, D.: Science **265**, 1212–1214 (1994)
75. Lau, K.T., Hui, D.: Composites **B33**, 263–277 (2002)
76. Calvert, P.: Nature **399**, 210–211 (1999)
77. Aamedov, M.A., Kotov, N.A., Prato, M., Guldi, D.M., Wicksted, J.P., Hirsch, A.: Nat. Mater. **1**, 190–194 (2002)
78. Zhan, G.D., Kuntz, J.D., Wan, J., Mukherjee, A.K.: Nat. Mater. **2**, 38–42 (2003)
79. Valentini, L., Biagiotti, J., Kenny, J.M., Santucci, S.: Appl. Polym. Sci. **87**, 708–713 (2003)
80. Bhattacharyya, A.R., Sreekumar, T.V., Liu, T., Kumar, S., Ericson, L.M., Hauge, R.H., Smalley, R.E.: Polymer **44**, 2373–2377 (2003)
81. Cadek, M., Coleman, J.N., Barron, V., Hedicke, K., Blau, W.J.: Appl. Phys. Lett. **81**, 5123–5125 (2003)

Diamond: Synthesis, Characterisation and Applications

Roslan Md Nor, Suriani Abu Bakar, Tamil Many Thandavan, and Mohamad Rusop

Abstract In this chapter we review some aspects of the synthesis and characterisation of chemical vapor deposited diamond. Chemical Vapor Deposited (CVD) diamond is arguably the first of the "new" carbon materials that has received extensive research attention due to its potential industrial applications. Intense research activities on CVD diamond that spanned over 30 years brought much progress in understanding and techniques on the synthesis and laboratory demonstration of applications. However, industrial scale applications are still elusive, mainly due to the many technical hurdles that must be overcome in order to fully benefit from the wonderful properties of diamond. Although CVD diamond has been superseded by fullerene in the 1990s, later carbon nanotubes and more recently the emergence of graphene, it is worth looking at this fascinating form of synthetic diamond which may yet make a comeback in years to come. Attention was given to the established techniques for the synthesis and characterisation of CVD diamond as well as issues related to the challenges of industrial applications of CVD diamond.

R.Md. Nor (✉) and T.M. Thandavan
Department of Physics, Faculty of Science, University of Malaya, 50603 Kuala Lumpur, Malaysia
e-mail: rmdnor@um.edu.my

S.A. Bakar
NANO-SciTech Center, Institute of Science, Universiti Teknologi MARA, 40450 Shah Alam, Selangor, Malaysia

Faculty of Electrical Engineering, Solar Cell Laboratory, Universiti Teknologi MARA, 40450 Shah Alam, Selangor, Malaysia

Physics Department, Faculty of Science and Technology, Universiti Pendidikan Sultan Idris, 35900 Tanjung Malim, Perak, Malaysia
e-mail: absuriani@yahoo.com

M. Rusop
NANO-SciTech Center, Institute of Science, Universiti Teknologi MARA, 40450 Shah Alam, Selangor, Malaysia
Faculty of Electrical Engineering, Solar Cell Laboratory, Universiti Teknologi MARA, 40450 Shah Alam, Selangor, Malaysia
e-mail: rusop8@gmail.com

N. Yahya (ed.), *Carbon and Oxide Nanostructures,* Adv Struct Mater 5,
DOI 10.1007/8611_2010_17, © Springer-Verlag Berlin Heidelberg 2010,
Published online: 6 July 2010

1 Introduction

Carbon is the only element in the periodic table that has four isomers from 0 dimensions to 3 dimensions. There are two well known polytypes, namely graphite with sp^2 hybridized bond and diamond with sp^3 hybridized bonds. During the last 30 years, a number of "new" carbon forms have received vast attention, all of which are variations of the two forms of carbon mentioned above. The "new" carbon forms are CVD diamond [1–3], which is the subject of this chapter, fullerenes, carbon nanotubes and graphene.

Fullerenes are a class of carbon materials with 0 dimensions which were first discovered in molecular beam experiments in 1985 by Harold W. Kroto, Robert F. Curl and Richard E. Smalley [4]. The most abundant of the fullerenes is C_{60}. It has the structure of hollow spherical cluster of 60 carbon atoms forming a complete molecule. In each molecule, 12 pentagons associated with sp^3 hybridization and 20 hexagonal formation of carbon atoms involving sp^2 hybridization [5, 6], joined together to form a complete spherical shape.

Carbon nanotubes were first discovered in 1991 created much attention as one dimensional nanostructured pure carbon materials with numerous useful properties which are of interest to both industrial applications and fundamental research [7]. The latest of the new carbon material is graphene which is one atom thick layer of two-dimensional carbon atoms with sp^2 bonding. Theoretical speculation have long showed that a single layer graphene sheet should be thermodynamically unstable but this was dispelled in 2004, when it was first fabricated by mechanically cleaving of Highly Oriented Pyrolytic Graphite (HOPG) [8].

2 Diamond

Diamond has been revered by humans since ancient times mainly as gemstone. Its scarcity, hardness and bright sparkle makes it a status symbol. Apart from being the much sought after gemstone, the many remarkable properties of diamond have similar appeal to material scientists and engineers. Among the properties are hardest known material, chemical inertness, and highest thermal conductivity at room temperature, least compressible and highest stiffness. When doped, diamond exhibit semiconducting properties with a larger band gap of 5.4 eV which can be useful for the fabrication of microelectronics devices for high frequency and high power applications.

Diamond which is carbon in the most concentrated form, involves sp^3 hybridization of the valence electrons in carbon to form chemical bonds. In diamond each carbon atom shares all four of its available electrons with adjacent carbon atoms, forming tetrahedral unit, with 1.54 Å in length [9]. This shared electron-pair bonding forms the strongest known chemical linkage, sp^3 hybridized electrons forming the covalent bond, which is responsible for the many superlative properties

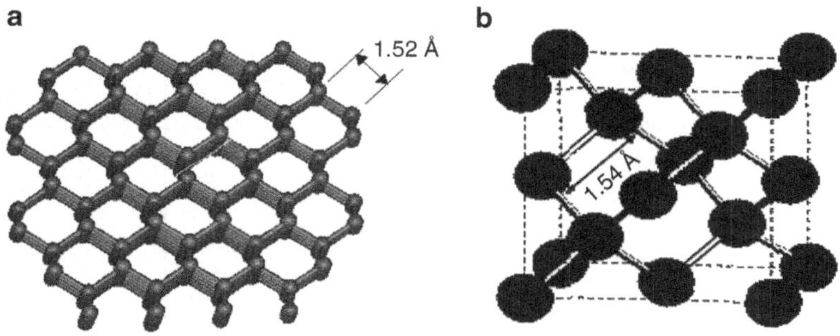

Fig. 1 (**a**) Diamond lattice (**b**) Face-centered cubic structure of diamond

of diamond. This bond form equal angles of 109°28′ separating each carbon atoms. The repeating structural unit of diamond consists of eight atoms, which are fundamentally arranged, in a cube [10] as shown in Fig. 1.

With this cubic form and its highly symmetrical arrangement of atoms, diamond crystals can develop in a variety of different shapes known as 'crystal habits'. The octahedron shape is the most common crystal habit. But diamond crystal can also form cubes, dodecahedra, and even combinations of these shapes.

2.1 Natural Diamond

Diamonds were first mined in India as early as 4,000 years ago. Up to until the eighteenth century mines of Golconda in India were the only known market for the trade of gemstones diamond. Then Brazil became the main producer after diamonds were found there in 1726. Later in 1867 diamond was discovered in Hopetown, South Africa [11], which up to this day remained a major supplier.

Natural diamonds were thought to be created under extreme conditions due geological processes at more than 150 km in the mantle of the earth. The diamond crystals formed were then transported to the surface of earth in molten rock, or magma in explosive volcanic eruptions. Although diamonds are not formed near the earth's surface, where graphite is the preferred stable form of carbon, some diamonds formed in extreme conditions in the crust of the earth survived at the surface because a temperature of about 600°C is required to decompose diamond to carbon dioxide.

2.2 Synthetic Diamond

Today, synthetic diamond can be fabricated using two different methods, with different conditions. The earlier and more established method was the High

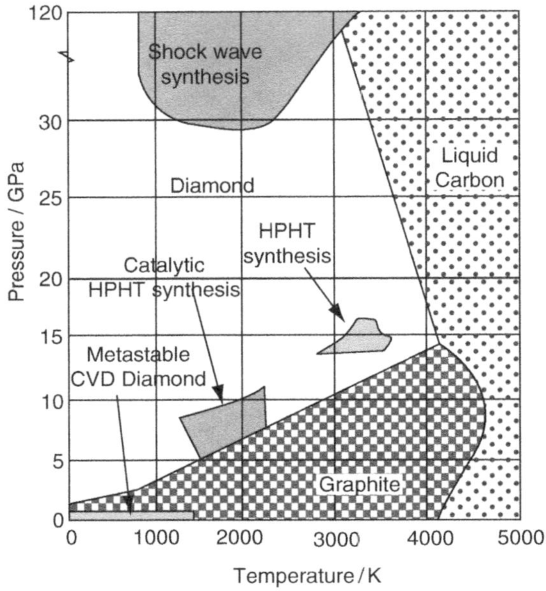

Fig. 2 Phase diagram of carbon [13] – gives a short description of how the phase-diagram was constructed from experimental data and calculations

Pressure High Temperature (HPHT) technique which duplicates the condition of natural diamond formation [9]. Using the HPHT technique synthetic diamond can be converted directly from graphite under specific conditions. At present, synthetic diamond formed using the high pressure high temperature technique is a major source of industrial diamond, mainly as abrasives and drill bits [12]. Alternatively, the Chemical Vapour Deposition (CVD) technique is the method to synthesis diamond at low pressure and low temperature conditions from gas phase precursors. The main attraction of CVD diamond is its film form where new potential applications such as coatings or optical mirrors are possible. The possibility of CVD diamond synthesis was first proposed in 1955, when Bundy et al. [13] presented the carbon phase diagram as illustrated in Fig. 2. The diagram shows the thermodynamic stability for graphite and diamond at different temperatures and pressures. The position of the diamond/graphite equilibrium line was established by thermodynamic calculations based upon the measured physical properties of graphite and diamond in the temperature range from 300 to about 1,200 K and by experiments on growing or graphitization of diamond. It has been proposed that a temperature, pressure phase diagram for carbon which included a wide temperature, pressure area at lower pressures and temperatures in which there would be in apparent equilibrium state or 'pseudo-equilibrium' between diamond and graphite [14]. The quantitative of free energies of graphite and diamond determination by Rossini et al. showed that their difference at one atmosphere, 0 K, is only about 2.5 KJ mol^{-1} [15].

2.3 High Pressure High Temperature (HPHT) Diamond

The carbon phase diagram illustrated in Fig. 2 shows that diamond is unstable with respect to graphite except at high temperatures (>1,300°C) and high pressures (>40 kbar).

In 1955, H. T. Hall and co-workers at General Electric developed the solvent-catalytic HPHT method [13, 16]. In this procedure graphite was compressed in a hydraulic press in the presence of a suitable molten metal catalyst/solvent until carbon dissolved and diamond crystallized. Sufficiently high temperature was provided by passing an electric current through the sample. Molten metals like Ni, Co or Fe are commonly used which dissolve graphite at elevated pressures of between 5 and 10 GPa, and temperatures in the range of 1,300–2,300°C. The dissolved carbon re-crystallizes into diamond at normal pressures and temperatures. The transition of graphite to diamond is accompanied by volume decrease of 43%, posing severe technical problems to maintain suitable pressure, being a vital condition [17]. If the pressure is not maintained, the formed diamond would convert itself back to graphite immediately. Synthetic diamond crystals with sizes ranging from few nanometres to millimetres were recovered from the process by dissolving the metal catalysts in acid [18].

There are applications of HPHT diamond materials. Diamond grit are used as cutting tool in industries and the main bulk of diamond used here are synthetic more viable than the use of natural diamond. Russell Hemley at Carnegie institution had found that most HPHT synthetic diamond is yellow and most CVD diamond is either milky white or greyish, limiting their optical applications. Colourless diamond are costly to produce using the HPHT method and this situation had limits the general applications of these diamonds as gems, in optics, and in scientific research. CVD diamond which is grown under non-thermal equilibrium conditions can be stable even at low pressure [19]. CVD diamond which can be produced in film form at this condition has similar properties as natural diamond and therefore provided tremendous prospective applications in industry.

2.4 Low Pressure Low Temperature CVD Diamond Synthesis

The notion of growing diamond from gas at pressures lower that atmospheric and temperatures lower than 1,000°C may seem counter intuitive, but the mechanism involved is totally alien of the HPHT process. The first reported success of low pressure low temperature was reported by Eversole in 1962, whereby the optimal temperature and pressures were 900°C and 0.1 Torr respectively [20]. In a patent they described a cyclic process for hydrocarbon pyrolysis onto a diamond surface followed by a separate graphite etching step using H atoms.

This was extended by Derjaguin et al. from the Soviet Union [10]. They performed a physical chemistry experiments such as thermal gravimetric analysis

(TGA) to observe weight gain in diamond powder with complete removal of residual graphite, from a CVD synthesis. They were partially successful in producing epitaxial films growth at a slow which was not completely free of graphitic carbon. These initial works demonstrated the feasibility of CVD diamond synthesis, albeit on a diamond substrate.

In 1968, Angus [21] showed that synthetic diamonds were identical with natural diamond and later showed that semiconducting diamond can be obtained by doping diamond grown from methane and hydrogen with boron [22, 23]. Japanese scientists at the National Institute for Research in Inorganic Materials (NIRIM) in Japan developed various diamond thin film deposition techniques such as CVD with the activation of carbon-containing gases, physical vapour deposition (PVD) from a solid carbon source and hybrid processes involving solids, carbon and catalytic materials and gases as the sources. These were done both on diamond or non-diamond substrates. Generation of atomic hydrogen near deposition surface was found to be the major common factor in diamond growth [11].

3 Mechanism of CVD Diamond Growth

Chemical vapor deposition refers to the process where solid materials are formed from the gas phase at some suitable temperature and pressure. Figure 3 shows a schematic diagram demonstrates the chemistry of hydrocarbon radicals and

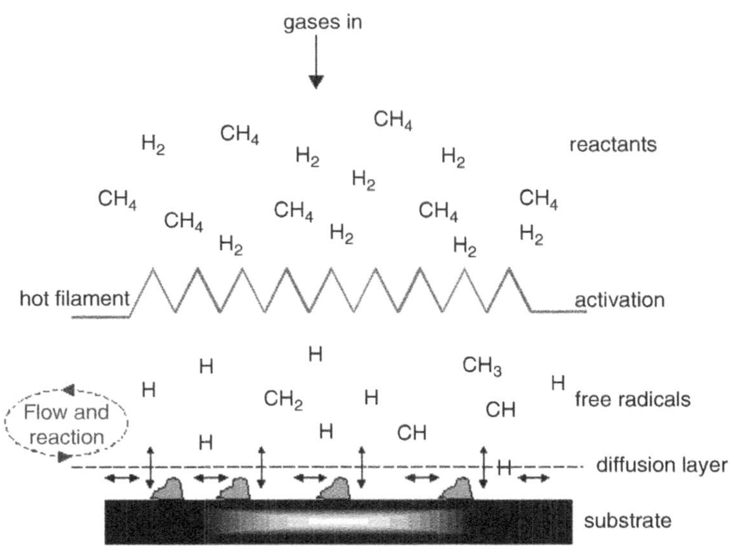

Fig. 3 Chemistry of hydrocarbon radicals and hydrogen atoms in a hot filament chemical vapor deposition method

hydrogen atoms generation in a hot filament, representing typical diamond CVD environment.

Synthesis of CVD diamond would involve some carbon containing gas (CH_4, C_2H_2) diluted in hydrogen. The precursor gas would not readily condense on the substrate to form diamond. A mechanism to dissociate the gas to generate active species such as CH_3, CH_2, CH, and C are necessary. One of the most common techniques is using a hot filament, where the gas molecules are thermally dissociated.

Typically the reactant gas mixture, for example low concentration of CH_4 in H_2 is introduced into an evacuated reactor chamber trough a hot filament above a substrate. With filament temperatures reaching 2,000°C, CH_4 and H_2 are decomposed upon collision on the surface of the hot filament wire to produce a variety of hydrocarbons radicals and H atoms.

The radicals then reacted chemically either in gaseous phase or on selected substrate materials. The general opinion is that the hydrocarbon methyl radical CH_3 is the main growth species in a CVD growth condition [24–26]. To picture this simplistically, in a deposition environment, CH_3 radical diffuses to substrate surface and is deposited on an active surface site, retaining its tetrahedral form with the three hydrogen atoms sticking away from the substrates. The hydrogen atoms can be removed from the surface by H atom to form H_2 or a CH_3 radical to form CH_4, leaving a dangling bond to receive another CH_3 radical or to form bridging bond with another surface carbon atom. The process will continue, building up the diamond nucleus with the addition of CH_3 radicals and the removal of surface H atoms and the addition of CH_3 radical to dangling bonds on the surface.

Other active species such as CH_2, CH and C although at a much lower concentration at the substrate surface, contributed to the growth of diamond [27]. The CH_2 radical upon condensation on the surface have a dangling bond ready for the addition of another radical or for bridging bond with another carbon atom on the surface. Similarly, the condensation of CH and H radicals may lead to carbon with diamond structures based on similar processes of addition and abstraction of actives radicals.

The chemistry of hydrocarbon species and hydrogen atoms has already been studied by many researchers and scientist in the past 10–15 years ago. Bachmann et al. [27] introduced the first C H O phase diagram providing a common scheme for all major diamond chemical vapor deposition (CVD) methods used to date. They had discovered that at low pressures, good quality diamond films can be synthesized from different gas mixtures by thermal CVD, Hot Filament CVD, combustion flames and various plasma deposition techniques.

3.1 Choice of Substrate for Diamond Growth

It is most desirable to deposit diamond on any substrate of choice. However, there are many restrictions. Ferrous metals are unsuitable due to the high solubility of carbon. Materials with melting points lower that the deposition temperatures are also not suitable.

For good adhesion the selected substrates are required to have a comparable thermal expansion coefficient with natural diamond which is 1.18×10^{-6} K^{-1} [28]. This is important because the substrate will expand at the high temperature and during deposition and upon cooling to room temperature; the substrate and the diamond film will contract unevenly and at different rates due to the different coefficient of expansions resulting in compressive or tensile residual stresses which could lead to cracks and delaminating of diamond film. For homoepitaxial growth, another consideration is the matching of the lattice of the substrate with that of diamond. This condition, while not affecting the growth of diamond can significantly affect adhesion of the diamond film to the substrate.

The most efficient substrate is diamond itself. Silicon is the next most efficient and the most studied substrate for CVD diamond growth. The lattice mismatch between silicon and diamond is about 52% where the lattice constant of silicon is 5.4301 Å and for diamond is 3.5667 Å. Furthermore, there a large difference in the coefficient of thermal expansion where for silicon it was two orders of magnitude smaller at 2.1×10^{-8} K^{-1} [29]. Despite these differences, efficient growth of CVD diamond with good adhesion to silicon substrates was routinely reported. The reason for this was that the growth of diamond was not directly on the silicon substrate but on a carbide layer formed prior to diamond nucleation [30]. Similarly, other carbide forming metals such as molybdenum, tungsten and titanium, and the carbides of this metal have been utilized as substrates. The degree of adhesion varied with different materials where the fabrication of free standing CVD diamond would favor substrates with the least adhesion. The effect of carbide formation can be demonstrated with the aid of Fig. 4, which shows the XRD spectra of CVD diamond on silicon and molybdenum substrate deposited using the MWPCVD at different temperatures. Notice that significant carbide formation is evident in the sample on molybdenum, which inevitably affect adhesion of the diamond film.

3.2 Substrate Pretreatment Methods and Diamond Nucleation

CVD diamond formation on non-diamond substrates occur by a two steps mechanism, that is, nucleation of diamond seeds and further growth of diamond on the seeds. The efficiency of the nucleation in terms of density and quality will inevitably influence the growth rate and morphology of the film. The nucleation process occurred in a carbon saturated state which facilitated the nucleation process [23]. The initial surface condition of substrate material is very important for diamond nucleation and growth rate. Even with silicon, the strongly favored substrate material, diamond nucleation is very poor on mirror-polished silicon substrates [31]. Therefore pretreatment of substrate are commonly applied to enhance diamond films growth on silicon.

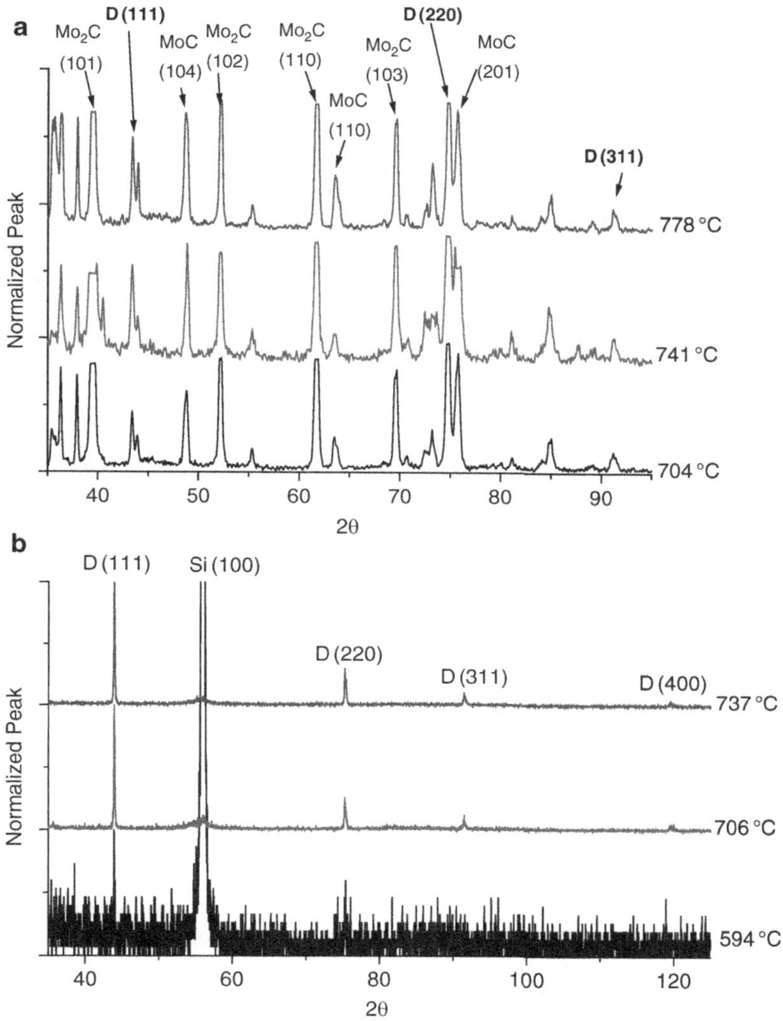

Fig. 4 XRD spectra of CVD diamond deposited at different temperatures on (**a**) Silicon substrates (**b**) molybdenum substrates. Notice the rich carbide peaks in the diamond sample on the molybdenum substrate

The nucleation density of diamond could be significantly enhanced by scratching or abrading the silicon substrate surface with micrometer sized diamond abrasive [32, 33], SiC sand paper [34], cubic BN [35] and stainless steel paste [36]. Among these abrasives, diamond pastes gave the best nucleation density and subsequently the best growth rates. This phenomenon is demonstrated with the aid of Fig. 5 which shows the SEM micrographs showing the effect of scratching on the nucleation of

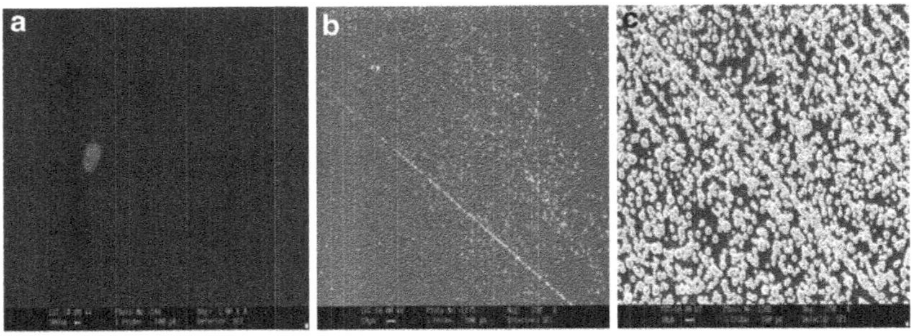

Fig. 5 Effect of scratching the substrate surface (silicon) with (**a**) unscratched (**b**) sand paper (**c**) SiC abrasive (**d**) diamond particle (15 μm)

CVD diamond grown using the microwave plasma deposition technique. Here, three mirror polished Si (100) substrates, the first was unscratched (Fig. 5a), second scratched with SiC abrasives (Fig. 5b) and the third with diamond particles of average grit size of 15 μm (Fig. 5c) were compared. Deposition was done on all three substrates simultaneously to keep the conditions similar. It is evident from the SEM micrographs that scratching Si surface with diamond significantly enhanced the nucleation of diamond, both in terms of quality and quantity. Notice that there were preferential growth in the scratch lines of the substrates abraded with SiC (Fig. 5b) and diamond grits (Fig. 5c).

This phenomenon can be explained as follows. Scratching removes atoms from the surface and leaves behind dangling bonds which are active sites for hydrocarbon radicals to react and condense on. When scratched with diamond abrasives, some of the nanosized diamond fragments may be trapped on the silicon surface and acts as seeds for efficient diamond nucleation and enhancement of growth. The formation of grooved lines facilitated turbulent flow of the precursor gas during deposition which somehow promoted nucleation of diamond on silicon.

Alternative techniques such as ion beam sputtering to create sub-micrometer scale craters using focused 25 keV Ga^+ beams [37], etching in KOH to create micrometer scale *V* grooves [38], coating with hydrocarbon oil of a low vapor pressure and high thermal stability [39], coating with thin layer of evaporated carbon [40], applying substrate bias voltage [41, 42] and 25 keV Si^+ ion beam implantation [43] are some of the pretreatment methods that have been applied to enhance the nucleation of diamond films. Yang et al. [44] had reported that ion implantation did not make substrate surfaces as rough as mechanically scratched or chemically etched process but still enhanced nucleation.

It is believed that by applying pretreatment process on silicon substrate, the resulting surface stresses have some influence on diamond nucleation. The surface energy of the substrate may increase due to an atomic displacement process which influences the diamond growth process. It is also possible that surface treatment cause reduction in the surface energy difference between Si and diamond enhance the nucleation. Pretreatment process allows changes in surface morphology because it favors the formation SiC which functioned as the diamond growth layer [44].

In a conventional method of hot filament chemical vapor deposition environment most of the species are neutral and positively and negatively charged species are present with different number density and with the overall density expected to be very much less than, for example, in a microwave plasma [45]. Electrically biasing silicon substrates during deposition, using either negative or positive voltage has been shown to be effective in enhancing nucleation rates. In the case of negative bias voltage, the effect may be attributed to a sputtering effect due to an increase in the kinetic energy of incident positive ions. When the substrate is positively biased negative ions and electrons bombard the substrate surface to bring about sputtering and the creation of active nucleation sites.

George et al. [45–47] reported that negative bias definitely enhances the amount and crystalline of the deposits which was identified by XRD and AFM studies. Cui and Fang [48] have reported that biasing influenced the distribution of hydrocarbon species and the electron temperature. They reported that nucleation density increased by a factor of two at a 120 V bias meanwhile at 200 V it was enhanced more than an order of magnitude. Sawabe et al. [49] found that positive biasing method can enhance nucleation density and Chao et al. [50] observed that with positive bias much smoother surface was obtained. This can be due to the significantly small number densities of negative ions, so that the surface was essentially bombarded by electrons which resulted in surface activation with little damage. Stoner et al. [51] reported that negative biasing in MWCVD have also increased the nucleation density of diamond. An investigation of DC and AC bias voltage by Kromka et al. outlined some interesting features [52]. At higher DC bias voltage degradation of nucleation process was reported compared to low DC bias voltage which was due to excessive sputtering during the deposition process. In AC bias voltage higher voltage is required to obtain dense packed and well-faceted diamond grains.

A slight increase of hydrogen atom percentage with biasing are not expected to affect the gaseous component mean temperatures and hydrocarbon species concentration distributions. However, methane and hydrogen decomposition can be accelerated by electron bombardment adding growth species and H atoms to enhance nucleation. Although the electron density depends on the filament temperature, the numbers of electron impinge on substrate surface will be increased due to biasing and create active sites for nucleation. In another study, Mahajan et al. reported coated SnO_2 on silicon substrates and observed nucleation density increased irrespective of whether the substrate surface is pretreated or not, prior to SnO_2 deposition [53]. The issue enhancing of nucleation density of CVD diamond is still unresolved. Although many published reports have provided some understand basic underlying mechanisms, a truly efficient technique have yet to be developed for industrial scale production of CVD diamond. Furthermore, nucleation enhancement may be influenced by the formation of diamond structured stable clusters of from hydrocarbon in the gas phase which condensed on the substrate surface as preferred nucleation sites [54].

3.3 The Role of Atomic Hydrogen in CVD Diamond Growth

As mentioned earlier atomic hydrogen (H) is one of the key constituent in the growth of CVD diamond. They were produced heterogeneously by the decomposition of hydrogen gas (H_2).

$$H_2(g) \rightarrow H + H \tag{1}$$

The condensation of all the carbon containing radicals basically adds carbon to the substrate surface. It is well known that the formation of non-diamond carbons (graphite and amorphous carbon) is much more efficient than that of cubic diamond. The atomic hydrogen formed preferentially etches sp^2 graphite and amorphous carbon at a much higher rate than diamond [55–57]. Another important role of hydrogen atom is to stabilize the growing sp^3 bonded carbon on the substrate whereby resulting in surface termination which prevented cross-linkage, and subsequent reconstruction of the surface to graphite. During deposition dangling bonds might be created by thermal desorption and abstraction of the surface hydrogen atoms. Thus a large number of reactive hydrogen atoms are required to make bonds to the excess dangling bonds, avoiding surface graphitization. Spatial distribution of atomic hydrogen is strongly dependent on parametric conditions [57].

Atomic hydrogen accelerates the gas phase reaction and promotes the decomposition rate of CH_4 or C_2H_2 into active radicals.

$$CH_4 + \bullet H \rightarrow CH_x + H_2 \quad (x = 1 \text{ to } 3) \tag{2}$$

$$C_2H_2 + H \rightarrow C_2H + H_2 \tag{3}$$

4 Plasma Enhanced Techniques for Chemical Vapour Deposition Diamond Synthesis

4.1 Microwave Plasma Enhanced CVD (MWPECVD)

Scientists at the National Institute for Research in Inorganic Materials (NIRIM), Japan, demonstrated the first working microwave plasma enhanced CVD diamond system [58] where polycrystalline diamond growth was demonstrated on a silicon substrate. It is known that molecules are efficiently dissociated in microwave plasmas giving rise to a cocktail of neutral and charged particles as well as electron. The generation of plasma in a microwave system involves the oscillation of charged particles at the microwave frequency. Collision of electrons with gas atoms and molecules generates high ionization fractions. Microwave plasmas are typically described as having "hot" electrons and "cool" ions and neutrals where the electron temperature can be in the hundreds of eV but the ions are essentially at room temperature. Under such conditions, high flux of energetic electrons on the substrate surface functioned as growth activators as a result of active sites formed.

Microwave energy is introduced into the reaction chamber using antennas or wave guides which convert a rectangular microwave signal into a circular mode, later pass through a silica microwave window (vacuum barrier) into a cylindrical cavity which is a PECVD process chamber. The location of the luminous plasma ball is controlled by the cavity tuning as well as the pressure, power, and gas phase composition of plasma. The luminous ball will further contract in size as the pressure in the chamber increases. The luminous ball will increase in diameter as the microwave power is increased.

This technique of plasma generation exhibits good stability and high density plasmas and the generation of the plasma does not require any electrode, thus eliminating the possibility of contamination from particles sputtered from the electrode. The plasma is also confined to the center of the deposition chamber to form ball-shaped plasma, thus carbon deposition on the walls of chamber is minimal. The main disadvantage the MWPECVD is the difficulty of scaling up to industrial scale.

4.2 Direct Current (DC) Plasma Enhanced CVD

In DC plasma CVD, a negative or positive DC voltage is applied to the substrate platen or to an auxiliary electrode to generate plasma in rarefied gas mixture. Intermediate electrodes may be used to guide the plasma and to alter the plasma position and properties. DC plasma enhanced CVD shows both advantage and disadvantage. The diamond deposition area is only limited by the size of the electrodes and the DC power supply which is technically relatively easy to overcome. The disadvantage of DC plasma enhanced CVD is that it may self extinguish if a very thick film is grown since a biased platen will change its bias if coated by a dielectric diamond film.

Japanese scientists developed the DC arc PECVD methods which created thermal plasmas capable of growing diamond films at rates $\geq 20\ \mu m\ h^{-1}$. Kurihara et al. at Fujitsu Laboratories have developed a DC plasma jet process called the DIA-jet process [59] which utilized a gas injection nozzle composed of a rod cathode concentric with a tube anode. Gases passed between the cathode and anode and were sprayed out from an orifice in the anode. The gas mixture is CH_4 and H_2 with a carrier gas (Argon or Helium); the total pressure is between 4.1 and 41 kPa (30–300 torr) were typical. The DC arc is sustained by 80–150 V and 10–50 A (power of 0.8–7.5 kW); the substrate was water cooled to maintain the substrate temperature between 520 and 1,220°C. Typical growth rates reported to be as high as 80 $\mu m\ h^{-1}$. DC arc methods have demonstrated high deposition rates and ability to synthesize high quality diamond and are gaining acceptance as a commercially viable manufacturing technology for diamond film synthesis; De Beers's industrial diamond developing CVD diamond using 10 kW DC Arc jet system.

4.3 Radio Frequency Plasma Enhanced CVD

Radio Frequency plasmas are widely used in the fabrication of silicon based microelectronics mainly in dry etching systems. As a result, rf plasma systems technology are the most developed. There have been sparse reports on the use of rf plasma to fabricate diamond thin films. In most reports, inductively coupled rf plasmas were used since these are simple systems capable of generating high density plasmas. Furthermore, rf power coupling to the gas using conductive coils located outside the reactor chamber, thus eliminating contamination from sputtered coil materials. Rudder et al. [60] and Bozeman et al. [61] reported the deposition of CVD diamond using rf plasmas with the methane–hydrogen system and verified the existence of cubic diamond based on Raman spectroscopy and X-ray diffraction. Although existence of cubic diamond was proven, Raman spectra showed significant inclusion of graphitic carbon. Furthermore, less facetted diamond crystallites based on their SEM analysis again indicated the inclusion of graphitic carbon, at least on the surface. An important point reported was that diamond growth in an rf plasma was feasible at pressures >1 torr and high input power. Noda et al. [62] reported using ICP rf plasma to deposited diamond using methanol-hydrogen-water system at relatively low pressure of 80 mtorr. They reported enhanced diamond growth with increasing water partial pressure of up to 40 mtorr.

Studies by our group [63, 64] showed that significant silicon substrate damage occurred at pressures lower than 20 mtorr based on the observation of pits and no deposition. This is especially true when the plasma was operated in the H-mode where the plasma density is much higher than the E-mode. This is mainly due to the nature of inductively coupled rf plasmas which is high ion and electron temperatures. Furthermore, the formation of the sheath layer at the interface created significant electric field which gives kinetic energy to positive ions in the plasma. The ions impinge on the substrate surface as a consequent of its kinetic energy and also due to the fact the substrate is also more negatively charged that the environment, as a result of faster electrons. Ion temperatures of tens of eV and electron temperatures in the hundreds of eV are common in inductively coupled plasmas. Thus, the needs for higher pressures where ion and electron densities and temperatures are reduced due to gas phase collisions where diamond liked carbon are formed. Our study on a hybrid rf plasma and hot filament system showed that the effect of rf plasma activation and hot filament activation were independent of each other [65] where results from increasing rf power from 0 to 400 W reduced the amount of cubic diamond and enhanced graphitic carbon with increasing rf power. Excessive surface sputtering in the rf plasma due to the high density and energetic ion species seemed to be the main inhibitor of diamond growth.

5 Doping of CVD Diamond

The many excellent mechanical, chemical, optical and chemical properties of CVD diamond coupled with it is a material with a wide band gap makes it a prime candidate for specialized microelectronics materials. Being a material of negative electron affinity(NEA) conducting diamond films can function as robust field electron emitters, Doping with boron, which has an acceptor level of E_v +0.37 eV [66] has been investigated using the Cold Implantation Rapid Annealing (CIRA) technique [67]. Besides boron as a p-type dopant, n-type dopants like sulphur, lithium, sodium, and phosphorus were used in doping of diamond with varying degrees of success. The doping process can be done either during deposition (in situ doping) or post deposition (ex situ doping).

Li et al. [68] reported that in microwave plasma assisted CVD, sulphur incorporation was enhanced by the presence of boron but the donor level energy was reported to decrease from 0.52 to 0.39 eV. Borst and Weis investigated the electrical properties doped diamond film and found that Li, Na and P doped films had very high resistivity of over 10^9 Ωcm^{-1} [69]. They reported boron doped film showed activation energy of electrical resistivity in the range of 0–0.43 eV compared to 0.16 eV for lithium doped. Suzuki et al. [70] demonstrated that the p-type conduction existed for B doped CVD diamond based on (I–V) and (C–V) measurements. The barrier height Φ_b for the Schottky junction was reported to be 1.77 eV. They also reported the proportional increase of net acceptor concentration with boron concentration based on C–V measurements. Boron-doped CVD diamond electrodes fabricated by Latto et al. [71] showed that at an oxygenated electrode surface, two time constants were observed in impedance plots where the electron transfer process was thought to be mediated by surface states.

Bohr et al. [72] suggested that phosphorus addition during diamond growth influence the growth kinetics primarily due to surface reaction rather than to changes in the gas activation. They used 5% ultrapure phosphine 99.998% mixed in 94% hydrogen and 1% CH_4. Koizumi et al. [73] successfully fabricated phosphorus doped n-type diamond thin films with the activation energy of carriers at 0.46 eV in a higher temperature region and the Hall mobility was 28 $cm^2\ V^{-1}\ s^{-1}$ for a sample with 600 ppm PH_3 content in the deposition mixture. There has been report that increasing the amount of dopant source gas during diamond growth enhanced the growth rate for example, Tsang et al. reported enhancement with addition up to 2,000 ppm but decreased when PH_3 concentration reached above 3,000 ppm [74].

Sulphur doping in diamond film using the MWCVD system was reported by Petherbridge et al. [75]. They showed that sulphur incorporation into the diamond film are directly proportional to the H_2S concentration in gas phase at number densities about 0.2%. The sulphur incorporation was further supported by the four point probe measurement. The group compared HFCVD with MWCVD deposition of sulphur doping into diamond [76]. They reported little effect on film morphology or growth rate was observed for HF grown diamond films, even at high doping levels (1% H_2S in the gas phase), and little or no evidence was seen for S incorporation into these films. In contrast, deposition from MW plasmas yielded diamond films of which the morphology, degree of S incorporation were strong variation of H_2S incorporation. Nishitani-Gamo et al. [77] reported producing sulphur doped homoepitaxial (001) diamond which showed n-type conduction by Hall Effect measurements in the temperature range 250–550 K. The mobility of electrons at room temperature was reported at 597 $cm^2 V^{-1} s^{-1}$.

6 Characterization of CVD Diamond

One of the main issues related to the utilization of CVD diamond as an industrial material is the purity of the diamond film. As it is well known that sp^2 bonded carbon formation competes with the formation of diamond which is sp^3 bonded carbon. Here, we discuss three of the most common techniques, namely scanning electron microscopy, X-ray diffractometry and Raman spectroscopy.

6.1 Scanning Electron Microscopy (SEM)

Scanning electron microscopic images is undoubtedly the most appealing analysis technique when applied to CVD diamond. SEM gives the morphology of the crystal structure. For example, in Fig. 6, the crystal particle structure of CVD diamond deposited using the hot filament technique is clearly visible. Useful as it is in visually evaluating diamond film growth; the SEM technique by itself is inadequate in determining CVD diamond quality in terms of diamond and non-diamond composition of the sample. Well facetted crystal structures may harbor layers of amorphous carbon on its surfaces.

6.2 X-Ray Diffraction

X-ray diffraction method is one of the non destructive methods used for both lattice parameter measurement and crystallinity measurement. The phenomenon of X-ray diffraction by crystals results from a scattering process in which X-rays are scattered by the electrons of atoms without any changes in wavelength. A diffracted beam is produced by coherent or Bragg scattering only when certain geometrical conditions are satisfied. This may be expressed in either of two forms, Bragg's diffraction law or the Laue equations. From the Bragg equation for the cubic system

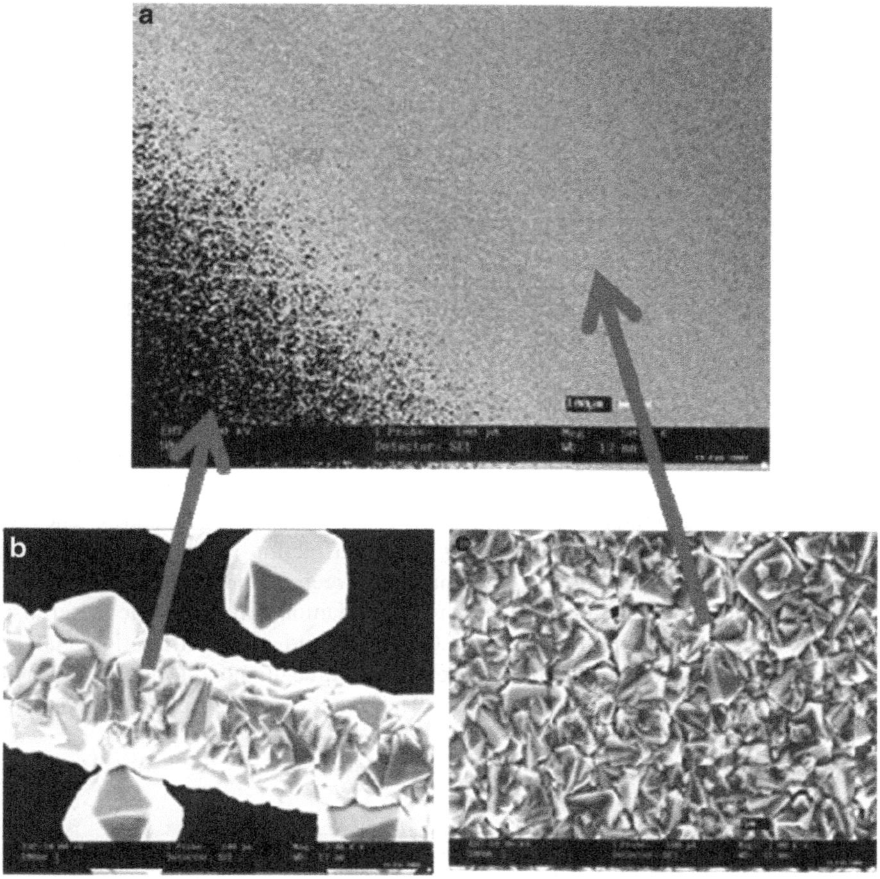

Fig. 6 Scanning electron microscopy (SEM) images (**a**) diamond film layer on silicon substrate (**b**) diamond crystal at the edge of film and (**c**) polycrystalline diamond film in the center position

$$n\lambda = \frac{2a}{\sqrt{h^2 + k^2 + l^2}} \sin\theta \tag{4}$$

$$n\lambda = 2d\sin\theta \tag{5}$$

where d equals to $a/\sqrt{h^2 + k^2 + l^2}$, is the inter-planar spacing, θ is the diffraction angle, n is an integer and λ is the incident wavelength. The h, k and l are the Miller indices of the peaks and a is the lattice parameter of the elementary cell.

Figure 7 shows typical X-ray diffraction spectra of CVD diamond where the crystalline nature of the films is evident. It is obvious that the film is polycrystalline and the peak high distribution can be compare to standard values for powder diffraction, which can reveal preferences in crystal orientation. Also, peak positions

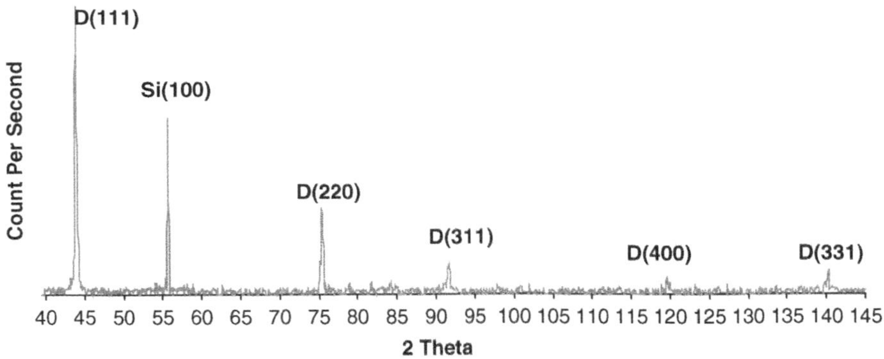

Fig. 7 Typical X-ray diffraction pattern of CVD diamond

which are normally shifted to higher or lower values than that of standards are signs of compressive and tensile stress respectively. Residual stress which is an important feature of CVD diamond on non-diamond substrates determines the adhesiveness of the film to the substrates. Studies using X-ray diffraction revealed that the main reasons can be attributed to lattice mismatch between diamond and substrate, and the difference in the coefficient of expansion of diamond and substrate materials.

In terms of elucidating the quality of CVD diamond, X-ray diffraction analysis has the capability to confirm the existence of crystalline diamond forms and to some extend crystalline graphitic carbon forms. The presence of amorphous carbon may not be efficiently detected.

6.3 *Raman Spectroscopy*

In the case of single crystal diamond, the carbon atoms are bonded to their neighbors by strong covalent sp^3 bonds, forming cubic structure belonging to the O_h^7 (Fd3m) space group. The diamond structure has only one triply generate optical mode at the centre of the Brillouin zone (T_{2g} symmetry) which appears as a sharp line at 1331.9 cm^{-1} [78]. Under ambient conditions, the graphite structure with strong in-plane sp^2 bonding is the most stable phase and the crystal structure belongs to the D_{6h}^4 (P63/mmc) hexagonal space group. The graphite crystal exhibits two Raman active modes, namely the E_{2g2} mode at 1,582 cm^{-1} and under special conditions, the E_{2g1} mode at 42 cm^{-1}.

The peculiarities of Raman spectra from CVD diamond can be demonstrated with aid of Fig. 7. For natural diamond, a sharp peak at about 1,332 cm^{-1} formed the signature for cubic diamond. The sharpness of the peak is due to symmetry of the carbon atoms bonded tetrahedrally in the diamond crystal. Typical spectra for CVD diamond shown in Fig. 7b illustrate a number of characteristics. Firstly, looking at the best diamond film which is the upper trace in Fig. 8b, there a

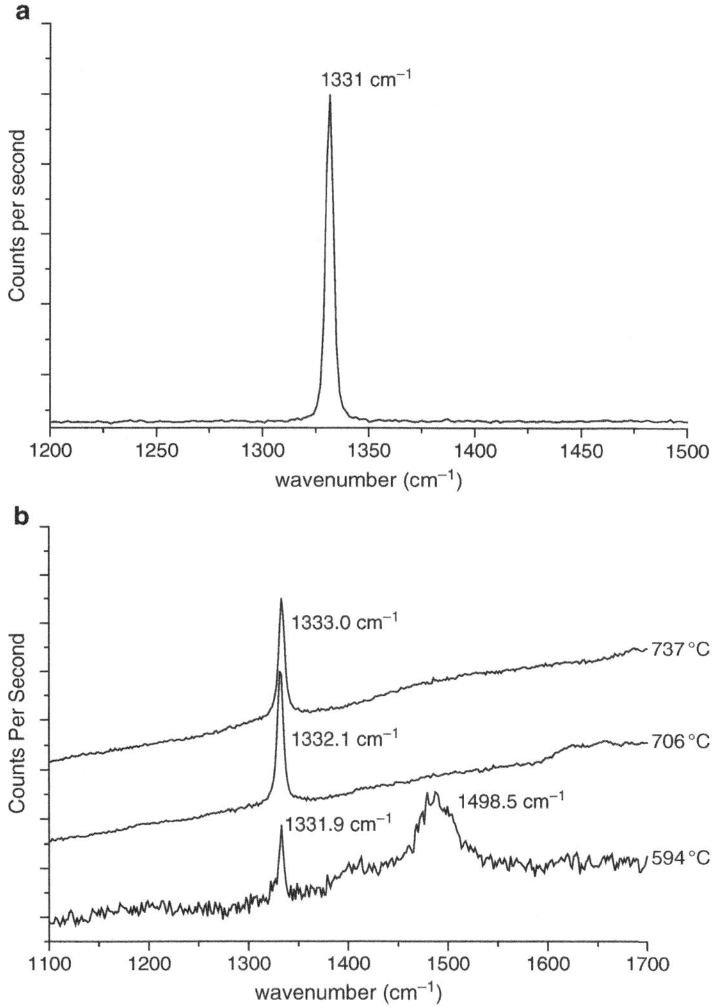

Fig. 8 Raman spectra of diamond (**a**) a clean spectrum from natural diamond (**b**) spectra obtained from CVD diamond synthesized using the microwave plasma technique on silicon substrates with 1% CH_4 and 99% H_2 at varying temperatures

luminescent background is evident. This was due to continuous emission as a result of crystal defects expected in CVD diamond. Another possible reason is due to the inclusion of impurity atoms, mainly nitrogen which is expected to be present in a CVD environment. By comparing the three spectra in Fig. 8b, it is clear that the film deposited at about 600°C significant peaks due to graphitic carbon around 1,400 cm^{-1} and 1,500 cm^{-1}. It is known that the Raman shift due to graphitic carbon is about 50 times more efficient than that of diamond [76], Raman spectroscopy is an excellent technique for monitoring the presence of graphitic carbon in the film. Notice that the peak position of the diamond line increased with deposition temperature. This is due to the increase in the residual stress of films deposited at increasing temperatures, as a result of the difference between the values of the coefficient of expansion of diamond and silicon.

7 Potential Applications of CVD Diamond

Although it has been about 30 years the potential applications of CVD diamond have been enthusiastically promoted in numerous publications, grant submissions and talks it still remained in a potential. However it is worth a revisit and assesses the realities of these potentials. As mentioned earlier, the main strength of diamond as a material is it mechanical, chemical, optical and electrical properties. Optically diamond is transparent from the uv to the far infrared, making an excellent optical windows especially in hard environments. As medical radiation detectors, diamond is attractive because it is stable, non-toxic and atomic number value close to that of biological tissues. Vittone et al. [79] compared the thermoluminescence response of polycrystalline diamond with standard LiF TLD100 under beta irradiation and reported similar response for one of their diamond sample.

However, the most promising potentials are with conducting diamond films. Although, as microelectronics devices the issues related to wafer size single crystal fabrication and n-type doping still persist, applications in the field of electrochemistry seemed the most promising especially as electrodes for operations in harsh chemical environment. Such applications require the film just to be conducting without the stringent specifications of the electrical properties. Vinokur et al. has reported that boron doped diamond films exhibited unsurpassed properties for such applications [80].

Another promising application of doped diamond is as field electron emitters. However undoped diamond seemed to be more efficient emitters than conductive boron doped samples. This was attributed to emission from graphitic carbon present at the grain boundary of diamond particles in polycrystalline diamond samples [81].

8 Challenges of Industrial Scale CVD Diamond Production

After the much progress in the science and technology of the synthesis of CVD diamond the realization for actual industrial applications is still elusive [82]. The main obstacle remains the slow growth rate of a few μm per hour renders the process and the product too expensive. Furthermore, issues on the homogeneity of film quality remain an issue. Substrate surface temperature being a key parameter in the deposition process changed significantly after a layer of diamond is formed due to the high thermal conductivity of diamond. For the application of CVD diamond as wear resistant coatings, serious issues related to substrate adhesion, substrate compatibility and residual stresses are still unresolved. Also, with the relatively high deposition temperature of present techniques most material surfaces are unsuitable for coating. As for potential applications for the fabrication of microelectronic devices, two main technical hitches have to be overcome. Firstly, efficient and cheap techniques have to be developed to fabricate wafer size single crystal diamond. Secondly, these single crystal diamond wafers will have to be doped. While techniques of p-type doping mainly with boron have been quite developed, the progress in n-type doping still lags behind. Taking the silicon experience as a guide, many attempts on doping diamond with phosphorus have been reported but with no real breakthrough in terms of producing n-type diamond suitable for microelectronic devices. As such, the field of CVD diamond is still wide open waiting for a breakthrough for the excitement to return and possible some real industrial applications to be realized.

9 Summary

In this chapter some aspects of the issues related to the synthesis and characterization of CVD diamond is briefly revisited. CVD synthesis techniques using the hot filaments and plasma are discussed as demonstration for the CVD synthesis of diamond as a whole. Characterization of CVD diamond films based on evaluating film quality is discussed in the light of the scanning electron microscopy, X-ray diffraction and Raman spectroscopic techniques. The challenges of producing industrial scale CVD diamond was appraised in the light of technical difficulties highlighted by the numerous work during the golden age of CVD diamond.

References

1. Yarbrough, W.A., Messier, R.: Science **247**, 688–696 (1990)
2. Angus, J.C., Hayman, C.C.: Science **241**, 913–921 (1988)
3. Spear, K.E., Dismukes, J.P.: Synthetic Diamond: Emerging CVD Science and Technology. Wiley, New York (1994)
4. Saito, R., Dresselhaus, G., Dresselhaus, M.S.: Physical properties of carbon nanotubes. Top. Appl. Phys. (1998)
5. Saito, S., Oshiyama, A.: Phys. Rev. Lett. **66**, 2637–2640 (1991)
6. Zhu, X.D., Xu, Y.H., Naramoto, H., Narumi, K., Miyashita, A., Miyashita, K.: J. Phys. Condens. Matter **15**, 2899–2906 (2003)
7. Iijima, S.: Nature **354**, 56–58 (1991)
8. Novoselov, K.S., Geim, A.K., Morozov, S.V., Jiang, D., Zhang, Y., Dubonos, S.V.: Science **306**(5696), 666–669 (2004)
9. Davis, R.F.: Diamond Films and Coatings. Noyes Publications, Park Ridge (1992)
10. Derjaguin, B.V., Fedoseev, D.V., Lukyanovich, B.V., Spitsyn, A.V., Ryanov, A.F., Lavrentyev, A.V.: J. Cryst. Growth **2**, 380–384 (1968)

11. Prelas, M., Popovicci, G., Bigelow, K. (eds.): Handbook of Industrial Diamonds and Diamond Films. Marcel Dekker, New York (1997)
12. Brand, J., Gadow, R., Killinger, A.: Surf. Coat. Technol. **180**, 213–217 (2004)
13. Bundy, F.B., Hall, H.T., Strong, H.M., Wentorf Jr., R.H.: Nature **176**, 51–55 (1955)
14. Bundy, F.B., Bassettand, W.A., Weathers, M.S., Hemley, R.J., Mao, H.K., Goncharov, A.F.: Carbon **34**, 141–153 (1996)
15. Rossini, F.D., Jessup, R.S.: J. Nat. Bur. Stds. **C21**, 491–497 (1938)
16. Bundy, F.P.: J. Chem. Phys. **38**, 631–643 (1963)
17. Bundy, F.P.: J. Geophys. Res. **85**, 6930–6936 (1980)
18. May, P.W.: Philos. Trans. R. Soc. Lond. A **358**, 473–495 (2000)
19. Simu, U., Karlsson, J., Bjorkman, H., Rangsten, P., Hollman, P.: Limits in Micro Replication of CVD Diamond by Moulding Technique, UPTEC 97 135R (1997)
20. Eversole, W.G.: Synthesis of diamond. U.S. Patents Nos 3030187 and 3030188, 17 April 1962
21. Angus, J.C., Will, H.A., Stanko, W.S.: J. Appl. Phys. **39**, 2915–2922 (1968)
22. Chung, D.D.L.: Materials for Electronic Packaging. Butterworth-Heinemann, Oxford (1995)
23. Haenen, K.: PhD thesis, Limburgs Universitair Centrum, Belgium (2002)
24. Harris, S.J.: Appl. Phys. Lett. **56**, 2298–2300 (1990)
25. Goodwin, D.G., Butler, J.E., Prelas, M.A.: In: Handbook of Industrial Diamonds and Diamond Films. Marcel Dekker, New York (1998)
26. Mankelevich, Y.A., May, P.W.: Diam. Relat. Mater. **17**, 1021–1028 (2008)
27. Bachmann, P.K., Leers, D., Lydtin, H.: Diam. Relat. Mater. **1**, 1–12 (1991)
28. Goss, J.P., Hourahine, B., Jones, R., Heggie, M.I., Briddon, P.R.: J. Phys. Condens. Matter **13**, 8973 (2001)
29. Okaji, M.: Int. J. Thermophys. **916**, 1101–1109 (1998)
30. Wedlake, R.J.: In: Field, J.E. (ed.) The properties of diamond. Academic, London (1979)
31. Jiang, X., Klages, C.P., Zhacai, R., Hartweg, M., Fusser, H.: J. Diam. Relat. Mater. **2**, 407–412 (1993)
32. Iijima, S., Aikawa, Y., Baba, K.: Appl. Phys. Lett. **57**, 2646–2648 (1990)
33. Sawabe, A., Inuzuka, T.: Thin Solid Films **137**, 89–99 (1986)
34. Suzuki, K., Sawabe, A., Yasuda, H., Inuzuka, T.: Appl. Phys. Lett. **50**, 728–729 (1987)
35. Singh, B., Arie, Y., Levine, A.W., Mesker, O.R.: Appl. Phys. Lett. **451**, 526–528 (1988)
36. Chang, C.P., Flamm, D.L., Ibbotson, D.E., Mucha, J.A.: J. Appl. Phys. **63**, 1744 (1988)
37. Kirkpatrick, A.R., Ward, B.W., Economou, N.P.: J. Vac. Sci. Technol. B **7**, 1947–1949 (1989)
38. Denning, P.A., Stevenson, D.A.: Appl. Phys. Lett. **59**, 1562–1564 (1991)
39. Morrish, A.A., Pehrsson, P.E.: Appl. Phys. Lett. **59**, 417–419 (1991)
40. Maeda, H., Ikari, S., Masuda, S., Kusakabe, K., Morooka, S.: Diam. Relat. Mater. **2**, 758–761 (1993)
41. Stoner, B.R., Ma, G.H.M., Wolter, S.D., Glass, J.T.: Phys. Rev. B **45**, 11067–11084 (1992)
42. Jiang, X., Klages, C.P., Zhacai, R., Hartweg, M., Fusser, H.: J. Appl. Phys. Lett. **62**, 3438–3440 (1992)
43. Yang, J., Su, X., Chen, Q., Lin, Z.: Appl. Phys. Lett. **66**, 3284–3286 (1995)
44. Azaroff, L.V., Brophy, J.J.: Electronic processes in materials (1963)
45. Dua, A.K., George, V.C., Friedrich, M., Zahn, D.R.T.: Diam. Relat. Mater. **13**, 74–84 (2004)
46. George, V.C., Das, A., Roy, M., Dua, A.K., Raj, P., Zahn, D.R.T.: Thin Solid Films **419**, 114–117 (2002)
47. George, V.C., Das, A., Roy, M., Dua, A.K., Raj, P., Zahn, D.R.T.: Appl. Surf. Sci. **212–213**, 287–290 (2003)
48. Cui, J.B., Fang, R.C.: J. Phys. D Appl. Phys. **29**, 2759–2762 (1996)
49. Sawabe, A., Inzuka, T.: Appl. Phys. Lett. **46**, 146–147 (1985)
50. Chao, C.H., Popeviei, G., Charison, E.J., Charlson, E.M., Meese, J.M., Prelas, M.A.: J. Cryst. Growth **140**, 454 (1994)
51. Stoner, R.J., Maris, H.J., Anthony, T.R., Banholzer, W.F.: Phys. Rev. Lett. **68**, 1563–1566 (1992)

52. Kromka, A., Janík, J., Balon, F., Kubovic, M., Cerven, I., Dubravcova, V.: Thin Solid Films **433**(1–2), 73–77 (2003)
53. Mahajan, J.R., More, M.A., Patil, P.P., Sainkar, S.R.: Mater. Sci. Eng. **B97**, 117–122 (2003)
54. Shao, Q., Fang, R., Wang, G., Xue, Z.: Prog. Cryst. Growth Charact. Mater. **30**, 221–226 (2000)
55. Dong, L., Zhang, Y., Ma, B., Fu, G.: Diam. Relat. Mater. **11**, 1648–1652 (2002)
56. Johnsson, E., Carlsson, J.O.: Diam. Relat. Mater. **4**, 155–163 (1995)
57. Ashfold, N.R., May, J.P., Petherbridge, W.R., Rosser, K.N., Smith, J.A., Mankelevich, Y.A., Suetin, N.V.: Phys. Chem. Chem. Phys. **3**, 3471–3485 (2001)
58. Hirakuri, K.K., Kobayashi, T., Nakamura, E., Mutsukura, N., Friedbacher, G., Machi, Y.: Vacuum **63**, 449–454 (2001)
59. Knotek, O., Bosch, W., Leyendecker, T., Munz, W.D., Falkenstein, R.: 8th ICVM, Linz, vol. C11.2, p. 340 (1985)
60. Rudder, R.A., Hudson, G.C., Posthill, J.B., Thomas, R.E., Hendry, R.C., Malta, D.P., Markunas, R.J., Humphreys, T.P., Nemanich, R.J.: Appl. Phys. Lett. **60**, 329–331 (1992)
61. Bozeman, S.P., Tucker, D.A., Stoner, B.R., Glass, J.T., Hooke, W.M.: Appl. Phys. Lett. **66**, 3579–3581 (1995)
62. Noda, H., Nagai, H., Shimakura, M., Hiramatsu, M., Nawata, M.: J. Vac. Sci. Technol. A **16** (6), 3170–3174 (1998)
63. Ng, K.K., Liew, Y.S., Md Nor, R., Wong, C.S.: Jurnal Fizik Malaysia **23**(1–4), 51 (2001)
64. Liew, Y.S.: MSc thesis, University of Malaya (2002)
65. Tan, C.H., Ng, K.H., Md Nor, R., Wong, C.S.: Jurnal Fizik Malaysia **25**(3–4) (2004)
66. Menon, P.M., Edwards, A., Feigerle, C.S., Shaw, R.W., Coffey, D.W., Heatherly, L., Clausing, R.E., Robinson, L., Glasgow, D.C.: Diam. Relat. Mater. **8**, 101–109 (1999)
67. Vogel, T., Meijer, J., Zaitsev, A.: Diam. Relat. Mater. **13**, 1822–1825 (2004)
68. Li, R., Hu, X., Shen, H., He, X.: Mater. Lett. **58**, 1835–1838 (2004)
69. Borst, T.H., Weis, O.: Diam. Relat. Mater. **4**, 948–953 (1995)
70. Suzuki, M., Yoshida, H., Sakuma, N., Ono, T., Sakai, T., Ogura, M., Okushi, H., Koizumi, S.: Diam. Relat. Mater. **13**, 198–202 (2004)
71. Latto, M.N., Roley, D.J., May, P.W.: Diam. Relat. Mater. **9**, 1181–1183 (2000)
72. Bohr, S., Haubner, R., Lux, B.: Diam. Relat. Mater. **4**, 133–144 (1995)
73. Koizumi, M., Kamo, Y., Sato, S., Mita, A., Sawabe, A.: Diam. Relat. Mater. **7**, 540–544 (1998)
74. Tsang, R.S., May, P.W., Ashfold, M.N.R., Rosser, K.N.: Diam. Relat. Mater. **7**, 1651–1656 (1998)
75. Petherbridge, J.R., May, P.W., Fuge, G.M., Rosser, K.N., Ashfold, M.N.R.: Diam. Relat. Mater. **11**, 301–306 (2002)
76. Petherbridge, J.R., May, P.W., Fuge, G.M., Robertson, G.F., Rosser, K.N., Ashfold MNR, J.: Appl. Phys. **91**(6), 3605–3615 (2002)
77. Mikka, N.G., Yasu, E., Xiao, C., Kikuchi, Y., Yushiwaza, K., Sakaguchi, T., Suzuki, T., Ando, T.: Diam. Relat. Mater. **9**, 941–947 (2000)
78. Harshavardhan, K.S., Vijayarahavan, M.N., Chandrabhas, N., Sood, A.K.: J. Appl. Phys **68**(7), 3303–3306 (1990)
79. Vittone, E., Manfredotti, C., Fizzotti, F., Lo Giudice, A., Polesello, P., Ralchenko, V.: Diam. Relat. Mater. **8**, 1234–1239 (1999)
80. Vinokur, N., Miller, B., Avyigal, Y., Kalish, R.: J. Electrochem. Soc. **143**, L238 (1996)
81. Zhu, W., Kochanski, G.P., Lin, S., Seibles, L.: J. Appl. Phys. **78**, 2707–2711 (1995)
82. May, P.W.: Science **319**, 1490–1491 (2008)

Versatility of ZnO Nanostructures

Muhammad Kashif, Majid Niaz Akhtar, Nadeem Nasir, and Noorhana Yahya

Abstract Development of novel devices depends on the size, structure and controlled morphology of nanomaterials. Understanding the growth parameters and growth mechanism of nanostructured materials is essential. ZnO is one of the most promising and important semiconductor materials for its semiconducting characteristics. Variety of ZnO nanostructures such as nanowires, nanorods, nanotubes, nanorings, nanohelixes, nanosprings, nanobelts can be prepared by using a solid–vapour method, vapour liquid solid method and hydrothermal methods under specific growth conditions. ZnO clearly demonstrates its versatility in its structures and characteristics. This chapter also reviews the novel nanostructures of ZnO synthesized by solid–vapour method, vapour–liquid–solid method and hydrothermal method and their growth mechanisms. The applications of ZnO nanostructures as gas sensing, field effect transistors, solar cell, piezoelectric and EM detector is discussed.

1 Introduction

ZnO is one of the most important semiconductor material due to its wide band gap (3.37 eV) and large exciton binding energy (60 meV) at room temperature. ZnO has unique properties and versatile applications such as nanolaser, resonators, biosensors, optoelectronic materials and gas sensors [1–3]. ZnO nanostructures such as

M. Kashif (✉), M.N. Akhtar and N. Nasir
Department of Electrical and Electronic Engineering, Universiti Teknologi PETRONAS, Bandar Seri Iskandar, 31750 Tronoh, Perak, Malaysia
e-mail: kashifmughal79@hotmail.com; majidniazakhtar@gmail.com; nadeemntu@hotmail.com

N. Yahya
Department of Fundamental and Applied Sciences, Universiti Teknologi PETRONAS, Bandar Seri Iskandar, 31750 Tronoh, Perak, Malaysia
e-mail: noorhana_yahya@petronas.com.my

N. Yahya (ed.), *Carbon and Oxide Nanostructures,* Adv Struct Mater 5,
DOI 10.1007/8611_2010_18, © Springer-Verlag Berlin Heidelberg 2010,
Published online: 14 July 2010

nanorods, nanowires, nanohelix, nanorings, and nanobelts are synthesis by different methods [4]. These nanostructures show novel optical, electrical, mechanical and chemical properties. The size, shape, crystal structure and surface structure of the nanomaterials effect the novel properties therefore an understanding of the preparations methods and mechanisms by which size and shape of nanostructures can be control is required [5, 6]. ZnO can be prepared by several methods such as metal organic chemical vapour deposition (MOCVD), thermal evaporation through vapour liquid solid, vapour solid mechanism, template technology, and chemical solution route such as hydrothermal, sol gel methods. High quality ZnO nanostructure can be obtained by using metal organic chemical vapour deposition (MOCVD), Molecular Beam Epitaxy (MBE), and RF magnetron sputtering methods, however these methods are expensive. Chemical solution methods are easier and cheaper method to obtain variety of ZnO nanostructures [7]. This chapter presents versatility of ZnO nanostructures prepared by different methods and their potential applications in the future.

2 Crystal Structure of ZnO

Zinc oxide is an oxide of group II metal Zinc that belongs to P63mc space group. Zinc is in the transition metal row which has $3d^{10}$ moments and hence it does not have any unpaired electron orbiting around the nucleus [5, 8, 9]. Zinc oxide is a semiconductor material with a hexagonal wurtzite [10] crystal structure (Fig. 1). It has been reported that ZnO has very large exciton binding energy of 60 meV [5, 8, 9, 11] at room temperature which makes it a promising candidate for short wavelength Light Emitting Diode (LED) [12].

The tetrahedral coordination in ZnO results in piezoelectric and pyroelectric properties [5, 9]. Tetrahedral ZnO (T-ZnO) nanostructures have been reported by Dai et al. [5].These nanostructures were synthesized by using thermal evaporation method without the presence of catalyst at 850–950°C. It could be observed that the ZnO structures are typically of tetrahedral in shape with four legs. ZnO nanostructures are wurtzite structure with lattice constant of a = 0.324 nm and c = 0.519 nm.

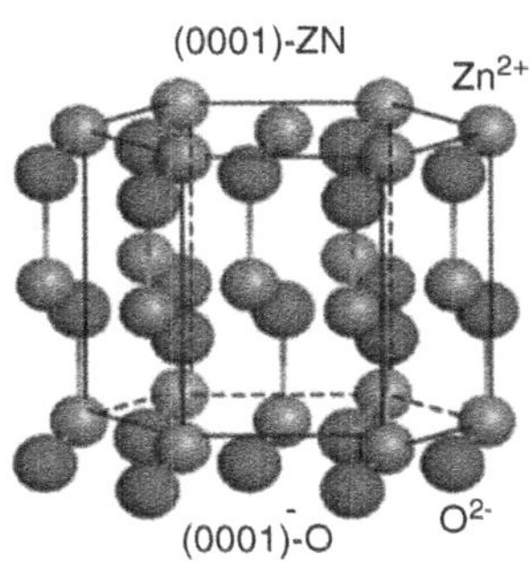

Fig. 1 The wurtzite structure model of the ZnO (adapted from [10])

The major peak observed is at (101) plain nanotubes (NT) was also studied by Zhang et al. [13] which could potentially be used for industrial applications in magneto optic devices. The ZnO nanotubes has wurtzite crystal structure with lattice parameter as a = 0.325 nm, c = 0.52 nm, $\alpha = \beta = 90°$ and $\gamma = 120°$. It was reported that the T-ZnO structure has two stages: nucleation the initial stage and are crucial role which induces diffusion, collisions of atoms and reaction between the vapour molecules in the formation of the tetrahedral of the ZnO structure. The second stage is the growth stage which will occur after super saturation ZnO tetrahedron nanostructures were also reported to have a link to the photoluminescence emission.

3 Synthesis Techniques

Different types of ZnO nanostructures can be prepared by two main techniques

1. Gas phase method
2. Solution phase method

In solution phase method liquids were used to get ZnO nanostructures where as in gas phase method different gases were used in vacuum chambers.

3.1 Gas Phase Method

Preparation of ZnO nanostructures can be done by using gas phase method which can be carried out at a very high temperature (500°C to 1,500°C). There are many methods for gas phase synthesis of ZnO nanostructures such as

1. Vapour–solid (VS) and vapour–liquid–solid (VLS) methods
2. Chemical vapour deposition (CVD) technique
3. Physical vapour deposition (PVD) technique
4. Evaporation of Zn and ZnO mixtures
5. Microwave technique

3.1.1 Vapour–Solid (VS) and Vapour–Liquid–Solid (VLS) Methods

The famous method for the synthesis of variety of ZnO nanostructures is vapour transport method. Due to variety of different nanostructures vapour transport method were classified into two methods vapour solid method and vapour liquid solid method. In vapour solid method ZnO nanostructures of different morphology produce without catalyst structure. In VLS method nanostructures are produced in the presence of the catalyst such as Au, Ag etc. In VS method evaporation of the source powder take place at elevated temperature and convert to vapours. These vapours under certain condition such a temperature, pressure, atmosphere and substrate condenses to form desired nanostructures products. A typical vapour solid process can be briefly described as follows: Pure zinc powder was put in an alumina boat, which was then transferred into a horizontal alumina tube placed in a horizontal tube furnace. The reaction chamber was heated to high temperature under a flowing argon atmosphere at suitable pressure.

In vapour phase method, ZnO nanostructures are grown by using ZnO and graphite powders as source materials. ZnO powder is reduced by carbon and carbon monoxide to low melting point Zn and suboxides of ZnO. These Zn and suboxides of ZnO were transferred to low temperature region by N_2/O_2 carrier gas where condensation took place. Nanodroplets were formed and these nanodroplets combine with oxygen to form nanostructures of ZnO on a substrate. Oxygen is used to form ZnO nuclei and the growth began until reactant flow is available. In the VLS method, ZnO vapour were dissolved in the catalyst to form eutectoid of Zn–Au and oxidized with oxygen from the ambient gas in the growth zone. Growth of ZnO nanostructures in VLS/VS depend on the various synthesis conditions namely temperature, growth pressure, starting reagents and flow rate [14]. Experimental set up for vapour solid method is shown in Fig. 2.

3.1.2 Metal Organic Chemical Vapour Deposition (MOCVD)

Metal organic chemical vapour deposition technique can be used to grow ZnO nanostructures. The ZnO crystal growth without catalyst depends on the substrate temperature, hydrogen flow rate, pressure of the precursor and substrate position as reported by [15]. ZnO nanowires can be prepared by MOCVD without catalyst [16]. It was found that high quality nanowires were clearly observed as compared to other methods which use catalyst. Schematic diagram of MOCVD method is shown in Fig. 3. The advantages of this technique are high quality film can be produced and these films can be used as industrial mass applications [16]. Yadouni et al. [15]

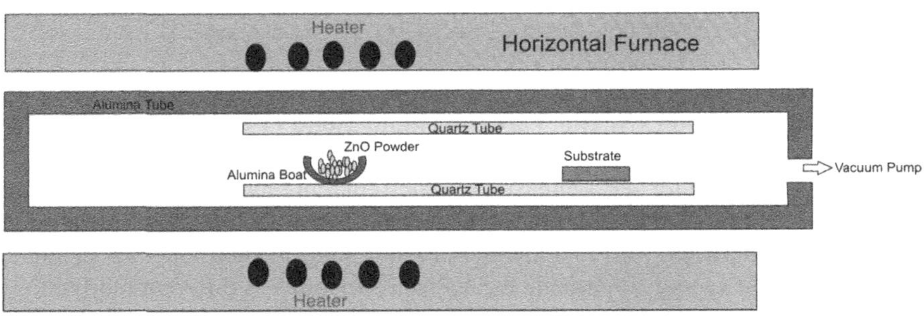

Fig. 2 Experimental set up for vapour solid method

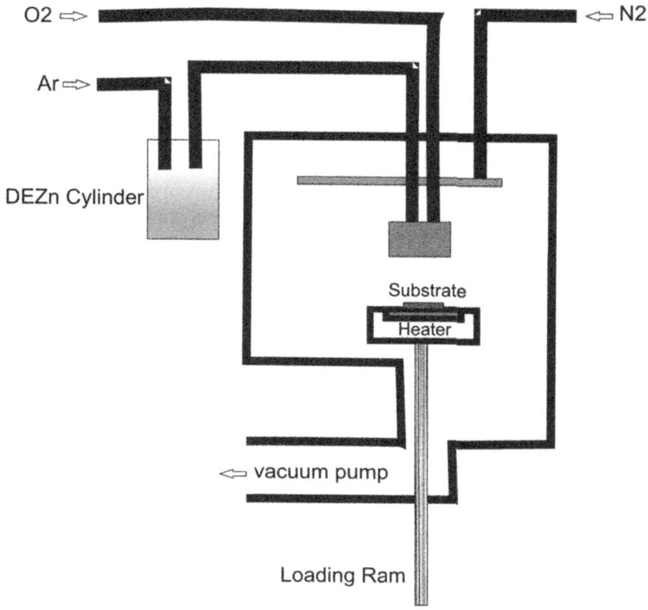

Fig. 3 Schematic illustrations of the experimental set-ups used for MOCVD

reported that a good quality ZnO film can be prepared by MOCVD method on the sapphire substrate at 270–450°C temperature with total flow rate of 5.5 L/mm. In this work, the pressure employed was P_{DEZn} at 14 Pa, $P_{t\text{-}but}$ at 70 Pa, the total pressure is equal to atmospheric pressure and substrate position was 3 cm from the gas inlet.

3.2 Solution Phase Method

Solution phase preparation method uses aqueous solutions for the growth of ZnO nanostructures. Most of the researchers reported the solution phase preparation methods which are as follows:

1. Sodium hydroxide (NaOH) and zinc chloride ($ZnCl_2$) were added in deionized water to produce ZnO nanostructures by hydrothermal process [17].
2. Zinc acetate [Zn $(CH_3COO)_2{\cdot}2H_2O$] was dissolved in methoxyethanol (MEA) solution and a glass substrate was used to obtained the thin films of ZnO by sol gel synthesis [18].
3. Zinc acetate dihydrate [Zn $(CH_3COO)_2{\cdot}2H_2O$] was mixed in deionized water and lithium hydroxide was added in the zinc acetate which was stirred by using a magnetic stirrer. ZnO nanostructures were formed using this method [19].

3.2.1 Hydrothermal Method

Hydrothermal method is a well known solution phase method for the growth of ZnO nanostructures. Different types of ZnO nanostructures with different morphology can be obtained by changing the conditions such as chemicals and their concentration and temperature as reported by Sun et al. [20]. In this synthesis method most of the ZnO nanostructures consist of nanorods and nanowires. For the growth of ZnO nanorods a seed layer is coated by pulse laser deposition (PLD) on the substrate and ZnO vertically aligned nanorods are grown on the seed layer. Many kinds of seed layers are used for the growth of ZnO nanostructure such as spin coated ZnO nanoparticles [21], sol–gel seed layer [22], RF sputtered [23], and ZnO based seed layer. Hydrothermal method is environment friendly easy to make low production cost as compared to other ZnO preparation techniques. Faster nucleation growth as compared to water can be done in alcoholic medium by using zinc nitrate hexahydrate. Baruwatie et al. reported the synthesis of ZnO nanoparticles by hydrothermal method in an autoclave by using Zinc nitrate hexahydrate. The distilled water and ammonium hydroxide solutions were mixed and stirred at 120°C temperature at a pH of 7.5 for the period of 6–24 h. It was then dried at 80°C to get nanoparticles [24].

4 Morphological Studies of Nanostructures ZnO

ZnO nanostructures are in the form of nanorods, nanowires, nanohelix, nanorings, nanocombs, nanobelts. These nanostructures produce due to different growth mechanisms which depend on synthesizing method and different growth conditions [25]. Some nanostructures growth mechanisms are well understood while other methods require investigation. One dimensional nanostructures of ZnO have been synthesizing by VS/VLS, electrochemical deposition and hydrothermal synthesis.

4.1 Nanowires and Nanorods

There is an increasing interest in studying nanowires and nanorods due to their applications. Aligned arrays of ZnO nanowires and nanorods are used in the optoelectronics, dyesensitized solar cell (DSSC), piezoelectric nanogenerator, field emission, and gas sensing [26].

The most famous method for the production of aligned ZnO nanowires is vapour liquid solid method in which metal catalyst is used to initiate and guide the growth of nanowires and nanorods. In this method Au, Fe, Sn, Ag, Cu is used as a catalyst but the most commonly used catalyst is Au. A systematic study of ZnO nanowire growth mechanism was carried out by studying the effect of Zinc vapour pressure and catalyst such as Au, Pt and Ag. It was observed these parameters play an effective role for ZnO nanowires growth [27]. Aligned nanorods was prepared without catalyst and at a low temperature by using vapour transport method [28]. The diameter of nanorods range from 80 to 900 nm and length of 12 μm. Figure 4 shows SEM images of nanorods on 6H-SiC substrate grown at 750°C and 10 mbar.

Micheal et al. [29] reported the effect of Au catalyst on the diameter of the ZnO nanowires by using vapour liquid solid method. ZnO nanowires were grown on the gold coated silicon substrate by heating the mixed powder of ZnO graphite to 900–950°C under constant flow for 30 min. It has been found that there is direct correlation between the size of catalyst and particles resulting from the diameter of the nanowires. A further decrease in nanowires diameters can be achieved through hydrogen reduction when ZnO nanowires were treated in hydrogen flow at 525–575°C for 30 min.

The formation of ZnO nanowires takes place after four steps, which are given below [30, 31]

1. Deposition of metal thin film
2. Formation of catalyst nanoparticles
3. Nucleation of the ZnO
4. Growth of ZnO nanowires

At the reaction temperature, reduction of ZnO will take place by graphite and CO (g). The chemical reaction is given below [30, 31]

$$\mathrm{ZnO(s) + C(s) \rightarrow Zn(g) + CO(g) at \sim 900°C} \quad (1)$$

$$\mathrm{CO(g) + ZnO(s) \rightarrow CO_2(g) + Zn(g) at \sim 900°C} \quad (2)$$

Fig. 4 SEM images of ZnO nanorods on 6H-SiC substrate grown at 750°C and 10 mbar (**a**, **b**) and at 750°C. Growth rate is higher in "lines" (*scratches*) on the initial SiC wafer caused by SiC wafer polishing (adapted from [28])

The deposition and condensation of these gaseous products take place at the catalyst particles and a following reaction take place

$$Zn(g) + CO(g) \rightarrow ZnO(g) + CO(s) + C(s) \tag{3}$$

$$Zn(g) + O(g) \rightarrow 2ZnO(s) \tag{4}$$

$$C(s) + CO_2(g) \rightarrow 2CO(g) \tag{5}$$

According to (3) and (4), O_2 and CO provided the oxygen source of the ZnO nanowires.

4.2 *Nanotubes*

A systematic study of the growth of hexagonal ZnO nanotube arrays using a chemical solution method by varying the growth temperature, time and solution concentration was carried out by [32]. GaN thin film was grown on the sapphire surface. Zinc nitratehexahydrate and hexamethylietetraamine (HMTA) in 1:1 ratio was used to make the precursor solution. The solution for reaction was kept at 95°C for 2–3 h and at 50°C for 3–48 h after that allows cooling at room temperature. The diameter of these nanotubes ranges from 500 to 800 nm and wall thickness 50–100 nm, length of nanotubes increase with growth time.

Growth of the nanotubes can be explained by the following reactions

$$(CH_2)6N_4 + 6H_2O \rightarrow 4NH_3 + 6HCHO_n \tag{6}$$

$$NH_3 + H_2O \rightarrow 4NH_3 \cdot H_2O \tag{7}$$

$$NH_3 \cdot H_2O \rightarrow 4NH^{4+} + OH^- \tag{8}$$

$$Zn^{2+} + 2OH^- \rightarrow Zn(OH)_2 \tag{9}$$

$$Zn(OH)_2 \rightarrow ZnO + H_2O \tag{10}$$

Growth of the ZnO nanotubes take place after two stages (1) growth of the ZnO nanorods (2) etching of the ZnO nanorods. In the first stage precipitation of ZnO nanorod take place. At the early stage of the growth, ${ZnO_2}^{2-}$ adsorbed on the positive polar face of ZnO nuclei surface and faster growth of the nanorods take place in (0001) direction. In the second stage dissolution of the ZnO nanorod take place. This dissolution of the nanorods produces non polar hollow structure. Single crystalline ZnO nanotubes were grown by MOCVD method at temperature 400°C on the sapphire substrate [33]. Nanotubes were hexagon-shaped and perpendicular to the substrate. Inner and outer bases of the hexagonal shape of nanotubes were 250 and 400 nm in length and had 130 nm wall thicknesses, respectively as shown in Fig. 5.

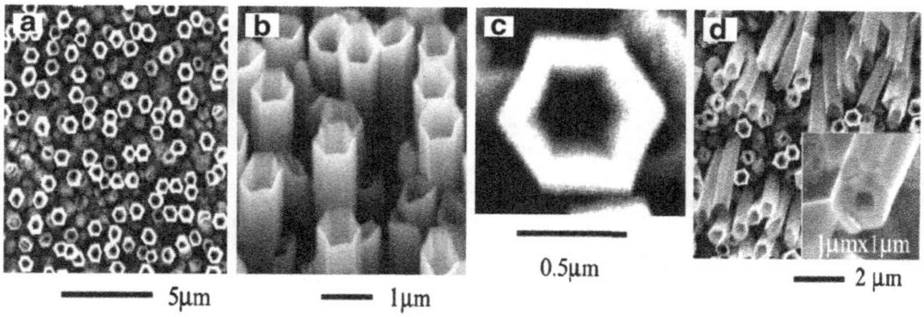

Fig. 5 SEM images of a sample surface taken along the surface normal (**a**) and with a 30° inclination (**b**). All of the ZnO tubes had exact hexagon-shaped cross sections and the same in-plane orientations. (**c**) Image of the top of a ZnO tube. (**d**) Images of a few ZnO tubes that were broken and lying on the sample surface. Inset in (**d**) shows the bottom image of a broken tube, indicating a hollow structure (adapted from [33])

4.3 *Nanobelts*

One of the most famous nanostructures of ZnO is nanobelt [34]. Nanobelts are free from the dislocation and other line defects due to which nanobelts are important for electronic and optoelectronic applications. Nanobelts are used in the fabrications of FET, Gas sensors nanoresonators and nano cantilevers [35, 36]. Nanobelts have rectangular shape with 10–50 nm thickness and width of 3–10 nm. Figure 6 show the TEM image of ZnO nanobelts [34].

Nanobelts are structurally and morphologically controlled because it has controlled properties. There are two factors that affect the growth of ZnO nanobelts (1) surface energy and (2) kinetic of the growth. Mostly ZnO nanobelts are grown by thermal evaporation method. ZnO nanobelts are grown in the lower energy and nonpolar surface such as (0 1 1 0) and (2 1 1 0) due to difference in surface energy (0 0 0 1), (2 1 1 0) and (0 1 1 0) surfaces [34].

Nanobelts can be grown using solid vapour method with catalyst [37] and at ZnO source decomposition temperature of 1,350°C. The nanobelts were grown on the alumina substrate at 400–500°C under 250 torr. Each nanobelt had uniform size distribution and the width of the nanobelt in the range of 10–60 nm and 5–20 nm and length up to several hundreds of micrometer. Electron diffraction show the nanobelt grows along [2 1 1 0] axis with top/bottom surface ± (0 0 0 1) and side surface ± (0 1 1 0). The nanobelts are single crystalline and dislocation free as revealed by the TEM and geometrically uniform [37].

4.4 *Nanorings and Nanohelix*

ZnO nanorings and nanosprings are used in actuators, sensors, resonators, piezoelectric for chemical and biological detection [38].The formation of nanorings and

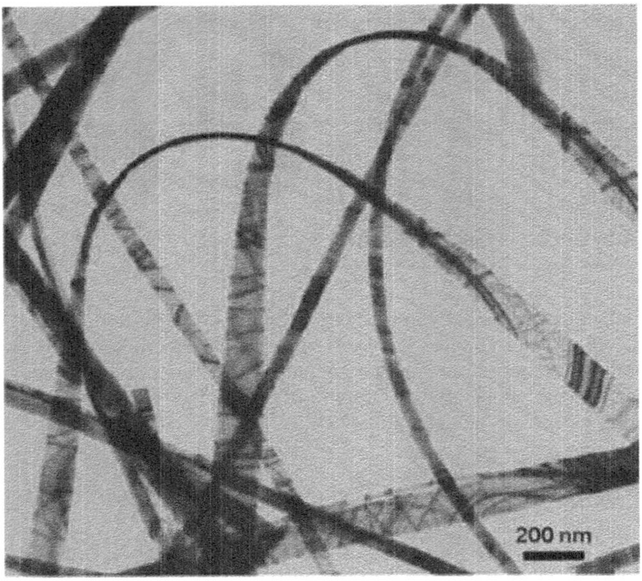

Fig. 6 TEM image of the as-synthesized ZnO nanobelts (adapted from [34])

nanosprings may take place due to rolling of single crystal nanobelts [38, 39]. Single crystal ZnO nanorings were grown by solid vapour method [39]. The source materials decompose at 1,400°C at low pressure (10^{-3} torr). Ar gas was later introduced at a flux of 50 sccm. Deposition of Si substrate at 200–400°C and Ar pressure of 500 torr were able to composed many free standing nanorings with diameter of 1–4 μm and wide shell of thickness 10–30 nm. SEM images show uniform, flat and circular shape of the nanorings. Two types of nanorings structures were found both have single crystal. Type I have non-uniform deform along the circumference but the type II has uniform deformation along the circumference ZnO nanorings growth can be explained by considering its polar surfaces of nanobelt. ZnO nanobelt has polar charges on its top and bottom surface positively charge plane (0 0 0 1) and negatively charge plane (−0 0 0 1). During the growth of the nanobelt, uncompensated surface charges tend to fold in the nanobelt and its length gets longer to minimize the area of the polar surface.

The interface of positively and negatively charge surface results the neutralization of the local polar changes which reduce the surface area to form a loop with an overlapped end. Initially the folding of nanobelt determines the radius of the loop and reduces the elastic deformation energy. After the ring formation, the electrostatic interaction is the force in subsequent growth of nanorings. Natural attraction of nanobelts on the rim of the ring continues the growth parallel to rim of the nanoring to neutralize the surface charges and surface area. This result in the formation of self coiled-coaxial uniradius, multiloopes nanoring structures as shown in Fig. 7 [10], complete neutralization of surface charges inside the nanorings may produce slinky shape. During the growth of the nanorings the repulsive force between the charges of surface stretches while deformation pulls the nanorings. The balance between them produces nanohelix /nanosprings and nanospiral is shown in Fig. 7 [10]. The ZnO nanobelts originated from the nanohelix have uniform shape with radius of 500–850 nm and evenly distributed pitches [40].

4.5 ZnO Nanoflowers

Rose-like structure of ZnO can be prepared by using chemical vapour deposition technique (CVD) [41]. This nanoflower structure was prepared at a temperature of 600°C by using SiC as a substrate. ZnO Nanoflower structure has the size of 1–2 μm.

Flower-like ZnO nanostructure consists of ZnO nanosticks obtained by solution process was reported by Wang et al. [42]. ZnO nanosticks consists of hexagonal nanorods was achieved by zinc acetate dehydrate, sodium hydroxide and polyethylene glycol at temperature of 180°C for 4 h. XRD results show that as synthesized product is ZnO (Fig. 8). SEM micrographs shows ZnO nanosticks have diameter of 350–450 nm and length of 1–2 μm (Fig. 9) and TEM image of the grown ZnO nanosticks are shown in Fig. 10.

Growth of ZnO nanoflowers can be synthesized by using zinc nitrate hexahydrate and hexamethylene tetramine at a temperature of 95°C on Si substrate [43]

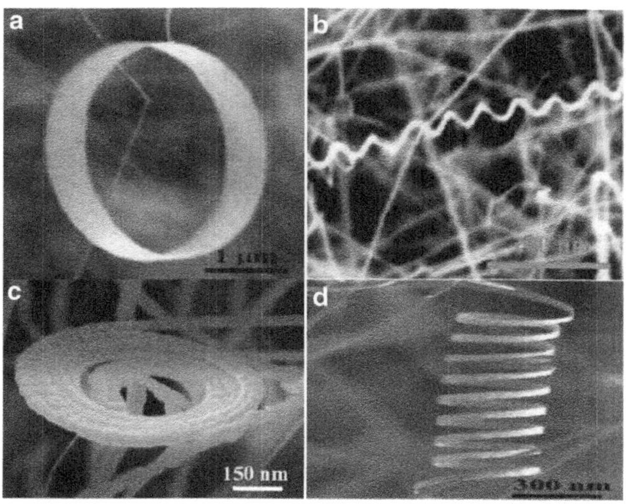

Fig. 7 (**a**) Single crystal seamless nanoring formed by loop-by-loop coiling of a polar nanobelt (**b**) deformation-free nanohelixes as a result of block-by-block self assembly; (**c**) spiral of a nanobelt with increased thickness along the length; (**d**) nanosprings (adapted from [10])

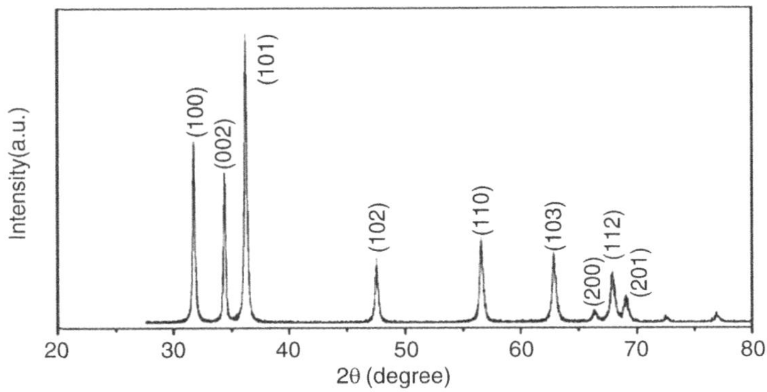

Fig. 8 Typical X-ray diffraction (XRD) pattern synthesized nanostructure: the indexed peaks correspond to the wurtzite hexagonal phase (adapted from [42])

Fig. 9 SEM images of ZnO: (**a** and **b**) overall product morphology and (**d** and **c**) detailed view on an individual flower (adapted from [42])

and the size of the nanoflower depend on growth time. ZnO nanoflowers structure can also be achieved with hydrothermal method by using zinc acetate and sodium hydroxide with different ionic liquids (ILs) [44]. The growth of ZnO branched flower like nanostructures can be obtained with a new technique has been reported by Sounart et al. [45] ZnO nanostructure crystal growth depends on the surface energy of the specific crystals planes by using organic diamine additives with

Fig. 10 (**a**) Low magnification TEM image of the grown ZnO nanosticks, (**b**) HRTEM image showing the difference between two lattice fringes, which is about 0.257 nm and corresponding FFT pattern (inset) is consistent with the HRTEM observation and (**c**) side elevation of the grown ZnO nanosticks. 257 nm and the corresponding FFT pattern (adapted from [42])

different range of chain length and concentration. New nucleation sites can be created by etching the surface and induction time. Growth rate can be increased by increasing the concentration of diamine.

4.6 Quantum Dots

ZnO nanoparticles and quantum dots show quantum confinement effects which increase the efficiency of optical devices. Quantum dots nanostructure can be achieved by different techniques such as sol–gel, electromechanical, vapour phase transport, RF magnetron sputtering, metal organic chemical vapour deposition method and laser ablation etc [46, 47].

ZnO quantum dot nanostructure prepared by sol–gel method was reported by Yatsui [46]. Zinc acetate were added into boiling ethanol at atmospheric pressure and the solution was cool down to 0°C and $LiOH{\cdot}H_2O$ dissolved in ethanol at room temperature in ultrasonic bath which also cool down to 0°C. Hydroxide solution was added dropwise in the zinc acetate suspension with constant stirring at 0°C. 0.1 g of LiOH was added to stop the particle growth.

Figure 11 shows the TEM images of ZnO quantum dots nanostructure inside the white dashed circle by using the sol–gel method. TEM image shows the monodispersed single crystalline ZnO nanostructure.

Synthesis of ZnO quantum dots with and without annealing at different temperatures has been reported [47]. When the temperature was increased from 873 to 1,273 k at 400 Pa some irregularity can be observed (Fig. 12). Figure 12a shows that quantum dots nanostructure of ZnO having irregular shape and some of them are spherical in shape prepared without annealing. Figure 12b indicates that QDs of ZnO nanostructure is still irregular even though the produced quantum dots are heat treated at 873 K. The less than 873 K temperature was not able to remove the

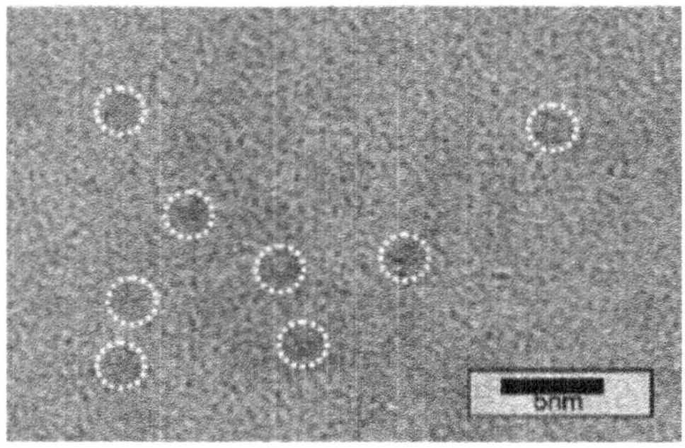

Fig. 11 TEM image of the ZnO QD. The dark areas inside the white dashed circles correspond to the ZnO QD (adapted from [46])

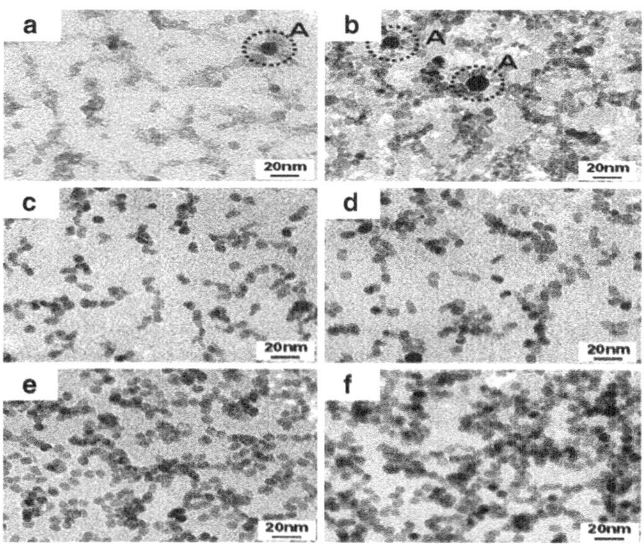

Fig. 12 TEM bright field images of ZnO quantum dots (**a**) without annealing and with annealing at temperatures of (**b**) 873 K, (**c**) 973 K, (**d**) 1,073 K, (**e**) 1,173 K, and (**f**) 1,273 K with a size classification at nominal size of 10 nm under a gaseous pressure of 400 Pa (adapted from [47])

irregularity in the quantum dot nanostructure of ZnO. The irregularity in the shape of QDs ZnO nanostructure can be removed at temperature greater than 973 K (Fig. 12c). However this irregularity remains at 1,073 K (Fig. 12d). ZnO QDs nanostructure annealed at temperature of 1,173 K and 1,273 K were totally converted into spherical in shape as shown in Fig. 12e, f.

5 Applications of ZnO Nanostructures

Nowadays, ZnO semiconductor has attraction due to their different nanostructures such as nanowires, nano rods, nanobelts, nanohelix, and nanospheres etc. Versatile ZnO nanostructures have been used in different applications such as

1. Gas sensors
2. Solar cell
3. Field effect transistor
4. Piezoelectric application
5. Electromagnetic (EM) detector

5.1 Gas Sensors

ZnO nanostructures have potential applications for gas sensing due to high surface to volume ratio [48, 49]. Many researchers used ZnO nanowires for gas sensing [50–53]. Liao et al. reported the growth of ZnO nanowires and carbon nano tubes on the Si (100) substrate. ZnO nanowires have diameter of 150–200 nm with length of several microns where as CNTs has length of 10 μm and diameter of 50 nm. A nanowire sensor was consisted of two electrodes one electrode consist of nanowires and other copper plate and separation distance between anode and cathode was kept 100 μm. By using ZnO nanowires as anode at 480 V in air, a current discharge of 150 μA was produced and CNTs at anode at breakdown voltage of 292 V produced current density of 280 μA. Similarly, by inserting metal electrode, when breakdown voltage was 870 V the current discharge produced was 72 μA. ZnO nanowires at the anode have moderated voltage in air than metal electrode and CNTs electrode has lower breakdown voltage than ZnO nanowires electrode. Anode ionization device with ZnO nanowires is used for detection of different gases such Co, NO_2, H_2, He, air, and O_2. ZnO nanowires gas sensor shows a very high sensitivity and response within a very small amount of time [54].

Comini et al. reported the application of ZnO nanowires and nanocombs for gas sensing .These ZnO nanowires and nanocombs were synthesized by thermal evaporation method [55]. Response of zinc oxide nanowires as a function of operating temperature for 5 ppm of nitrogen dioxide, 1 ppm of ammonia, 50 ppm of acetone,

100 ppm of carbon monoxide and 500 ppm of ethanol is shown in Fig. 13. Ethanol and acetone show highest response and increases with operating temperature.

The sensitivity of zinc oxide nanowires as a function of the concentration of acetone and ethanol at temperature of 400°C is shown in Fig. 14.

5.2 *Solar Cell*

The solar energy can be converted into electrical energy using photovoltaic (PV), concentrating solar power (CSP), and dye-sensitized solar cell (DSSC) etc [56]. In Photovoltaic devices at the interface of two materials charge separation takes place. A monolayer of the charge transfer dye is attached to the surface of the nanocrystalline

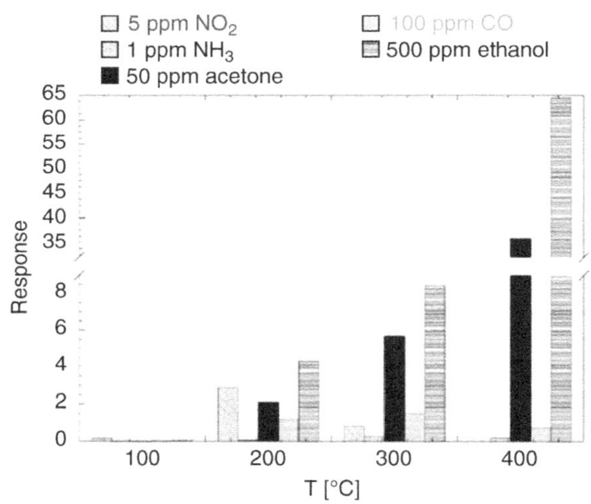

Fig. 13 Response of zinc oxide nanowires as a function of operating temperature for 5 ppm of nitrogen dioxide, 1 ppm of ammonia, 50 ppm of acetone, 100 ppm of carbon monoxide and 500 ppm of ethanol. The highest response is registered for ethanol and acetone (adapted from [55])

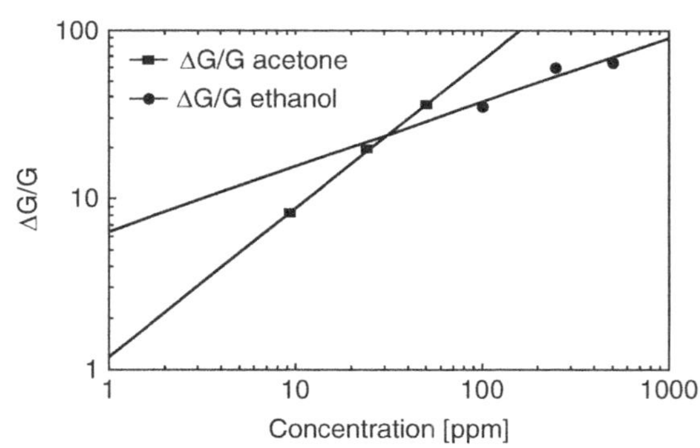

Fig. 14 Response of zinc oxide nanowires towards acetone and ethanol as a function of the concentration at an operating temperature of 400°C. The response follows the power behavior and reveals a very low detection limit (adapted from [55])

film. The dye is then used by electron donation from the electrolyte. The voltage produced under light due to difference between Fermi level of the electron in the solid and the potential of the electrolyte. Dye sensitized solar cells with silicon acts as source of photoelectrons, causes electric field for charge separation and creates current. The semiconductor materials such as ZnO, ITO_2 were used for the transportation of charges. Photosensitive dye was used to provide Photoelectrons. The molecules of dye are so small, so in order to capture maximum amount of light, nanomaterial need to be introduced with dye molecules. Schematic diagram for dye sensitized solar cell is shown in Fig. 15.

ZnO nanostructures were studied for their applications in photonics. Photoluminescence (PL) spectra of ZnO nanostructures give information about the excitonic emissions [57]. ZnO nanowires green emission density increases when their diameter decreases. The green-yellow bands give emission peak due to oxygen vacancy present in ZnO nanostructures. The band to band transition gives strong emission peak than green yellow emission bands. At the red luminescence band, emission peak mainly occurs due to ionized oxygen vacancies [58]. The morphology of ZnO nanostructures would directly effect on the photovoltaic of a dye-sensitized solar device (DSSC). Blue shift was observed in the near UV emission peak in ZnO nanobelt. ZnO nanowire is the best nanostructures for UV emission, where as ZnO nanorods are better for optical waveguides due to large refractive index ($\sim$2) [59, 60].

Absorption spectra of dye-absorbed ZnO nanorods and nanoparticles were reported Amaratunga et al. The ZnO nanorod based solar cell shows higher cell efficiency (η) of (1.32%) than ZnO nanoparticle based solar cell efficiency (0.87%). It is observed that the intensity of the absorption peak of the nanoparticles with N_3 dye is less than nanorods. An increase in the conversion efficiency of the solar cell was observed by morphology of ZnO nanostructures [61].

Min et al. studied the photovoltaic performances for dye-sensitized solar cell (DSSC) of the ZnO nanowires synthesize by a microwave-assisted aqueous route at low temperature of 120°C [62]. The *I–V* characteristics curve of DSSC using the nanowires (Cell-a) and commercially ZnO powder (Cell-b), respectively is shown in Fig. 16.

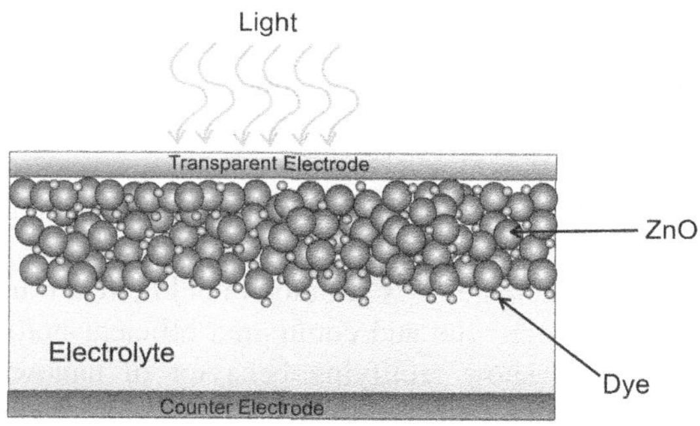

Fig. 15 Schematic diagram for dye sensitized solar cell

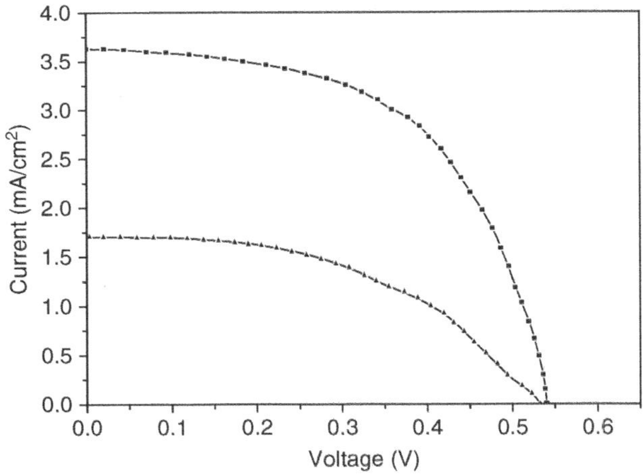

Fig. 16 Current-voltage characteristics of dye-sensitized solar cells using the as-prepared nanowires (*filled square* Cell-a) and commercial ZnO powder (*filled triangle* Cell-b) (adapted from [62])

5.3 Field Effect Transistor

Nanostructures of ZnO semiconductor have made revolutionary changes in the fabrication of electronics devices and industry as well. Nanowires and nano rods of ZnO nanostructures were used for the electric transport measurements [63, 64]. ZnO nanostructures can be use for fabrication of field effect transistors (FETs). Figure 17 shows schematic diagram of FET configured by using nanowire. A layer of SiO_2 with p-type silicon back gate was used for the development of field effect transistor. Two metal electrodes which acts as source and drain were fixed at the two ends of the nanowire. Electrical properties of the nanowire FET was carried out by using current versus source drain voltage and current versus gate voltage [65].

Field effect transistor (FET) was made-up from ZnO having nanostructures by many procedures. ZnO nanowires have n-type semiconductor behavior due to oxygen vacancies and defects. ZnO thin films transistors considered as high (7 cm^2/v) electron field effect mobility but on the other hand single crystalline ZnO nanowires gives high (80 cm^2/N-s) [66]. It was also reported that coating of polyimide on the ZnO nanowires would increase the electron mobility up to 1,000 cm^2/V-s [67]. ZnO has versatile applications in photonics and electronics due to different nanostructures and its semiconductor behaviour. But, due to difficulty in p-type doping, many applications are still restless. Many researchers are trying to make efficient p-type doping in ZnO nanostructures. P-type ZnO nanowires creates P-N junction in field effect transistor. PN junction was combined with logic circuits and configured efficient field effect transistor (FET). IV characteristics show rectifying behavior of nanowires due to PN junction. It was observed that when vertically aligned ZnO nanowires was used as field emission

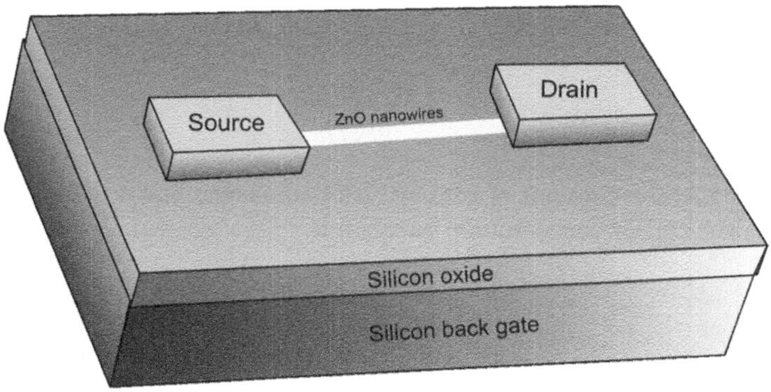

Fig. 17 Schematic of a single ZnO nanowire based FET

measurement, field of 18 V/μm was found by applying a current density of 0.01 $\mu A/cm^2$ with emission current of 0.1 mA/cm^2. ZnO nanowires prepared at low temperature had produced better results than ZnO nanowires/nanoneedles at high temperature [68].

Nanobelts were also used to fabricate field effect transistor (FET). Ethanol was used in fabrication of ZnO nanobelts. ZnO nanobelts were distributed in ethanol by ultrasonication process. After dispersing ZnO nanobelts dried them and deposit on SiO_2/Si substrate for AFM analysis. Gold electrode arrays were used with ZnO dispersed nanobelts for the configuration of field effect transistor (FET). Electrode gaps were taken as large as 6 μm and 100 nm as small [69].

Field effect transistors (FET) were configured by using ZnO nanowires as reported by Goldbrger et al. ZnO nanowires were prepared by using gold thin film as catalyst via carbothermal reduction process at 900°C. It was observed that ZnO nanowires show single crystals with wurtzite structure. Nanowires have diameter ranging from 50 to 200 nm with length of 10 μm. ZnO nanowires dispersed in VLSI grade 2-propanol and transferred onto prefabricated Cr/Au probe with Si (100) wafers. Electron beam lithography was used for the connection between probe pads and ZnO nanowires, Measurements of field effect transistor (FET) with nanowires were done by using probe station with HP 4156B parameter analyser. Threshold voltage for current measured was −12.5 V for nanowire FET where as it varies from −4.5 to −28 V for all samples [70].

At more negative voltage, there was no current in ZnO nanowire FET. From plots, mobility of ZnO nanowires FET can be calculated [70]

$$\frac{dI_{sd}}{dV_G} = \frac{\mu C}{L^2} \tag{11}$$

Where, C is the capacitance. μ is the carrier mobility, and L is the length of nanowire channel.

The capacitance is given by

$$C = \frac{2\pi\varepsilon_o\varepsilon_{SiO_2}L}{\ln\left(\frac{4h}{d}\right)} \tag{12}$$

ε_{SiO2} is dielectric constant of gate SiO_2, h is thickness and d is the diameter of nanowire. The carrier concentration equation is

$$n_e = \frac{V_{th}C}{q\pi\left(\frac{d}{2}\right)^2 L} \tag{13}$$

The mobility of 17.1 $cm^2\ V^{-1}\ s^{-1}$ with a carrier concentration of 1.1×10^7 for FET nanotransistor was observed. The average carrier concentration and average mobility $5.2 \pm 2.5 \times 10^{17}\ cm^{-3}$ and $13 \pm 5\ cm^2\ V^{-1}\ s^{-1}$ was found for FET nanotransistor. I_{ON}/I_{OFF} for FET wass ranges from 10^5 to 10^7 for FET nanowire transistor [70].

Chen et al. fabricated the field effect transistor by hydrothermally synthesize ZnO nanotubes [71]. ZnO nanotubes show typical n-type semiconducting behavior. The I_{ds}–V_{ds} curves obtained at different gate voltages is shown in Fig. 18a. The I_{ds}–V_g curve is shown in Fig. 18b. For a given V_{ds}, I_{ds} decreases with increasing negative Vg.

5.4 *Piezoelectric Application*

ZnO nanostructures are potentially used for the enhancement of piezoelectric properties. Due to unique nanostructures ZnO can be used for the piezoelectric

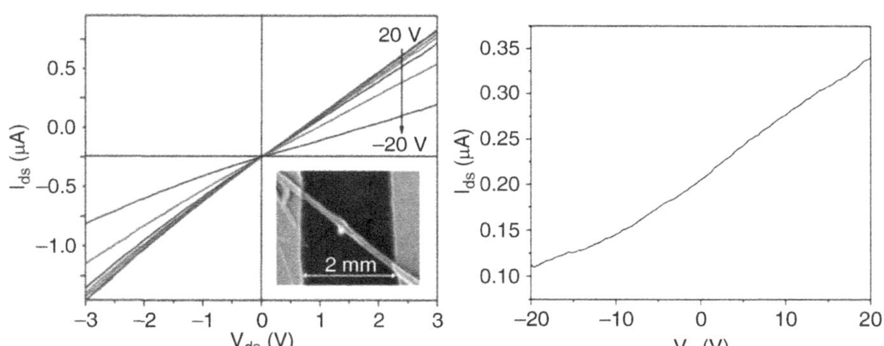

Fig. 18 (**a**) Typical I_{ds}–V_{ds} curves obtained at different gate voltages and (**b**) I_{ds}–V_g curve of a single ZnO nanotube FET. The inset is a SEM image of the device (adapted from [71])

devices due to its non central symmetry, conductivity at room temperature, higher band gap and large excitation energy [10].

The piezoelectric effect could be understand by considering an atom consists of positive charge and surrounded by an ions as shown in Fig. 19.The negative charges center of gravity lies at the centre of the tetrahedron when pressure was applied the distortion will take place and the center of gravity of the negative charges will not coincide with the position of the positive atom which lies at the centre of the tetrahedron. As a result of this pressure, electric dipole created as shown in the Fig. 19a.

Wang [10] measured the piezoelectric coefficient of the ZnO nanobelts by using atomic force microscope. ZnO nanobelts were spread on the conductive surface which was 100 nm Pd coated (1 0 0) Si wafer. After that to avoid electrostatic effect and to achieve uniform electric field the whole surface was again coated with 5 nm Pd. Piezoresponse force microscopy (PFM) was used to measure the piezoelectric coefficient of individual ZnO nanobelt. Comparison of ZnO piezoelectric coefficient with ZnO bulk is shown in the Fig. 19b. The piezoelectric coefficient for ZnO nanobelts changes from 14.3 pm/V to 26.7 pm/V with the change of frequency while the bulk (0 0 0 1) of ZnO have 9.93 pm/V as shown in Fig. 19b. This result shows that ZnO nanobelt is strong candidate for the piezoelectric applications.

Zhang et al. [72] reported the hydrothermally growth of ZnO nanowire array with Zn foil and ammonia solution at 95°C for the controlled lengths and surface to volume ratio of aligned ZnO. Surface to volume ratio can be changed by changing the concentration of aqueous ammonia solution and the controlled length of nanowires can be adjusted by the reaction time. Piezoelectric property of ZnO nanowires were studied by fabricating nanogenerator. Nanogenerator was made by two pieces of stalked ZnO nanowires (NW) and penetrates to each other. One piece is coated with gold for conductive nanotip. Nanogenerator (NG) was driven by ultrasound waves as shown in Fig. 20. Piezoelectric current was measured and different characteristic curve were obtained at different conditions.

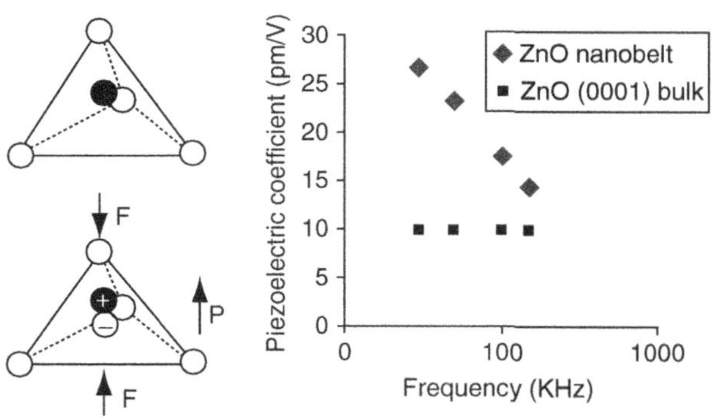

Fig. 19 (**a**) Schematics showing piezoelectric effect in tetrahedrally coordinated cation-anion unit. (**b**) Experimentally measured piezoelectric coefficient d_{33} for ZnO and its comparison to that of the bulk (adapted from [10])

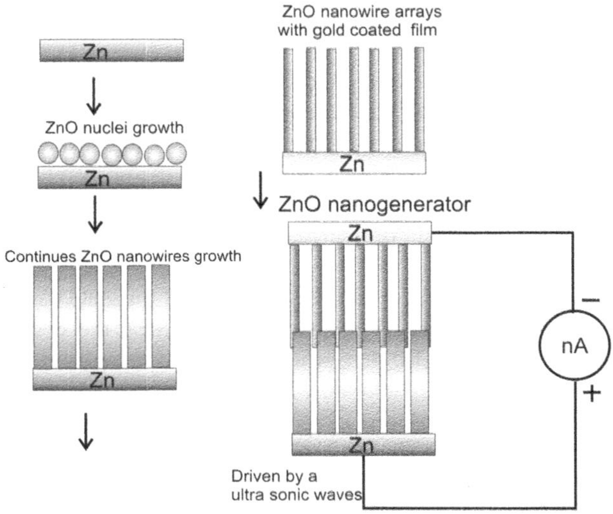

Fig. 20 Schematic illustration of growth mechanism of the ZnO NW arrays and the assembling procedures of a ZnO NG

Figure 21a shows the short circuit current for four samples prepared with different concentration of 4,7,10 and 15% and having different surface to volume ratio. The ultrasonic frequency of 41 KHz was used to analyze the output current. The output current was taken when the ultrasonic wave turned on. It was found that the average output current ~ 20, ~ 30, ~ 44 and 47 nA for sample 1–4 respectively. It has been concluded that by increasing the surface to volume ratio of ZnO nanowires the output current increases as shown in Fig. 21a. Figure 21b indicates the short circuit current with the change of length of nanowires of ZnO nanostructure ranges from 3.6 to 10 μm. The average short circuit current received at the output is ~ 7 ~25 and ~ 48 nA respectively. This result shows that with increasing length of ZnO nanowires the output current will increase. A short circuit current was measured for sample 4 having 15% concentration and 10 μm length with the change of frequency of the ultrasonic wave from 10 to 50 KHz as shown in the Fig. 21c. From results it has been concluded that there is no relation between piezoelectric out-put and the frequency because the variation for current is irregular.

5.5 Electromagnetic(EM) Detector

ZnO nanoparticles have been studied as electromagnetic (EM) detectors [73] potentially to be used for Seabed Logging applications. The ZnO nanoparticles were prepared by self combustion method and the XRD profile (Fig. 22a) indicates single phase with average particle size below 50 nm. The self combustion samples showed Raman shift at 437.67 cm^{-1} (annealed at 100°C and 200°C), 433.11 cm^{-1}

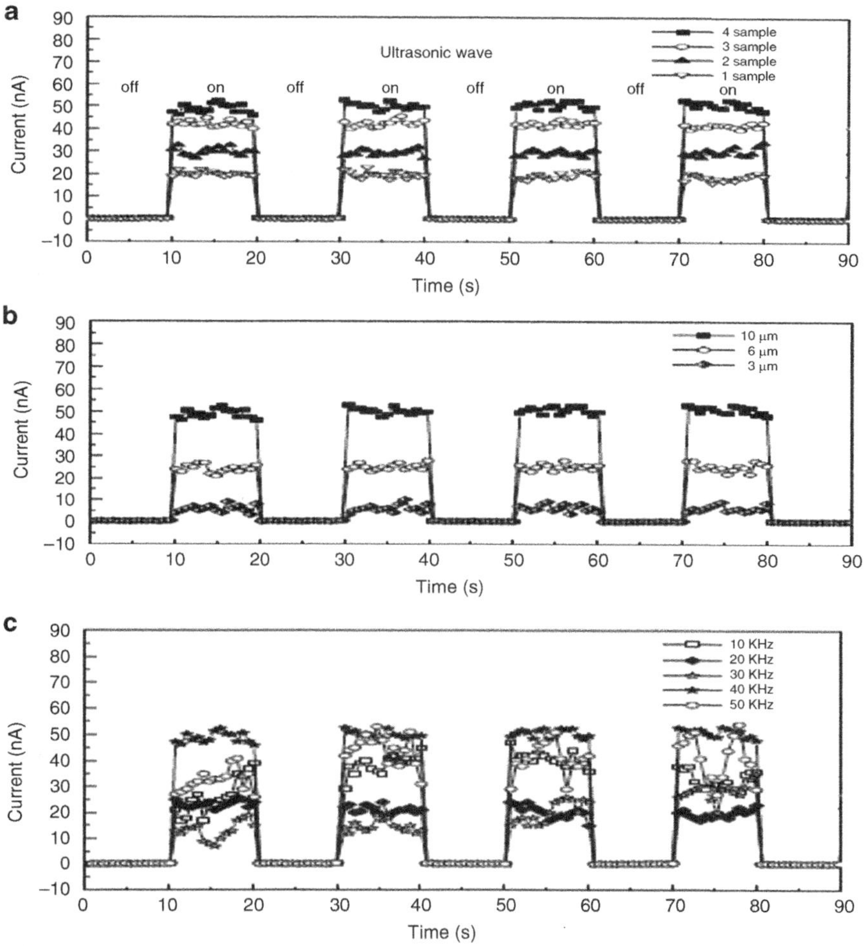

Fig. 21 Piezoelectric output current of different ZnO NW NG models (**a**) Short circuit current output from the different NGs with diverse surface-to-volume ratio NW arrays. The No. 1–4 samples were prepared under 4, 7, 10, and 15% (V/V) concentration conditions, respectively. (**b**) Short circuit current output measured from the NGs with different length NW arrays. The length of NW arrays is about 3, 6, and 10 μm, respectively. (**c**) Short circuit current output measured from identical NG driven by ultrasonic waves from 10 to 50 kHz. The length of the ZnO NW arrays is 10 μm and the sample was prepared in 15% aqueous ammonia solution (adapted from [72])

(annealed at 300°C) and 440.203 cm^{-1} (annealed 400°C). These two major shifts are known as second order Raman shift are in the same range of reported values which are 323–1,120 cm^{-1} shifts [73]. Figure 22b, c indicate that PVA + ZnO polymer sample had shown higher voltage peak to peak (Vpp) comparing to the PVA polymer without ZnO as filler. At 50 MHz, PVA polymer sample and PVA-ZnO polymer sample showed 31.1 mV and 110 mV, respectively. The voltage

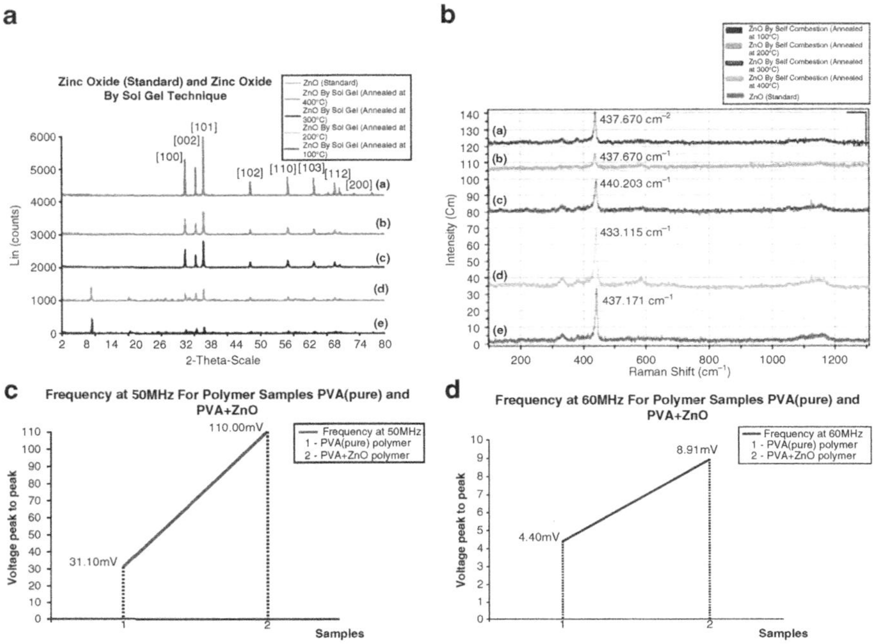

Fig. 22 (**a**) XRD patterns of as-synthesized zinc oxide samples with a major peak (101) (a) zinc oxide standard (b) 400°C and (c) 300°C (c) 200°C and (d) 100°C (**b**) Raman results at annealing temperature (a) 100°C (b) 200°C (c) 300°C (d) 400°C and (e) Standard Zinc Oxide Raman Shift (**c**) The peak to peak voltage (Vpp) of PVA(pure) and PVA + ZnO detectors at 50 MHz (**d**) The peak to peak voltage (Vpp) of PVA (pure) and PVA + ZnO detectors at 60 MHz

peak to peak was decreased to 4.40 mV for PVA (pure) polymer and 8.91 mV for of PVA + ZnO polymer sample. The different percentage of PVA (pure) polymer and PVA + ZnO polymer sample are 253.7% (at 50 MHz) and 102.5% (at 60 MHz) respectively. The results indicate that PVA-ZnO polymer can be used as EM detector. It should be noted that the EM detector is a polymer based composite which was prepared using casting technique.

6 Conclusion

In this chapter we presented ZnO nanostructures grown by different techniques namely hydrothermal, vapour–liquid–solid, vapour–solid and MOCVD. The various growth morphologies are summarized and their growth processes are discussed. Potential applications have been selected and discussed such as the field effect transistors, gas sensors, solar cell, piezoelectric and EM detector. ZnO could be one of the most important nanomaterials in the future due to its remarkable properties especially their versatility in morphology.

References

1. Riaz, M., Fulati, A., Zhao, Q.X., Nur, O., Willander, M., Klason, P.: Nanotechnology **19**, 415708 (2008)
2. Kar, J.P., Ham, M.H., Lee, S.W., Myoung, J.M.: Appl. Surf. Sci. **255**, 4087–4092 (2009)
3. Seung, Y.B., Chan, W.N., Ja, H.K., Park, J.: J. Phys. Chem. B **109**, 2526–2531 (2005)
4. Wang, Z.L.: Zinc Oxide Bulk, Thin Films and Nanostructures. Elsevier, Oxford, UK (2006)
5. Dai, Y., Zhang, Y., Wang, Z.L.: Solid State Commun. **126**, 629–633 (2003)
6. Jun, W., Shunxiao, Z., Jia, Y., Huijun, Y., Li, Z., Jing, X., Zhang, M.: Bull. Mater. Sci. **31**, 597–601 (2008)
7. Kyu, S.H., Yun, J.L., Seung, H.: J. Ceram. Process. Res. **8**, 305–311 (2007)
8. Sun, J.W., Lu, Y.M., Liu, Y.C., Shen, D.Z., Zhang, Z.Z., Li, B.H., Zhang, J.Y., Yao, B., Zhao, D.X., Fan, X.W.: J. Phys. D Appl. Phys. **41**, 155103 (2008)
9. Zhong, L.W.: J. Phys. Condens. Matter **16**, 829–858 (2004)
10. Wang, Z.L.: Appl. Phys. A **88**, 7–15 (2007)
11. Hasuike, N., Nishio, K., Katoh, H., Suzuki, A., Isshiki, T., Kisoda, K., Harima, H.: J. Phys. Condens. Matter **21**, 064215 (2009)
12. Jia, C.H., Chen, Y.H., Liu, G.H., Liu, X.L., Yang, S.Y., Wang, Z.G.: J. Phys. D. Appl. Phys **42**, 015415 (2009)
13. Chun, Z.F., Yong, Z.Z., Hu, Z.W., Feng, Y.J., Ni, Y.J.: Chin. Phys. Lett. **26**, 016105 (2009)
14. Yang, S.H., Chen, P.C., Hong, S.Y.: Curr. Appl. Phys. **9**, E180–E184 (2009)
15. Yadouni, A.E., Boudrioua, A., Loulergue, J.C., Vincent, S., Triboulet, R., Sallet Vand Triboulet, R.: Opt. Mater. **27**, 1391–1395 (2005)
16. Lee, W., Jeong, M.C., Myoung, J.M.: Acta. Mater. **52**, 3949–3957 (2004)
17. Elen, K., Rul, H.V.D., Hardy, A., Bael, M.K.V., D'Haen, J., Peeters, R., Franco, D., Mullens, J.: Nanotechnology **20**, 055608 (2009)
18. Masashi, O., Hiromitsu, K., Toshinobu, Y.: Thin Solid Films **306**, 78–85 (1997)
19. Wang, M., Ye, C.H., Zhang, Y., Wang, H.X., Zeng, X.Y., Zhang, L.D.: J. Mater. Sci. Mater. Electron. **19**, 211–216 (2008)
20. Sun, Y., Riley, D.J., Ashfold, M.N.R.: J. Phys. Chem. B **110**, 15186–15192 (2006)
21. Habibia, M.H., Sardashtia, M.K.: J. Iran. Chem. Soc. **5**, 603–609 (2008)
22. Lin, D., Lin, J.X., Hong, T.Y., Ying, M., Wei, L.J.: Trans Nonferrous Met. Soc. China **17**, S721–S724 (2007)
23. Song, J., Lim, S.: J Phys. Chem. C **111**, 596–600 (2007)
24. Baruwati, B., Kumar, D.K., Manorama, S.V.: Sens. Actuators B Chem. **119**, 676–682 (2006)
25. Wang, G.Z., Wang, Y., Yau, M.Y., To, C.Y., Deng, C.J., Ng Dickon, H.L.: Mater. Lett. **59**, 3870–3875 (2005)
26. Jianye, L., Qi, Z., Hongying, P., Henry, O.E., Luchang, Qin, Jie, Liu: J Phys. Chem. C **113**, 3950–3954 (2009)
27. Zhang, Z., Wang, S.J., Yu, T., Wu, T.: J Phys. Chem. C **111**, 17500–17505 (2007)
28. Mofor, A.C., Bakin, A.S., Elshaer, A., Fuhrmann, D., Bertram, F., Hangleiter, A., Christen, J., Waag, A.: Appl. Phys. A **88**, 17–20 (2007)
29. Michael, H.H., Yiying, W., Henning, F., Ngan, T., Eicke, W., Peidong, Y.: Adv. Mater. **13**, 113–116 (2001)
30. Li, L.Y., Jing, H.Y., Dan, D.W., Yong, J.Z., Ya, X.W., Hui, L.L., Hou, G.F., Ji, H.L.: Physica E **40**, 920–923 (2008)
31. Hejazi, S.R., Hosseini, H.R.M., Ghamsari, M.S.: J. Alloys Compd. **455**, 353–357 (2008)
32. Xi, Y., Song, J., Xu, S., Yang, R., Gao, Z., Hu, C., Wang, Z.L.: J. Mater. Chem. **19**, 9260–9264 (2009)
33. Zhang, B.P., Binh, N.T., Wakatsuki, K., et al.: Appl. Phys. A **79**, 1711–1714 (2004)
34. Wang, Z.L.: Chinese Sci. Bull. **54**, 4021–4034 (2009)
35. Wang, Z.L.: Adv. Mater. **15**, 432–436 (2003)

36. Zhou, Z.K., Hao, Z.H., Wei, M.Z., Gang, W.X., He, Y.S.: Chin. Phys. Lett. **26**, 024201 (2009)
37. Kong, X.Y., Wang, Z.L.: Nano Lett. **3**, 1625–1631 (2003)
38. Gao, P.X., Wang, Z.L.: Small **1**, 945–949 (2005)
39. Kong, X.Y., Ding, Y., Yang, R., Wang, Z.L.: Science **303** (2004)
40. Wang, Z.L.: Mater. Today **7**, 26–33 (2005)
41. Zhang, N., Yi, R., Shi, R., et al.: Mater. Lett. **63**, 496–499 (2009)
42. Wang, J., Zhang, S., You, J., Gaoa, G., Chena, G., Liu, X.: Bull. Mater. Sci. **31**, 597–601 (2008)
43. Vernardoua, D., Kenanakisa, G., Courise, S., Manikasf, A.C., Voyiatzisf, G.A., Pembleg, M. E., Koudoumasa, E.H., Katsarakis, N.: J. Cryst. Growth **308**, 105–109 (2007)
44. Yavari, I., Mahjoub, A.R., Kowsari, E., Movahedi, M.: J. Nanopart. Res. **11**, 861–868 (2009)
45. Sounart, T.L., Liu, J., Voigt, J.A., Huo, M., Spoerke, E.D., McKenzie, B.: J. Am. Chem. Soc. **129**, 15786–15793 (2007)
46. Yatsui, T., Jeong, H., Ohtsu, M.: Appl. Phys. B **93**, 199–202 (2008)
47. Suzuki, K., Inoguchi, M., Kageyama, K., Takagi, H., Sakabe, Y.: J. Nanopart. Res. **11**, 1349–1360 (2009)
48. Seiyama, T., Kato, A.: Anal. Chem. **34**, 1502 (1962)
49. Basu, S., Dutta, A.: Sens. Actuators B **22**, 83 (1994)
50. Wan, Q., Li, Q.H., Chen, Y.J., He, X.L., Li, J.P., Lin, C.L.: Appl. Phys. Lett. **84**, 3654 (2004)
51. Wan, Q., Li, Q.H., Chen, Y.J., Wang, T.H., He, X.L., Gao, X.G., Li, J.P.: Appl. Phys. Lett **84**, 3085 (2004)
52. Suchea, M., Christoulakis, S., Moschovis, K., Katsarakisa, N., Kiriakidis, G.: Thin Solid Films **515**, 551–554 (2006)
53. Hsueh, T.J., Hsu, C.L., Chang, S.J., Chen, I.C.: Sens. Actuators B **126**, 473–477 (2007)
54. Liao, L., Lu, H.B., Shuai, M., Li, J.C., Liu, Y.L., Liu, C., Shen, Z.X., Yu, T.: Nanotechnology **19**, 175501 (2008)
55. Wei, D., Unalan, H.E., Han, D., Zhang, Q., Niu, L., Amaratunga, G., Ryhanen, T.: Nanotechnology **19**, 424006 (2008)
56. Comini, E., Faglia, G., Ferroni, M., et al.: Appl. Phys. A **88**, 45–48 (2007)
57. Fan, Z., Lu, J.G.: J. Nanosci. Nanotechnol. **5**, 1561–1573 (2005)
58. Fan, Z., Cheng, P., Walter, E.C., Penner, R.M., Lin, C.H., Lee, H.P.: Appl. Phys. Lett **85**, 6128 (2004)
59. Yang, P., Yan, H., Mao, S., Russo, R., Johnson, J., Saykally, R., Morris, N., Pham, J., He, R., Choi, H.J.: Adv. Funct. Mater. **12**, 323 (2002)
60. Johnson, J.C., Yan, H., Yang, P., Saykally, R.J.: J. Phys. Chem. B **107**, 8816 (2003)
61. Amaratunga, Ryhanen, T.: Nanotechnology **19,** 424006 (2008)
62. Min, C., Shen, X., Sheng, W.: Appl. Phys. A **96**, 799–803 (2009)
63. Chik, H., Liang, J., Cloutier, S.G., Kouklin, N., Xu, J.M.: Appl. Phys. Lett. **84**, 3376 (2004)
64. Liu, C.H., Yiu, W.C., Au, F.C.K., Ding, J.X., Lee, C.S., Lee, S.T.: Appl. Phys. Lett. **83**, 3168 (2003)
65. Shen, G., Chen, D.: Front. Optoelectron. China. (2009). doi: 10.1007/S12200**,** 010-0001-4
66. Park, W.I., Kim, J.S., Yi, G.C., Bae, M.H., Lee, H.J.: Appl. Phys. Lett. **85**, 5052 (2004)
67. Chang, P., Fan, Z., Tseng, W., Chiou, W.A., Hong, J., Lu, J.G.: Chem. Mater. **16**, 5133–5137 (2004)
68. Fan, Z., Lu, J.G.: J. Nanosci. Nanotechnol. **5**, 1561–1573 (2005)
69. Arnold, M.S., Avouris, P., Pan, Z.W., Wang, ZL.: J. Phys. Chem. B **107**, 659–663 (2003)
70. Goldberger, J., Sirbuly, D.J., Law, M., Yang, P.: J. Phys. Chem. B **109**, 9–13 (2005)
71. Chen, H., Wu, X., Gong, L., Ye, C., Qu, F., Shen, G.: Nanoscale Res. Lett. (2009). doi: **10.1007/S11671**, 009–9506–4,
72. Zhang, J., Li, M.K., Yu, L.Y., Liu, L.L., Zhang, H., Yang, Z.: Appl. Phys. A **97**, 869–876 (2009)
73. Yahya, N., Daud, H., Tajuddin, N.H., Zaid, H.M., Shafie, A., Puspitasari, P.: J. Nano Res. **11**, 25–34 (2010)

Supported Nanoparticles for Fuel Synthesis

Noor Asmawati Mohd Zabidi

Abstract This chapter has been written as an introductory on the preparation method of supported iron and cobalt oxides nanocatalysts. It begins with an overview on the gas-to-liquid (GTL) process. Emphasis is given on the catalysis for Fischer–Tropsch (FT) synthesis. Both the iron and cobalt-based catalysts have their own merits and features of these catalyst systems are highlighted. The spherical-model catalyst approach has been adopted as it can bridge the gap between the well-defined single crystal surfaces and those poorly-defined complex industrial catalysts. The synthesis methods for the oxide-supported nanoparticles of iron and cobalt oxides described in this chapter include the colloidal, reverse microemulsion, ammonia deposition, impregnation, precipitation and strong electrostatic adsorption. The applications of electron microscopy techniques on the morphological characterization of supported nanoparticles are illustrated in this chapter.

1 Introduction

The conversion of natural gas to liquid (GTL) fuel is an attractive option for monetizing stranded natural gas. The GTL process consists of the following steps [1]:

1. Production of synthesis gas (a mixture of carbon monoxide and hydrogen) from natural gas
2. Synthesis of hydrocarbon via the Fischer-Tropsch synthesis (FT) reaction
3. Product upgrading of the synthesized hydrocarbon

N.A. Mohd Zabidi
Fundamental and Applied Sciences Department, Universiti Teknologi PETRONAS, Bandar Seri Iskandar, 31750 Tronoh, Perak, Malaysia
e-mail: noorasmawati_mzabidi@petronas.com.my

N. Yahya (ed.), *Carbon and Oxide Nanostructures*, Adv Struct Mater 5,
DOI 10.1007/8611_2010_19, © Springer-Verlag Berlin Heidelberg 2010,
Published online: 25 June 2010

The Fischer–Tropsch (FT) reaction, shown in (1), produces clean gasoline and diesel fuels.

$$nCO + 2nH_2 \rightarrow C_nH_{2n} + nH_2O \quad (1)$$

Catalysts used for the Fischer–Tropsch reaction are generally based on iron (for high temperature process) and cobalt (for low temperature process). Iron has been the traditional catalyst of choice for FT reaction. It is reactive and the most economical catalyst for synthesis of clean fuel from the synthesis gas mixture. Cobalt has higher activity for Fischer–Tropsch reaction but more expensive compared to iron [2]. Catalyst supports that have been utilized include silica, alumina, titania, zirconia, magnesium, carbon and molecular sieves [1]. The cost of catalyst support, metal and catalyst preparation contributes to the cost of FT catalyst, which represents a significant part of the cost for the FT technology. Fundamental understanding of the relationship between the catalyst performance and its physical properties, such as particle size, surface area and porosity is vital. The deactivation behavior of cobalt has been linked to its crystallite size, therefore, control of crystallite size is of importance [3]. A very stable and active catalyst is required to ensure the catalytic system is economically attractive. A model catalyst consists of well-defined catalytically active metal deposited on non-porous support [2]. Spherical model catalysts can be used to bridge the gap between the poorly-defined porous industrial catalysts and the well-defined single crystal surfaces. Knowledge on the relation between the rate of the reaction to the composition and morphology of the catalyst is still lacking [4]. Thus, characterization of model catalysts can relate the physical properties, such as size and shape of particles, to the catalytic behavior of the catalytic materials.

This chapter focuses on the preparation methods and the morphological characterization of supported iron and cobalt nanoparticles using electron microscopy. The term nanoparticle is used for particles having diameters ranging from 2 to 50 nm with variable crystallinity whereas well-defined crystalline nanoparticles are classified as nanocrystals [5]. The commercially applied iron catalyst is in the fused form comprising large iron particles and therefore difficult to investigate using microscopy. The loss of catalyst activity is associated with changes of iron into a mixture of iron oxide and iron carbide during the Fischer–Tropsch synthesis reaction. The relation between deactivation and changes in composition and morphology are not fully understood for iron and cobalt catalysts [4]. The application of electron microscopy techniques on the supported nanoparticles are well suited to investigate the morphology of supported catalysts and morphological changes that occur during Fischer–Tropsch synthesis. The knowledge on the effect of particle size on the product selectivity and yields for Fischer–Tropsch reaction is still lacking. It has been reported [1] that particle sizes of the catalyst material have an effect on the pressure drop in the reactor, and can influence the product distribution. Iron and cobalt nanoparticles of sizes less than 10 nm are expected to improve the kinetics of the Fischer–Tropsch reaction, selectivity for gasoline and stability of the catalyst.

Niemantsverdriet [6] described the principles of electron microscopy and its application in catalysis. Transmission electron microscopy (TEM) and scanning electron microscopy (SEM) are among the popular techniques in catalysis as these techniques can yield information on the size and the shape of supported particles. These electron techniques use electron beam with energy ranging between 100 to 400 keV and magnification of up to 10^6 which can reveal detail information about the surface structure. Atomic resolution of about 0.1 nm can be achieved using a TEM instrument.

2 Preparation of Catalyst Support

The industrial catalyst systems are complex system which posed difficulty for carrying out fundamental level studies on the active sites of the catalyst [2].The active metal particles are often hidden in the pores of the support. To overcome this problem, simplified catalyst model can be designed. Using this approach, the active metal can be deposited on the external surface of a nonporous support which can facilitate profile views of the supported nanoparticles using electron microscopy. SiO_2 is a commonly used support for FT catalysts. Non-porous SiO_2 spheres can be synthesized using the Stöber's method [7] and this support can be used to design the spherical catalyst model for studying Fischer–Tropsch catalysis. The synthesis of non-porous SiO_2 spheres began using two solutions as follows: Solution A comprised 76 mL of NH_4OH (25%) in 600 mL of absolute ethanol. Solution B was prepared by stirring 64 mL of tetraethylorthosilicate (TEOS, 98%) with 260 mL of absolute ethanol [8]. Solution B was added to solution A and the milky white mixture was stirred for 24 h. Ethanol was removed from the mixture in a rotary evaporator at 75°C. The precipitate was dried in an oven at 110°C for 16 h and then calcined in an oven at 500°C for 1 h to remove the ammonia. The FESEM image (Fig. 1) shows spherical-shaped SiO_2 particles having diameters ranging from 100 to 230 nm [8]. These non-porous SiO_2 spheres are useful catalyst support for carrying out fundamental level investigation using spherical model catalyst approach.

3 Preparation and Microscopic Characterization of Oxide-Supported Iron Nanoparticles

3.1 Preparation of Iron Nanoparticles

The synthesis of nanoparticle is of importance for applications in catalysis. Some of the concerns of nanoparticle synthesis are the size and the uniformity of the particle size, control of particle shape, crystallinity and reproducibility of synthesis methods [5]. Iron nanoparticles can be prepared by several methods as described in the following sections.

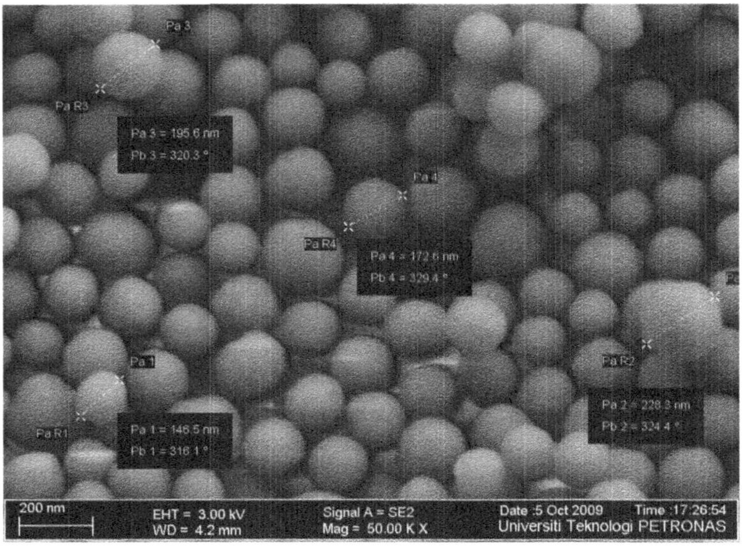

Fig. 1 FESEM image of non-porous Stöber SiO_2 spheres. The BET surface area of the SiO_2 spheres is 17.8 $m^2\ g^{-1}$ and the cumulative pore volume is 0.098 $cm^3\ g^{-1}$. (Adapted from [8])

3.2 Colloidal Method [9]

Colloids are synthesized in the presence of surfactants which disperse and stabilize the nanoparticles in an organic solvent. Some of the approaches include polyol method, ethylene glycol method, modified coordination capture method and pseudo-colloidal method. The polyol process involves heating a mixture of catalyst precursor in surfactants, such as oleic acid and oleyl amine in a high-boiling solvent, such as diphenyl ether. Sun and Zeng [10] have reported that the high temperature alcohol reduction of iron (III) acetylacetonate metal precursor resulted in monodispersed iron nanoparticles. This synthesis process is also called "heating-up" process [9]. The size of the nanoparticles is controlled by changing the concentration of the precursor, the amount and type of surfactant, the aging time and temperature of the reaction. Another synthetic method that produces uniform nanocrystals that is comparable to the "heating up" process is called the "hot injection" method. The "hot injection" method induces high supersaturation and leads to fast homogeneous nucleation reaction followed by diffusion-controlled growth process, which control the particle size distribution.

Using the modified colloidal synthesis method, non-porous silica spheres were sonicated in a mixture of oleylamine, oleic acid and cyclohexane for 1 h and then heated and stirred in a multi-neck quartz reaction vessel [11]. A liquid mixture of iron(III) acetyl acetonate, oleylamine, oleic acid, 1,2 hexadecanediol, and phenyl ether was slowly added to the stirred SiO_2 suspension once the reaction temperature reached 150°C. The reaction mixture was refluxed under nitrogen atmosphere at 265°C for

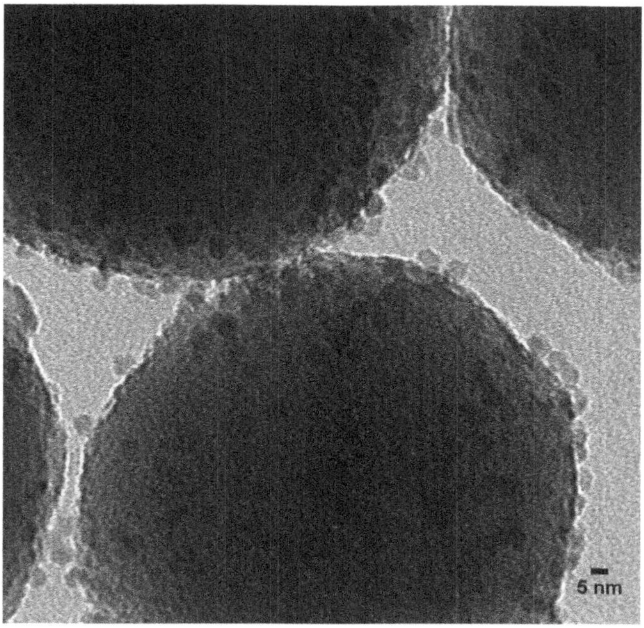

Fig. 2 TEM image of spherical model catalysts 6% FeO_x/SiO_2 prepared by the modified colloidal synthesis method. Non-porous Stöber SiO_2 spheres were used a catalyst support. (Adapted from [12])

30 min. Figure 2 shows the TEM image of SiO_2-supported iron oxide nanoparticles prepared via the modified colloidal synthesis method [12]. The SiO_2 spheres were pre-treated in the oleic acid and oleylamine mixture in a bath sonicator at 33°C for 3 h prior to the synthesis. The long contact of the SiO_2 spheres in the surfactant mixture might have led to a good coverage of the surfactant molecules on the SiO_2 spheres, hence increased its affinity towards the surfactant-stabilized iron nanoparticles. Spherical-shaped iron oxide nanoparticles with average diameters of 6.2 ± 0.9 nm were formed via the modified colloidal synthesis method and the nanoparticles were almost evenly dispersed on the SiO_2 surfaces. An equimolar mixture of oleylamine and oleic acid was used in the colloidal synthesis approach and these surfactants were able to prevent the agglomeration of the iron oxide nanoparticles. Iron oxide nanoparticles were anchored on the SiO_2 surfaces and did not lie in between the SiO_2 spheres, as shown in Fig. 2, thus suggesting that nucleation occurred heterogeneously. The iron loading was kept at 6 wt% as it was discovered that increasing the iron loading resulted in highly agglomerated nanoparticles [11]. The size of the nanoparticle is influenced by temperature, aging time, surfactant, amounts of metal precursor as well as the ratio of the metal precursor to the surfactant [5].

Combination of the modified colloidal method and the seed-mediated growth [13] has been attempted in order to produce larger nanoparticles [11]. The procedure is similar to the modified colloidal synthesis method, but iron precursor reagent was

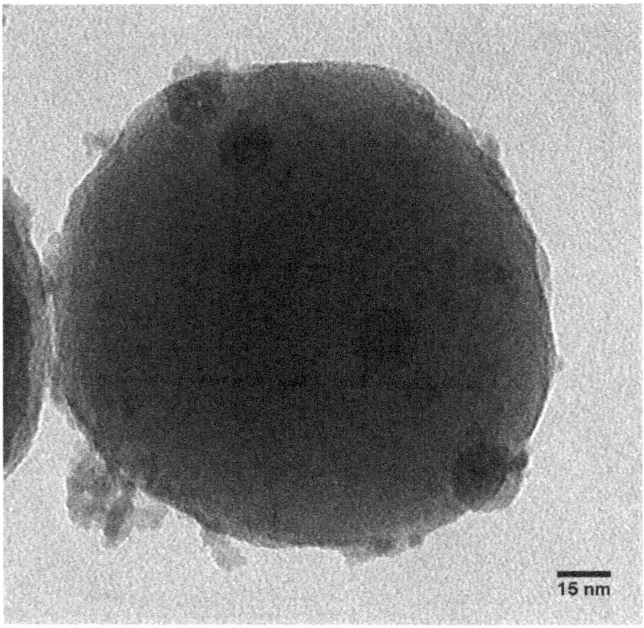

Fig. 3 TEM image of spherical model catalysts 6% FeO_x/SiO_2 prepared by the combination of modified colloidal synthesis method and the seed-mediated growth method. (Adapted from [11])

added in two stages. Upon completion of the first reaction stage, the reaction mixture was cooled to 150°C, then a second portion of precursor reagent was added and the reaction mixture was heated again to 265°C for 30 min. The combined synthesis method had resulted in larger nanoparticles, as shown in Fig. 3.

3.3 *Reverse Microemulsion Method [9, 14]*

A microemulsion is a liquid mixture of water, a hydrocarbon and a surfactant. A surfactant is a molecule that possesses both the polar (hydrophilic head) and the non-polar (hydrophobic tail) groups. When the concentration of the surfactant exceeds the critical micelle concentration, molecules aggregate to form micelles. When micelles are formed in an organic medium, the aggregate is referred as a reverse micelle, in which the polar heads are in the core and the non-polar tails remain outside to maintain interaction with hydrocarbon. The reverse microemulsion synthesis method consists of preparing two microemulsions containing the metal salt and the reducing agent. The precursor metal salt and reducing agent are dissolved in the aqueous phase whereas the surfactant is prepared in an organic medium. Mixing these two microemulsions caused percolation. The reducing agent reduced the metal salt inside the micelles to metal particles.

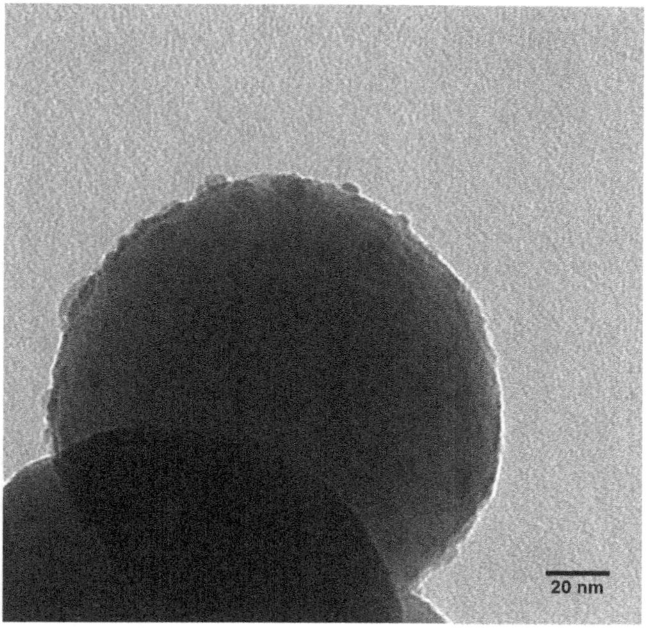

Fig. 4 TEM image of FeO_x on SiO_2 spheres prepared using the reverse microemulsion method, AOT surfactant, 6 wt% Fe loading, calcined in O_2/Ar flow at 500°C for 3 h. (Adapted from [12])

Mohd Zabidi synthesized the SiO_2-supported iron nanoparticles using the reverse microemulsion method [11]. The first reverse microemulsion consisted of $Fe(NO_3)_3{\cdot}9H_2O$ (aq) and sodium bis(2-ethylhexyl) sulfosuccinate (AOT, ionic surfactant) in hexanol and the second reverse microemulsion was prepared by mixing an aqueous hydrazine solution (reducing agent) with the AOT solution. SiO_2 spheres were added to the mixture and the slurry was stirred for 3 h under nitrogen environment. Figure 4 shows the TEM micrograph of a spherical model catalysts prepared using the reverse microemulsion method. The reverse micro-emulsion method produced spherical-shaped iron oxide nanoparticles with average diameters of 6.3 ± 1.7 nm, however, the coverage of the SiO_2 surfaces was found to be less than that obtained using the colloidal synthesis approach [12].

3.4 Ammonia Deposition Method [15]

The ammonia deposition method was proven successful for the preparation of SiO_2-supported cobalt catalyst. This method involves adding ammonia to cobalt (II) nitrate solution followed by reaction on the surface of the silica support. The application of the ammonia deposition method for synthesis of supported iron nanoparticles has not been so successful, as it resulted in extensive agglomeration of the iron nanoparticles, as depicted in Fig. 5.

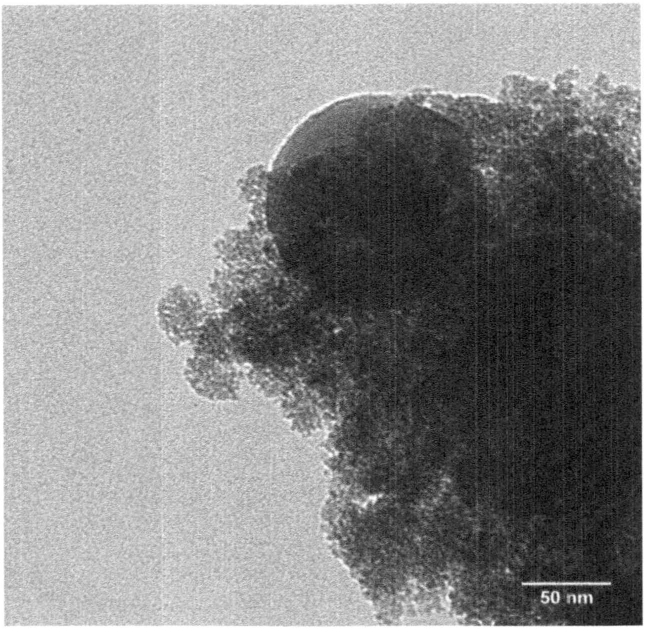

Fig. 5 TEM image of 6% FeO_x/SiO_2 prepared by the ammonia deposition method. (Adapted from [12])

3.5 *Impregnation Method*

The impregnation method seems to be the simplest synthesis method as it utilizes a metal precursor $Fe(NO_3)_3{\cdot}9H_2O$ dissolved in aqueous phase, followed by drying and calcination process [16]. The metal loading influenced the particle size distribution. Figure 6 shows the TEM image for the SiO_2-supported iron-based catalyst prepared by the impregnation method for 6% FeO_x/SiO_2, which has particle size ranged from 4 to 10 nm [17]. However, increasing the iron loading to 15% resulted in a broader particle size distribution and revealed bimodal distribution (Fig. 7).

4 Preparation and Microscopic Characterization of Oxide-Supported Cobalt Nanoparticles

4.1 *Preparation of Cobalt Nanoparticles*

An extensive review on the development of cobalt catalyst was presented by Khodakov et al. [9]. The performance of the cobalt catalyst in Fisher–Tropsch reaction is greatly influenced by the catalyst preparation method. The variables

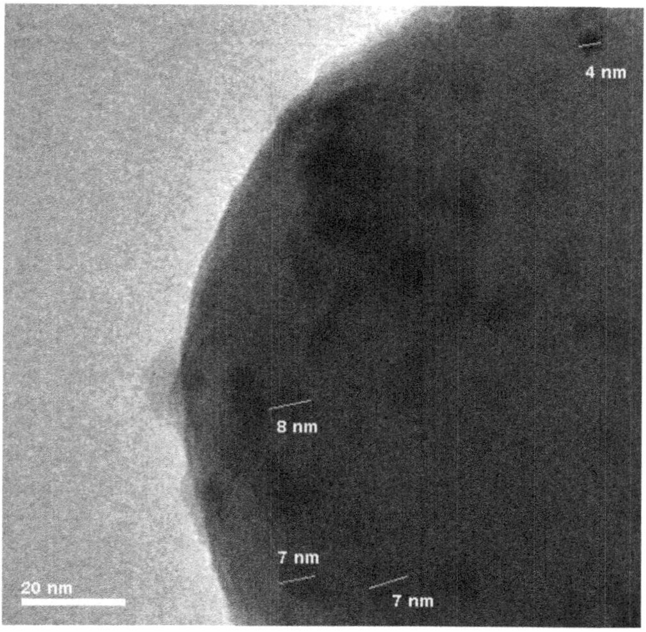

Fig. 6 TEM image of 6% FeO_x/SiO_2 prepared via the impregnation method. (Adapted from [17])

include suitable support, deposition method of the cobalt metal precursor, catalyst promoter, and the subsequent thermal treatments [2]. Cobalt has shown better resistant to deactivation and attrition, but it is also much more expensive compared to ferum. Therefore, well-dispersed cobalt on the catalyst support is highly desired to gain economic attractiveness. The reactivity in FTS is correlated to the number of cobalt metallic particle exposed to the syngas molecules [9]. This factor in turns depends on the cobalt loading, dispersion of cobalt species and its reducibility. Hence, an ideal supported catalyst would have uniformly distributed cobalt species that undergoes complete reduction forming cobalt metallic particle at optimum size of 6–8 nm, where high dispersion guarantees optimum use of cobalt without jeopardizing the FTS performance. Some of the common preparation methods involving supported cobalt catalysts are discussed in the following sections. Similar preparation methods as those for iron synthesis have also been applied in cobalt preparation procedures such as the impregnation, reverse microemulsion and precipitation methods.

4.2 Impregnation Method

The incipient wetness impregnation method is a commonly used method for preparing supported cobalt catalyst. Typically required amount of the precursor

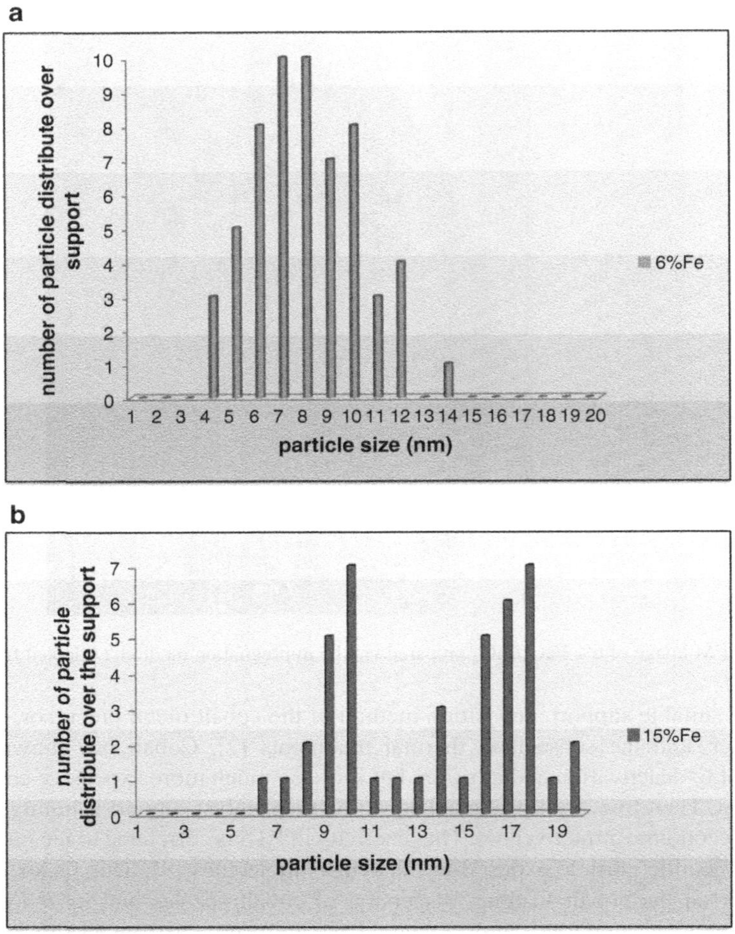

Fig. 7 Particle size distribution at iron loadings of (**a**) 6% (**b**) 15% on SiO_2 spheres. (Adapted from [17])

salt i.e. Co $(NO_3)_2{\cdot}6H_2O$ is dissolved in deionized water and added dropwise to the support under constant stirring, followed by drying in an oven at 120°C overnight and calcining at temperature 500°C [3]. Variables which can affect the resultant catalyst are the rate of addition of precursor solution, rate of drying, temperature and duration of heating. Figure 8 shows the TEM image of alumina-supported cobalt oxide sample [18]. The particle size distribution ranged from 5 to 20 nm. A morphology different from that of Fig. 8 was obtained when non-porous SiO_2 spheres were used as a support, as depicted in Fig. 9. Nevertheless, extensive agglomeration was observed even at 5 wt% loading of the cobalt [19]

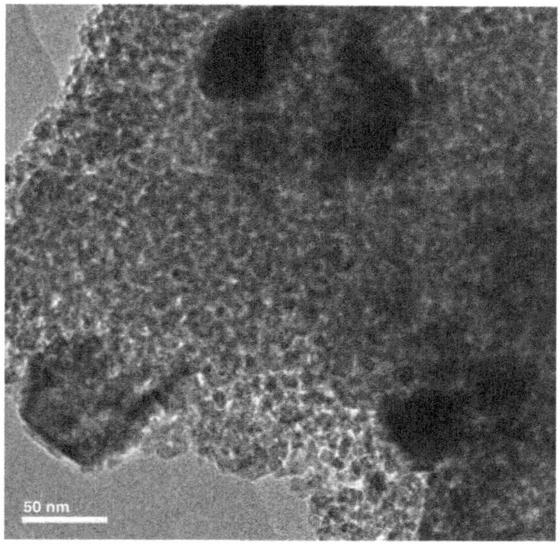

Fig. 8 TEM image of cobalt oxide supported on alumina, prepared via impregnation. (Adapted from [18])

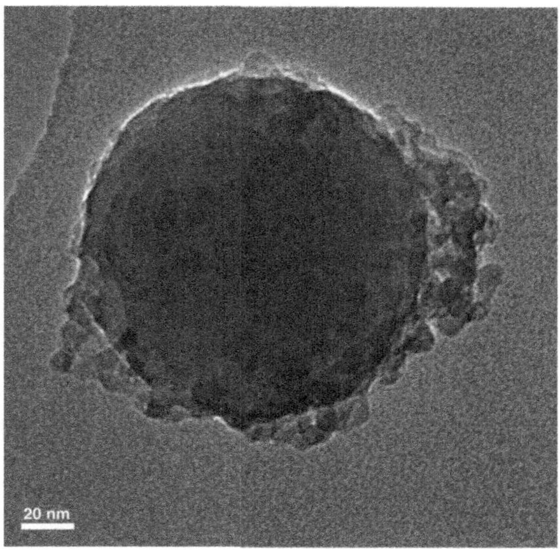

Fig. 9 TEM image of cobalt oxide supported on SiO_2, prepared via impregnation method (Adapted from [19])

4.3 Precipitation Method

The required amount of the precursor salt i.e. Co $(NO_3)_2{\cdot}6H_2O$ was dissolved in de-ionized water and added to the support with constant stirring followed by the addition of 200 ml of 25 vol% ammonia solution dropwise, then temperature was gradually raised to 90°C [18]. The resultant slurry was stirred at this temperature for 8 h. The slurry was then dried at 120°C overnight, ground and calcined at 500°C for 6 h. Particle size distribution for catalyst synthesized using the precipitation method was not as good as the one obtained using the impregnation method [18].

4.4 Strong Electrostatic Adsorption (SEA) Method

Synthesis of uniformly distributed Co particles remains a great challenge. Strong electrostatic adsorption is a catalyst preparation method which is based on basic concept of electrostatic attraction of oppositely charged particle. The mechanism for the electrostatic adsorption has been described Jiao and Regalbuto [20]. Silica and other metal oxides contain hydroxyl groups on its surface. Point of zero charge (PZC) is the pH value of a medium where the hydroxyl groups on the surface of the support remain neutral. In a pH $<$ PZC medium, the hydroxyl groups will protonate and become positively charged and thus attracting anions. When pH $>$ PZC, the hydroxyl groups will deprotonate and became negatively charged and attracting cations. In other words, pH value plays an important role in the deposition of metal precursor. The PZC of silica support was found to be 4.25 ± 0.25. Figure 10 shows the TEM image of SiO_2-supported cobalt oxide nanoparticles synthesized via the strong electrostatic adsorption method [8]. This synthesis method produced

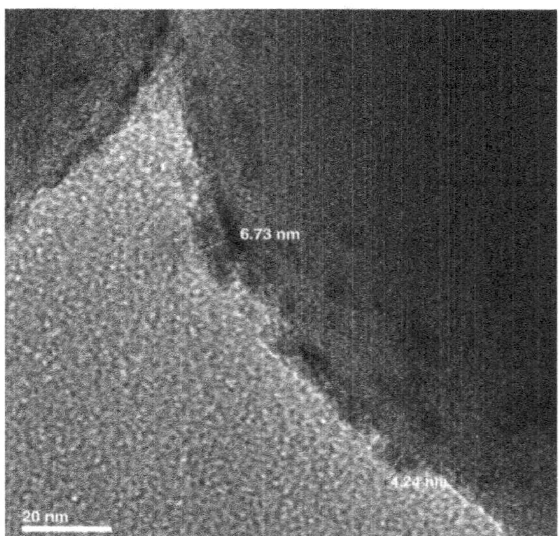

Fig. 10 TEM image of cobalt oxide supported on SiO_2, prepared via the strong electrostatic adsorption method. Cobalt loading on SiO_2 is 5 wt% (Adapted from [8])

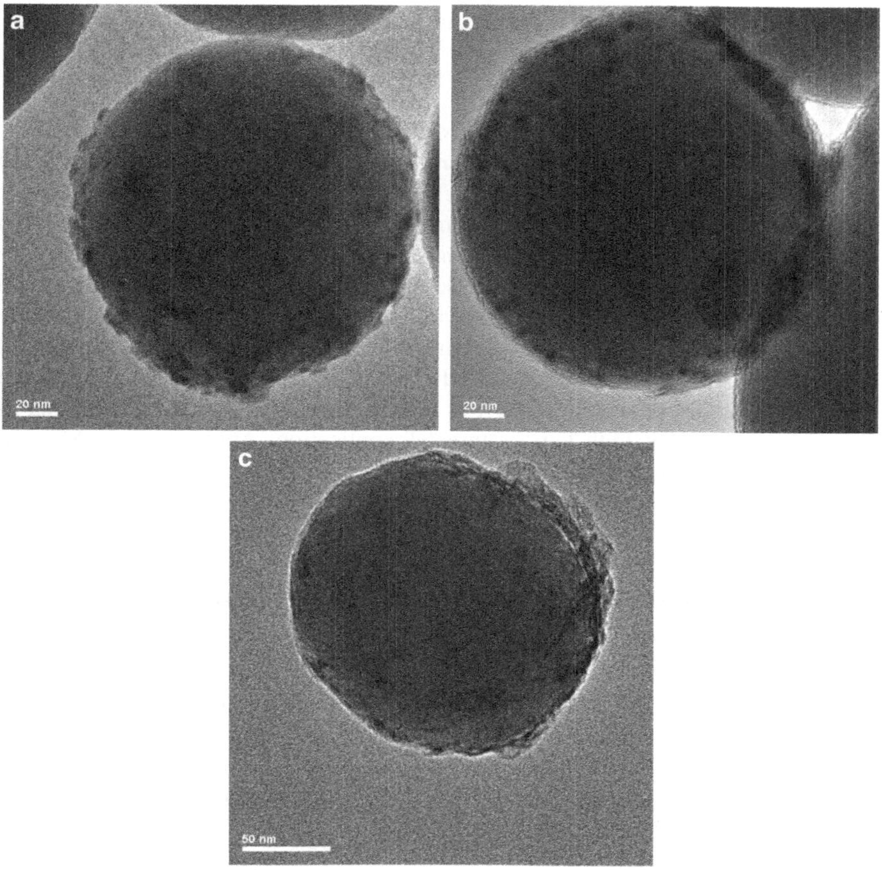

Fig. 11 TEM images of cobalt-oxide supported on SiO_2 at cobalt loadings of (**a**) 5 wt% (**b**) 10 wt% (**c**) 20 wt% (Adapted from [8])

uniform particle size and shape and most of the cobalt oxide nanoparticles have sizes less than 10 nm. The size distribution of cobalt oxide on SiO_2 support prepared using the strong electrostatic adsorption method is better than those obtained using other synthesis methods.

Besides the preparation method, the metal loading on SiO_2 support also influenced the size and the distribution of the nanoparticles on the support as illustrated in the TEM images in Fig. 11. At low cobalt loading (1 wt%) SiO_2 surface was not covered uniformly by the nanoparticles whereas at high cobalt loading (20 wt%) extensive agglomeration of cobalt oxide nanoparticles was observed. It was discovered that 93% of the cobalt oxide crystallites were in the optimum size range of 6–10 nm with minimal agglomeration at 10 wt% cobalt loading on SiO_2 [8].

5 Preparation and Microscopic Characterization of CNT-Supported Iron and Cobalt Nanoparticles

Tavasoli and co-workers [21] have investigated the Fischer–Tropsch synthesis using cobalt particles supported on carbon nanotubes (CNT). Whereas Zaman et al. [22] discovered MgO-modified CNTs enhanced the product selectivity of C_{5+} due to the synergetic effect between CNTs and MgO in the Fischer–Tropsch reaction. Carbon nanotubes (Fig. 12) have several advantages to be used as a catalyst support owing to their acid/base-resistant, inertness and stability at high temperature. Their inertness avoids the formation of mixed compound which are difficult to reduce under normal reduction condition. The decrease in the strength of the metal-support interaction enhanced the reducibility of the CNT-supported cobalt catalyst. These special properties have given CNTs increasing potential as a catalyst support for high temperature FT application. Applying the CNT-supported Co catalysts in FT reaction increased C_{5+} selectivity by more than 10% and also decreased the methane selectivity by nearly 30% [22].

5.1 Preparation of Cobalt and Iron Nanocatalysts Supported on CNTs

Monometallic catalysts can be prepared by a single step impregnation method in which the required amounts of metal precursor salts $Fe(NO_3)_3{\cdot}9H_2O$ and $Co(NO_3)_2{\cdot}6H_2O$ are dissolved in deionized water and added to the treated CNTs dropwise, followed by ultrasonication and drying in a rotary evaporator at 90°C [23].

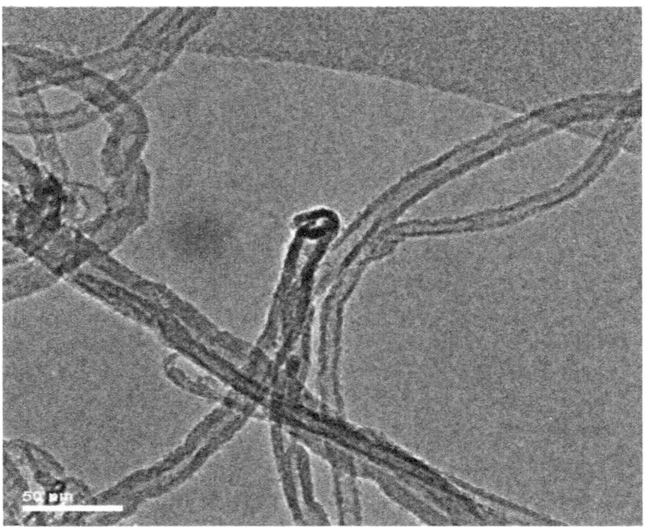

Fig. 12 TEM image of pure multi-walled CNTs (Adapted from [23])

These catalysts are further dried in an oven at 120°C overnight and calcined at 350°C for 3 h. Bimetallic nanocatalysts are prepared by sequential impregnation method using similar steps as those for monometallic in which cobalt precursor is impregnated first onto the support followed by iron salt.

FESEM and TEM images of iron oxide nano-particles on CNTs are shown in Figs. 13 and 14, respectively. Iron oxide nano-particles were well dispersed not only inside the CNTs but also on the outer walls of the CNTs. Nanoparticles encapsulated inside the CNTs were more uniform than those at the outer CNTs. The channels of CNTs restrict the growth of the particles encapsulated inside it, which resulted in smaller nanoparticles than the ones attached to the outer walls of the CNTs.

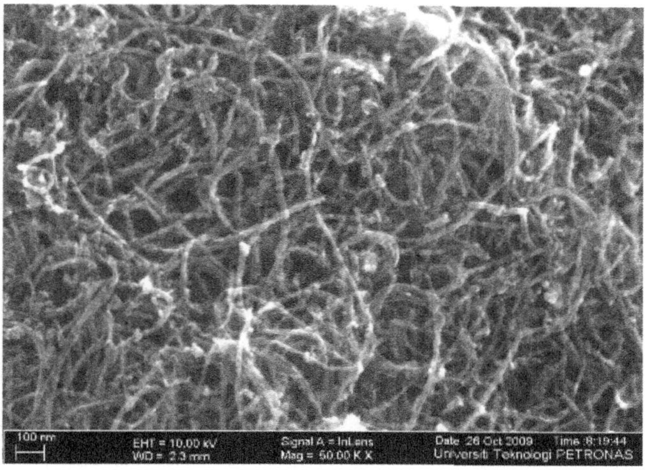

Fig. 13 FESEM image of CNT-supported iron oxide nanoparticles (Adapted from [23])

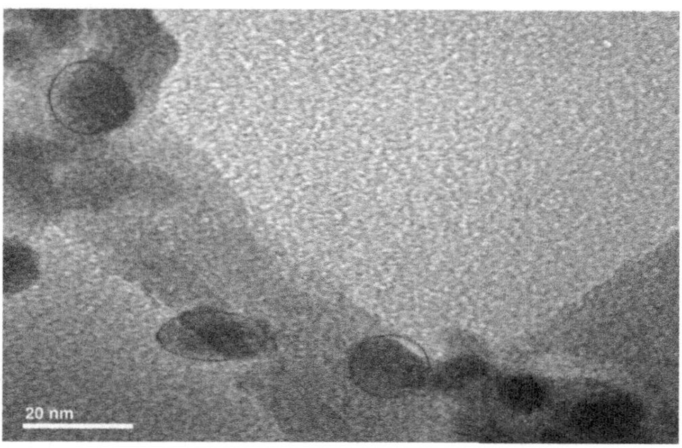

Fig. 14 TEM image of iron oxide nanoparticles on CNTs. (Adapted from [23])

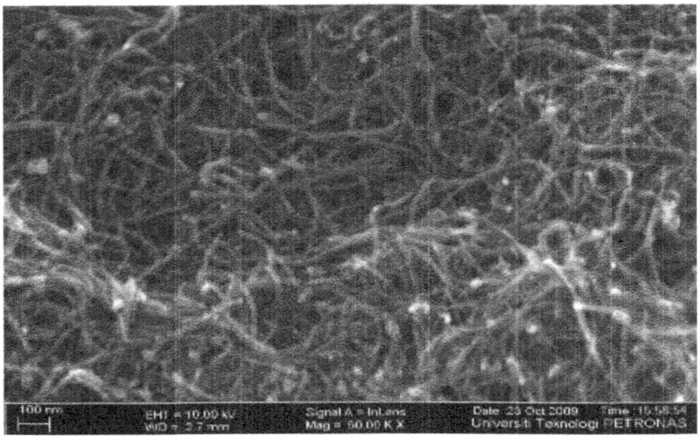

Fig. 15 FESEM image of 50:50 Fe:Co bimetallic on CNTs. (Adapted from [23])

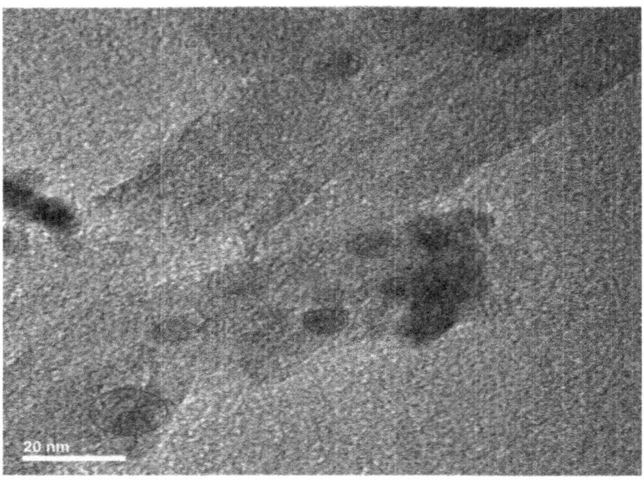

Fig. 16 TEM image of 50:50 Fe:Co bimetallic on CNTs (Adapted from [23])

The presence of cobalt in the 50:50 Co:Fe bimetallic on CNTs did not change the morphology of the particles significantly, as shown in Figs. 15 and 16.

6 Conclusion

This chapter provides an overview on the preparation methods of supported iron and cobalt oxides nanocatalysts. The discussion was focused on iron and cobalt as both materials are the most commonly-used catalysts for fuel synthesis via the Fischer Tropsch (FT) reaction. The synthesis methods for the nanoparticles reviewed in this chapter include the colloidal method, reverse microemulsion, impregnation, precipitation, ammonia deposition and strong electrostatic adsorption methods. The particle size and distribution of the nanoparticles are greatly influenced by the synthesis methods. The colloidal synthesis method resulted in fairly uniform-sized iron oxide nanoparticles on spherical SiO_2 support. Well-defined cobalt oxides nanoparticles can be prepared via the strong electrostatic adsorption synthesis method (SEA). Both TEM and FESEM are powerful tools for investigating morphology of catalysts materials. This chapter illustrated the microscopic characterization of nanoparticles on SiO_2, Al_2O_3 and CNTs supports. Spherical models Fe/SiO_2 and Co/SiO_2 nanocatalysts can be well-characterized using TEM and FESEM. The application of both microscopic techniques on supported nanocatalysts facilitates the size-dependent studies of both cobalt and iron-based catalysts in Fischer Tropsch synthesis (FT) reaction.

References

1. Tijm, P.J.A.: Gas to Liquids, Fischer-Tropsch Catalysis, Reactors, Products and Process (2006)
2. Moodley, D.J.: On the deactivation of cobalt-based Fischer-Tropsch synthesis catalysts. PhD thesis, Eindhoven University of Technology, The Netherlands (2008)
3. Saib, A.M., Borgna, A., van de Loosdrecht, L., van Berge, P.J., Geus, J.W., Niemantsverdriet, J.W.: J. Catal. **239**, 326–339 (2006)
4. Sarkar, A., Seth, D., Dozier, A.K., Neathery, J.K., Hamdeh, H., Davis, B.H.: Fischer Tropsch synthesis. Catal. Lett. **117**, 1–17 (2007)
5. Hyeon, T.: Chemical Communications (Royal Society of Chemistry 2004), 927–934 (2003)
6. Niemantsverdriet, J.W.: Spectroscopy in Catalysis, 3rd edn. Wiley-VCH, Weinheim (2007)
7. Stöber, W., Fink, A.: J. Colloid Interface Sci. **26**, 62–69 (1968)
8. Chee, K.L., Mohd Zabidi, N.A., Mohan, C.: Presentation at the 10th postgraduate symposium, Malaysia, 2010
9. Khodakov, A.Y., Chu, W., Fongarland, P.: Chem. Rev. **107**, 1692–1744 (2007)
10. Sun, S., Zeng, H.: J. Am. Chem. Soc. **124**, 8204–8205 (2002)
11. Mohd Zabidi, N.A.: Sabbatical Report. Eindhoven University of Technology, The Netherlands (2007)
12. Mohd Zabidi, N.A., Moodley, P.A., Thüne, P.C., Niemantsverdriet, J.W.: Presentation at the international conference on nanoscience and nanotechnology, Kuala Lumpur, 2008
13. Yamamuro, S., Ando, T., Sumiyama, K., Uchida, T., Kojima, I.: Jpn. J. Appl. Phys. **43**, 4458–4459 (2004)
14. Martinez, A., Priesto, G.: Catal. Commun. **8**, 1479–1486 (2007)
15. Barbier, A., Hanif, A., Dalmon, J.A., Martin, G.A.: Appl. Catal. A **168**, 333–343 (1998)
16. Saib, A.M.: Towards a cobalt Fischer-Tropsch synthesis catalyst with enhanced stability: a combined approach. PhD thesis, Eindhoven University of Technology, The Netherlands (2008)
17. Tasfy, S.H.F., Mohd Zabidi, N.A., Subbarao, D.: Presentation at the regional conference on solid state science & technology, Penang, 2009
18. Ali, S., Mohd Zabidi, N.A., Subbarao, D.: Presentation at the international conference on nanotechnology research and commercialization, Langkawi, 2009
19. Mohd Zabidi, N.A., Salleh, S.B., Tasfy: Presentation at the Nanotech Malaysia Conference, Kuala Lumpur, 2009

20. Jiao, L., Regalbuto, J.R.: J. Catal. **280**, 329–341 (2008)
21. Tavasoli, A., Irani, M., Nakhaeipour, A., Mortazavi, Y.,Khodadadi, A. A., Dalai, A.J.: Iran J. Chem. Eng. **28**, 37–48 (2009)
22. Zaman, M., Khodadi, A., Mortazavi, Y.: Fuel Proc. Technol. **90**, 1214–1219 (2009)
23. Ali, S., Mohd Zabidi, N.A., Subbarao, D.: Synthesis and characterization of bimetallic catalysts. Presentation at the 10th postgraduate symposium, Malaysia, 2009

Nanotechnology in Solar Hydrogen Production

Balbir Singh Mahinder Singh

Abstract The projected increase in human population has triggered the comprehensive search for suitable renewable energy related electrical power generation technologies. The efforts to exploit these technologies dates back to the last century, but breakthroughs certainly fall short in terms of competition with the current fossil fuel based energy systems. One of the strong points that allow solar energy to remain competitive is the fast deterioration of the environment and the accompanying natural disasters linked to the extensive usage of fossil-fuels. Concepts such as energy efficiency and energy conservation must be converted to strategies and initiatives, leveraging on nanotechnology as one of the important elements in solar hydrogen production. Although solar hydrogen production concept is not new, but the issues such as effective energy balance and management have been hindering the implementation process. The practicality and total energy management studies must be able to facilitate the sustainable implementation of solar hydrogen related electrical power generation systems. In this chapter, an Integrated Nano-Solar Hydrogen Production Scheme is discussed.

1 Introduction

The dynamic growth of human population has somehow accelerated the search for renewable energy resources. Although efforts moving in this direction dates back to the last century, breakthroughs certainly fall short in terms of competition with the current fossil fuel based energy systems. One of the disadvantages noticed, that is fast gaining momentum is pertaining to the polluting nature of the current main stream energy systems, and the results can be noticed by observing the fast

B.S. Mahinder Singh
Department of Fundamental and Applied Sciences, Universiti Teknologi PETRONAS, Bandar Seri Iskandar, 31750 Tronoh, Perak, Malaysia
e-mail: balbir@petronas.com.my

N. Yahya (ed.), *Carbon and Oxide Nanostructures,* Adv Struct Mater 5,
DOI 10.1007/8611_2010_20, © Springer-Verlag Berlin Heidelberg 2010,
Published online: 25 June 2010

deterioration of the environment and the accompanying so-called natural disasters. Efforts are now parallel; with the on-going intensified energy-related research activities that are either trying to reduce the impacts on the environment or seeking opportunities to further develop renewable energy based systems. The deterioration in the environment can be felt by the sudden change in global climate and the accompanying natural disasters such as earthquakes, tsunamis and floods. The need to develop renewable based energy systems is inevitable and strategies to leverage directly on the main source of energy, which is our sun, must be put in place urgently, by combining both the current state of technologies and the conventional power generating systems. Concepts such as energy efficiency and energy conservation must be converted to strategies and initiatives that include nanotechnology as one of the important enablers for the solar hydrogen production roadmap [1]. The solar hydrogen production concept is not new, but the issues such as effective energy balance and management have been hindering the implementation process. The practicality and total energy management studies must be balanced, in order to facilitate the transfer of R&D efforts to the power generating industry [2]. The world's energy consumption is currently estimated to be around 15×10^{12} W [3]. Our earth receives approximately 170×10^{15} W of energy daily, and although only part of this reaches earth at any point of time, through global cooperation, energy from the sun can be collected continuously and shared across the globe. Governments of the world must meet and find solutions to enforce the Global Power Sharing Scheme, and implement Global Renewable Energy Grid [4]. The strong reason to initiate this move is the fact that fossil fuel based systems will face difficulties due to depletion and negative impact on the environment. The depletion will not happen overnight and the time available must be used for parallel efforts of developing renewable energy related technologies and policies that will allow for global cooperation. Due to the oil embargo in 1973 and some other similar shocks in the early eighties [5], R&D efforts for producing solar hydrogen and oxygen from water intensified. But after temporary stabilization of oil-based power generation, the activities slowed down, and currently, innovative approaches are being used in direct water dissociation.

In Malaysia, the government initiated the national depletion policy, to provide some kind of protection to the depleting oil reserves by capping the domestic crude oil production to about 650,000 barrels per day [6]. This is to control and prolong the life of fossil fuels, and at this controlled production rate, the reserves are projected to last for another 15 years. The policy also allowed for the inclusion of alternative energy resources, which was highlighted in the eighth Malaysia Plan [7] and subsequently, in the ninth Malaysia Plan. The need to ensure that sufficient and reliable energy supply to be intensified was addressed and this allowed the relevant parties to develop a roadmap linking solar, hydrogen and fuel cell research initiatives for sustainable development. The other governments in the world have also introduced similar policies and approaches, as the global oil and gas supply continues to deplete, timelines are forecasted to safeguard economies around the globe. The utilization of solar energy for hydrogen generation is viable, and hydrogen is perhaps the simplest element that has the highest energy content per unit of weight, as compared to any other fuels. The solar hydrogen system is rather promising and a strategic move towards leveraging on this combination for sustainable power generation is rather sensible and viable [8].

2 Solar Hydrogen Generation

The focus of this chapter is on the use of nanotechnology for solar hydrogen generation. The motivation is to develop a sustainable and pollution-free energy system that will allow the world population to improve their living standards by utilizing an economically viable method to produce large quantities of hydrogen from water using solar energy [9]. The main fundamental process to be exploited in order to produce solar hydrogen is by using the electrolysis approach [10], coupled with clean electrical power generating system. There are many initiatives currently being used for producing hydrogen, as shown in Fig. 1 and the process of electrolysis will form the underlying basis for discussion that will lead towards the initiative of producing nano-solar hydrogen. Although the focus of this chapter is on the use of electrolysis as the main method for hydrogen production, it will also be useful to look at the process of reformation. The two typical reformation processes are steam reformation and gasification [11]. In the case of steam reformation, the feedstock can be natural gas, namely methane or ethane. These gases are combined with high-pressurized steam, at temperatures between 650 and 950°C, where, with the use of a catalyst, the chemical bonds are broken, thus producing hydrogen, which is stored, while carbon monoxide and carbon dioxide are removed. The so-called residuals are removed by using suitable techniques, and hydrogen is stored. Apparently steam

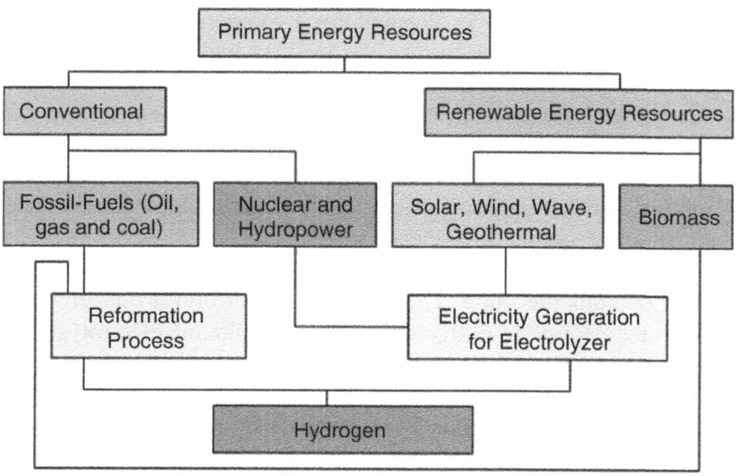

Fig. 1 Utilisation of primary energy resources and methods for producing hydrogen

reformation of natural gas is the popular choice and is used for producing around 80% of hydrogen globally. The major setback of using this approach is linked to the fact that natural gas is part of the depleting fossil fuels that will jeopardize the economies across the globe.

The other reformation method is known as the gasification process. Gasification is a method used for extracting energy from many different types of organic materials, where at high temperatures, with a controlled amount of oxygen and/or steam, are converted into carbon monoxide and hydrogen. The organic materials used can be coal, petroleum and biomass and the gaseous mixture is referred to as synthesis gas (syngas) [12]. One of the advantages of gasification is that syngas is potentially more efficient than direct combustion of the original fuel, and can be used in internal combustion engines. The other advantage of gasification is that it can also be used with fuels derived from biomass and organic waste. In the next sub-section, the focus will be on the process of electrolysis, a hydrogen production method that is certainly not new, but will be the fundamental technology needed to move forward.

2.1 *Electrolysis*

The electrolysis process requires the use of electrolyzers, and there is a need to use electrical power. The invention of battery by an Italian physicist, Alessandro Giuseppe Antonio Anastasio Volta [13] in the 1800 provided the necessary support that allowed William Nicholson and Anthony Carlisle to discover the electrolysis process and it was in the following century, that the technology was transferred to the industry. The industrial revolution certainly played an important role in accelerating the development of electrolyzers, as hydrogen became an important industrial feedstock. The increase in demand for hydrogen allowed steam methane reformation related technology to take the lead, as hydrogen production via this method was relatively cost effective and sustainable. The research opportunities are available for any technology that can produce hydrogen that is clean and sustainable to be developed. The immediate improvement to electrolysis related technology is to search for new alternatives in generating electrical power for the electrolyzers and strongly motivated the increase in prices of fossil fuels which caused chaotic economical situations across the world again recently.

There are many different variants of electrolyzers, although the basic of producing hydrogen remains the same. The first water electrolyzer conceived utilised a tank and alkaline electrolyte design approach. The alkaline electrolyzer was filled with 25% solution of potassium hydroxide in pure water and operated at voltages between 1.85 and 2.50 V. The electrolysis process involved the use of direct current to split water into its basic elements of hydrogen and oxygen, and ideally produced hydrogen and oxygen that are 99.995% pure [14]. The negatively charged cathode and positively charged anode were placed in the electrolyte. The reduction reaction occurred at the cathode, and electrons were donated to hydrogen cation to produce hydrogen gas. The oxidation reaction took place where electrons mobility towards the cathode ensured that the circuit is complete. The half-reactions were balanced by using acids or alkalines and the combination produced the same overall decomposition of water into oxygen and hydrogen.

$$\text{Cathode(reduction):}\ \ 2H^{+}(aq) + 2e^{-1} \rightarrow H_2(g) \tag{1}$$

$$\text{Anode(oxidation):}\ \ 2H_2O(liquid) \rightarrow O_2(g) + 4H^{+}((aq) + 4e^{-}) \tag{2}$$

$$\text{Balanced reaction:}\ \ 2H_2O(liquid) \rightarrow 2H_2(g) + O_2(g) \tag{3}$$

The considerably good performance by the electrolysis process can be transformed into clean-energy technology, by supplying renewable electricity [15]. The process of splitting can be further improved, by enhancing the chemical reactions. In fact the idea of injecting the renewable based conversion was first mooted by John B.S. Haldane, in 1923 where he proposed that the electrolysis process should be able to capitalize on electricity generated by using wind turbines [16]. The hydrogen and oxygen that is produced can be stored, and can be used based on demand. This is certainly an interesting concept of harnessing renewable based energy, but the implementation must take the total energy management system into consideration.

2.2 *Electrolysis Evolution*

Although there seems to be a race towards finding viable solutions to further improve the electrolysis process, the focus still revolves around the fact that electrolyzers are powered by electricity obtained mainly from the grid to produce hydrogen and oxygen from water. In Malaysia, grid electricity is mainly generated through combustion of fossil fuels, and in the quest to produce clean energy via hydrogen generation will definitely leave behind carbon footprints. One immediate solution to reduce the carbon footprints is to utilize renewable electricity with the electrolyzer. The one identified setback in using solar cells to power electrolyzers is cost, as it is much cheaper and sustainable to use SMR approach [17]. However, lab-scale research activities focusing at the use of renewable energy based splitting of water for hydrogen and oxygen production is currently being intensified. Since the focus of this chapter is on the utilization of solar energy, the immediate attention is directed towards the processes of thermolysis, and photoelectrolysis. The thermolysis process involves the splitting of water by using high temperature thermal solar collecting elements, where else photoelectrolysis involves the splitting of water in a liquid solar semiconductor cell. This cell is referred to as photoelectrochemical cell (PEC) where photoelectrolytic reactions take place when photons are absorbed from solar insolation. In Fig. 2, the possible different paths of utilizing

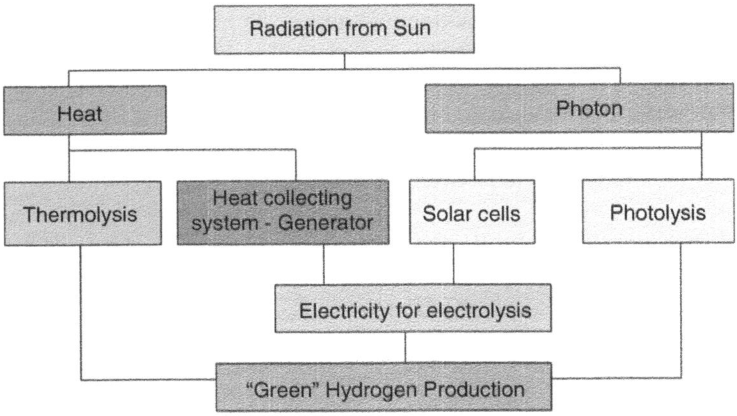

Fig. 2 Various different pathways to produce hydrogen by using solar energy

solar energy for producing hydrogen shows that there are many approaches that can be stand-alone, or hybrid systems can also be formulated. The thermal energy can be harnessed by using high-end concentrating systems, whereby by using direct steam electricity generating systems, electrolyzers can be powered up to produce up 100 MW of power.

3 Nano-Solar Hydrogen

The advancements in nanoscience have been proven to be the way forward in enhancing currently available energy conversion and harnessing technologies, such as solar cells. The efforts are geared towards searching new novel devices, methods and apparatus for solar hydrogen generation and the use of nanoscience to get the enabling technology is certainly inevitable, as the worldwide demand for energy is predicted to be approximately 30×10^{12} W by 2050 [3]. The two nano-related technologies that are worth describing are related to the processes known as photolysis and photocatalysis [18]. Photolysis can occur by utilizing solar energy to chemically decompose water by utilizing electromagnetic radiation around the visible region. An effort to expand the absorption band to capitalize on solar ultraviolet radiation is being carried, with the aid of nanotechnology. The medium for harnessing solar energy for hydrogen generation here can be solar cells directly, usually referred to as PEC (photoelectrochemical). These cells can convert incoming visible light photons to electricity and basically use semiconductor photoanode and metal cathode, immersed in an electrolyte. In certain cases, PEC cells are used to generate hydrogen, adopting the process of electrolysis in a single cell. In PEC cells, the semiconductor photoanode is used to absorb photons of light, in order to enable the reduction–oxidation (REDOX) chemical reaction to take place in the electrolyte. The PEC cell will function as an electricity-generating cell, if the net products and reactants of the reaction are unchanged. One such example is the dye-sensitized solar cells, where these cells use dye-adsorbed highly porous nanocrystalline titanium oxide to produce electrical energy [19]. The PEC cells can produce hydrogen if there is a net change in reactants. The type of process depends on the nature of the chemical reaction that takes place, whether it is exothermic or endothermic. The photoelectrolysis process involves endothermic net change of reactants, and hydrogen is produced. The chemical reactions occurring in a photoelectrolytic cell is fundamentally similar to that of electrolysis, except that it involves holes and electrons as charge carriers. Since it is an endothermic reaction, the harnessed energy is stored in hydrogen and the REDOX reactions are as given below:

$$\text{Photoanode: } H_2O(liquid) + 2holes^+ \rightarrow 2H^+(aq) + \frac{1}{2}O_2(g) \quad (4)$$

$$\text{Cathode: } 2H^+(aq) + 2e^- \rightarrow H_2(g) \quad (5)$$

As for the photocatalyst process, the initiatives that were started by Fujishima and Honda, both Japanese researchers, in the 1970s, used this approach for photocatalytic hydrogen production [20]. The photon absorption based photocatalyst is used in a cell containing nano-titanium oxide photoelectrode and platinum counter electrode, both immersed in iron based electrolyte. The oxidation process will take place at the photoelectrode when it is irradiated by ultraviolet radiation and light, which leads to reduction of water at the platinum electrodes. Motivated by their findings, current research activities are intensively looking into the use of nano-photocatalyst as well as nanocrystalline coatings, to produce high photo current densities, in order to improve conversion efficiency, within a cell [21]. In the next section, the integrated approach will be discussed in an effort to combine these technologies with the available solar harnessing technologies. An energy balance audit must be carried out in order to ensure that the implementation is proper and will not be just a net zero process that is not all an energy efficiency project [22].

3.1 Applications of Nanotubes in Solar Collectors

The hydrogen generation described in the earlier section concentrates on direct use of solar energy. In the next section, an integrated effort will be introduced, whereby proper energy management studies indicate the need to use both solar cells and solar heat collectors, in order to sustain the production of hydrogen. Both solar cells and solar heat collectors are widely available, but due to the low efficiencies and high cost, the opportunities for enhancing the performance by applying nanotechnology exist. There is ongoing research at the solar labs at Universiti Teknologi PETRONAS that is looking into the improvement of dye sensitised solar cells

(DSSC) performance. These solar labs are fully equipped to produce DSSC, and have the advantage of synthesising nanomaterials in the Nanotechnology laboratories. Typically, DSSC conversion efficiency is around 10%, and 20 nm nanoparticles layer is used for collection of energised electrons. The interconnection between the nanoparticles affects the electron mobility, and due to this, the efficiency is low. The use of uniformly ordered TiO_2 nanotubes arrays can be one of the ways to increase the electron transportation due to the enhanced inter-tube connections that improves the electron percolation. The other effort undertaken is to improve the performance of solar heat collection elements. The incorporation of carbon nanotubes in polymers to produce solar heat collector surfaces is expected to increase the absorptance, and lower the reflectance. The material produced will have significantly higher thermal conductivity as well, and is expected to lower the production cost.

4 Design of an Integrated Solar-Nano Hydrogen System

New innovative technologies will surface quickly, as the world approaches towards the speculated and much anticipated reduction in fossil-fuel supply. In order to ensure successful implementation of these technologies, integrated approaches are highly recommended, whereby proper energy management simulation study that incorporates the fundamentals related to the solar energy principles are taken into consideration. The proposed integrated system in this chapter revolves around Fig. 3. The use of both solar cells and solar heat collection elements to boost the production of hydrogen in a single system can be considered as an important step towards optimizing all the current efforts.

The study involves a comprehensive energy management process, and simulation studies are important to determine the system's overall efficiency. Before the efficiency can be determined, there is a need to look at the solar geometry aspects as well. Usually, a standard value of 1,000 W/m^2 is used when the sizing of the solar related energy-harnessing devices are carried out. This value is not reliable, as the meteorological conditions are highly transient. In Fig. 3, the integrated hydrogen generation paths via utilizing energy from the sun is shown schematically, and can be translated into a framework that can be used to develop methods that will allow the issues of availability and sustainability to be addressed. The possibility of using nanotechnology to improve the processes can be visualized by using Fig. 3 as the scheme clearly provides the necessary opportunities for an integrated nano-solar hydrogen pilot-plant to be realized. The incoming solar radiation, which consist both of direct and diffuse radiation will be captured, both in terms of photons and heat. The energy contained in the photons will be converted directly to electricity, while the thermal energy can be used to generate electricity, or just to increase the temperatures. The immediate need to carry out an energy balance audit is to look at the quality and quantity of solar radiation, which is actually an intermittent source of energy.

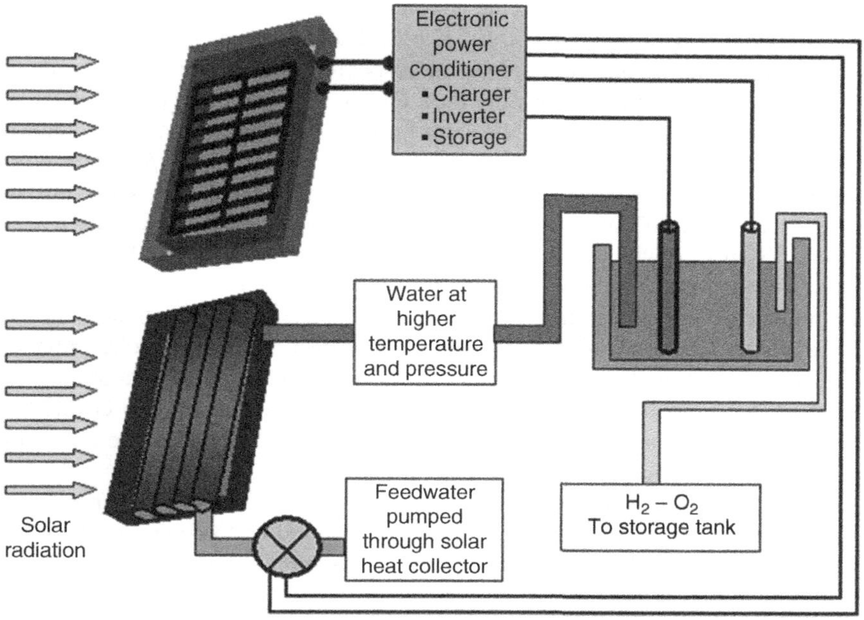

Fig. 3 The integrated nano-solar hydrogen production scheme [4]

4.1 Solar Radiation

Solar radiation, or at times referred to solar insolation, is referring to the electromagnetic radiation emitted by the sun, where most of the radiation is in the broadband solar radiation wavelength region of 280–4,000 nm. Solar radiation has a wavelength distribution, where the range of wavelengths can describe the different regions of the solar spectrum. The region of the spectrum that is visible to all of us is in the wavelength range of about 380–720 nm, where wavelengths of the solar spectrum are related to different energy levels [23]. The solar and terrestrial radiation falls between 0.15 and 120 μm, where the radiation of practical importance to be converted to useful energy lies between 0.15 and 3.0 μm [24].

The solar radiation originates from the sun, which can be taken as a sphere made of intensely hot gaseous matter. The sun's radius is estimated to be around 6.960×10^8 m, which is around 109 times more than the radius of earth. The distance between the earth and the sun varies as the earth rotates around the sun in an elliptical orbit, where earth is approximately around 1.47×10^{11} m away from the sun on 4th January and the distance is approximately 1.53×10^{11} m on 4th July. Although there is a variation in the distance between earth and sun, this does not have a significant effect on the amount of solar radiation that reaches earth. The theoretically calculated mass of sun is estimated to be 1.991×10^{30} kg, which is about 330,000 times more than earth's mass. Although the outer surface effective temperature of the sun is 5,762 K, the innermost core temperature is estimated to be in the region of 8×10^6 to 40×10^6 K [25]. Solar constant is the rate at which solar radiation strikes the earth's upper atmosphere where it is the average amount of energy received in a unit of time on a unit of area perpendicular to the sun's direction at the mean distance of the earth from the sun [26]. The average intensity of solar radiation reaching the upper atmosphere is about 1,353 W/m^2. The World Radiation Centre in Switzerland has adopted a different value for the solar constant, which is around 1,367 W/m^2 [27]. The amount of this energy that actually reaches the earth's surface will vary according to atmospheric and meteorological conditions. The solar rays that move through the atmosphere can be absorbed, scattered and reflected by air molecules, water vapour, clouds, dust and pollutants in fact. Global solar radiation refers to the sum of the direct and diffuse solar radiation. Figure 4 shows how the atmospheric conditions can influence the amount of solar radiation that is received at the surface of earth. The thick clouds that covered the sun for a period of approximately 6 min caused a sharp decline in the amount of solar radiation that reached the surface of the earth.

An analysis based on Fig. 4 shows that the highest value of solar radiation received on the surface of earth at Bercham, Ipoh around 12.13 noon was 560 W/m^2, which is approximately around 41% of the solar constant valued at 1,353 W/m^2. As the clouds moves past the sun, the average fraction that arrived was around 15.2%, which clearly shows the significant impact of cloudiness. A whole system analysis of solar cells and solar thermal collectors requires simulation to be carried out, where design specifications are changed and simulated under the same meteorological conditions. The locality is also important and the place, as shown in Fig. 5, where selection is based on whether there is a stable and high amount of solar radiation received at that particular locality.

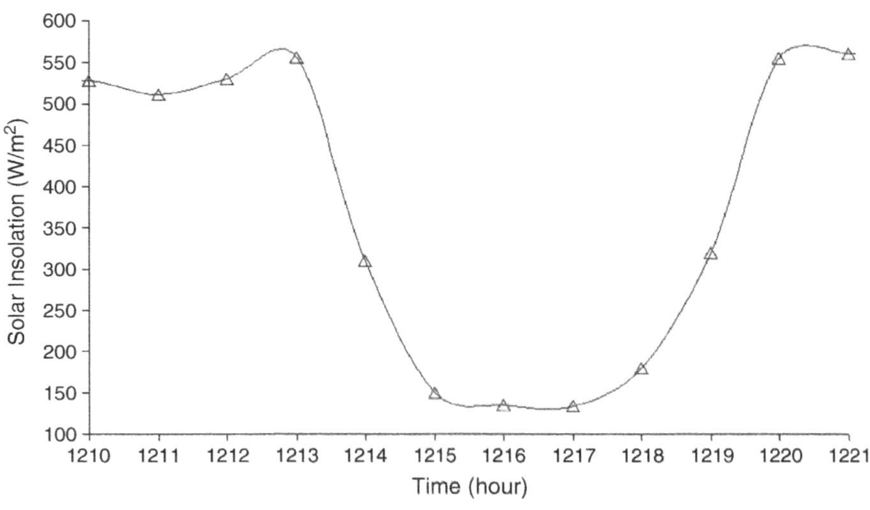

Fig. 4 Variation in solar insolation for a period of 1 h

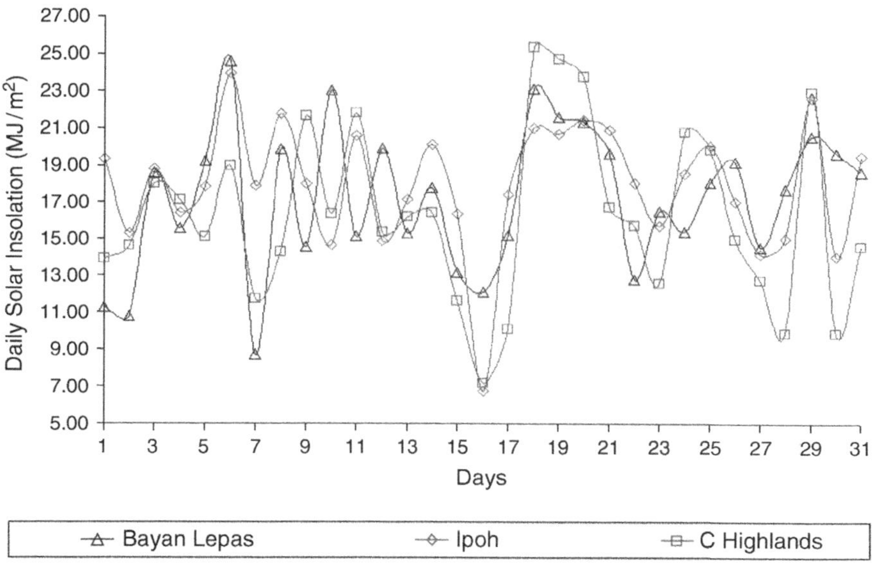

Fig. 5 Monthly daily average global solar radiation for three different places in Malaysia

Therefore, for simulation purposes, a large database is required, just to store the measured solar radiation data, and a good data compression method is necessary. One of the compression methods that have a correlation of up to 99.85% is given as (6), based on Fig. 6. Equation (1) can be easily incorporated into simulation software to determine the hourly global radiation, I_G in MJ/m^2, where t is time in hours, without the need to create and callback large amount of data from the solar radiation database.

$$I_G = 1.3720 + [1.3764 \cos(0.4796t + 0.3748)] \quad 6 \leq t \leq 19 \qquad (6)$$

4.2 Solar Cells

Solar cells are devices that can generate electricity by using the photovoltaic (PV) effect. PV modules have no moving parts and operate silently without any emissions of dangerous gases, although significant carbon footprint during production is recorded. There are different generations of solar cells that are currently being developed, but PV cells that uses semiconductor technology to capture the energy in sunlight remains popular, due to relatively higher efficiency and availability. Crystalline silicon is formed into crystals and silicon wafers are cut from grown ingots. A conventional solar cell consists of a silicon wafer with a thickness of 0.05 cm, and typical cells that are 10 cm in diameter produce about 1 W of power,

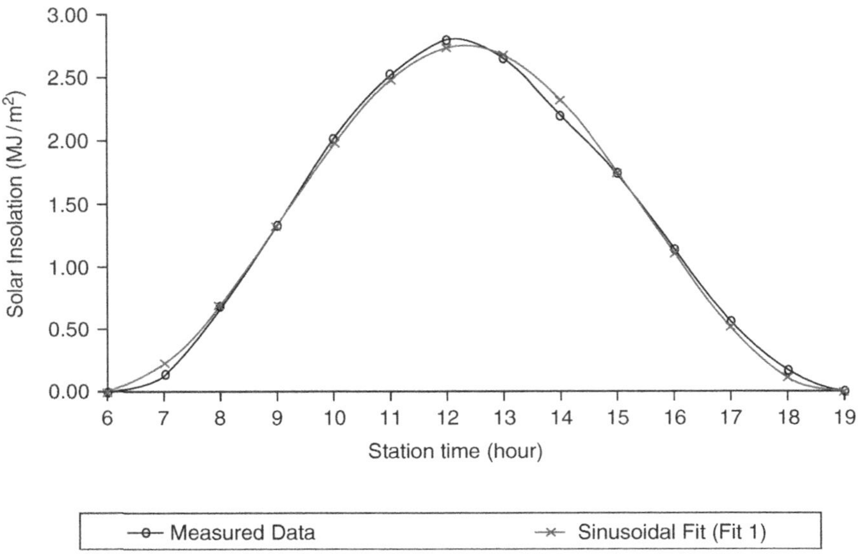

Fig. 6 The average of the hourly solar insolation data measured at Ipoh station is compared to a sinusoidal curve fit

and are grouped into modules of dozens of cells. Modules are further grouped into panels and then arrays, which may produce several kilowatts of power. The first generation PV cells have efficiencies of 5–15%, to convert the solar energy into usable energy. Efficiency is constantly increasing with the use of new materials and manufacturing processes are developed.

Most cells in operation today are single crystal silicon cells. Silicon cells provide a good balance of cost effectiveness, reliability, and efficiency. A number of other metals can be transformed into semiconductors and used in photovoltaic cells such as copper indium diselinide, cadmium sulfide, cadmium telluride, gallium arsenide and indium phosphide. It is predicted by the Centre for third Generation Photovoltaic at University of New South Wales, Australia that PV in the future will have energy conversion efficiencies between 30 and 60%, based on high efficiency thin film technology [28]. The effect of PV has been observed since 1839 and the practical application only came about in 1958, when PV modules were placed on Vanguard I, the second US satellite launched into orbit. PV technology is the ideal source of electrical energy to overcome electrification problem in the rural areas, where extending the utility's electricity grid is expensive or impossible. In Malaysia, Tenaga Nasional Berhad installed PVs on a kampung house located at Bukit Apit Felcra settlement in Malacca, at a cost of RM 5,000. The solar kit that was used came along with a battery that had a capability of offering up to eight power points. Similar facility was also extended to provide electricity to 49 orang asli houses in Kampung Bukit Panjang, Melekek, also in Malacca [29]. PV systems are easy to operate, rarely need maintenance and do not pollute the environment, but the implementation of these proven technology for large-scale electricity generation is still premature, perhaps as the third and fourth generation PVs are conceived it would be economically feasible to hybridize it with solar hydrogen technologies.

4.3 Solar Thermal Collector

Solar collector is a device used to harness the solar energy to thermal energy. The wide application of solar collectors in urban areas in Malaysia is for domestic water heating purposes. So the minds of the people are confined to that usage, although there is a wide potential, including using it for drying agricultural products [30]. Figure 7 summarizes the type of collectors available currently. The most commonly used collector that is widely used is the flat plate collector. In the case of a flat-plate collector, the attractive features are noticed to be the simplicity in design; no tracking requirements and the need for periodical maintenance is much lesser as compared to the concentrating devices. Concentrating devices are complicated and require scheduled maintenance as it involves moving parts and expensive optical systems. But yet, for high temperature applications such as to operate solar thermal electric power plants, the concentrating devices form an inseparable part that will ensure reliable and sustainable operation.

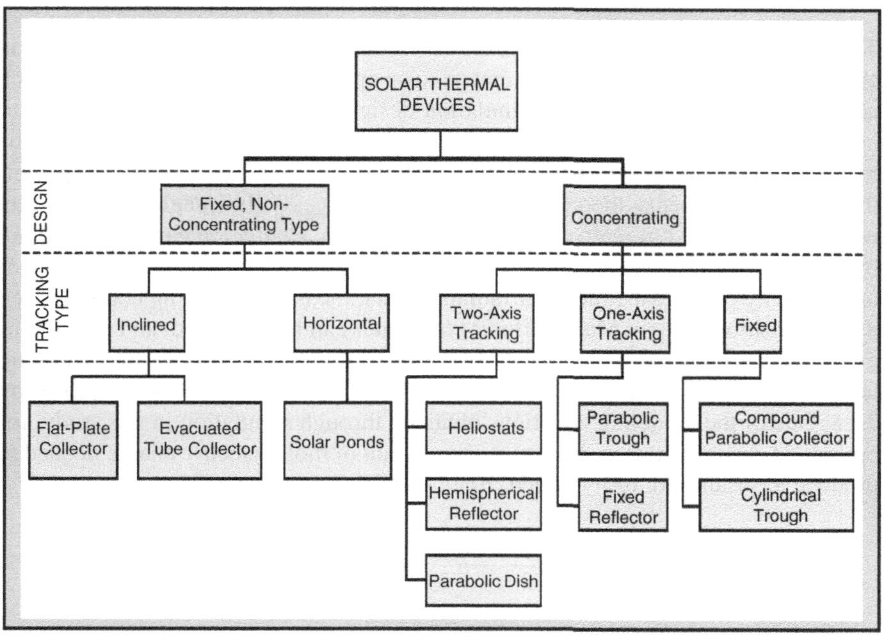

Fig. 7 Different types of solar thermal collectors

4.4 Simulation Approach

In line with the current technology development, the escalating complexity of energy and environmental systems are dealt with computer modelling and simulation. These tools are emerging as a viable approach to design and performance evaluation. There is a need to understand the theoretical and operational principles underlying this technology, as the solar thermal processes are dynamic in nature. The traditional way of designing is by developing a prototype using some basic calculations and data is then collected for a certain period of time. If the performance is not as expected, then continuous refinements are carried out and the whole process of testing is repeated. Usually, after a few refinements, if the prototype fails, it is discarded. Computer simulation is one of the most powerful tools currently used for analysing and designing complex systems. According to [31], simulation is the process of developing a simplified model of a complex system that is used to analyse and predict the behaviour of a real system.

Simulation is carried out because real-life systems are often difficult to be analysed due to their complexity. It is generally possible to develop a model that can be used to predict the behaviour of the real system as accurately as possible.

4.5 Simulation Results

The simulation and experimental results must be organized based on a systematic approach, allowing the design parameters to be optimized. The systematic approach used was based on Fig. 8. The simulation design approach in Fig. 8 was used to integrate the different components, declared as modules. The modules can be represented by individual mathematical models, and can be integrated together, through comprehensive theoretical development process. The integrated design can be further subjected to realistic design data, whereby evaluation can be carried out by using measured meteorological data [22]. The simulation process can be further improved by using real-time solar radiation data, measured by using computerized data acquisition system. Many different outcomes can be measured, and one of it as given in Fig. 9. It can be observed from Fig. 9 that by increasing the operating temperature of the electrolyzer [32] the amount of hydrogen produced increases by 23%. One of the modeling equations obtained through simulation in this study, for optimized performance; to determine the amount of molar electric charge needed to produce a certain mol of hydrogen, $I(t)/n$ in A/mol, is given as (7).

$$\frac{I(t)}{n} = \frac{192970}{t} - 1.1122 \times 10^{-10} \tag{7}$$

Equation (7) is specifically derived to be used in this simulation studies, and can be refined, to be used with other integrated models. The flexibility is given, as the

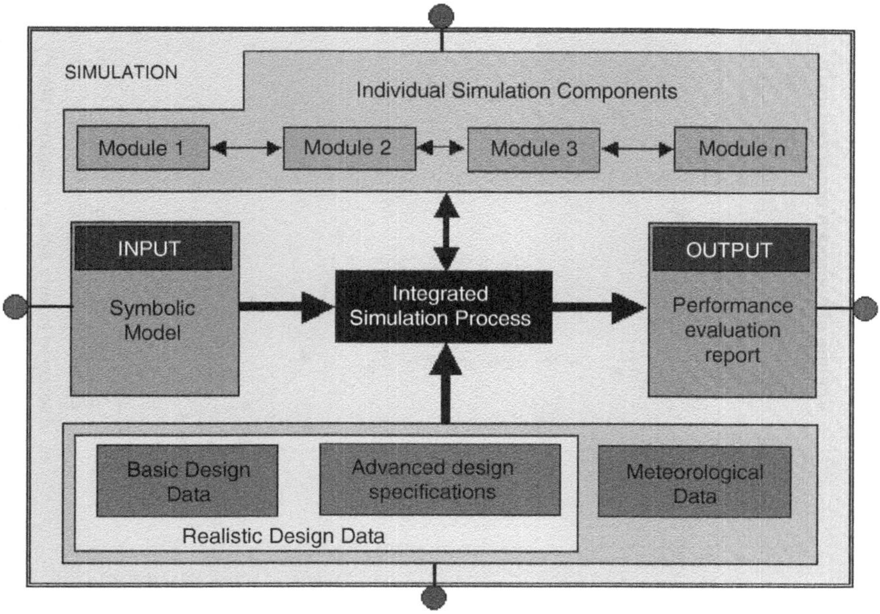

Fig. 8 Simulation process flow diagram developed by the author

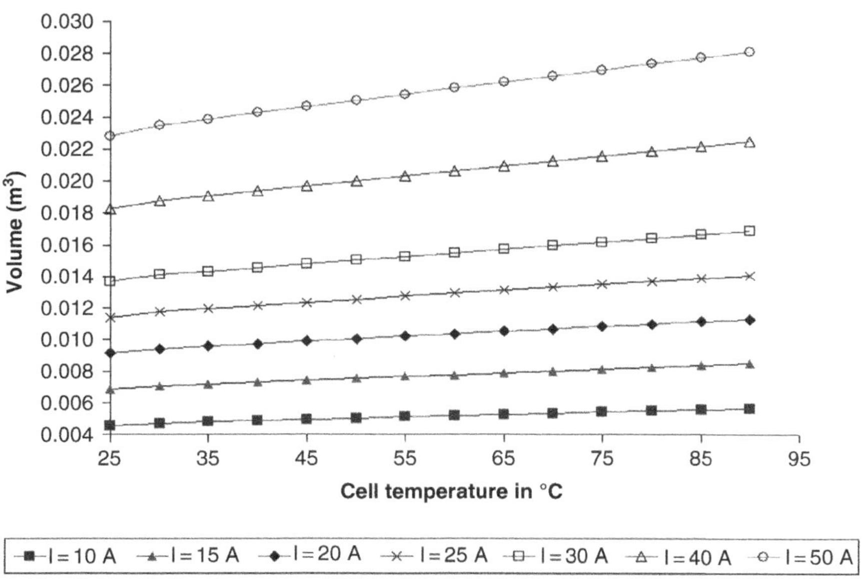

Fig. 9 Simulation results for the volume of hydrogen produced in an electrolysis cell, where the cell temperature is increased at different power density, while the cell voltage is kept constant at 1.68 V, for a duration of 1 h

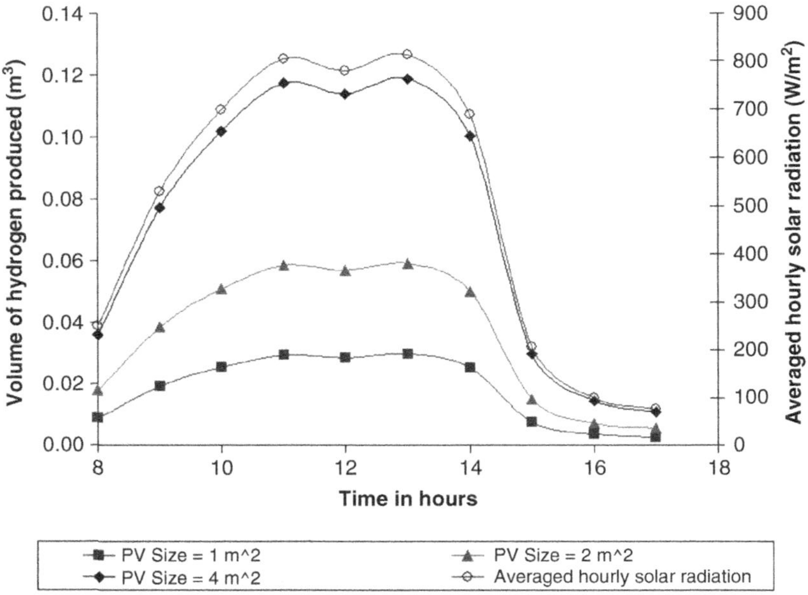

Fig. 10 Simulation outcomes based on averaged solar radiation received in Ipoh, for three different PV sizes and optimized system based on setup as shown in Fig. 3

modeling equation can be declared as one of the modules. The equation was used to determine the PV system and storage system sizing, while the outcomes as shown in Fig. 9 was used to size the solar thermal system. The overall system evaluation via simulation was executed based on optimized parameters, and one of the integrated outcomes for different PV sizing is as shown in Fig. 10, when subjected to the local weather conditions in Ipoh, Malaysia.

The same setup can be used, but the meteorological data can be changed. The output from solar thermal is also determined with the incoming solar radiation. The simulation results can be used to optimise the system, and fuel cell module can be included to gauge the output electrical power that can be generated, when subjected to the averaged hourly solar radiation. The improvement to the system via nanotechnology can be carried out on the solar harnessing devices, as well as the electrolyzer. In fact, the storage system for hydrogen can also be improved, with the use of nanotechnology, and can be declared as one of the modules as well.

5 Conclusion

The implementation of any solar related processes faces the issue of sustainability, and reliability, as the intermittent availability reduces the practicality of the systems. The need to utilize all parts of earth, to tackle the issue of intermittency requires global attention. In order to increase the operating temperature of electrolyzers, high performing solar thermal concentrators are required. The quality of solar radiation becomes an important factor, and deserts located in America, Africa, Australia and Asia can be strategically used for this purpose. A global energy management system is required to plan for effective power generation systems, especially renewable energy based systems. A Comprehensive Global Renewable Energy Management System (CGREM) [4] is a solution of tomorrow's energy needs, whether it is going to be implemented via nanotechnology or any other technical technology that will be made available in the future. There is bound to be many other initiatives developed to produce hydrogen via clean energy sources, whereby it can be distributed and stored in a variety of ways and has the potential to replace fossil fuels to provide electricity and transportation fuels, and hence leads to energy independence. The need for giant economies to participate and develop this promising "zero-polluting" system is also the way forward in reducing the impact on the environment. The race against time to develop other production techniques such as photobiological [33] and plasmatron methods, are intrinsically motivated towards meeting the future energy demands.

References

1. Balat, M.: Potential importance of hydrogen as a future solution to environmental and transportation problems. Int. J. Hydrogen Energy **33**, 4013–4029 (2008)
2. Marban, G., Valdes, S.T.: Towards the hydrogen economy? Int. J. Hydrogen Energy **32**, 1625–1637 (2007)
3. Energy Information Administration (EIA), http://www.eia.doe.gov, retrieved on 21 January 2010
4. Balbir Singh: Comprehensive global renewable energy management system (CGREM), Intellectual Property Pending, February 2010. Universiti Teknologi PETRONAS, Malaysia, 2010
5. Kadir, K.A.: Renewable energy: west should help. Business Times. 31 July 1996, Malaysia, p. 24, (1996)
6. Economic Planning Unit – Ninth Malaysia Plan (2006–2010). (2006)
7. Jaafar, M.Z., Kheng, W.H., Kamaruddin, N.: Greener energy solutions for a sustainable future: issues and challenges for Malaysia. Energy Pol. **31**, 1061–1072 (2003)
8. Nowotny, J., Sorrell, C.C., Sheppard, L.R., Bak, T.: Solar-hydrogen: environmentally safe fuel for the future. Int. J. Hydrogen Energy **30**, 521–544 (2005)
9. Bak, T., Nowotny, J., Rekas, M., Sorrell, C.C.: Photoelectrochemical hydrogen generation from water using solar energy – materials-related perspective. Int. J. Hydrogen Energy **27**, 991–1022 (2002)
10. Stojic, D., Marceta, M., Sovilj, S., Miljanic: Hydrogen generation from water electrolysis – possibilities of energy saving. J. Power Sources **118**, 315–319 (2003)
11. Xu, J., Yeung, C., Ni, J., Meunier, F., Acerbi, N., Fowles, M., Tsang, S.C.: Methane steam reforming for hydrogen production using low water-ratios without carbon formation over ceria coated Ni catalysts. Appl. Catal. A **345**(2), 119–127 (2008)
12. Andrew, L.D.: The role of carbon in fuel cells. J. Power Sources **156**, 128–141 (2006)
13. Pancaldi, G.: Volta: Science and Culture in the Age of Enlightenment. Princeton University Press, Princeton (2003)

14. Konopka, A., Gregory, D.: Hydrogen production by electrolysis: present and future. Presented at 10th intersociety energy conversion engineering conference, 1975
15. Barbir, F.: PEM electrolysis for production of hydrogen from renewable energy sources. Sol. Energy **78**, 661–669 (2005)
16. Haldane, J.B.S.: DAEDALUS or Science and the Future. Cambridge (1923)
17. Thomas, L., Gibson, T.L., Kelly, N.A.: Optimization of solar powered hydrogen production using photovoltaic electrolysis devices. Int. J. Hydrogen Energy **33**, 5931–5940 (2008)
18. Armor, J.: The multiple roles for catalysis in the production of hydrogen. Appl. Catal. A **176**(2), 159–176 (1999)
19. Gratzel, M., McEvoy, A.J.: Hydrogen production by solar photolysis of water. American Physical Society Symposium, Basic Research for the Hydrogen Economy, Canada (2004)
20. Fujishima, A., Honda, K.: Electrochemical photolysis of water at a semiconductor electrode. Nature **238**, 37–38 (1972)
21. Zhang, F.J., Chen, M.L., Oh, W.C.: Fabrication and electro-photolysis property of carbon nanotubes/titanium composite photocatalysts for methylene blue. Bull. Korean Chem. Soc. **30** (8), 1798–1804 (2009)
22. Taylor, M.P., Zhang, W.D., Wills, V., Schmid, S.: A dynamic model for the energy balance of an electrolysis cell. Trans IchemE Part A 74, (1996)
23. Serway, R.A., Beichner, R.J.: Physics for Scientists and Engineers with Modern Physics, 5th edn. Saunders College Publishing: A Division of Harcourt College Publishers, USA (2000)
24. Garg, H.P., Prakash, J.: Solar Energy: Fundamentals and Applications. Tata McGraw-Hill, New Delhi (1997)
25. Duffie, J.A., Beckman, W.A.: Solar Engineering of Thermal Processes. Wiley, New York (1980)
26. Thekaekara, M.P.: Solar energy outside the earth's atmosphere. Sol. Energy **14**, 109 (1973)
27. Frohlich, C., Brusa, R.W.: Solar radiation and its variation in time. Sol. Phys. **74**, 209–215 (1981)
28. Green, M.A.: Third generation photovoltaics: solar cells for 2020 and beyond. Physica E **14**, 65–70 (2002)
29. Yatim, A.H.: Thirty six-year wait for electricity. New Straits Times, 12 December 2000, Malaysia, p. 16 (2000)
30. Sopian, K., Othman, M.Y.H., Yatim, B., Daud, W.R.W.: Sustainable and environment friendly solar drying technologies for agricultural produce. In: Proceedings of World Renewable Energy Congress VII (WREC 2002), Cologne, Germany (2002)
31. Aburdene, M.F.: Computer Simulation of Dynamic Systems. Wm.C. Brown Publishers, Dubuque (1988)
32. Mingyi, L., Bo, Y., Jingming, X., Jing, C.: Thermodynamic analysis of the efficiency of high-temperature steam electrolysis system for hydrogen production. J. Power Sources **177**, 493–499 (2008)
33. Rene, A.R., Hubertus, V.M.H., Gerrit, J.W.E., Sybrand, J.M., Cees, J.N.B.: Principle and perspectives of hydrogen production through biocatalyzed electrolysis. Int. J. Hydrogen Energy **31**, 1632–1640 (2006)

Fe–FeO Nanocomposites: Preparation, Characterization and Magnetic Properties

Jamshid Amighian, Morteza Mozaffari, and Mehdi Gheisari

Abstract To date, nano-magnetic materials have gain great attention by the research community due to their importance for future applications. A brief introduction of Fe–FeO nanocomposites in the form of particles and thin films is given in the first part of this chapter. This includes definition, magnetic properties, preparation, structure and applications. Different preparation methods of Fe–FeO are then introduced in the second part of the chapter. These include mechanical alloying, high energy ball milling, mechanochemical processing, DC magnetron sputtering, molecular-beam-epitaxy, plasma gas condensation. Among these preparation techniques, mechanochemical processing has been fully explained. Different techniques and instruments which have been used to characterize the samples have been explained. These include XRD, TEM, VSM, Superconducting Quantum Interferences Devices (SQUID), and Mössbauer. Magnetic properties of the nanocomposites especially Fe–FeO have been presented in the final part of the chapter. These include magnetization, coercivity, Mössbauer, hysteresis loops, exchange bias effect, vertical shift, spin glass phase, rotational hysteresis, FC and ZFC hysteresis loops.

J. Amighian (✉)
Islamic Azad University-Najafabad Branch, Najafabad, Isfahan, Iran
e-mail: jamighian@sci.ui.ac.ir

M. Mozaffari
Physics Department, Razi University, Taghbostan, Kermanshah, Iran
e-mail: mozafari@sci.ui.ac.ir

M. Gheisari
Islamic Azad University Aligoudarz Branch, Aligoudarz, Lorestan, Iran
e-mail: m.gh.godarzi@gmail.com

N. Yahya (ed.), *Carbon and Oxide Nanostructures*, Adv Struct Mater 5,
DOI 10.1007/8611_2010_22, © Springer-Verlag Berlin Heidelberg 2010,
Published online: 13 August 2010

1 Introduction

Composites have been generally classified based on the matrix such as metal-matrix composites (MMCs), ceramic-matrix composites (CMCs) and polymer matrix composites (PMCs). Nanocomposites are a new class of materials in which at least one of the phases (the matrix, the reinforcement or both) is of nanometer dimensions. Magnetic nanocomposites containing fine (single domain) magnetic particles, isolated electrically and magnetically by a nonmagnetic, nonmetallic component, also exhibit interesting magnetic properties and they are attracting increasing attention for their magnetic applications [1]. Recently, ceramic-matrix and/or metal-matrix nanocomposites have also received increased attention because of their unique mechanical, electrical and interesting magnetic properties [2, 3] such as high coercivity at room temperature [4, 5], giant magnetoresistance [6] and superparamagnetism [7]. Also, ceramic/Fe magnetic nanocomposites, such as Fe/Fe oxides, Al_2O_3/Fe, SiO_2/Fe [8–11], have shown high values of coercivity with respect to iron and therefore, interesting for application in recording media [8]. Nanocomposites are prepared using various physical [1, 12–15] and chemical [16, 17] methods. Among these methods mechanical alloying (MA) or high-energy ball milling (HEBM) [1, 15] is a powerful method to synthesize nanomaterials. Mechanochemical processing (MCP) is the term applied to powder processing in which chemical reactions and phase transformations take place due to application of mechanical energy. Most of the mechanochemical processing reactions studied in recent years have displacement reactions of the type: $MO + R = M + RO$ where the metal oxide (MO) is reduced by a more reactive metal (reductant, R) to the pure metal, M [1].

Single phase (ferrites) together with multiple phases (composites) of magnetic nanoparticles have been studied intensively for decades since they are of fundamental interest [18, 19] and have importance in technological applications, [20] particularly in the information storage industry [21]. An assembly of magnetic nanoparticles shows behaviors very different from their bulk counterpart because of the finite-size effect [22] and the surface effect [23] as well as the interparticle interactions [24] and may display magnetic phenomena, such as spin-glass-like and exchange bias effect behaviors which the latter will be describe here. Previous studies have shown that once the metallic nanoparticles are exposed to air, core (metal)/shell(oxide)- structured nanoparticles are formed [25].Furthermore, if the oxide layer is antiferromagnetic (AFM), the core/shell-structured magnetic nanoparticles usually show an exchange-bias effect [26] . The microscopic exchange interaction at the ferromagnetic (FM)/AFM interface is believed to be responsible for this macroscopic phenomenon. This exchange-bias effect makes the behavior of the magnetic nanoparticles even more complicated and is described in different magnetic nanocomposites below.

When magnetic nanocomposites such as ferromagnetic (FM)–antiferromagnetic (AFM) interfaces are cooled through the AFM Neel temperature (T_N), with the FM Curie temperature (T_C) larger than T_N, a shift in hysteresis loop along the applied magnetic field will be observed. This effect is known as exchange bias. Meiklejohn and Bean were the first researchers who observed this phenomenon when they investigated the low temperature magnetic behavior of partially oxidized cobalt fine particles (Co–CoO) [27]. They observed that field cooled magnetization curve of a magnetic system will exhibit a shift along the magnetic field axis. They attributed this phenomenon to the exchange interaction at the interface between the FM Co core and the AFM CoO shell. In addition to AFM–FM interfaces, exchange bias and related effects have also been observed in other types of interfaces, e.g. AFM-ferrimagnetic (Ferri), Ferri-FM.

The exchange bias nanostructured systems containing these interfaces can be found in different systems such as thin films, core-shell nanoparticles, inhomogeneous materials, lithographed nanostructures (such as: patterned wires, dots, rings) or FM nanoparticles embedded in AFM matrices [28]. Iron–iron oxides systems have been studied for a number of years in connection with exchange bias phenomenon. These systems are based on Fe–FeO, Fe-α-Fe_2O_3 and Fe-γ-Fe_2O_3 and Fe–Fe_3O_4 [10, 26].

In this chapter characterization and magnetic properties of Fe–FeO nanocomposites are fully described. There are four major iron oxide phases which consist of FeO, Fe_3O_4, γ-Fe_2O_3 and α-Fe_2O_3, and are named wüstite, magnetite, maghemite and hematite respectively [29]. In these oxides, wüstite is almost nonstoichiometric with some Fe deficiency and can be denoted as $Fe_{1-x}O$. Apart from wüstite, other iron oxide phases are very important technologically and have many applications in different industries. Nevertheless wüstite is very interesting on its own: for example, its unusual electronic properties. $Fe_{1-x}O$ is an interesting semiconductor whose carrier type changes from p to n type around $x = 0.08$ [30].On the other hand from the viewpoint of magnetic properties, stoichiometric FeO is an antiferromagnet with a Neél temperature of about 200 K and has a rock salt structure, with a closed-packed fcc O^{2-} lattice in which Fe^{2+} ions occupy the octahedral (B) interstitial sites [29].

In order to preserve the total crystal electroneutrality of $Fe_{1-x}O$, some of the Fe^{2+} ions give away another electron and become Fe^{3+}. This means that for a particular value of x, there are $2xFe^{3+}$ and $(1 - 3x)$ Fe^{2+} ions and this iron deficiency leads to the formation of some vacancies. These vacancies are partly located as Frenkel defects on interstitial tetrahedral sites [31–36].

Neutron-diffraction [31–33, 36] and x-ray-scattering [32] studies on quenched $Fe_{1-x}O$ powders and single crystal $Fe_{1-x}O$ indicate that the vacancies are not randomly distributed but clustered around Fe^{3+} ions [34]. These Fe^{3+} ions preferentially occupy the tetrahedral (A) interstitial sites.

Significant progress towards understanding the possible structure of a defect cluster has been achieved in the theoretical work of Catlow and Fender [34]. In this work they have suggested that the basic cluster is identified as a complex of four cation vacancies and one tetrahedral Fe^{3+} ion (or 4:1 cluster), Fig. 1a. This cluster is formed as a result of a large Coulomb energy term favoring the occupation of the tetrahedral site when all nearest-neighbor cations are vacant. The aggregation of these 4:1 clusters occurs by vacancy-sharing and the calculations suggest that the

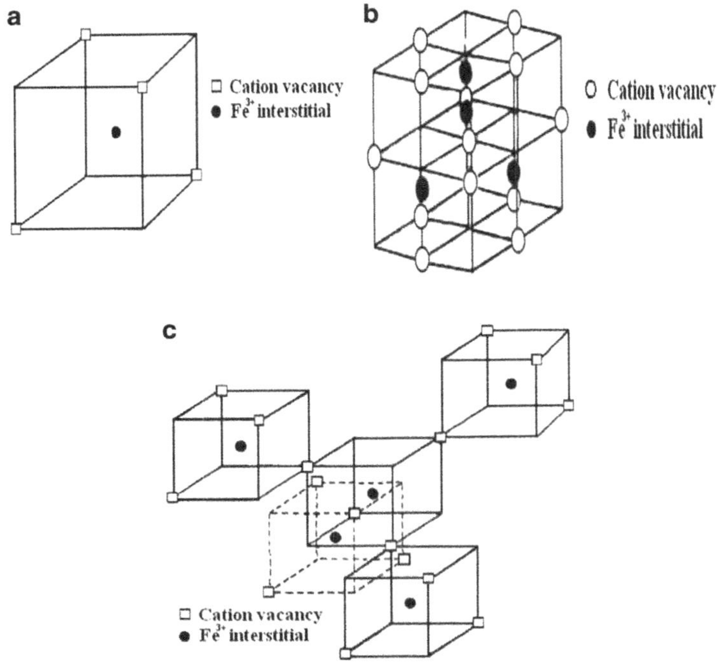

Fig. 1 (**a**) 4:1 Cluster; (**b**) Koch and Cohen cluster; (**c**) 16:5 spinel like cluster [34]

most stable small aggregates are formed by edge-sharing rather than corner-sharing (6:2 or 8:3 clusters). If more extended clusters are formed they are likely to involve corner-sharing, but the binding energy for a cluster closely related to the inverse spinel structure of Fe_3O_4 (16:5), Fig. 1b, is larger than that calculated for the Koch–Cohen cluster (13:4) [32]. Koch and Cohen [34] reported that by using x-ray studies, defects in $Fe_{1-x}O$ are aggregated to the order of 13:4 cluster (13 octahedral vacancies surrounding four tetrahedral Fe^{3+}) with a noticeable displacement of ions near the cluster and that the defect does not quite have the Fe_3O_4 structure (Fig. 1c). Catlow and Fender studies have been supported by the neutron diffraction studies of Chetham et al. and Battle et al. [33, 36].

Anomalous high magnetization and low temperature coercivity were found in sputtered $Fe_{1-x}O$ films [37, 38]. In other works, defect clusters in $Fe_{1-x}O$ films and their ferrimagnetism properties have been reported by Dimitrov et al. [38–40].

2 Preparation Methods of Fe–FeO

There are different forms of Fe–FeO nanocomposites, such as nanolayers, nanopowders and core- shell. These nanostructures are prepared by different methods, such as mechanical alloying, mechanical milling, high energy ball milling, mechanochemical processing, DC magnetron sputtering [41] molecular-beam-epitaxy [42, 43], fast evaporating [44], plasma gas condensation [45, 46].

One of the most important and easiest way to prepare Fe–FeO nanocomposites is mechanochemical processing, which will be explained here. Different authors used mechanochemical processing to prepare Fe–FeO nanocomposites. Ding et al. [3] have used a mixture of Fe and Fe_2O_3 powders in the stoichiometric composition and have mechanically milled in a Spex 8000 Mixer/Mill. The powders were first loaded together with 12 mm diameter steel grinding balls into a hardened steel vial. The ball to powder mass ratio was 10:1. To study the reaction kinetics, mechanical milling was performed for different times between 1 and 65 h. As-milled samples were annealed under vacuum at temperatures in the range of 200–900°C.

Figure 2 shows XRD patterns of as-milled powders after milling for different times. After milling for 1 and 2 h, the XRD patterns still showed a mixture of the starting phases, Fe_2O_3 and Fe. However, the diffraction peaks, particularly for F_2O_3 became broadened. After milling for 4 h, magnetite, F_3O_4, and wüstite, FeO, appeared, coexisting with F_2O_3 and Fe. Hematite disappeared after milling for 10 h. Wüstite (FeO) with a trace of Fe was found after milling for 20 h. After further milling, samples consisted of wüstite, with no other phases being evident on the diffraction pattern.

They have also used Mössbauer spectroscopy to study the formation of the magnetite and a wüstite phase quantitatively, as shown in Fig. 3. The results of Mössbauer spectroscopy showed that for the starting powder and the sample milled for 2 h, 65% of Fe atoms were found in the hematite phase and the rest in BCC-Fe. This result is expected for the starting composition of F_2O_3 + Fe. However after milling for 4 h, 30% of Fe atoms were found in the magnetite phase and 9% of Fe atoms in FeO, with the amount of significantly reduced. Fe_2O_3 disappeared after milling for 10 h. No Fe_3O_4 was found after milling for 20 h, and the fraction of Fe atoms in BCC-Fe was below 10%. Nearly single phase FeO was observed after milling for 30 h or longer [3].

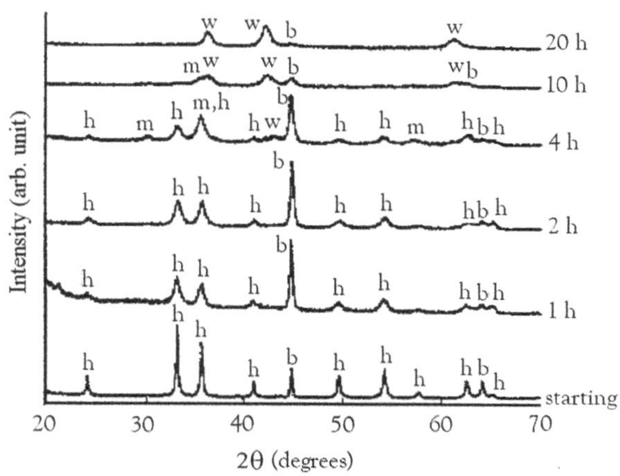

Fig. 2 XRD patterns of as-milled powders after milling for different times (b: BCC-Fe, w: wüstite, m: magnetite, h: hematite) [3]

Fig. 3 Fraction of Fe atoms in different phases as a function of milling time, using Mössbauer spectroscopy [3]

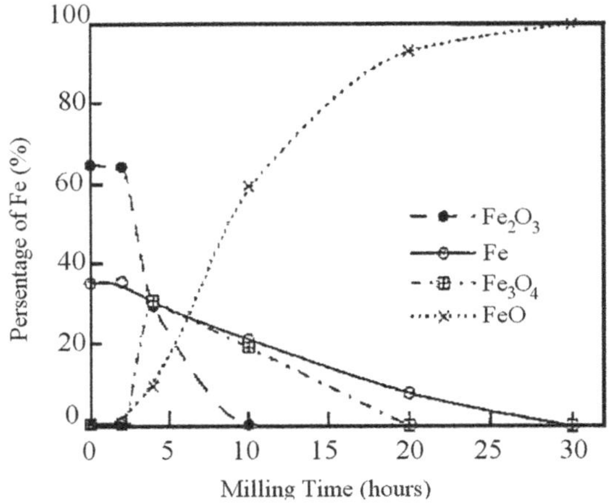

In another work Ding et al. [47] used a mixture of Fe (99.9%, 300 mesh) and Fe O (99.9%) < 200 mesh) powders in a nominal composition of xFe · (1 − x) Fe_2O_3 with x varied between 0 and 1 and mechanically milled them for 30 h. The as-milled powders were annealed at temperatures in the range of 200–900°C for 1 h under a vacuum of 10 Torr.

X-ray diffraction (XRD) measurements showed that the samples with x = 0.5 consisted of nearly single phase FeO. For higher values of x the samples consisted of a mixture of FeO and Fe. The fraction of FeO decreased with increasing x, and only α-Fe was present for x = 1.

The X-ray diffraction (XRD) results were also verified by Mössbauer measurements. The xFe · (1 − x)Fe_2O_3 sample with x = 0.5 showed a paramagnetic doublet, for milled and annealed sample at 700°C (Fig. 4) which is expected for FeO. Interesting however, is the fact that at 300°C a mixture of Fe and Fe_3O_4 was found (Fig. 4).

However it should be noted that for x = 0.5, the as-milled sample consisted of single phase FeO according to XRD and Mössbauer measurements [47]. Figure 5 shows the phases present in the as-milled samples and after annealing at different temperatures. The temperature boundaries correspond to the decomposition and reformation temperatures, respectively, as determined by the DSC measurements. The phases formed during heat treatment above 250°C correspond to the equilibrium Fe/Fe_2O_3 phase diagram. The decomposition and re-formation of FeO was clearly evidenced by Mössbauer measurements. As shown in Fig. 5, nearly single phase FeO was found in the as-milled sample with x = 0.5. After annealing at 300°C, FeO decomposed entirely into Fe_3O_4 and Fe. After annealing at 700°C, FeO re-formed and no Fe_3O_4 was detectable. A small fraction of α-Fe, containing about 5% of the Fe atoms, was also present after annealing at 700°C, in good agreement with the Fe/Fe_2O_3 phase diagram.

Bonetti et al. [48] used two different sets of specimens which have been synthesized, using as precursor materials: (a) Fe_3O_4 powder (purity 99.9%); (b) a mixture of 20 wt% metallic Fe (99.9%) and 80 wt% Fe_3O_4 powders. The grinding process has

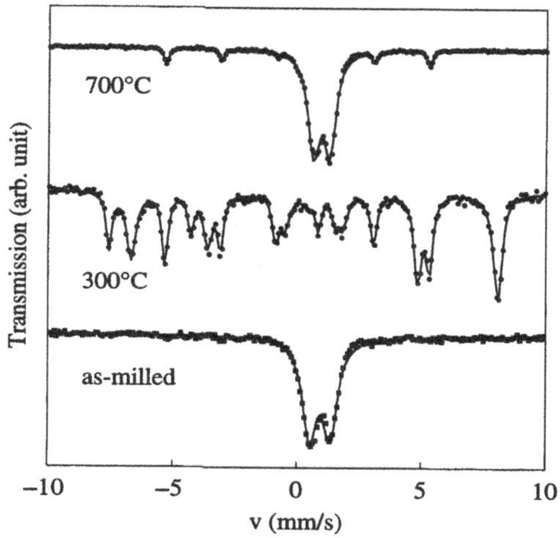

Fig. 4 Mössbauer spectra of $xFe \cdot (1-x)Fe_2O_3$ sample with $x = 0.5$ in the as-milled state, and after annealing at 300°C for 10 min and 700°C for 1 h, respectively [47]

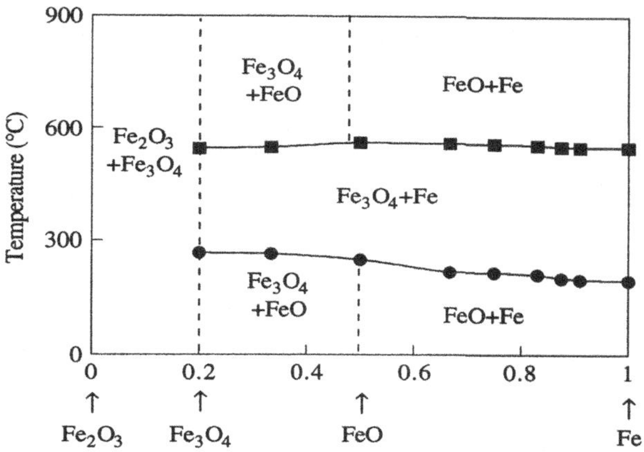

Fig. 5 Phases present in mechanically alloyed and heat treated $xFe-(1-x)FeO$ samples [47]

been carried out in an original planetary mill apparatus, operating in vacuum of 10^{-6} mbar and allowing a constant cooling of the vials by liquid nitrogen. In this case, the temperature of the vials was maintained at 230°C. The vials and balls were made of hardened steel and the ball-to-powder weight ratio was 7:1. The rotational speed of the vials was of 600 rpm and the milling was prolonged for 50 h. Then, the as-milled powders were annealed for 1 h, in flowing argon, at selected annealing temperatures in the range of 100–600°C.

The XRD pattern of the $Fe + Fe_3O_4$ sample annealed at 600°C showed that a high fraction of wüstite (FeO). Based on the Fe-O phase diagram, Bonetti et al. found that a eutectoid reaction proceeds at 570°C, which is reversible on cooling: $Fe + Fe_3O_4 \rightarrow FeO$ [48].

Gheisari et al. [49] have used mechanochemical processing to prepare wüstite nanoparticles. They used high purity hematite (Fe_2O_3) and iron (Fe) powders as the starting materials. Desirable Fe/Fe_2O_3 mole ratios (from 0.6 to 1 by a step of 0.1) together with 270 g hardened steel balls of different sizes were loaded into a 500 cc volume hardened steel vial. The milling was performed for 20 h in air in a high energy planetary mill (Fritsch, Pulverisette 6) with a rotational speed of 500 rpm. In order to determine iron wear in the course of milling, the weight of balls and vial were carefully weighed before and after milling.

In order to get a single phase wüstite different mole ratios of (Fe/Fe_2O_3) were milled, using a planetary mill. Based on the following chemical reaction: $Fe + Fe_2O_3 \rightarrow 3FeO$, it is necessary to choose an equimolar of Fe and Fe_2O_3 to get a single phase FeO. But XRD investigations of the as-milled powders with different mole ratios show that only the sample with a mole ratio 0.6 is a single phase wüstite, Fig. 6a and for higher mole ratios the products are iron-wüstite composites. Figure 6b shows the XRD pattern of the sample with a mole ratio 1, as a typical XRD pattern of samples with mole ratio higher than 0.6. The reason of using a mole ratio of 0.6 (obtained experimentally) instead of equimolar one, is due to iron uptake in the course of milling. Also according to the following relation [35]

$$a = 4.334 - 0.478x \tag{1}$$

and the unit cell parameter of $Fe_{1-x}O$ (a = 4.2998 Å) obtained from XRD pattern, the x value is 0.072. As can be seen the main peaks observed at diffraction angles of

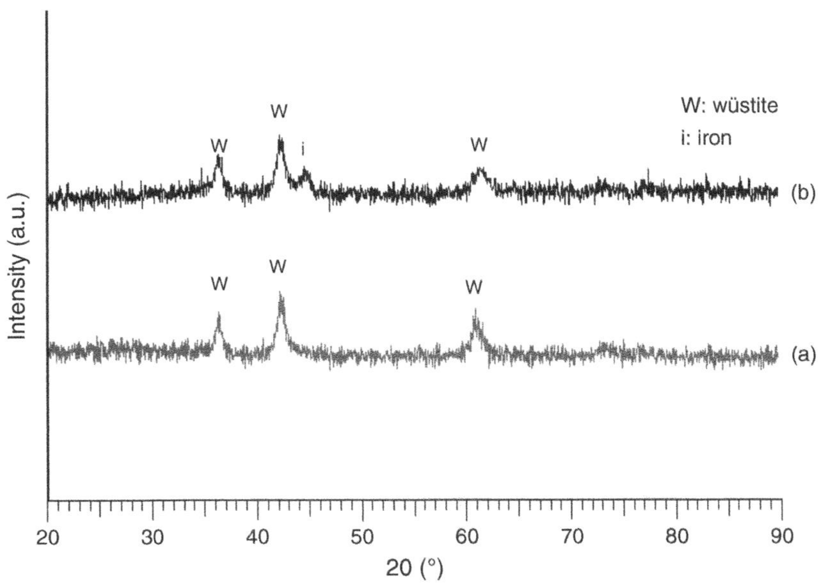

Fig. 6 XRD patterns of the samples with mole ratios (**a**) 0.6 (**b**) 1 [49]

36.1, 42.1 and 60.9 correspond to wüstite ($Fe_{1-x}O$). An average crystallite size of 13 ± 1 nm and microstrain of $\varepsilon = 0.008$ were obtained for the single phase sample, using Williamson and Hall formula [49].

Figure 7 shows the variations of pressure and temperature in the air-filled vial in the course of milling as a function of time for the sample with MR = 0.6. As can be seen, the pressure rises sharply in the first 20 min and reaches a maximum value of about 90.6 ± 0.1 kPa. There are two reasons for this pressure increase. First, it is related to the released oxygen due to the following chemical reaction [50]:

$$Fe_2O_3 \rightarrow Fe_2O_{3-\delta} + \frac{1}{2}\delta O_2 \qquad (2)$$

where δ refers to the extent of oxygen vacancy and second, is related to increase in temperature, mainly due to iron oxidation in the presence of the oxygen and conversion of mechanical energy into thermal energy. A sharp increase in temperature in the first 100 min, confirms this hypothesis. As can be seen in Fig. 7, in a time interval of 20–120 min, the pressure decreases sharply. This is due to the reaction of vacant iron oxide, oxygen and iron, based on the following reaction:

$$Fe_2O_{3-\delta} + \frac{1}{2}\delta O_2 + (1 - 3x)Fe \rightarrow 3Fe_{1-x}O \qquad (3)$$

which leads to formation of wüstite. As time is passing the thermal energy due to mechanical energy conversion leads to an increase in temperature and then it reaches a constant value after 5 h, due to a thermal equilibrium condition. As can be seen the pressure also has the same behavior.

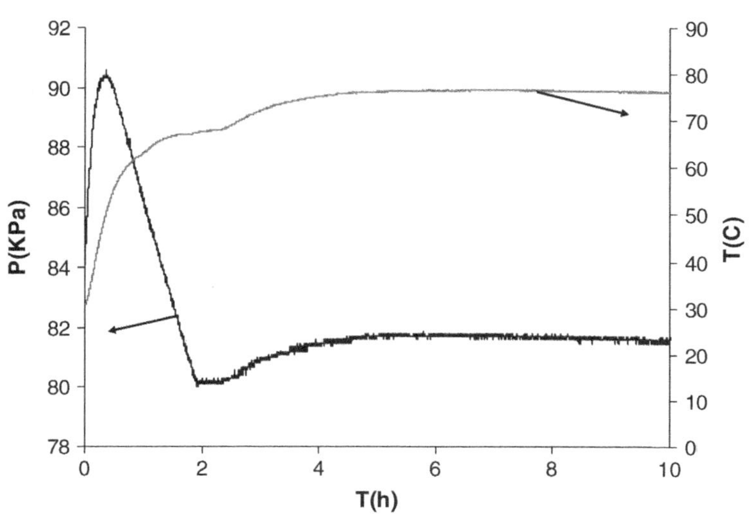

Fig. 7 The variations of temperature and pressure of the air filled vials as a function of time for samples with MR = 0.6 [49]

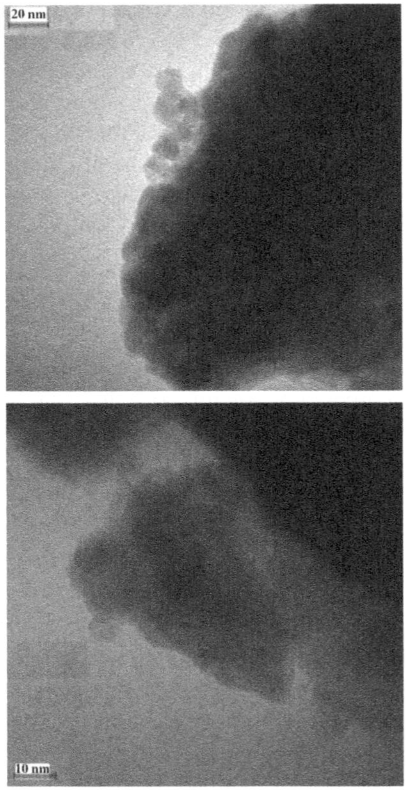

Fig. 8 TEM micrographs of the single phase wüstite [49]

Figure 8 shows the TEM micrographs of the single phase wüstite. The average size of the particles in the aggregates is in the range of the crystallite size obtained by Scherrer's formula [51].

Following the work by Gheisari et al. [49], Mozaffari et al. [52] have used mechanochemical processing to prepare iron-wüstite (Fe–Fe_yO) nanocomposites with other Fe/Fe_2O_3 mole ratios (MR) = 0.9, 2.3, 4.9 and 13.6. Figure 9 shows XRD patterns of the as-milled samples with different mole ratios (MR), as labeled on the figure. All mixtures with MRs higher than 0.6 resulted in iron-wüstite composites, except for MR = 13.6. As can be seen on the XRD pattern of the sample with MR = 13.6, there are no detectable peaks related to wüstite. The mean crystallite sizes of the iron and wüstite in the nanocomposites were obtained, using Scherrer's formula and were about 9 ± 1 and 7 ± 1 nm respectively. Also from Fig. 9 it can be seen (refer to vertical line drawn on the figure) that by increasing MR, the main diffraction peaks of wüstite have shifted to higher angles which is a result of reduction in its lattice parameter.

Figure 10 shows the variation of wüstite lattice parameters with respect to MR. This reduction can be due to: (a) nano sized particles and (b) iron deficiency [49].

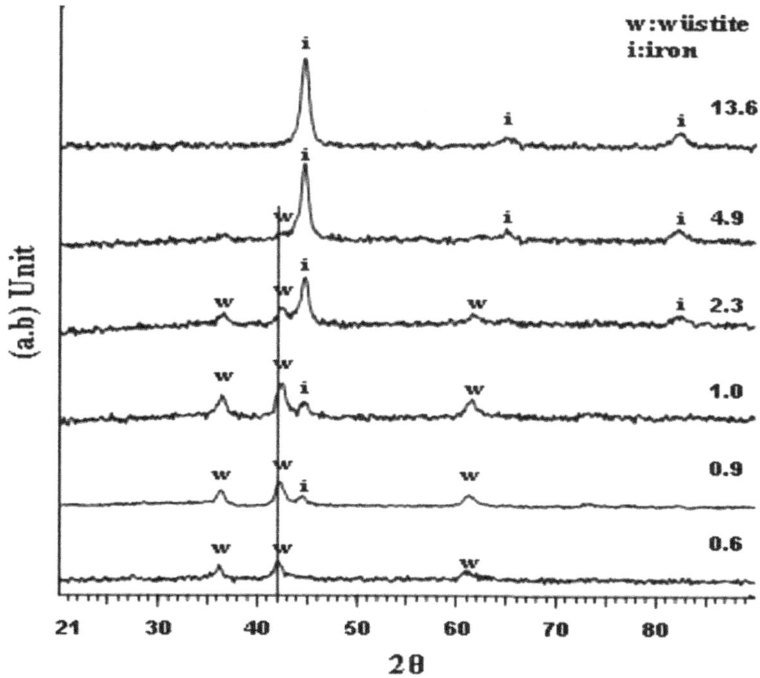

Fig. 9 XRD patterns of the samples with different mole ratios (Fe/Fe_2O_3), as labeled on the patterns. The *vertical line* at 42° has been drowning to guide eye [52]

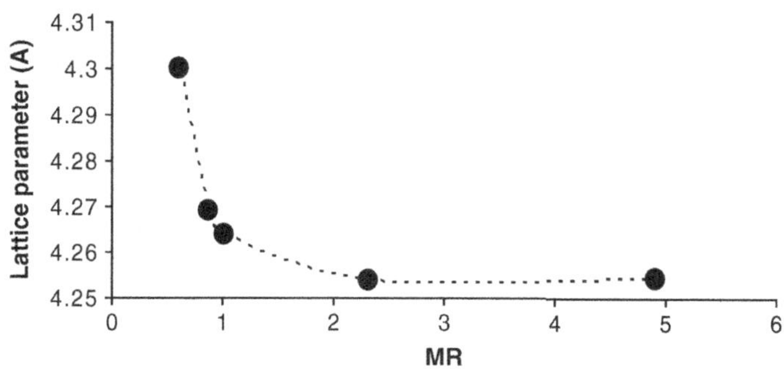

Fig. 10 The variation of the wüstite lattice parameters in nanocomposites as a function of MR [52]

As the average crystallite sizes of the wüstite in the nanocomposites have more or less the same values for all MRs the reduction in lattice parameter can be due to different iron deficiencies. Wüstite is almost nonstoichiometric with some Fe deficiency and can be denoted as Fe_yO [29, 49]. Using the formula (1), where y is the Fe content, a composition of $Fe_{0.93}O$ was estimated for the wüstite single phase, using lattice parameters for higher mole ratios, MR = 0.9–4.9, the same formula was used and the corresponding compositions were found to be $Fe_{0.87}O$ to $Fe_{0.83}O$. It should be noted that in order to achieve electroneutrality, an appropriate proportion of iron ions should be considered as Fe^{3+}, therefore one could consider wüstite as $Fe_{1-}{}^{2+}{}_{3x}Fe_{2x}{}^{3+}\square_xO$, where □ shows a vacancy [52].

Yagodkin et al. [53] have used mechanochemical processing with starting materials of α-Fe_2O_3 and Fe using different milling times. First they found a mixture of Fe_2O_3 + Fe_3O_4 + FeO + Fe but at higher milling time a mixture of FeO + Fe + amorphous phase was found according to the XRD results.

Nanocrystalline composite containing FeO, α-Fe and an amorphous phase were obtained as a result of high-energy ball milling. The average sizes of the crystallites in the produced composites were in the range of 15–20 nm. The amorphous phase was a mixture of oxygen and iron [53].

3 Magnetic Properties

3.1 *Saturation Magnetization and Coercivity*

Saturation magnetization and coercivity as a function of the composition, x, for xFe · (1−x)Fe_2O_3 is shown in Fig. 11 [47]. The low magnetization in Fig. 11 is partly due to the paramagnetic FeO and also indicates a possible presence of small amounts of magnetic phases, probably α-Fe. For x > 0.5, the as-milled samples consisted of mixtures of FeO and Fe, and the measured values of magnetization were in good agreement with the values expected for the respective compositions [47].

Samples based on ferrimagnetic Fe_3O_4 phase for x ≤ 0.5 had coercivities of 200–400 Oe. Higher coercivities of 500–600 Oe were measured for samples consisting of Fe and FeO, i.e. for x = 0.6–0.8. The coercivity then decreased with

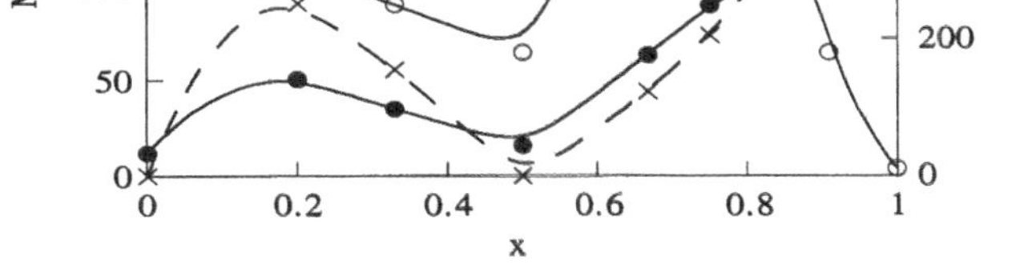

Fig. 11 Saturation magnetization, M_s, and coercivity, H_c, as a function of the composition, x, for xFe · (1 − x)Fe_2O_3 in the as-milled state. The dashed line is the theoretical magnetization calculated from the composition, x [47]

increasing x, as the fraction of paramagnetic FeO decreased. For $x = 1$ (pure α-Fe), the coercivity was less than 20 Oe. These measurements show that high values of coercivity can be obtained provided the α-Fe grains are separated by a non-magnetic phase (here paramagnetic FeO) as is the case for $0.5 < x \leq 0.85$. For larger values of x the α-Fe grains increasingly percolate [52] within the structure and the exchange interactions between the magnetic crystallites causes the coercivity to decrease.

In Fig. 12 measurements of saturation magnetization and coercivity for samples with $x = 0.67$ are plotted as a function of annealing temperature. The as-milled powder had a magnetization of 62 emu/g, which is in good agreement with the magnetization calculated for a two phase mixture of FeO and α-Fe with $x = 0.67$. After annealing at 200°C the magnetization increased due to the partial decomposition of FeO into Fe and Fe_3O_4. Samples annealed at 300 and 400°C possessed a magnetization of around130 emu/g. This value agrees with the magnetization of a mixture of Fe_3O_4 and Fe for $x = 0.67$ (Fig. 5). Annealing at higher temperatures caused a rapid decrease in magnetization due to the formation of FeO. The samples annealed at 700 and 900°C had values of magnetization close to that of the as-milled sample, indicating similar fractions of Fe and FeO [47].

Yagodkin et al. found that the milled powders ($Fe_2O_3 + Fe_3O_4 + FeO + Fe \rightarrow FeO + Fe +$ amorphous phase.) had properties, which are characteristic of hard magnetic materials [53]. Improvement of the magnetic properties was achieved by low-temperature annealing of the milled powders. An intrinsic coercive force $\mu_0 H_c \geq 0.05$ T at 300 K (Fig. 13) was achieved for the Fe_2O_3 + 50% Fe mixture already after 1 h milling. (Intrinsic coercive forces of hard magnetic materials should exceed 0.01 T.) After 3 h milling, B_r and $(BH)_{max}$ of this powder were about 0.38 T and 6 kJ/m^3, respectively. In the other powders these values were lower, which may be explained by a lower content of nanocrystalline α- Fe [53].

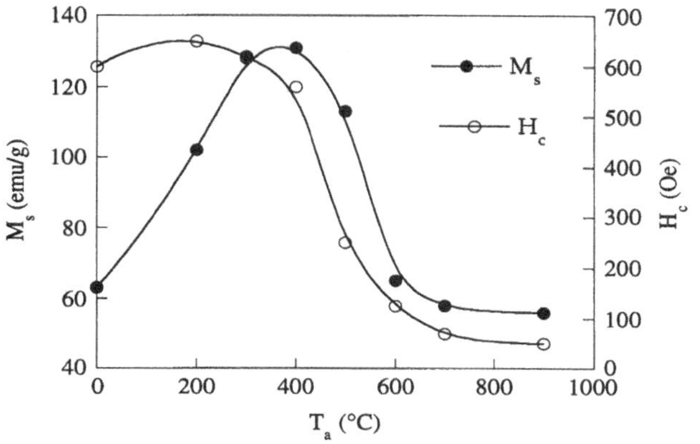

Fig. 12 Saturation magnetisation, M_s, and coercivity, H_c , for xFe – (1 – x)Fe O with $x = 0.67$ as a function of the annealing temperature, [47]

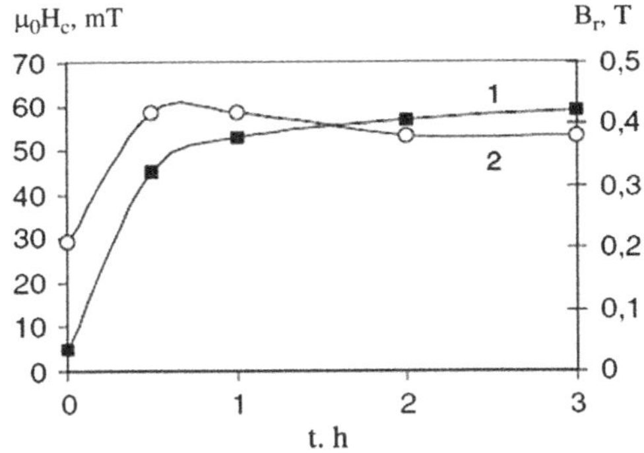

Fig. 13 μ_0H_c (1) and B_r (2) at room temperature as a function of a milling time [53]

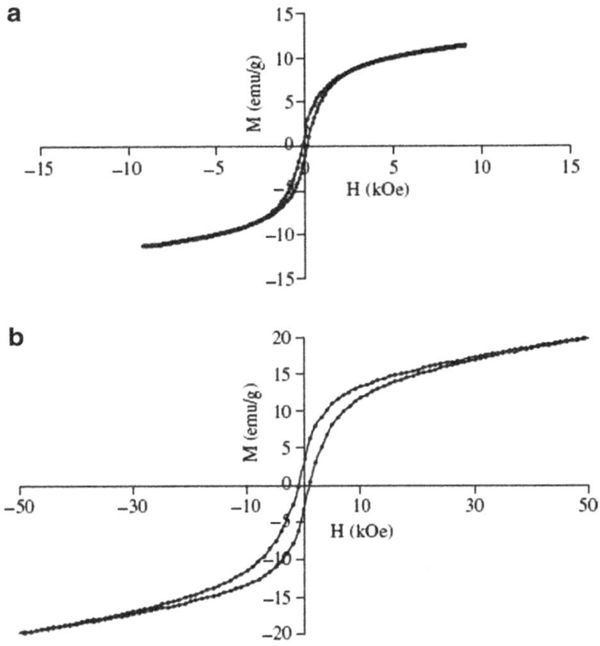

Fig. 14 Hysteresis loops of the wüstite single phase sample at (**a**) room temperature and at (**b**) 5 K [49]

Figure 14 shows the hysteresis loops of the single phase wüstite at room temperature and at 5 K, prepared by Gheisari et al. [49]. As can be seen both curves show non-zero coercivities and remanent magnetizations in the form of ferrimagnetic behaviour. The same behaviour has been observed for wüstite thin films at low temperatures (10 K) [13, 14]. The magnetizations have not been saturated even in a field of 50 kOe and at 5 K temperature. The magnetizations are 11 and 20 emu/g in applied fields of 9 and 50 kOe respectively. This is in contrast to the behaviour of bulk wüstite, which is antiferromagnetic at temperatures below 200 K (Neél temperature) [29].

According to Dimitrov et al. [38–40], the observed magnetization can be due to the spinel-type defect clusters, Fig. 1c. Based on these works, it is possible to consider that defect clusters in our sample may comprise of a Fe_3O_4 like phase coherently embedded in an ideal FeO matrix. In this case for a particular value of x in $Fe_{1-x}O$, there are x molecules of Fe_3O_4 and $(1 - 4x)$ molecules of FeO. Then in the single phase wüstite sample (MR = 0.6) there are 0.712 molecules of FeO and 0.072 molecules of Fe_3O_4 which the latter can be the cause of observed hysteresis loops.

The Mössbauer spectra of the Fe–FeO nanocomposites are shown in Fig. 15 [52]. The observed asymmetry of spectrum for the single-phase wüstite sample is due to two overlapping quadrupole doublets [54]. The doublet with $\delta = 0.95$ and $\Delta = 0.85$ mm/s is assigned to occupation of octahedral sites and to the feature of electronic exchange between Fe^{2+} and Fe^{3+} ions both in octahedral sites. Also another doublet with $\delta = 0.65$ and $\Delta = 0.57$ mm/s is assigned to Fe^{3+} ions on tetrahedral sites which interact with Fe^{2+} ions on octahedral sites [54]. For higher MR, Mössbauer spectra exhibit sextets, which show the existence of iron in the samples. From the area of the peaks, corresponding to each sample, the relative

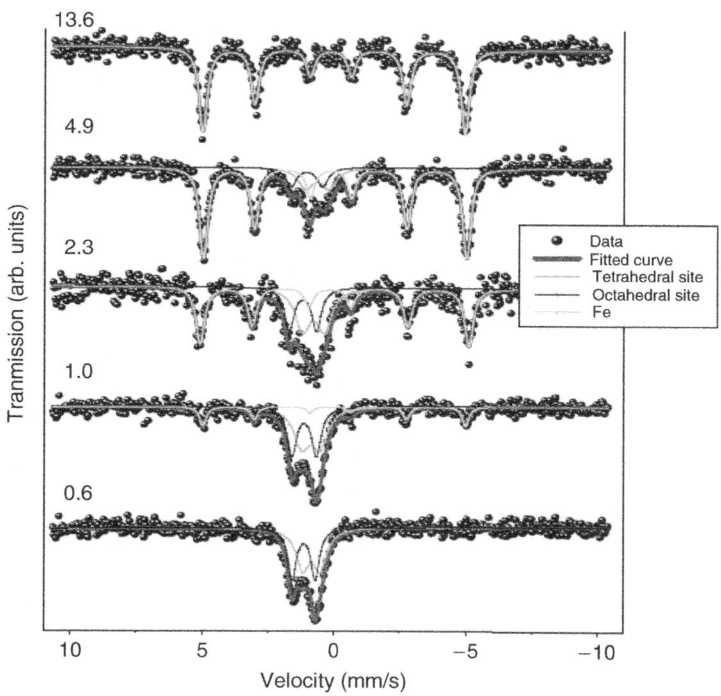

Fig. 15 Mössbauer spectra of the samples with different mole ratios (Fe/Fe_2O_3), as labeled on the spectra [52]

contents (%) of Fe and wüstite were determined and shown in Table 1. Mössbauer spectrum related to the sample with MR = 13.6, shows a single sextet which is related to α-Fe and confirms the XRD results. This also, is due to lack of high detectability in Mössbauer method.

Figure 16 shows room temperature hysteresis loops of the cold pressed powders with different MRs.

The variation of Ms values with respect to MR, that was obtained from VSM measurements together with those calculated based on Mössbauer data and chemical reaction are shown in Fig. 17. The Ms calculation based on Mössbauer data was performed by the following formula: $M_S = \alpha[M_S]_{Fe} + (1-\alpha)[M_S]_{wustite}$ where α is the Fe fraction in the nanocomposites and $[M_s]_{Fe}$ and $[M_s]_{wüstite}$ are the saturation magnetizations of Fe and wüstite, respectively, were obtained from the VSM measurements. A non-zero magnetization of 12 emu/g has been considered for wüstite phase, which is due to formation of spinel-like defect clusters, as they have already reported [49]. In addition the Ms value of Fe nanopowders, which were

Table 1 Calculated values of isomer shift (IS), quadrupole splitting (QS), hyperfine magnetic fields (H_{hf}) and the iron weight percents of each sample with different mole ratios[52]

MR	[IS] δ (mm/s)			[QS] Δ (mm/s)			H_{hf} (kOe)	(%) $Fe_{(obs)}$
0.6	A = 0.65	B = 0.95		A = 0.57	B = 0.85		0	0
1	A = 0.63	B = 0.95	Fe = 0.05	A = 0.68	B = 0.93	Fe = 0.05	306	20
2.3	0.57	0.97	0.03	0.78	0.97	0.03	305	45
4.9	0.43	0.83	0.03	0.85	1.1	0.03	307	75
13.6	0	0	0.02	0	0	0.02	307	100

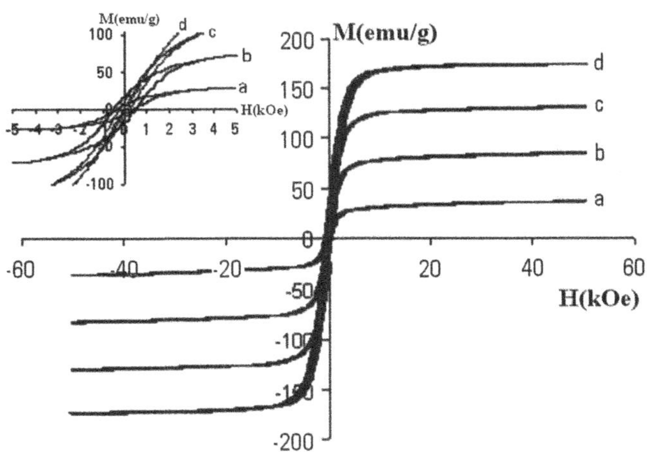

Fig. 16 Room temperature hysteresis loops of the powders with different MRs (**a**) 1.0, (**b**) 2.3, (**c**) 4.9 and (**d**) 13.6. *Inset* shows low field part of the loops [52]

obtained by VSM measurements is 170 emu/g, which is lower than the value of bulk Fe (217.2 emu/g) [55].

Ms calculation based on the chemical reaction was performed by:

$$M_S = \beta[M_S]_{Fe} + (1 - \beta)[M_S]_{wustite} \tag{4}$$

where $\beta = [55.9(MR + 2 - 3y)]/[55.9(MR + 2 - 3y) + 3(55.9y + 16)]$ and M_s of Fe and wüstite are defined as before. The chemical reaction used in this calculation is:

$$(MR)Fe + Fe_2O_3 \rightarrow 3Fe_yO + (MR + 2 - 3y)Fe \tag{5}$$

The measured Fe uptake in the course of milling was about 2 g and was taken into account.

The increase in Ms with respect to MR is due to the increase of α-Fe content in the samples. As can be seen, the calculated values of M_s are in good agreement with the experimental ones.

Figure 18 shows the variation of coercivity with respect to MR for the nanocomposite as-milled powders. As can be seen the values of H_c are not zero for the lowest MR (0.6) [23]. Also it can be seen that as MR increases from 0.6, the coercivity increases sharply to a value of 480 Oe, and drops off at the percolation threshold at about MR = 2.3 [52], in which the value of Fe content is 45% (Table 1). This behavior has been seen in other granular magnetic systems, exactly, when the percolation threshold is crossed. The decreases in the H_c for MR values greater than the percolation threshold is due to an increase in α-Fe phase in the sample, in which the low H_c value of Fe dominates the composite coercivity and reaches the H_c value of Fe less than 10 Oe. The same behavior has been reported in $Fe–Fe_3O_4$ systems [10].

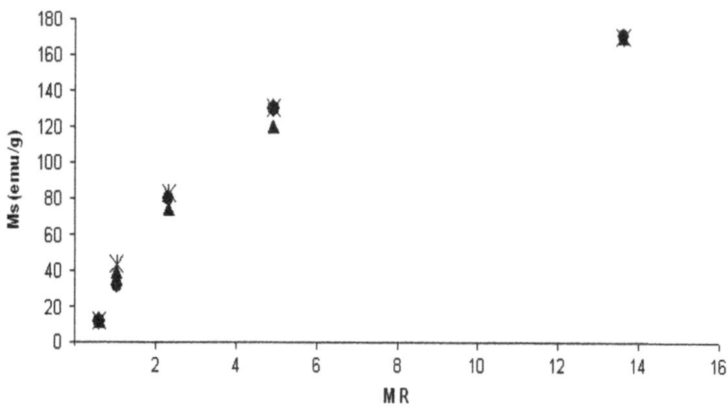

Fig. 17 The variation of saturation magnetizations with respect to MR, for VSM measurements (*filled circle*) and calculation based on Mössbauer data (*cross symbol*) and chemical reaction (*filled triangle*)[52]

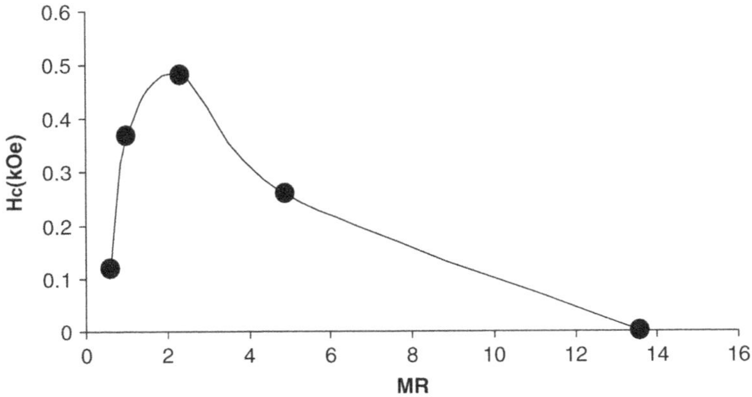

Fig. 18 Variation of coercivity as a function of MR, for the as-milled powders [52]

3.2 Exchange Bias Effect in Fe–FeO

Fe–FeO nanocomposite was the second system in which the phenomenon of exchange anisotropy (exchange bias) was investigated by Meiklejohn [56]. They measured rotational hysteresis of this system (Fig. 19) and saw that the Fe–FeO system has rotational hysteresis like the Co–CoO system, therefore has exchange anisotropy.

They took a more detailed look at the high field rotational hysteresis as a function of temperature. They found that rotational hysteresis should vanish at exactly the Neel temperature of particular oxide; this is shown for Co–CoO and Fe–FeO in Fig. 20. The Neel temperatures for CoO and FeO are 290 and 185 K, respectively, as shown by notation T_N in Fig. 20.

Fiorani et al. [57] prepared the nanogranular Fe/Fe oxide samples by cold-compacting oxide-layered Fe particles, by inert gas condensation and oxygen passivation. The Fe particle mean size, $D = (67 \pm 1)$ nm, and the Fe weight fraction, $xFe = (20 \pm 3)\%$, were estimated by X-ray diffraction through the Rietveld analysis method [58]. They have observed exchange bias in nanogranular Fe/Fe oxide samples (Fig. 21). As it is obvious the ZFC loop is symmetric about the origin while the FC loops are shifted towards the negative field values (Fig. 21, inset). The shift is related to the exchange field parameter $H_{ex} = (H_{right} + H_{left})/2$, whereas the coercivity is defined as $H_C = (H_{right} - H_{left})/2$, H_{right} and H_{left} being the points where the loop intersects the field axis.

FC hysteresis loops were measured at different temperatures below 250 K. The curves of H_{ex} vs. T for $H_{cool} = 4$ kOe and $H_{cool} = 20$ kOe are shown in Fig. 22. For both values of H_{cool}, H_{ex} is completely absent above T = 150 K and it appears only below such temperature, which corresponds to the freezing of most of the moments of the oxide regions. This indicates that the exchange bias effect originates from the exchange interaction at the interface between the metallic particles and the oxide

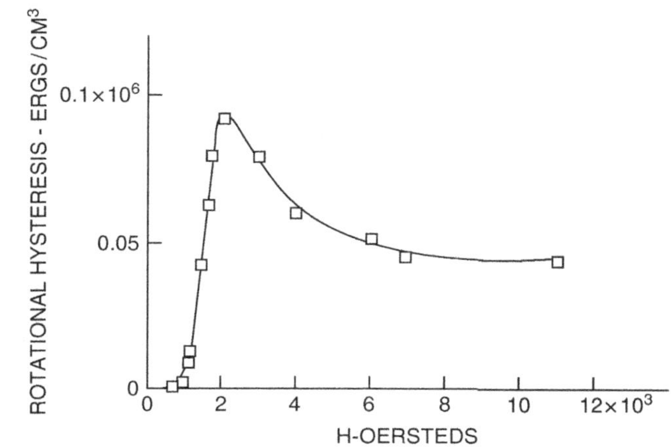

Fig. 19 Rotation hysteresis of 200 Å iron particles that have a ferrous oxide shel. Data were taken at 77 K [56]

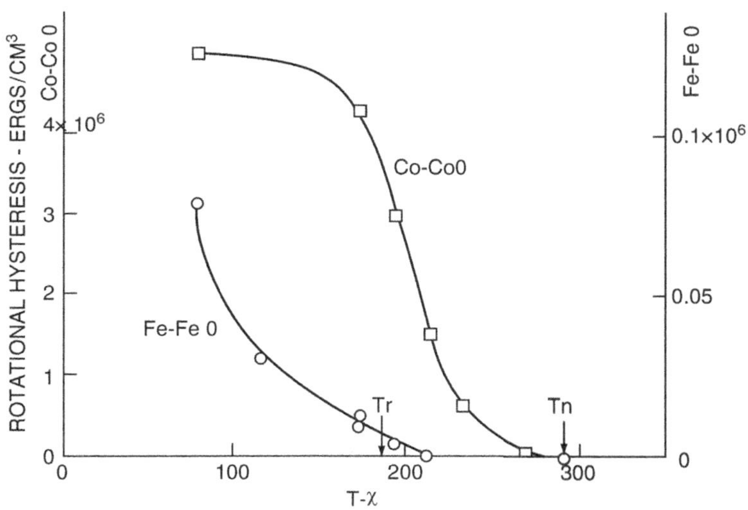

Fig. 20 Rotation hysteresis of the Fe–FeO and Co–CoO systems at 10,500 and 15,000 Oe, respectively [56]

matrix. With reducing T, H_{ex} increases because of the progressive freezing of a rising number of oxide region moments [57].

This effect has been also observed in $NiFe_2O_4$ [59] and γ-Fe_2O_3 [60] nanoparticles, where the ferrimagnetic core is surrounded by a disordered surface shell, freezing in a spin glass like state at low temperature.

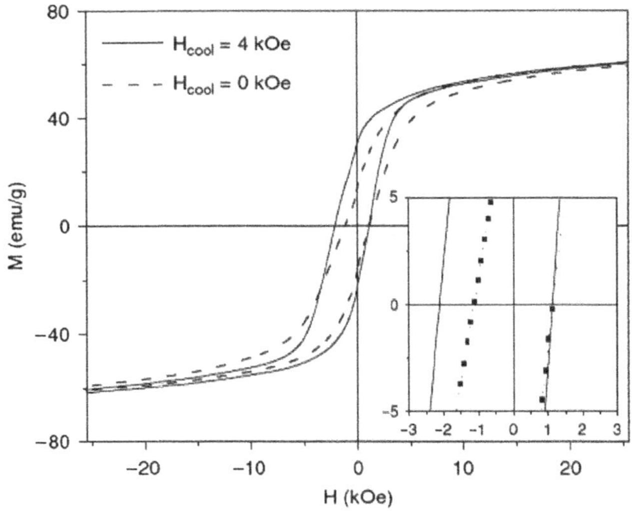

Fig. 21 Hysteresis loops at T = 5 K after field-cooling from T = 250 K in Hcool = 0 and 4 kOe. Inset: enlarged view of the central region of the loop at Hcool = 0 and 4 kOe [57]

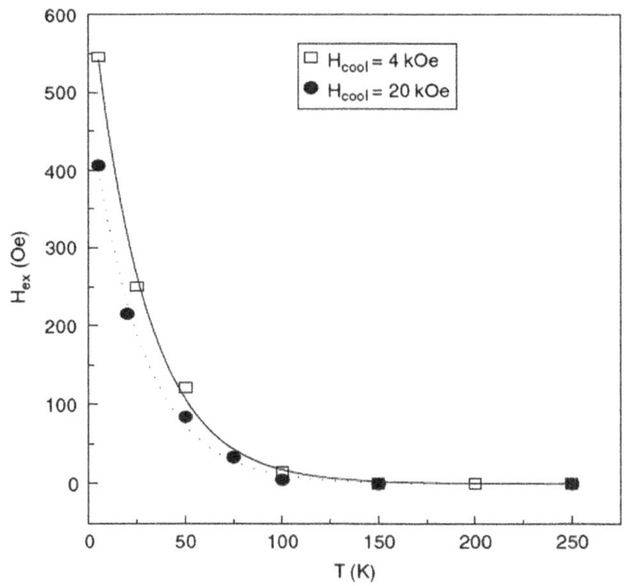

Fig. 22 Exchange bias field vs. temperature after field cooling at different fields (H = 4 kOe, *open symbol*; H = 20 kOe, *solid symbol*) [57]

The exchange bias in Fe nanoparticles with an oxide layer of γ -Fe_2O_3 has been studied by several groups [61, 62], and their results show that the bias field (H_{ex}) is generally below 3 kOe in this system. In this article, they report the giant exchange bias in γ-Fe_2O_3-coated Fe nanoparticles; that is, $H_{ex} = 6{,}000$ Oe at 2 K, which is much larger than that previously reported [61, 62] for the Fe nanoparticles. A simple model is proposed to interpret the giant exchange bias in this system. Zheng et al. [46] fabricated core/shell-structured Fe nanoparticles, in which the α-Fe core is about 5 nm in diameter and the γ-Fe_2O_3 shell is about 3 nm thick, and systematically studied their structural and magnetic properties. The magnetic hysteresis (M–H) loops, measured at low temperatures, after the particles were cooled from 350 K in a 50 kOe field, show significant shifts in both horizontal and vertical directions. It has been found that the exchange-bias field can be as large as 6.3 kOe at 2 K (Fig. 23), and that the coercive field is also enhanced greatly in the field-cooled (FC) loops.

The interesting feature in the M–H curves is that both the ZFC and FC loops remain open even in a 50 kOe field, known as high field irreversibility, which could be interpreted as being due to the existence of the spin-glass like phase [46]. More importantly, the coercive field has been greatly enhanced from 2.4 kOe for the ZFC loop to 6.4 kOe for the FC loop, and that the FC loop becomes asymmetrical. From the FC loop, the exchange bias has been easily extracted to be 6.3 kOe, which is much larger than the previous reported values in similar systems. The large exchange bias and vertical shifts of the FC loops at low temperatures may be ascribed to the frozen spins in the shells [46].

Muñoz et al. [63] have grown iron thin films by DC magnetron sputtering at controlled substrate temperatures. Magnetic hysteresis loops showed an exchange bias consistent with the air passivation of the samples. The obtained exchange bias in their samples is in agreement with the observed one in Fe–FeO samples, this is reasonable because of the sample passivation after preparation, when they go from the vacuum chamber to the room environment.

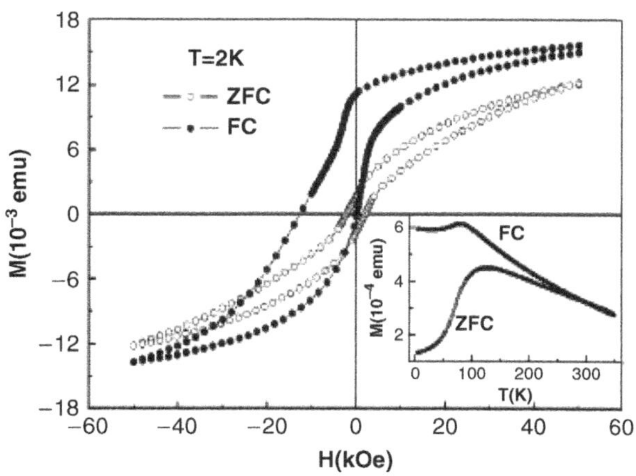

Fig. 23 The ZFC and 50 kOe FC magnetic hysteresis loops at 2 K. Both the horizontal and vertical shifts in the FC loop are apparent. The high field irreversibility up to 50 kOe is also seen clearly in both the ZFC and FC loops. Inset: the ZFC–FC curves measured in a 0.1 kOe field [46]

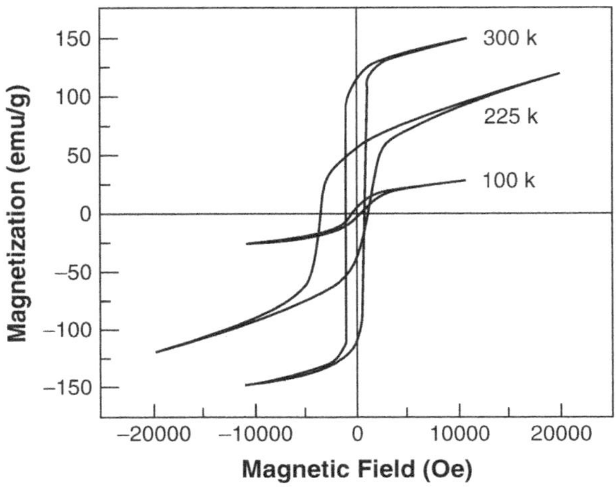

Fig. 24 Magnetic hysteresis loops at different temperatures, magnetic field have been applied parallel to the films. [63]

As can be seen from the Fig. 24 the exchange bias has its maximum values for the samples prepared at temperatures near 200 K [63]. Chen et al. [42] have also grown the iron thin films by molecular-beam epitaxy and studied their magnetic properties. The films were grown on (110) GaAs substrates and were allowed to develop a natural oxide. The iron oxide on the free surface is FeO and has an antiferromagnetic transition temperature around 200 K. This antiferromagnetic oxide provides an exchange bias for the iron film at low temperatures. Although it is usual to study exchange coupling with magnetization measurements they have used the low-temperature magneto transport properties of the films to study the exchange coupling and compare it to models of this phenomenon [42].

Gheisari [64] has measured exchange bias in Fe–FeO samples, prepared according to reference [52]. Figure 25 shows the hysteresis loops of the iron-wüstite nanocomposites after field cooling to 5 K in a 70 kOe field. As can be seen in Fig. 26 the FC loops are shifted towards the negative applied field. In addition to the horizontal shift, the vertical asymmetry of hysteresis loops was also observed, which can be interpreted as the existence of a spin-glass phase [46, 64]. The values of exchange field (H_{ex}), coercivity field (H_c), magnetization at maximum applied field (70 kOe) and vertical shift ΔM are listed in Table 2. With increasing the percentage of iron in nanocomposites the magnetizations of the samples are increased, but the value of H_{ex}, H_c and ΔM are decreased. This is obvious because Fe nanoparticles are soft ferromagnetic with a very low H_c and high magnetization. Also susceptibility measurement of the samples show that there is a spin -glass like phase in low Fe concentrations which supports the higher H_{ex}, H_c and ΔM in the nanocomposites.

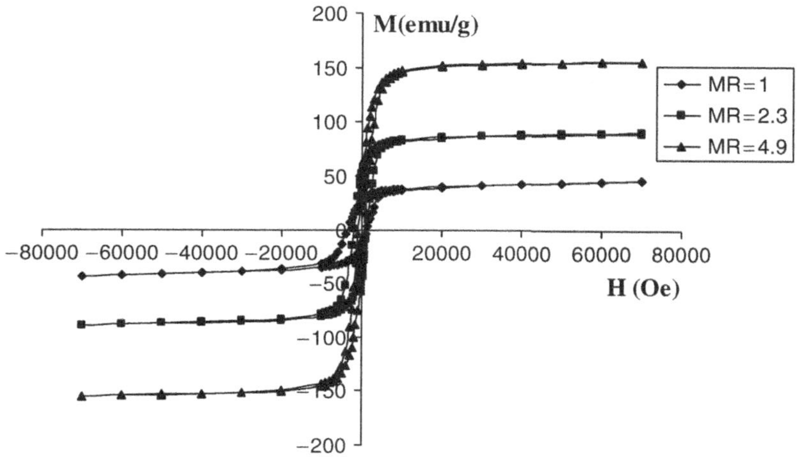

Fig. 25 Hysteresis loops of the iron-wüstite nanocomposites after field cooling to 5 K in a 70 kOe field, measured by SQUID [64]

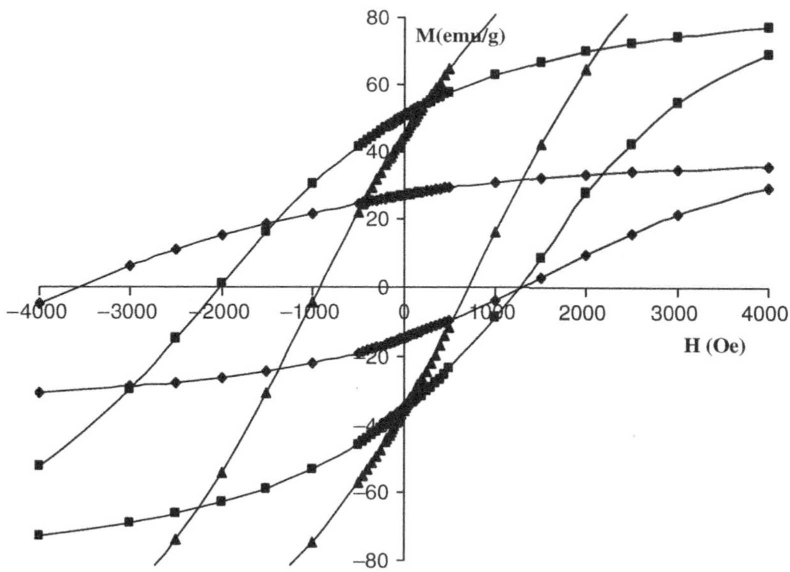

Fig. 26 Low field FC hysteresis loops of the iron-wüstite nanocomposites after field cooling to 5 K in a 70 kOe field, measured by SQUID [64]

Table 2 The values of exchange field (H_{ex}), coercivity field (H_c), magnetization at maximum applied field (70 kOe) and vertical shift ΔM [64]

MR	%Fe	He (Oe)	Hc (Oe)	ΔM	M(70 kOe)
1	20	1,138	2,430	1.8	44.8
2.3	45	375	1,645	1.6	89.5
4.9	75	100	800	0.7	155.4

4 Conclusion

In this chapter structure, preparation and characterization methods of Fe–FeO nanocomposites in the form of particles and thin films have been reviewed. The magnetic properties, namely saturation magnetization, coercivity and exchange bias effect have been discussed.

References

1. Suryanarayana, C.: Mechanical Alloying and Milling. Marcel Dekker, New York (2004)
2. Suryanarayana, C.: Mechanical alloying and milling. Prog. Mater. Sci. **46**, 1 (2001)
3. Ding, J., Miao, W.F., Pirault, E., Street, R., McCormick, P.G.: J. Magn. Magn. Mater. **177–181**, 933 (1998)
4. Sort, J., Nogues, J., Amils, X., Surinach, S., Munoz, J.S., Baro, M.D.: Phys. Rev. B **65**, 174420 (2002)
5. Giri, A.K.: Mater. Res. Bull. **32**, 523(1997)
6. Xiao, J.Q., Jiang, J.S.: Phys. Rev. B **49**, 3982 (1994)
7. Jacobs, I.S., Bean, C.P., Rado, G.T., Sul, H. (eds.): Magnetism, vol. 3. Academic, New York (1963)
8. Pardavi-Horvath, M., Takacs, L.: IEEE Trans. Magn. **28**, 3186 (1992)
9. Takacs, L.: Nanostruct. Mater. **2**, 241 (1993)
10. Ding, J., Miao, E., Street, R., McCormick, P.G.: Scr. Mater. **35**, 1307 (1996)
11. Barahama, P., Banerjee, S., Das, D., Mukhopadhy, P.K., Chatterjee, S., Nigam, A.K., Chakravorty, D.: J. Magn. Magn. Mater. **246**, 162 (2002)
12. Mauvernay, B., Presmanes, L., Bonningue, C., Tailhades, Ph: J. Magn. Magn. Mater. **320**, 58 (2008)
13. Löffler, J.F., Meier, J.P., Doudin, B., Ansermet, J., Wagner, W.: Phys. Rev. B **57**, 2915 (1998)
14. Parkin, S.S., More, N., Roche, K.P.: Phys. Rev. Lett. **64**, 2304 (1990)
15. Koch, C.C.: Nanostructured Materials: Processing, Properties, and Applications. William Andrew Publishing, New York (2006)
16. Koshizaki, N., Yasumoto, K., Terauchi, Sh, Umehara, H., Sasaki, T., Oyama, T.: Nanostruct. Mater. **9**, 587 (1997)
17. Zhang, D., Klabunde, K.J., Sorensen, ChM, Hadjipanayis, G.C.: NanoStruct. Mater. **12**, 1053 (1999)
18. Fiorani, D., Testa, A.M., Tronc, E., Lucari, F., D'Orazio, F., Nogués, M.: J. Magn. Magn. Mater. **226–230**, 1942 (2001)
19. Hansen, M.F., Koch, ChB, Morup, S.: Phys. Rev. B **62**, 1124 (2000)
20. Kodama, R.H.: Magnetic nanoparticles. J. Magn. Magn. Mater. **200**, 359 (1999)
21. Weller, D., Moser, A.: IEEE Trans. Magn. **35**, 4423 (1999)
22. Kodama, R.H., Makhlouf, S.A., Berkowitz, A.E.: Phys. Rev. Lett. **79**, 1393 (1997)
23. Kodama, R.H., Berkowitz, A.E.: Phys. Rev. B **59**, 6321 (1999)
24. Luo, W., Nagel, S.R., Rosenbaum, T.F., Rosensweig, R.E.: Phys. Rev. Lett. **67**, 2721 (1991)
25. Kwok, Y.S., Zhang, X.X., Qin, B., Fung, K.K.: J. Appl. Phys. **89**, 3061 (2001)
26. Nogues, J., Sorta, J., Langlais, V., Skumryev, V., Suriñach, S., Muñoz, J.S., Baro, M.D.: Phys. Rep. **422**, 65 (2005)
27. Meiklejohn, W.H., Bean, C.P.: Phys. Rev. **102**, 1413 (1956)
28. Meiklejohn, W.H., Bean, C.P.: Phys. Rev. **105**, 904 (1957)
29. Cornell, R.M., Schwertman, U.: The Iron Oxides. Wiley-VCH, Weinheim (2003)
30. Johnson, D.P.: Solid State Commun. **7**, 1785 (1969)

31. Roth, W.L.: Defects in the crystal and magnetic structures of ferrous oxide. Acta Crystallogr. **13**, 140 (1960)
32. Koch, F., Cohen, J.B.: The defect structure of Fe1-xO. Acta Crystallogr. B **25**, 275 (1969)
33. Cheetham, K., Fender, B.E.F., Taylor, R.I.: J. Phys. C Solid State Phys. **4**, 2160 (1971)
34. Catlow, C.R.A., Fender, B.E.F.: J. Phys. C Solid State Phys. **8**, 3267 (1975)
35. McCammon, C.A., Liu, L.: Phys. Chem. Miner. **10**, 106 (1984)
36. Battle, P., Cheetham, K.A.: The magnetic structure of non-stoichiometric ferrous oxide. J. Phys. C Solid State Phys. **12**, 337 (1979)
37. Kim, Y.K., Oliveira, M.: J. App. Phys. **75**, 431 (1994)
38. Dimitrov, D.V., Hadjipanayis, G.C., Papaefthymiou, V., Simopoulos, A.: IEEE Trans. Magn. **33**, 4363 (1997)
39. Dimitrov, D.V., Unruh, K., Hadjipanayis, G.C., Papafthymiou, V., Simopoulos, A.: Phys. Rev. B **59**, 14499 (1999)
40. Dimitrov, D.V., Unruh, K., Hadjipanayis, G.C., Papafthymiou, V., Simopoulos, A.: J. Appl. Phys. **87**, 7022 (2000)
41. Lin, X., Murthy, A.S., Hadjipanayis, G.C., Swann, C., Shah, S.I.: J. Appl. Phys. **76**, 6543 (1994)
42. Chen, Y., Lottis, D.K., Dahlberg, E.D.: J. Appl. Phys. **69**, 4523 (1991)
43. Kiewicz, S., Jiménez, V.J., Prieto, F.: J. Appl. Phys. **70**, 5822 (1991)
44. Slawska, W.A., Roig, A., Gich, M., Casas, L., Racka, K., Nedelko, N., Molins, E.: Phys. Rev. **70**, 054412 (2004)
45. Baker, C., Ismat Shah, S., Hasanain, S.K.: J. Magn. Magn. Mater. **280**, 412 (2004)
46. Zheng, R.K., Wen, G.H., Fung, K.K., Zhang, X.X.: J. Appl. Phys. **95** (2004)
47. Ding, J., Miao, W.F., Pirault, E., Street, R., McCormick, P.G.: J. Alloys Compd. **267**, 199 (1998)
48. Bonetti, E., Del, B.L., Singnoretti, S., Tiberto, P.: J. Appl. Phys. **89**, 1806 (2001)
49. Gheisari, M., Mozaffari, M., Acet, M., Amighian, J.: J. Magn. Magn. Mater. **320**, 2618 (2008)
50. Park, J.C., Kim, D., Lee, C.S., Kim, D.K.: Bull. Korean Chem. Soc. **20**, 1005 (1999)
51. Cullity, D.B., Stock, S.R.: Elements of X-Ray Diffraction. Prentice Hall, Englewood Cliffs, NJ (2001)
52. Mozaffari, M., Gheisari, M., Niyaifar, M., Amighian, J.: J. Magn. Magn. Mater. **321**, 2981 (2009)
53. Yagodkin, Y.D., Lileev, A.S., Grishina, E.N., Reissner, M., Steiner, W.: J. Mater. Sci. **39**, 5255 (2004)
54. Elias, D.J., Linnett, J.W.: Oxidation of metals and alloys, Part 3 Mössbauer spectrum and structure of wüstite. Trans. Faraday Soc. **65**, 2673 (1969)
55. Cullity, B.D.: Introduction to Magnetic Materials. Addison-Wesley, Menlo Park, CA (1972)
56. Meiklejohn, W.H.: J. Appl. Phys. **29**, 454 (1958)
57. Fiorani, D., Binaco, L.D., Testa, A.M.: J. Magn. Magn. Mater. **300**, 179 (2006)
58. Mozaffari, M., Manouchehri, S., Yousefi, M.H., Amighian, J.: J. Magn. Magn. Mater. **322**, 383 (2010)
59. Kodama, R.H., Berkowitz, A.E., McNiff, E.J., Foner, J.S.: Phys. Rev. Lett. **77**, 394 (1996)
60. Martinez, B., Obradors, X., Balcells, L.L., Rouanet, A., Monty, C.: Phys. Rev. Lett. **80**, 181 (1998)
61. Prados, C., Multigner, M., Hernando, A., Sanchez, J.C., Fernandez, A., Conde, C.F., Conde, A.: J. Appl. Phys. **85**, 6118 (1999)
62. Loffler, J.F., Meier, J.P., Doudin, B., Ansermet, J.P., Wagner, W.: Phys. Rev. B **57**, 2915 (1998)
63. Munoz, M.A., Prietoa, C., Ocala, C., Martíneza, J.L.: Scr. Mater. **43**, 919 (2000)
64. Gheisari, M.: Preparation and investigation of magnetic properties of iron–iron oxide nanocomposites. PhD thesis, Department of Physics, University of Isfahan, Isfahan, Iran (2009)

Nanostructured Materials Use in Sensors: Their Benefits and Drawbacks

Aleksandra Lobnik, Matejka Turel, Špela Korent Urek, and Aljoša Košak

Abstract The development of nanoscale materials for optical chemical sensing applications has emerged as one of the most important research areas of interest over the past decades. In this chapter we firstly present some general aspects of nanostructured materials and give a description on the analytical aspects of sensors and sensing principles. The broad variety of nanomaterials as well as sensors' design made us to limit our presentation, which concentrates on nanomaterials, such as quantum dots, polymer- and sol-gel-based particles. The benefits and drawbacks of the properties of these nanomaterials used in optical sensing applications are given, and the recently developed optical chemical sensors and probes based on photoluminescence are overviewed. Finally, some future trends of the nanomaterial-based optical chemical sensors are given.

1 Introduction to Nanostructured Materials and Sensing Principles

1.1 General Aspects of Nanostructured Materials

Microtechnology and microfabrication technology are key terms which continue to dominate discussions in all branches of sensors research and development. Microfabrication has reached a stage of serious application and is accepted as a good alternative to classical "macroscopic" technologies. It has provided us with the

A. Lobnik (✉), M. Turel, Š. Korent Urek, and A. Košak
Faculty of Mechanical Engineering, University of Maribor, Centre of Sensor Technology, Smetanova ulica 17, 2000, Maribor, Slovenia
e-mail: aleksandra.lobnik@uni-mb.si, matejka.turel@uni-mb.si, spela.korent@uni-mb.si, aljosa.kosak@uni-mb.si

N. Yahya (ed.), *Carbon and Oxide Nanostructures*, Adv Struct Mater 5,
DOI 10.1007/8611_2010_21, © Springer-Verlag Berlin Heidelberg 2010,
Published online: 19 August 2010

means of producing sensors characterized by high sensitivity, small size, enhanced optics and low cost [1, 2].

New developments in solid-state physics, the technological application of quantum effects, material research and optical technology have opened the door to the world of nanoscience that will probably gain importance in all fields of sensor application over the next 10–20 years. All the innovative production, characterization and modification methods suitable for nanotechnology are oriented consistently towards the idea of “engineering on the atomic and molecular level”.

Nanoscience is a field of knowledge of the properties of matter in the nanostate. The subject of nanoscience is investigation of fundamental mechanisms of structure formation, structural organization and transformation at a nanolevel and involves complex interdisciplinary investigations of the physical and chemical properties of nanoscale objects [2–7].

Nanoscience serves as a basis for nanotechnology. The main goal of the latter is to develop economically and environmentally efficient methods for the design of novel nanostructured materials and highly disperse systems, preparation of films and coatings, fabrication of functional nanostructures and elements of nanoelectronic devices that are promising for applications in various fields from the information and telecommunication systems, sensors, optoelectronics and catalysis to medicine and bioengineering.

Nanostructured materials cross the boundary between nanoscience and nanotechnology and link the two areas together, so these definitions are very appropriate. Although nanotechnology is widely talked about, there is little consensus about where the nano-domain begins. It is recognized that the size range that provides the greatest potential and, hence, the greatest interest is that below 100 nm; however, there are still many applications for which larger particles can provide properties of great interest. Therefore, for the purposes of this review chapter, we have arbitrarily taken nanoparticles to be discrete particles that have a diameter of 100 nm or less. To properly understand and appreciate the diversity of nanomaterials, the most typical way of classifying nanomaterials is to identify them according to their dimensions. As shown in Fig. 1, nanomaterials can be classified as zero-dimensional (0-D) (nanoparticles,), one-dimensional (1-D) (nanowires, nanorods, and nanotubes), two-dimensional (2-D), and three-dimensional (3-D). This classification is based on the number of dimensions, which are not confined to the nanoscale range ($<$100 nm) [8–10].

Materials at the nanoscale lie between the quantum effects of atoms and molecules and the bulk properties of materials, where many physical properties of materials are controlled by phenomena that have their critical dimensions at the nanoscale. A change from macro- and micro- to nano-scale qualitatively modifies the most physicochemical properties of materials. When the sizes of materials are reduced in one or more dimensions their physical and chemical properties can change dramatically due to extraordinary increasing of surface-to-volume ratio. As a result, the larger surface area of nanoparticles compared to their volume plays a significant role in dictating these materials’ important properties. These changes affect their optical properties (Stokes’ shift, resonance, etc.), chemical properties

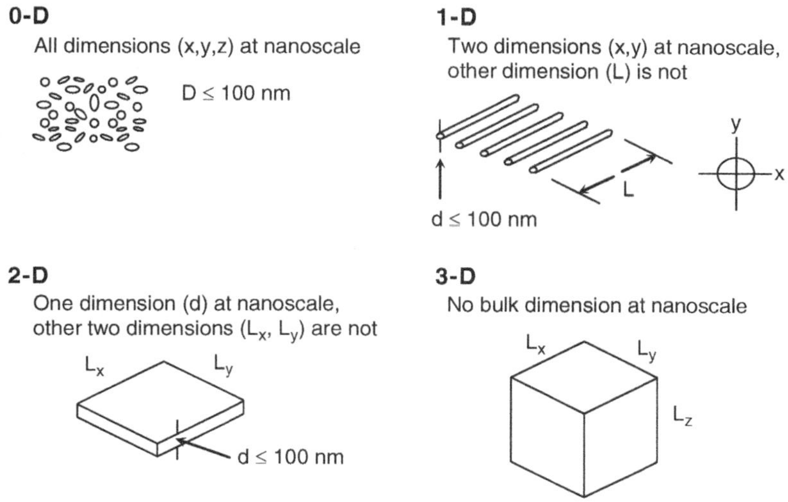

Fig. 1 Classification of nanomaterials according to their dimensions (adapted from [8])

(chemiluminescence, surface functionalisation, etc.) electromagnetic (electronic, magnetic, dielectric, etc.), mechanical (lattice dynamics, mechanical strength, etc.), and thermal properties (Seebeck coefficient, thermal resistance, etc.), causes the resulting compounds and materials to display properties lacking in macro- and micro-scopic objects of the same chemical nature [9–11].

Large surface-to-volume ratio and its effects on material properties is a key feature of nanoscience and nanotechnology. For these reasons, a nanomaterial's shape is of great interest because various shapes will produce distinct surface-to-volume ratios and therefore different properties. Thus, nanoparticles are characterized by a nonmonotonic dependence of the properties on their size, shape and composition. The upper limit of the size of nanomaterials or nanosystems is that whose further increase (or increase in number of atoms in the cluster) does not modify the final properties of the material (there is no further transition of quantitative to qualitative changes) [8, 9, 11]. Qualitatively, the special properties of nanomaterials are both due to the extraordinarily developed surface of their constituting particles and to the electronic and quantum effects exhibited by nanoparticles. Furthermore, in materials with one or more nanoscale dimensions, these properties can be purposefully engineered, enhancing and tailoring the performance of sensors.

1.2 Analytical Aspects of Sensors

In general, the sensor requirements are defined by the specific application, where sensing always involves an interaction between the target and the system that is able to detect it (the sensor). The following features are of particular importance for all types of sensors, and thus, they must be carefully considered in R&D of sensors [12–19]:

- *Sensitivity*: Change in measurement signal per concentration unit of the analyte, i.e. the slope of the calibration graph
- *Selectivity*: Ability to discriminate in detection the target from its close analogs that can be also present in the tested system. It is the basic characteristic of the analytical method that determines the accuracy of results
- *Limit of detection (LOD)*: Lowest concentration of an analyte that the analytical process can reliably detect. The LOD concentration is statistically distinguishable from the blank or background signal
- *Dynamic range*: Concentration range between the detection limit and the upper limiting concentration (the lower limit is determined by the sensor sensitivity and the higher limit appears due to the effects of saturation)
- *Reversibility*: Ability of the sensor to respond dynamically (reversibly) to changes in sample concentration in the course of measurement
- *Linearity*: Relative deviation of an experimentally determined calibration graph from an ideal straight line. Usually values for linearity are specified for a definite concentration range
- *Response time*: Time for a sensor to respond from zero concentration to a step change in concentration. Thus, e.g. the value of t_{95} represents the time necessary to reach 95% of the ful-scale output
- Robustness and reliability
- Small size
- Low cost

In the area of analytical and clinical chemistry, the importance and power of chemical sensors has been recognized for many years. Classical analytical procedures are usually performed by means of sophisticated instrumentation which cannot be easily moved away from laboratory, requiring thus the transport of the sample to the lab. In contrast to such methods, chemical sensors provide a possibility of real-time analysis, which can be accomplished directly in the field, plant, home, or in the hospital. Ideally, such a sensor can be stuck directly into the sample and the result of the measurement is displayed within a couple of seconds. The ultimate power of the ideal chemical sensor is the ability to provide the spatial and temporal distributions of a particular molecular or ionic species in real time [20].

1.2.1 Definition and Classification of Chemical Sensors

According to the definition given by IUPAC Commission on General Aspects of Analytical Chemistry, a chemical sensor *is a device that transforms chemical information, ranging from the concentration of a specific sample component to total composition analysis, into an analytically useful signal. The chemical information may originate from a chemical reaction of the analyte or from a physical property of the system investigated.* According to the Cambridge definition [21], one of the most appropriate definitions is that *Chemical sensors are miniaturized devices which can deliver real-time and on-line information on the presence of specific compounds or ions in even complex samples.*

The usual aim of a chemical sensor is to produce a measurable signal which can be correlated to the concentration of a specific compound present in the immediate environment of the sensor [22]. Typically, a chemical sensor consists of a *chemical recognition* phase (sensing element or receptor) coupled with a *transduction element* (Fig. 2). The receptor identifies a parameter (e.g. concentration of a given compound, pH, etc.) and gives a signal proportional to the magnitude of this parameter. The receptor function is fulfilled in many cases by a thin layer which is able to interact with analyte molecules, catalyse a reaction selectively, or participate in a chemical equilibrium together with the analyte. The transducer translates the signal produced by the receptor into a measurable signal, which is amenable to processing by amplification, filtering, recording, display, etc. The transducer operates according to the physicochemical nature of the signal appearing when the molecular recognition occurs [18, 21–24].

Chemical sensors may be classified according to the operating principle of the *transducer element*, as listed bellow [18].

- *Optical sensors*; they are based on absorbance, reflectance, luminescence, fluorescence, refractive index, optothermal effect and light scattering. Examples of transducers used: fiberoptics connected to optoelectronic systems, photodiode. As an example, a luminescent sensor can be constructed by associating a sensing element, which emits light when in contact with a specific analyte, with a photodiode which converts the energy of the incident light into a measurable signal.
- *Electrochemical sensors*; among them voltametric and potentiometric devices, chemically sensitized field effect transistor (CHEMFET) and potentiometric solid electrolyte gas sensors. Examples of transducers used: amperometric and potentiometric detection-based electrodes and semiconductors.

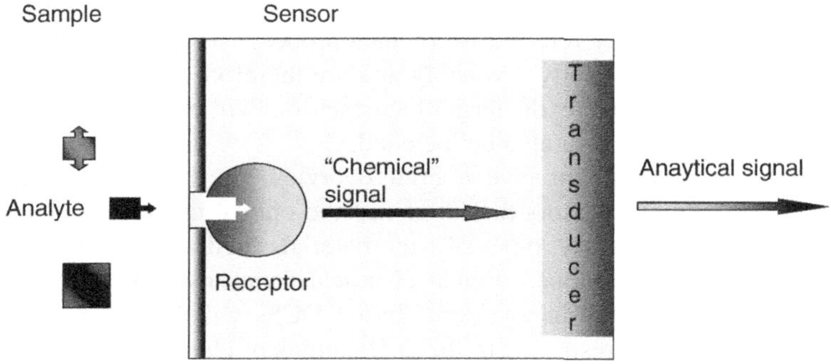

Fig. 2 Schematic representation of the composition and function of a chemical sensor

- *Electrical sensors*; including those with metal oxide and organic semiconductors as well as electrolytic conductivity sensors.
- *Mass sensitive sensors*, i.e. piezoelectric devices and those based on surface acoustic waves.
- *Magnetic sensors*; (mainly for oxygen) based on paramagnetic gas properties.
- *Thermometric sensors*; based on the measurement of the heat effect of a specific chemical reaction or adsorption which involves the analyte. Example of transducer used: thermistor.

Alternative sensor classification schemes follow the receptor principles [18, 25]:

- *Physical*; here no chemical reaction takes place. Typical examples are sensors based on measurement of absorbance, refractive index, conductivity, temperature or mass change.
- *Chemical*; in which a chemical reaction with participation of the analyte gives rise to the analytical signal.
- *Biochemical*, in which a biochemical process is the source of the analytical signal. The biochemical principle may be regarded as a subgroup of the chemical one.

Sensors having a receptor part based on a biochemical principle are usually called *biosensors*. Selectivity and sensitivity provided by nature have been utilized in such sensors, frequently by immobilizing the biologically active compounds, such as enzymes and immunoglobulins, within a receptor part of the sensor. The effective way of obtaining the biological selectivity is the combination of cell cultures, tissue slices, organs and sometimes of whole living organisms with the transducer.

As it can be seen, many different types of chemical sensors are possible to create, and the consideration of all of them would make this chapter too long. We will therefore limit our contribution to optical chemical sensors (OCSs), which are among the most important types of sensors which have recently been utilizing the advantages of a variety of nanomaterials for the continuous, real-time monitoring of diverse analytes. Furthermore, the topics covering optical biosensors, sensors based on surface plasmon resonance, dendrimers, and those applying optical sensing molecular imprinting principles will not be included. It is also important to note, that many devices described in the literature as nanosensors act virtually irreversible and would be better referred to as 'nanoprobes'. 'True nanosensors' are in fact devices that respond fully reversibly and are therefore capable of continuous monitoring (=sensing). Here, the term of nanosensors will be used less strictly and examples of nanoprobes will also be included.

The following pages will give a brief overview on the basic principles of OCSs and on the major factors that influence the optical chemical (nano)sensors characteristics. Subsequently, types of nanomaterials used in sensors and applications of the recently developed optical chemical nanosensors will be presented. Future perspectives of the nanomaterial-based OCSs will be finally given. The extent of this chapter is restricted to the applications of chemical nanosensors and nanoprobes based on optical principle, such as photoluminescence (PL), and to the sensing nanomaterials (nanoparticles) that have their dimensions $\leq$100 nm. The term photoluminescence (also referred to as luminescence, fluorescence or phosphorescence) is related to the detected signal changes caused by the electromagnetic radiation (EM).

1.2.2 Opto-Chemical Sensor (opt(R)odes)

Optical sensors, or *optrodes*, represent a group of chemical sensors in which electromagnetic radiation is used to generate the analytical signal in a transduction element. The optical sensor system itself comprises the active part, the light source and the read-out device. These sensors can be based on various optical principles (absorbance, reflectance, luminescence, fluorescence), covering different regions of the spectra (UV, Visible, IR, NIR) and allowing the measurement not only of the intensity of light, but also of other related properties, such as lifetime, refractive index, scattering, diffraction and polarization [26].

The optical sensing techniques, in comparison to electrical methods, have some advantages, such as selectivity, immunity to electro-magnetic interference, and safety during work with flammable and explosive compounds. They are also sensitive, inexpensive, non-destructive, and have wide capabilities. Optrodes do not require a reference cell as in potentiometry. They can easily be miniaturized and allow multiple analysis with a single control instrument at a central site [27].

Besides a number of advantages, optical sensors also exhibit disadvantages: the ambient light can interfere, the long-term stability is limited due to indicator leaching or photobleaching, there may be a limited dynamic range, selectivity may be poor, and a mass transfer of the analyte from the sample into indicator phase is necessary in order to obtain an analytical signal [28].

Fiber-Optic Chemical Sensors

Fiber-optic chemical sensors (FOCSs) represent a subclass of chemical sensors in which an optical fiber is commonly employed to transmit the electromagnetic radiation to and from a sensing region that is in direct contact with the sample. The spectroscopically detectable optical property can be measured through the fiber optic arrangement, which enables *remote sensing*. In addition to advantages in terms of cheapness, ease of miniaturization, obtaining safe, small, lightweight, compact and inexpensive sensing systems, a wide variety of sensor designs are made possible [26–28].

The most common classification of FOCs distinguishes between *intrinsic* and *extrinsic* type of sensors [28–30].

- In *intrinsic* type of FOCs, the sensing principle is based on the change in light-transmission characteristics due to the change occurred in a fiber property (e.g., refractive index or length) upon the interaction with the analyte or the system being studied. The optical fibre itself has sensory characteristics. This type of sensor is mainly applied to measure physical or physicochemical parameters, such as pressure, temperature, or enthalpy of reactions.

- In *extrinsic* type of FOCs, the optical fiber acts as a transporting media by means of guiding the radiation from the source to the sample or from the sample to the detection system. Extrinsic sensors can be subdivided into (a) distal and (b) lateral types. The most common are distal-type sensors, in which the indicator is immobilized at the distal end (tip) of the optical fibre. Alternatively, in lateral sensor, the sensing chemistry can be immobilized along a section of the core of the optical fibre to make an evanescent field sensor.

Optical Detection Principles

The most commonly applied methods in optical sensing are those based on light absorption or light emission. However, compared to absorption based methods, molecular emission (fluorescence and phosphorescence, generally speaking luminescence) is particularly important because of its extreme sensitivity and good specificity. The sensitivity of luminescence methods is by about 1,000 times greater than that of most spectrophotometric methods; as well, lower LODs for the desired analytes can be achieved [31]. Therefore, the photoluminescence based techniques will be briefly described in this section, as these are mostly applied also in the recently published nanomaterial-based OCSs and because luminescence offers a variety of techniques to be used. The most important among them are the steady-state measurements (emission intensity measurements), decay time (lifetime)-based measurements, ratiometric measurements and fluorescence resonance energy transfer (FRET).

Measuring the emission intensity is most popular because the instrumentation needed is very simple and low cost. However, measuring the light emission intensity has some disadvantages compared to emission lifetime measurements, in which the sample is excited only by a pulse of EM rather than via continuous illumination which is the case of intensity-based methods. Precision and accuracy of luminescence intensity-based schemes are greatly affected by fluctuations of the light-source's intensity, detector sensitivity, inner filter effects, indicator concentration (bleaching and leaching), sample turbidity, and sensing layer thickness. Some of these problems can be minimized or even overcome by measuring luminescence lifetimes instead of intensities. However, the drawbacks of lifetime measurements are the complexity and high costs, typically associated with instrumentation for lifetime measurements, along with a limited number of indicator dyes available that show significant analyte-dependent changes in lifetime [32–35].

Another way to reduce the problems associated with intensity as well as with lifetime detection principles is the use of ratiometric measurements. The technique employs dual emission or dual excitation indicators or mixtures of two luminophores, exhibiting separated spectral areas with different behaviour. For example, the ratio of two fluorescent peaks is used instead of the absolute intensity of one peak. The sensors therefore typically contain a reference dye; the advantage of this approach is that factors, such as excitation source fluctuations and sensor concentration will not affect the ratio between the fluorescence intensities of the indicator and reference dye [36–38].

Another important process that occurs in the excited state is the Förster or fluorescence resonance energy transfer (FRET). This process occurs whenever the emission spectrum of one fluorophore, which is the donor, overlaps with the absorption spectrum of another molecule, which is the acceptor. The acceptor must absorb the energy at the emission wavelength(s) of the donor, but does not necessarily have to remit the energy fluorescently itself. The transfer of energy leads to a reduction in the donor's fluorescence intensity and excited state lifetime, and an increase in the acceptor's emission intensity. The rate of energy transfer from donor to acceptor is highly dependent on many factors, such as the extent of spectral overlap, the relative orientation on the transition dipoles, and, most importantly, the distance between the donor and acceptor. Due to its sensitivity to distance, FRET has been used to investigate molecular interactions [39].

1.2.3 Key Factors Influencing the Optical Sensor Characteristics

An optical detection system may be based either on (a) direct sensing or (b) indicator-mediated sensing. In a direct optical sensor, the analyte is detected directly via some intrinsic optical property such as, for example, absorption or luminescence. In an indicator-mediated system, a change in the optical response of an intermediate agent, usually an analyte-sensitive dye molecule (indicator) is used to monitor analyte concentration [21, 40]. On the principle of immobilized indicators relies a large group of optical chemical sensors, because the measuring analytes mostly have no intrinsic optical property or this property is not convenient for their detection.

The active component of an optical sensor is a thin polymer film or membrane in which the sensitive indicator is immobilized. This "smart" material responds to the species of interest by altering its optical properties [28, 41]. For example, pH is measured optically by immobilizing a pH indicator on a solid support and observing changes in the absorption or fluorescence of the indicator as the pH of the sample varies with time [26].

Indicators

The basic principle of the indicator chemistry (immobilized in or on the polymer matrix) in an optical chemical sensor is in transforming the measuring concentration of the analyte into a measurable analytical signal. The analyte concentration is measured indirectly, through the alteration of the indicator optical properties.

Various types of indicators are used in optical chemical sensing, such as colorimetric (based on light absorption [25]) and luminescent (based on light emission,

[42–47]) indicators. However, the latter are of primary importance due to their high interdisciplinarity, great sensitivity, and applicability to different detection principles. A great number of fluorescent synthetic organic products is nowadays available so that a researcher can easily select the proper dye (indicator) corresponding to a particular sensing application in terms of spectroscopic properties and chemical reactivity. On the other hand, also the basic organic chemistry offers a great potential as it enables synthesizing tailor-made indicators for specific applications [47, 48].

Immobilization Techniques

A method for indicator immobilization into a suitable polymer matrix has also an important influence on the sensing characteristics. The following possibilities are usually applied:

- *Impregnation* – the indicator is immobilized in the polymer matrix through physical adsorption, chemisorption or electrostatic bonding. The polymer thin film is dipped into a saturated indicator solution and the solvent is then let to evaporate [49].
- *Covalent bonding* – the indicator is covalently bonded to polymer matrix. This may be achieved by (a) choosing the indicator that contains a functional group for covalent bonding to polymer, which is at the same time insensitive for the target analyte, or (b) polymerizing the indicator to certain monomers to form a copolymer [50].
- *Doping* – the indicator is entrapped in the matrix during the polymerization process, where the indicator is simply added to the starting polymer solution [51].

Covalent immobilization enables the sensor having good stability (no leaching, crystallization and evaporation of components) and longer operational lifetime. The disadvantage is that the covalent bonding often lowers the sensitivity for the analyte and prolongs the response time of the sensor [52]. Although the impregnation technique is widely used and low-priced, it is used first of all for test strips and in gas sensors due to its low stability (indicator leaching). Doping is one of the most used immobilizations as it is not restricted to certain indicators and polymers. The sensor stability (in terms of indicator leaching) is better compared to impregnation and worse compared to covalent bonding. The response time is better than in covalent immobilization.

Polymers

Polymer chemistry is an extremely important part of the optical sensor technology. Both the light guide (including its cladding and coating) and the sensing chemistry of indicator-mediated sensors are made from organic or inorganic polymer [40, 41]. The choice of polymer is governed by the permeability of the polymer for the analyte, its stability and availability, its suitability for dye immobilization, its compatibility with other materials used in the fabrication of optrodes, and its compatibility with the sample to be investigated. The polymer microenvironment has a strong effect on the spectral properties of the immobilized indicator, pK_a value, luminescence lifetime, binding constant, etc. [53, 54]. Consequently, the choice of polymer material has a pronounced influence on the sensor performance and its characteristics, such as selectivity, sensitivity, working range, calibration, response time, (photo)stability [55, 56]. The response time, for example, will be governed by the diffusion coefficients of gases or liquids, and the quenching efficiency by the solubility of the gas in the polymer.

However, although most authors have compiled a considerable amount of data on various polymers, numerous new materials are available for which no data are known. It is also known that copolymers and polymer mixtures do not necessarily display the properties that may be expected from averaging the data of the pure components. On the other hand, nano-sized (polymer) materials pose new technological and analytical challenges in many different sensor designs to improve industrial process monitoring (air and water quality), food quality surveillance, and medical diagnostics, and to reliable, real-time detection of chemical, biological, radiological and nuclear hazards for military and anti-terrorism applications – all this by enabling improved sensor characteristics, such as sensitivity, selectivity and response time, along with dramatically reduced size, weight and power requirements of the resulting monitoring devices compared to conventional, macroscaled alternatives.

1.2.4 Effect of Nanodimensions on Sensor Characteristics

The sensor characteristics can be tuned not only by the choice of the indicator and polymeric support but also by merely reducing the size (<100 nm). This is because materials that are smaller than the characteristic lengths associated with the specific phenomena often display new chemistry and new physics that lead to new properties that depend on size. Perhaps one of the most intuitive effects is due to the change in the surface-to-volume ratio. When the size of the structure is decreased, this ratio increases considerably and the surface phenomena predominate over the chemistry and physics in the bulk. Therefore the sensor characteristics, such as sensitivity [57] and response time [58] can be dramatically improved.

Literature [59, 60] defines an optical nanosensor as a device with dimensions smaller than 1 μm that is capable to detect chemical or biological parameters by optically transforming the information into an analytically useful signal. However, the nanotechnology deals with the study and application of structures of matter with at least one dimension of the order of less than 100 nm [61]. Here we will use the term nanosensors strictly for less than 100 nm sized particles that are used in sensing applications. This is also due to the fact that beads bigger than 100 nm behave similarly to bulk film optrodes. In contrast, sensing beads smaller than 100 nm show improved sensing characteristics [62].

The nanosensors consist of an inert, for intracellular measurements biofriendly matrix in which a sensing component and/or an optical reporter are entrapped. The transduction method of choice is currently fluorescence because of the high sensitivity and relative ease of measurement. Most of the nanosensors reported so far have used fluorophores that are sensitive and selective to the chosen analyte as the sensing component. More recently there have been signs that more complex sensing schemes are being developed, these utilize the capacity of the sensor matrix to accept encapsulation of more than one component, thus allowing a synergistic approach to be employed [60]. Nanoparticles containing indicator dyes can either be used directly as nanosensors or as components of optical sensor materials. Most of nanosensors known as PEBBLEs (Probes encapsulated by biologically localized embedding) have been designed for making quantitative measurements in the intracellular environment as they are small enough to be inserted into living cells with a minimum of physical perturbation [37, 63]. PEBBLES have many advantages over widely used fluorescence dye based methods, such as: (a) increased number of analytes that can be measured because nanosensors are not limited to using a single fluorophore and can utilize cooperative interactions between ionophores, enzymes, reporter dyes etc., (b) the matrix protects the intracellular environment from any potentially toxic effects of the sensing dye, (c) the matrix protects the sensing dye from potential interferences in the cellular environment, e.g., non-specific binding proteins and organelles, (d) no selective sequestration of nanosensors into cellular compartments or leaking from, or being pumped out of, cells, (e) enhanced ability to carry out ratiometric measurements, (f) the in-vitro calibration of nanosensors is valid for in-vivo measurements [60]. When the dyed nanoparticles are used as components of optical sensor, multi-analyte sensing becomes possible since several types of beads (e.g. pH-sensitive and oxygen-sensitive) can be incorporated into one polymer.

Nanosensors pose advantages, such as improved sensitivity, response time and ability to perform in-vivo measurement. However, the dark side of using "free" nanosensors for in-vivo measurements needs to be considered. Nanoparticles are very different from their everyday counterparts, so their adverse effects cannot be derived from the known toxicity of the macro-sized material. The prime concern is the retention of these particles in the body and the harmful effect in the long run [64] since nanoparticles can be responsible for a number of material interactions that could lead to toxicological effects [65].

By being able to fabricate and control the size, morphology and surface characteristics of nanoparticles, it could be possible to influence the resulting properties and, ultimately, design materials to give desired properties. The optical properties that can be controlled at nanoscale are of great interest in the field of optical sensor designing [66]. Some optical chemical nanosensors rely on quantum dots [67], metal beads [11] and other materials; however, most of the them make use of indicators embedded in polymer beads [68–75] and sol-gels [38, 76–81]. Figure 3 gives an overview of common nanosensor concepts [59].

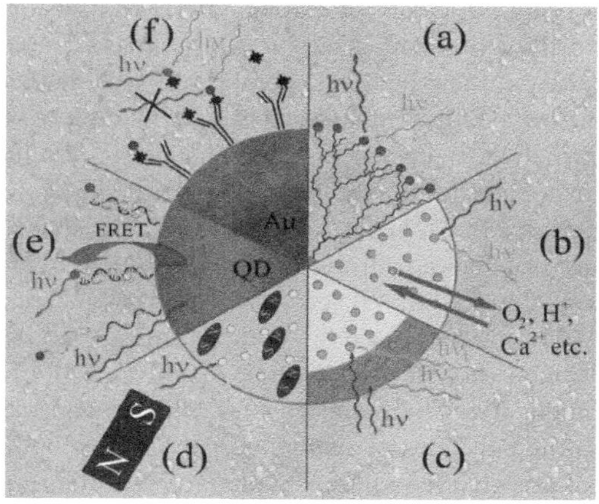

Fig. 3 Schematic representation of optical nanosensors: (**a**) macromolecular nanosensors; (**b**) NSs based on polymer materials and sol-gels; (**c**) multi-functional core-shell systems; (**d**) multi-functional magnetic beads; (**e**) NSs based on quantum dots; (**f**) NSs based on metal beads (Adapted from [59])

2 Nanomaterials Used in Sensors

In optical chemical sensing it is desirable to employ appropriate nanomaterials to achieve highly sensitive and predictable responses over wide analyte concentration ranges within the shortest possible time. A fairly broad spectrum of nanomaterials is used in sensors. Some recent publications [1–11] suggest that chemical sensors utilized nanomaterials with different structures, which can be classified as follows:

- Nanoparticles, nanoclusters, nanocrystals used in optical sensors (biochemical, immune sensors, electrochemical sensors) (0D)
- Nanotubes, nanorods, nanotapes, and nanowires used mainly in electrochemical and in optical biochemical sensors (1D)
- Nanosized assembled film structures (Langmuir–Blodgett films and self-assembled mono- and multi- layers) used mainly in optical sensors (2D)

Most frequently nanosensors utilize so called (0D) nanomaterials, i.e., nanoparticles, nanocrystals, nanoclusters, and quantum dots. These are ensembles of several hundreds or thousands of single-charged atoms or molecules, several nanometers in size, with discrete energy levels. Such nanoparticles are smaller than the de Broglie wave of electron, which renders them capable of intensively absorbing the electromagnetic radiation in the visible or near UV regions.

Nanosensors utilize nanoparticles with different chemical nature and physical properties [11]:

- Semiconducting materials (CdS, CdSe, CdTe, ZnS, ZnSe, ZnO, PbS, PbTe, etc.)
- Semiconducting materials doped with lanthanides (Eu^{3+}, Sm^{3+}, Tb^{3+}, Gd^{3+}, etc.)
- Noble metals (Au, Ag)
- Magnetic materials (Fe_3O_4, Fe_3S_4, γ-Fe_2O_3, $MO{\cdot}Fe_2O_3$, where $M = Ni^{2+}$, Co^{2+}, Zn^{2+}, Mn^{2+}, Mg^{2+})
- Polymer- and silicon-based particles doped with dyes or containing surface-grafted functional groups

High diversity of nanomaterials and sensor types differing in the operation and analytical signal generation principles, coupled with a broad spectrum of media analyzed and analytes, complicate systematization of the published data on nanosensors. Different authors classify nanosensors based on nanomaterial geometry, in which case the sensing elements are represented by tubes, rods, tapes, wires [82], nanoparticles, nanotubes, and porous silicon [61, 83, 84], nanosized films [85, 86], spheres, rods, cubes, or prisms [87, 88], etc. Other authors emphasize the application of nanomaterials in analysis of environmental objects [61] with the use of optical [86, 88, 89] sensors. Some authors concentrate on only one specific type of nanomaterials used in optical sensors, i.e. quantum dots [89] or dye-modified silicon dioxide nanoparticles [90]. One brief review cannot cover all types and applications of nanosensors, which makes us to restrict ourselves to selected principal types of nanomaterials used in optical chemical sensors. The emphasis of this chapter will be devoted to properties and applications of recently developed optical chemical nanosensors based on semiconducting materials (quantum dots) and sensors based on polymers and sol-gel materials, all operating on the photoluminescence principle.

2.1 Semiconducting Quantum Dots

Semiconducting nanomaterials have attracted significant attention in research and applications in area of optical sensing during the last decade. The reason for this interest is based on the very high surface-to-volume ratio that could significantly change optical properties of semiconductor nanomaterials. The recent results obtained in studies of the optical and other physico-chemical properties were difficult to interpret for long owing to polydispersity, presence of surface defects, and a low degree of crystallinity of the nanoparticles synthesized [2–4, 9, 11].

Inorganic semiconductors include the periodic group (IV) elements silicon and germanium; compounds such as GaN, GaP, GaAs, InP, and InAs (collectively the III–V materials); and ZnO, ZnS, ZnSe, CdS, CdSe, and CdTe (II–VI materials). Solid solutions of many of these semiconductors can be made, and the band gap of the resulting solid solutions is intermediate between the two end-members; thus GaP has a room-temperature band gap of 2.3 eV ($\lambda \sim 540$ nm), GaAs has a room-temperature band gap of 1.4 eV ($\lambda \sim 890$ nm), and GaP_xAs_{1-x} has a band gap energy that depends nearly linearly on phosphor content (x) (10) [91].

By binding more and more atoms together, the discrete energy levels of the atomic orbitals merge into energy bands for a semiconducting material. Therefore semiconducting material can be regarded as a hybrid between small molecules and bulk material (Fig. 4) [92].

In the case of semiconductor bulk materials, an electron may be excited from the valence band to the unfilled conduction band if an incident photon has energy greater than the band gap of the material. Under these conditions, the photon is absorbed while a hole is left in the valence band when the electron jumps to the

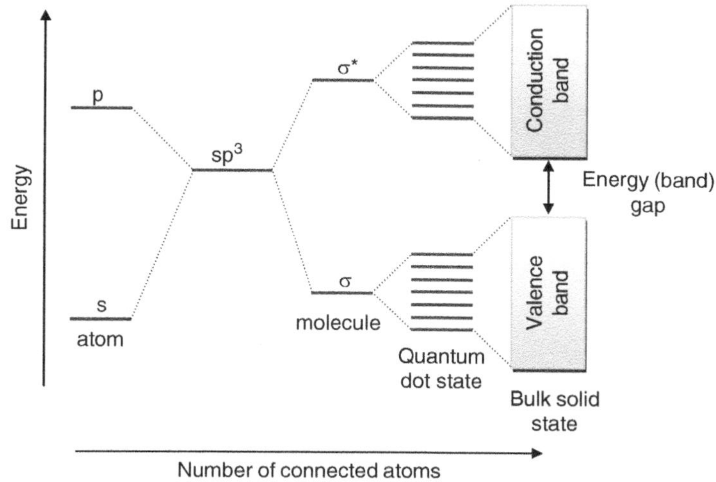

Fig. 4 Electronic energy levels depending on the number of bound atoms (Adapted from [92])

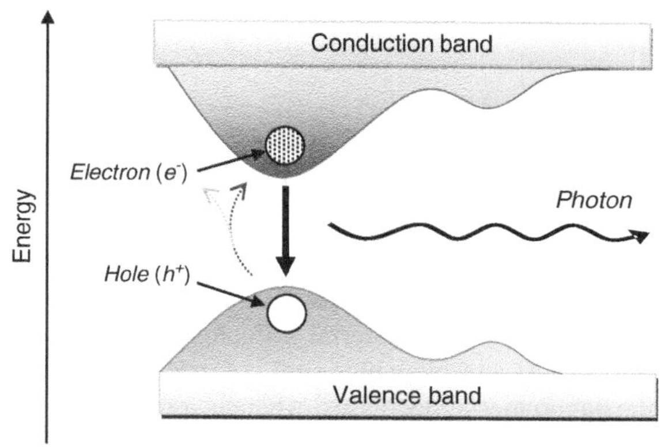

Fig. 5 Emission of a photon upon recombination of an electron-hole pair

conduction band. Inversely, if an electron in the conduction band returns to the valence band and recombines with a hole, a photon is released with energy equal to the band gap of the semiconductor (Fig. 5).

However, at low temperatures, bulk semiconductors often show optical absorption just below the energy gap. This process is associated with the formation of an electron and hole bound to each other, which is called an exciton. The exciton has mobility and thus can move freely through the material.

In general, the effects of nanoscale on optical absorption are associated with the density of states in the valence and conduction bands, the quantized energy levels of the nanostructure, and the influence of excitonic effects. As shown in Table 1, the exciton radius has nanoscale dimensions [92].

Table 1 Exciton Bohr diameters and band gap energies for various semiconductors

Material	Exciton (nm)	Band gap energy (eV)
CuCl	1.3	3.4
CdS	8.4	2.58
CdSe	10.6	1.74
GaAs	28	1.43

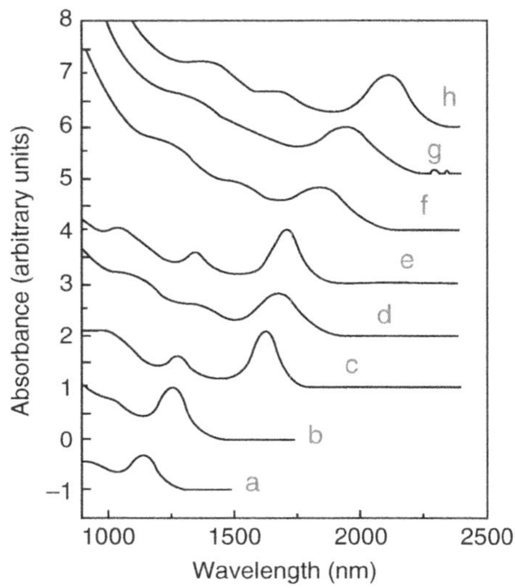

Fig. 6 Room temperature optical absorption spectra of PbSe nanocrystals with diameters (**a**) 3 nm, (**b**) 3.5 nm, (**c**) 4.5 nm, (**d**) 5 nm, (**e**) 5.5 nm, (**f**) 7 nm, (**g**) 8 nm, and (**h**) 9 nm (Adapted from [92])

As the dimension of semiconductor nanomaterial changes from 3D to 0D (the size of nanoparticles decreases) the density of states becomes more quantized and the band gap of the material shifts toward higher energies and shorter wavelengths [93, 94]. As a result, a blue shift is expected in the absorption spectrum as the size of the nanomaterial decreases, whereas a red shift occurs for an increase in size.

This effect is visible in Fig. 6, which shows the absorbance spectrum of PbSe nanocrystals as a function of their size. The highest energy (shortest wavelength) absorption region, called the absorption edge, is shifted toward the blue as the nanoparticle size is reduced.

Optical emission may also occur if the electron and the hole recombine, leading to the generation of a photon. If the photon energy is within 1.8–3.1 eV, the emitted light is in the visible range. Such special spectroscopic properties of semiconducting nanocrystals, appear due to an effect called »*quantum confinement*« that emerges when the particle size is smaller than the so called Bohr exciton radius. Because of the quantum confinement, the emission of visible light can be tuned by varying the nanoscale dimensions. The typical trend is the shift of the emission peak toward shorter wavelengths (blue shift) as the size of the nanoparticles decreases [92]. This phenomenon is clearly visible in Fig. 7, which shows the photoluminescence spectra for CdSe of different particle sizes [95, 96].

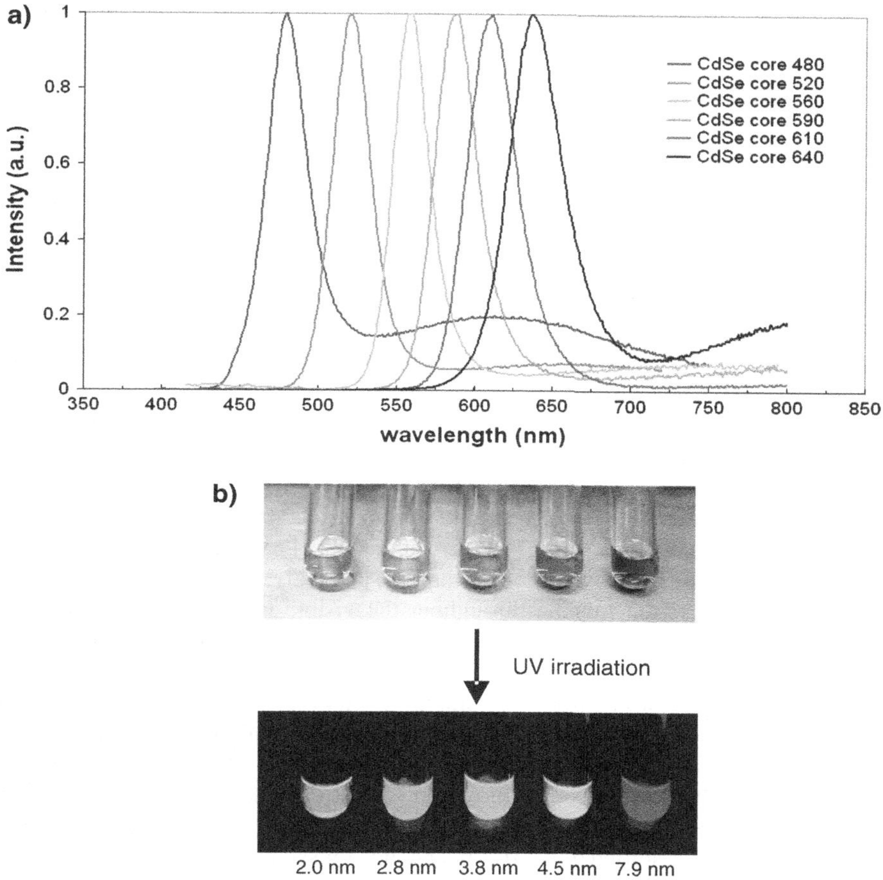

Fig. 7 (**a**) Emission spectra of CdSe quantum dots with different sizes
Source: http://www.nanotechnologies.com, and (**b**) CdSe quantum dots of different sizes in solution emitting light at different wavelengths (Adapted from [96])

At a sufficiently small size (bellow 10 nm) of semiconducting nanocrystallites, so called "quantum-dots", the density of states of conduction electrons can take only particular discrete values, which is responsible for the quantum size effect [94]. As a result, quantum dots possess unique optical properties not observed for corresponding bulk material [97–99]. The quantum-dots (QDs) (Table 2) are luminescent nanocrystallites emerging as a new class of fluorescent reporters with properties and applications that are not available with traditional organic dyes or nanocomposite structures doped with them [47, 100]. Their novel properties have opened new possibilities in many areas including ultrasensitive chemical analysis and cellular imaging. Thus, the major advantages of these materials over organic dyes include high photoresistance, possible directed control of the fluorescence

Table 2 Nanocrystalline semiconducting materials (quantum dots) and range of tunability

QD material	Fluorescence-emission range (nm)	QD-diamater range (nm)
Zinc sulfide (ZnS)	300–410	–
Zinc selenide (ZnSe)	370–430	
Cadmium sulfide (CdS)	355–490	1.9–6.7
Cadmium telluride (CdTe)	620–710	
Lead sulfide (PbS)	700–950	2.3–9.0
Lead selenide (PbSe)	1,200–2,340	
Lead tellurite (PbTe)	1,800–2,500	

wavelength, a small (15–40 nm) half-width of the fluorescence spectra, and a high quantum yield which affords extensive application of quantum dots in optical sensors. So these nanoparticles absorb light and emit fluorescence without the involvement of electronic transitions, which is a completely different way from that of traditional dyes [47, 91, 101–103].

The most popular choices of quantum dot materials for chemists are CdSe and CdS. Their bulk band gaps are 1.7 eV for CdSe and 2.4 eV for CdS, corresponding to absorption at $\sim$720 nm and $\sim$520 nm, respectively, which means that their absorption energies are tunable throughout the visible region [102]. Along with changes in the particle size, the shape of the fluorescence spectrum, i.e. the change in the color of the emitted light, is also affected. For example, 2.5 nm and 7 nm CdSe particles exhibit green and red fluorescence [47, 102, 104], respectively. In the case of CdS, is also valid that absorption bands are shifted as the particle size varies [105]. As the particle size decreases, the absorption edge is structurized and shifted toward the short wavelength region so that the CdS particles smaller than 2.2 nm become colorless. However, at a particle size of less than 1 nm the absorption peak is shifted to 280 nm and no emission is observed [106–108]. Particle size variation also affects the lifetime of luminescence [109].

Figure 8 shows (a) the transmission electron microscopy (TEM) image [110] and (b) the high resolution transmission electron microscopy (HRTEM) of monodisperse colloidal CdSe nanocrystals synthesized under kinetic size control.

In optical chemical sensing applications QDs have been utilized in a variety of approaches, including measurements based on emission intensity or on shifts of emission maxima, for detecting different parameters, mostly heavy metal ions. Probes, rather than sensors were presented, in which the quantum dots used were either functionalized or nonfunctionalized [111–121].

2.1.1 Semiconducting "Core-Shell" Systems

Knowledge of the surface chemistry of semiconducting nanocrystallites (quantum dots) is needed to understand their optical properties and to manipulate them chemically for a desired application. For sensing applications, it is highly desired

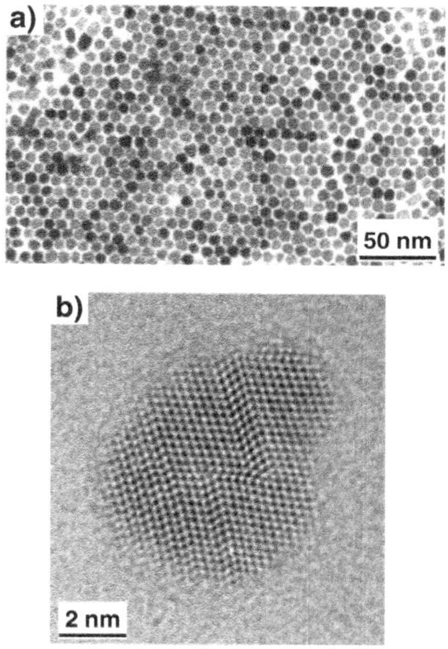

Fig. 8 (**a**) Transmission electron microscopy (TEM) image of monodisperse colloidal CdSe nanocrystals [110], (**b**) High resolution transmission electron microscopy (HRTEM) image of CdSe nanocrystal
Source: http://www.cnano-rhone-alpes.org

to be able to control and alter the properties and functionalities of semiconducting nanomaterials with greater flexibility and possibility.

Uncoated semiconducting nanocrystallites usually exhibit hydrophobic surface effects and are therefore insoluble in water and in other highly polar solvents. However, the prerequisite for the development of semiconducting nanocrystallites for effective sensing in different media and with better photophysical properties is to gain access to photostable, compatible and water-soluble nanocrystals. Thus, it is accepted that control of the composition and structure of the surface of nanocrystallites plays the key role in fabrication of nanomaterials with desired optical properties.

One approach to control and alter the properties and functionalities of semiconducting nanocrystallites is to use the so-called "core-shell" structures, which contain more than one single component and have properties different from those of single component nanomaterials [64].

In such a multiple component material all nanoparticles, even ultrasmall ones, are comprised of at least two fragments, the core and the shell. The cores may be any kind of colloidal particles, i.e. polymers, metals, insulators, and all classes of semiconductors. Likewise, the shells may consist of any sort of materials including inorganic and organic ones. In the latter case, the semiconducting nanocrystallites could be capped with some appropriate organic functional ligands used as stabilizing agents including mercaptopropionic acid, L-cysteine, thioglycerol, etc. [116, 122, 123] or polymers (polystyrene, polyvinyl alcohol, polyethylene glycol, polyacrylamid, etc.). Such coatings usually protect the core from the action of external factors and can significantly influence the physical properties of the material. They can alter the charge, functionality, and reactivity of the semiconductor nanocrystallite surfaces, their thermal and chemical stability, improve solubility, make them less cytotoxic and allow conjugation of other molecules to these particles [64]. Besides, the shell can also provide a steric barrier to prevent the nanoparticles from agglomerating, enhances excitation and emission intensities, improves selectivity and gives the desired surface functionality [47, 113, 124–126].

Semiconductor nanocrystallites having an inorganic, i.e. semiconducting, shell (ZnS, CdS, PbS, etc.) are generally complex [127]. In these systems the influence of the nanocrystallite surface on photoluminescence can be understood in terms of "trap" states, caused by main structural defects of the surface (bond angle and bond length distortions, nonstoichiometry of composition, etc.) and impurities (atoms of the environment). For bulk semiconductors, surface coating is a well-known phenomenon that decreases the possibility of charge carriers residing in traps. For semiconducting nanocrystallites, i.e. quantum dots, surface coating has most frequently been achieved by higher band-gap semiconductors, to obtain "quantum-shell" [47, 103].

In the majority of these cases, the electronic structure of the "core-shell" particles is as follows. The core material having a certain band-gap is capped by a material with a larger band-gap, the conduction band energy of the capping material being higher (more negative) than that of the core material and the valence band energy of the capping material being lower (more positive on the electrochemical energy scale) than that of the core material. The main consequence of this capping is that the exciton photogenerated in the core is prevented from spreading over the entire particle, and by this it is forced to recombine while spatially confined to the core. In most cases this is accompanied by enhanced luminescence [128].

While the absorption spectrum of semiconducting "core-shell" tend to be close to the simple sum of the spectra of the two individual components, at least when their electronic interaction is not strong, their photoluminescence properties often change significantly, especially in terms of intensity, either quenched or enhanced depending on interactions between the two components. When one component is substantially dominant in size or weight over the other, the optical properties of such composite system tend to be primarily determined by the predominant component.

Many semiconducting quantum dots are the composites made of a narrow dispersed highly crystalline CdSe core overcoated with a shell of a few atomic layers of a material with a larger band gap, such as ZnS and CdS, on top of the nanocrystal core. The typical so-called solid–solid "core-shell" structures are formed (Fig. 9) [95]. Such constructions are beneficial because the surface defects in the crystal structure act as temporary "traps" for the electron or hole, preventing their radiative recombination and thus reducing the quantum yield. The luminescence of the surface coated material becomes relatively insensitive to its local environment, and quantum dots can therefore be used as large, inorganic analogues of fluorescent dyes. With the proper shell design, it becomes possible to obtain photoluminescence quantum yields close to 90% and to increase photostability by several orders of magnitude relative to conventional dyes [47].

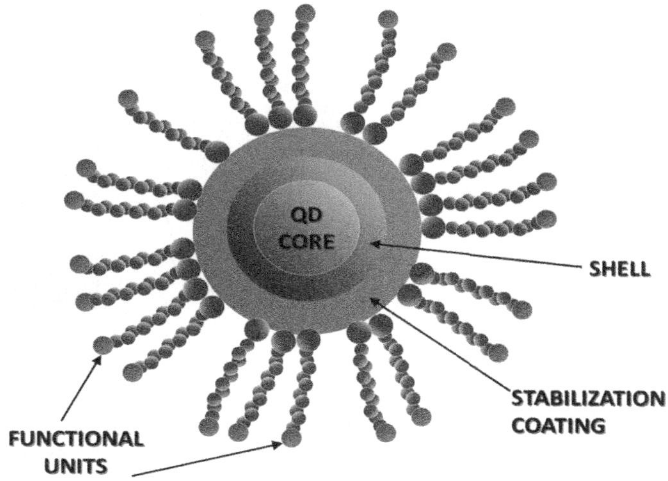

Fig. 9 Overall structure of a quantum dot nanocrystal conjugate. The layers represent the distinct structural elements

The "core-shell" approach may significantly change optical, chemicals and photocatalytic properties of semiconductor nanocrystallites, i.e. quantum dots [127, 129], which in general lead to the following effects:

- It may enhance their excitonic state and defect emission by blocking nonradiative electron/hole (e^-/h^+) recombination defect sites (traps) on the surface of the semiconductor ("activation" of fluorescence) [130–132].
- It may create new traps on the surface of the nanoparticle leading to the apperance of new emission bands [133, 134].
- It may enhance the photostability of semiconductor nanoparticles [130]: the photoresistance of the crystal core is increased and this prevents surface quenching of excitons and aggregation of the articles, thereby additionally increasing the quantum yield of fluorescence.
- It may enhance the selectivity and efficiency of light-induced reactions occurring on the surface of semiconductor nanoparticles [127, 135, 136] and it may lead to the formation of a layer of another semiconductor on the semiconductor nanoparticle (i.e., CdS/HgS and HgS/CdS [134], CdS/ZnS [137], ZnSe/CdSe [128], PbS/CdS [138, 139], TiO_2/SnO_2 and SnO_2/TiO_2 [140], CdS/CdSe and CdSe/CdS [141], etc.) and in some cases to the formation of a three-layered structure-quantum dot quantum well like CdS/HgS/CdS [142, 143].

Many other examples of semiconducting "core-shell" systems include various materials in which the semiconducting crystal core of CdSe or CdTe is capped with a thin layer of the other semiconductor having a broader energy gap, e.g. CdS, ZnS, or ZnTe. This challenge has been utilized in a wide range of combinations and systems, like CdS/Ag_2S, CdS/ZnS, AgI/Ag_2S, ZnS/CdSe, ZnSe/CdSe, CdS/HgS,

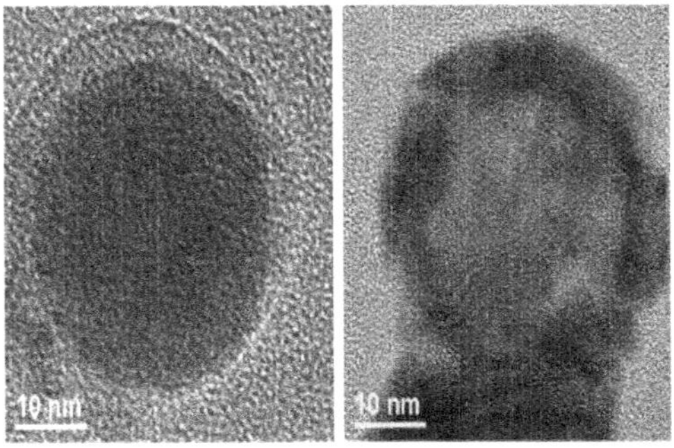

Fig. 10 Transmission electron microscopy (TEM) image of CdSe/ZnS (*left*) and ZnS/CdSe (*right*) core-shell nanocrystals
Source: http://www.ph.utexas.edu/~laser/projects/shelled.html

CdS/PbS, HgS/CdS, CdSe/ZnSe, CdSe/ZnS, CdSe/CdS, CdS/TiO_2, CdS/Au, TiO_2/Pt, TiO_2/Au, TiO_2/CdSe, CdSe/SiO_2, CdTe/SiO_2, and Au/SiO_2. Many of these core-shell materials have been already reviewed [104, 129, 144–158].

Using a transmission electron microscopy (TEM) the spherical ZnS coated CdSe-cores and CdSe coated ZnS-cores can be observed (Fig. 10).

2.2 Polymers and Sol-Gel Materials

2.2.1 Polymers

Plastic materials are quite different from particles built up from metal sulfides so-called quantum dots, other semiconductor materials, metal nanoparticles (mainly gold) or glass and its modifications including certain sol-gels. Polymer nanoparticles (NPs) are usually obtained by microemulsion polymerization. Microemulsions are clear, stable, isotropic liquid mixtures of oil, water and surfactant, sometimes in combination with a cosurfactant. Microemulsion polymerization of monomers may be achieved by incorporating a monomer in any of the water and oil phases of the system [159]. The two basic types of microemulsions, direct (oil dispersed in water, o/w) and reversed (water dispersed in oil, w/o) are frequently used. In w/o microemulsion nanodroplets of oil surrounded by surfactant are dispersed in the continuous bulk water phase. The oil nanodroplets serve as confined reactor for polymerization of discrete nanoparticles. The size of the synthesized particles is determined by the size of those droplets [160–165].

NPs can be also prepared by precipitation method which is based on the use of two miscible solvents [62, 166]. The nanobeads are formed by diluting the polymer solution with of poor solvent. Gradually, evaporation of the good solvent at room temperature causes precipitation of the polymer solute as fine particles. By using this method, nano-particles can be prepared from variety of polymers (e.g., engineering plastics, biodegradable polymers and electro-conductive polymers, etc.). The diameter of particles can be controlled by changing the concentration of the solution, the mixing ratio of the good solvent and the poor solvent, respectively. It should be emphasized that this process does not require the addition of surfactants (and their subsequent removal) as in the case of nanosensors prepared via polymerization [62]. For the preparation of nanosensors water is commonly used as poor solvent since the nanosensors are usually designed for measurements in aqueous medium.

An indicator can be added into the mixture of monomers to be entrapped in the bead during polymerization. Both physical [62, 73, 75, 167] entrapment and covalent [38] coupling are used. Physical entrapment of an indicator in NPs is preferred because of its simplicity and reproducibility. In contrast to bulk sensor films (typically several microns thick), in NSs many indicator molecules are located close to the surface so that leaching can become a serious problem. To avoid leaching, covalent binding can be used. However, in this case, both the dye and the beads require having a reactive group through which a covalent bond can be formed between the polymer and the indicator. This situation is often undesirable because excess reactive groups on the surface of beads may compromise their properties and often make them more prone to aggregation [168]. Swelling is another widespread method for the encapsulation of indicators in NPs [169, 170]. This method is only useful for hydrophobic materials which are not swellable in water. It is essential to use water insoluble indicators for this method otherwise leaching can occur over time.

NSs based on hydrophobic materials even allow the monitoring of hydrophilic species with acceptable response times (in contrast to monolithic films based on the identical composition), due to small diffusion distances. Polymers with polar properties (such as polyacrylonitrile) have a large surface-to-volume ratio and are therefore especially useful for ion sensing. Indicators of amphiphilic nature are often located onto the surface of the bead, allowing response even to hydrophilic analytes.

2.2.2 Sol-Gel Materials

Sol-gels (inorganic silica beads and organically modified silica – Ormosils) are very popular materials for designing optical NSs [43, 45, 52, 54, 56, 171–176]. This is due to the fact that the beads can easily be manufactured, are porous to allow an analyte to diffuse freely inside, are robust, and are biocompatible, making them suitable for intracellular measurements. Compared with polymer nanoparticles, silica nanoparticles possess several advantages. Silica nanoparticles are easy to separate via centrifugation during particle preparation, surface modification, and other solution treatment processes because of the higher density of silica (e.g. 1.96 g/cm^3 for silica vs. 1.05 g/cm^3 for polystyrene). Silica nanoparticles are more hydrophilic and biocompatible, they are not subject to microbial attack, and no swelling or porosity change occurs with changes in pH [177].

Nanoparticles based on sol-gel materials can be prepared by two general synthetic routes: the Stöber [178–180] and reverse microemulsion processes [181–186].

In a typical Stöber-based protocol, a silica alkoxide precursor (such as tetraethyl orthosilicate, TEOS) is hydrolyzed in an ethanol and ammonium hydroxide mixture. The hydrolysis of TEOS produces silicic acid, which then undergoes a condensation process to form amorphous silica particles. In general, the lower the concentration of water and ammonia, the smaller are particles. Indicators are typically entrapped inside the pores of the beads. However, that does not always prevent them from leaching into solution. Therefore, similar to the polymer beads, covalent coupling is often preferred.

Dye-doped silica nanoparticles can also be synthesized by hydrolyzing TEOS in a reverse-micelle or water-in-oil (w/o) microemulsion system. In a typical w/o microemulsion system, water droplets are stabilized by surfactant molecules and remain dispersed in bulk oil. The nucleation and growth kinetics of the silica are highly regulated in the water droplets of the microemulsion system, and the dye molecules are physically encapsulated in the silica network, resulting in the formation of highly monodisperse dye-doped silica nanoparticles [187]. Polar dye molecules are used in the w/o microemulsion system to increase the electrostatic attraction of the dye molecules to the negatively charged silica matrix, so that dye molecules are successfully entrapped inside the silica matrix. Water-soluble inorganic dyes, such as ruthenium complexes, can be readily encapsulated into nanoparticles by this method [100, 169, 188–190]. Leakage of dye molecules from the silica particles is negligible, probably because of the strong electrostatic attractions between the positively charged inorganic dye and the negatively charged silica. To synthesize organic-dye-doped nanoparticles, various trapping methods have been used, such as introducing a hydrophobic silica precursor [191], using water-soluble dextran-molecule-conjugated dyes, and synthesizing in acidic conditions [100]. These alternative methods aid in trapping hydrophobic dye molecules into the silica matrix. The unique advantage of the w/o microemulsion method is that it produces highly spherical and monodisperse nanoparticles of various sizes. It also permits the trapping of a wide variety of inorganic and organic dyes as well as other materials, such as luminescent quantum dots.

2.2.3 The Use of Core-Shell Systems for Sensing Chemistry

Besides core-shell nanostructures based on QDs, also the core-shell nanostructures based on polymers and sol-gel materials are well known [64, 78, 170, 192]. The versatility matrix allows the sensing chemistry to be incorporated either into the core or into the shell of a bead which for some nanosensors can slightly tune the sensitivity and helps to minimize interferences [73]. Moreover, the encapsulation of two different fluorescent dyes in the core-shell structure, one analyte-sensitive dye and other analyte-insensitive incorporated into the core of the bead used as reference, allows ratiometric detection. This method overcomes the problems (fluctuations in light source intensity of the instrument, photobleaching of the analyte-sensitive dye, drifts in the optoelectronic setup and background fluorescence) associated with intensity-based measurements [193–201].

3 Applications of Optical Chemical Nanosensors

In the following section the recently developed photoluminescence-based optical chemical nanosensors will be described, separately for sensors based on quantum dots and those based on polymers and sol-gel materials. The applications highlight the most important sensor characteristics, such as the type of quantum dots and the indicator and matrix used, dynamic range, response time, limit of detection, indicator leaching, operational lifetime, interferences, etc. and give some their critical viewpoints.

3.1 Sensors Based on Quantum Dots

Quantum dots (QD) are inorganic semiconductor nanocrystals with the diameter of 1–12 nm [9, 10]. Originally, quantum dots were applied only in optical sensors whose operation was underlain by the fluorescence phenomenon [89], and more recently, also in sensors with electrochemical detectors [3, 4, 6, 10]. The fluorescent particles play a passive role, in which only their presence in the system is detected. This cannot satisfy the researchers that try to find possibilities to make QDs directly sensitive to intermolecular interactions. The problem here is an insignificant knowledge of QD properties. Very little is known on the variation of the different parameters of their emission in response to different kind of intermolecular interactions in view of various possibilities of forming their surface shells. The properties of QDs are not studied in such detail as that of organic dyes and no focused attempts have been made to date to increase this response by optimizing the composition of core and coating layers.

Several authors have reported on sensing small molecules and ions by using quantum dots with surface either being unmodified or functionalized with a ligand. Especially, the determination of heavy metals in the aquatic environment and biological systems is of tremendous interest owing to their deleterious effects on the ecosystem and human health depending on the dose and the toxicity.

In recent years, QDs have attracted considerable attention as novel luminescence probes [202–204]. Quantom dots used in sensor applications are summarized in Table 3.

Isarov and Chrysochoos [205] observed that the copper ions bound onto the surface of CdS quantum dots led to formation of a new, red-shifted, luminescence band. Chen and Rosenzweig [206] have studied the selectivity of luminescent CdS quantum dots to certain ions by changing the capping layer of the quantum dots. Polyphosphate, L-cysteine, and thioglycerol were used as ligands. Surface ligands play a critical role in ion selectivity. Polyphosphate-capped CdS QDs were sensitive to nearly all mono- and di-valent cations, showing no ion selectivity. Conversely, thioglycerol-capped CdS QDs were sensitive to only copper and iron ions. Similar concentrations of physiologically relevant cations, such as zinc, sodium,

Table 3 Nanosensors/probes based on QDs

QDs	Analyte	Ligand	LOD	Interferences	Ref.
CdS	Cu^{2+}	Thioglycerol	0.1×10^{-6} M	Fe^{2+}	[206]
CdSe	Cu^{2+}	2-Mercaptoethane sulphonic acid	3.15×10^{-9} M	no	[207]
CdTe	Cu^{2+}	3-Mercaptopropionic acid	2.9×10^{-9} M		[208]
CdSe	Cu^{2+}	Mercaptosuccinic acid	Not given	Fe^{2+}	[114]
CdSe/ZnS		Mercaptosuccinic acid	Not given	Fe^{2+}, K^{+}	
ZnS:Mn	Cu^{2+}	–	7.3×10^{-9} M	–	[209]
SdS	Cu^{2+}	Chitosan	1.2×10^{-9} M	–	[115]
CdTe	Ag^{+}	Thioglycolic acid	5.0×10^{-8} M	Cu^{2+}, Hg^{2+}	[111]
CdSe	Ag^{+}	Mercaptoacetic acid and bovine serum albumin	7×10^{-8} M	–	[210]
CdSe	CN^{-}	*Tert*-butyl-*N*-(2-mercaptoethyl)-carbamate	1.1×10^{-7} M	–	[211]
CdSe	CN^{-}	2-Mercaptoethane sulfonate	1.1×10^{-6} M	I^{-}, SCN^{-}	[212]
CdTe	Zn^{2+} Co^{2+}	Thioglycolic acid	0.2×10^{-6} M 0.2×10^{-6} M	– Ni^{2+}	[214]
CdSe/ZnS	Zn^{2+}	Cyclen	1.4×10^{-6} M	Fe^{2+}, Fe^{3+}, Cu^{2+}, Co^{2+}	[215]
CdSe/ZnS	Hg^{2+}	Sulfur calixarene	15×10^{-9} M	Pb^{2+}	[100]
SdSe	Hg^{2+} I^{-}	Triethanolamine	1.9×10^{-7} M 2.8×10^{-7} M	No	[216]
CdS	Hg^{2+}	Mercaptoacetic acid	4.2×10^{-9} M	Cu^{2+}	[113]
CdTe	Pb^{2+}	Thioglycolic acid	2.7×10^{-7} M	Cu^{2+}, Ag^{+}	[117]
CdSe/ZnS	pH	1,3 Oxazine ring	–	–	[217]
CdSe/ZnS	pH	Squaraine dye	–	–	[118]
CdTe	pH	Thioglycolic acid	–	Mg^{2+}, Ca^{2+}, Ag^{2+}, Fe^{3+}, Hg^{2+}	[119]
CdTe	pH	Thioglycolic acid	–	Mg^{2+}, Ca^{2+}, Ag^{2+}, Fe^{3+}, Hg^{2+}	[119]
CdSe/ZnS	K^{+} Ca^{2+}	Schiff base	–	Ca^{2+} K^{+}	[121]
CdSe	NO	Triethanolamine	3.02×10^{-7} M	–	[112]
CdS	Sulfadiazine	Thioglycolic acid	0.8×10^{-6} M	Cu^{2+}, Pb^{2+}	[116]
CdSe	Spironolactone	–	0.48×10^{-6} M	No	[120]

potassium, calcium, and magnesium ions did not affect the luminescence of thioglycerol-capped CdS QDs. On the other hand, L-cysteine-capped CdS QDs were sensitive to zinc ions and insensitive to other physiologically important cations, such as copper, calcium, and magnesium ions. The detection limits were 0.8 μM for zinc and 0.1 μM for copper ions. Maria Teresa Fernandez-Arguelles et al. [207] investigated a water-soluble luminescent CdSe quantum dots (QDs), whose surface was modified with 2-mercaptoethane sulphonic acid and with 2-mercaptoacetic acid, for the sensitive and selective determination of copper (II) ions in aqueous solutions. Higher sensitivity was obtained for the sulphonic-modified CdSe QDs.

Detection limits for Cu(II) of 0.2 μg/l and a dynamic range up to 30 μg/l. Sulphonic-modified QDs showed a high selectivity. Bo and Ping [208] reported on a new kind of functionalized CdTe nanocrystals for the quantitative and selective determination of Cu(II) ions. Nanocrystals were prepared and modified in aqueous solution by 3-mercaptopropionic acid (MPA). This method was also successfully applied to the determination of Cu(II) ions in environmental (industrial effluent samples) and human hair samples [208]. CdSe and CdSe/ZnS quantum dots surface functionalized with mercaptosuccinic acid ligands were prepared [114]. The presence of the ZnS shell has a significant effect on the both the binding constant, the range over which quenching occurs and sensitivity. Silica-coated ZnS:Mn nanoparticles were synthesized by coating hydrophobic ZnS:Mn nanoparticles with silica shell through microemulsion [209]. Quenching of fluorescence intensity of the silica-coated nanoparticles allows the detection of Cu^{2+} concentrations as low as 7.3 nM. Silica-coated nanoparticles have improved luminescent intensity as well as good photostability compared to uncoated ZnS:Mn nanoparticles. Moreover, silica-coated nanoparticles can endure large variations in pH and keep stable luminescent intensity with wide pH value. Fluorescent probe for Cu^{2+} based on the decrease of relative fluorescence intensity of CdS QDs modified by chitosan (CTS) has been reported [115]. The authors presumed that quenching mechanism is due to a chemical displacement of surface Cd^{2+} ions by Cu^{2+} ions to form CuS particles on the surface of QDs core, which leads to the enhancement of electron and whole transfer, and the quenching of fluorescence signals of CTS-QDs. Under the optimum conditions, the relative fluorescence intensity was linearly proportional to the concentration of Cu^{2+} in range 8.0 nM to 3.0 μM, and the limit of detection was 1.2 nM. Probe was insensitive to Pb^{2+}, Hg^{2+}, Ag^{+} and other physiologically important cations even if their concentrations were 100 times higher than that of Cu^{2+}.

A probe for the detection of Ag^{+} ions based on red-shift of the emission band of CdS quantum dot surface modified with thioglycolic acid (TGA) was developed [111]. Under optimal conditions, a linear relationship does exist between the red-shift of the emission and the concentration of Ag(I) in the range from 0.1 to 0.15 μM, with a detection limit of 50 nM. The interference of alkali, alkaline earth ions, Co^{2+}, Mn^{2+} and Ni^{2+} are very weak. Cd^{2+}, Mg^{2+}, Zn^{2+}, Pb^{2+} and Ba^{2+} have a relative fluorescence wavelength blue-shift effect on probe at relative higher concentration. Cu^{2+}, Hg^{2+} also reveal certain red-shift on TGA-capped CdTe QDs fluorescence wavelength; however, when the same concentration of Cu^{2+} or Hg^{2+} are added in the QDs-Ag^{+} system, the red-shift of emission wavelength of probe has no obvious increment in comparison with the redshift caused by Ag^{+}. Jian et al. [210] reported on water-soluble and biocompatible CdSe quantum dots (QDs) that have been surface modified with mercaptoacetic acid and bovine serum albumin (BSA) to improve fluorescence intensity and stability in water solution. They found the quench of fluorescence signals of the functionalized CdSe QDs in the 543 nm wavelength and enhancement of them in the 570–700 nm wavelength range by Ag(I) ions at pH 5.0. The response was linearly proportional to the concentration of Ag(I) between 4.0×10^{-7} and 1.5×10^{-5} mol/l, with the limit of detection 7.0×10^{-8} M.

Wei et al. [211] prepared luminescent surface-modified CdSe semiconductor quantum dots (QDs), with nanoparticle (NP) size distribution in the order of 2–7 nm, for optical determination of cyanide ions. The nanoparticles that have been functionalised with *tert*-butyl-*N*-(2-mercaptoethyl)-carbamate (BMC) groups, exhibited a strong fluorescent emission at about 580 nm with rather long fluorescence lifetimes (several hundred nanoseconds) in aerated methanolic solution. The functionalisation of the nanocrystals with carbamate ligand allowed a highly sensitive determination of free cyanide via analyte-induced changes in the photoluminescence (fluorescence quenching of intensity at 580 nm) of the modified quantum dots (excited at 400 nm). A detection limit of 1.1×10^{-7} M (2.9 μg/l) of cyanide ions was obtained, while the interfering effect of other inorganic anions (including NO^{3-}, Cl^- or SCN^-) was negligible even at 200-fold level concentrations in excess of cyanide. Wei et al. [212] prepared surface modified CdSe quantum dots with 2-mercaptoethane sulfonate for the selective determination of free cyanide in aqueous solution with high sensitivity. The limit of detection was unaltered after more than 2 months storage. The quenching of the fluorescence from the mercaptoethane sulfonate coated CdSe QDs by cyanide anions was rather selective. The influence of other anions (including $SO_4{}^{2-}$, $SO_3{}^{2-}$, $NO_2{}^{2-}$, $NO_3{}^-$, SCN^-, Cl^-, Br^-, I^- and acetate) on the luminescence emission from the synthesized QDs showed that most of these common anions did not exhibit any significant effect on the photoluminescence emission of the QDs, even at relatively high concentration levels. Only, concentrations of I^- and SCN^- higher than 2×10^{-4} M produced a measurable quenching of the luminescence of the QDs.

Moore and Patel [213] found that the quantum dots CdS photoluminescence activation could also be induced by Zn^{2+} and Cd^{2+} salt introduction. Li et al. [214] synthesized water-soluble luminescent thiol-capped CdTe QDs that are sensitive to cink and cobalt ions. They found that zinc ions enhanced the luminescence emission and cobalt ions quenched it. However, other divalent metal ions (e.g., Ca^{2+}, Mg^{2+}, Mn^{2+}, Ni^{2+} and Cd^{2+}) also quenched luminescence. Zn^{2+} selective nanosensors based on PET fluorophore-ligand have been designed using azamacrocycles, a class of nitrogen-containing analogues to crown ethers, conjugated via an amide link to 3-Mercaptopropionic acid-CdSe/ZnS QDs [215]. Three azamacrocycles were used as receptors: TACN (1,4,7-triazacyclononane), cyclen (1,4,7,10-tetraazacyclododecane), and cyclam (1,4,8,11-tetraazacyclotetradecane). The QDs conjugated with cyclen showed the highest sensitivity towards Zn^{2+} ions. The response time is from 1 to 3 min depending on the tested concentration. Other physiologically important cations that exist at high concentration in living cells, such as Ca^{2+}, Mg^{2+}, Na^+, and K^+, did not give rise to any changes in the emission of the QD–azamacrocycle conjugates. On the other hand, Cu^{2+}, Fe^{3+}, Fe^{2+}, and Co^{2+} produced the quenching of QD–azamacrocycle conjugate luminescence.

Sensor for Hg^{2+} has been developed based on a selective fluorescence quenching of CdSe/ZnS QDs modified with sulphur calixarene. This system showed a detection limit in the nM range and the influence of other metal ions (Na^+, K^+, Ca^{2+}, Mg^{2+}, Zn^{2+}, Mn^{2+}, Co^{2+}, Ni^{2+}) was very weak with the exception of Pb^{2+} at higher concentrations that produced a measurable quenching of QDs fluorescence [100].

Triethanolamine (TEA)-capped CdSe QDs showed to be responsive through an effective fluorescence quenching only in the concomitant presence of Hg^{2+} and I^- [100, 216]. The adequate experiments showed that iodine (I) anions could bridge between TEA–CdSe-QDs and Hg^{2+} to form a stable complex (QDs-I^--Hg^{2+}) and the following effective electron transfer from the QDs to the Hg^{2+} could be responsible for the fluorescence quenching of QDs. Water soluble CdS quantum dots surface functionalized with mercaptoacetic acid were presented as Hg^{2+} probes [113]. Under the optimum conditions, the fluorescence intensity of CdS QDs is linearly proportional to the mercury ion concentration in the range 0.005–0.4 μM with a detection limit of 4.2 nM. The effect of various coexisting ions on the fluorescence of probe was studied by mixing Hg(II) ion and interfering ions. Only Cu^{2+} ion appears to cause some interference.

Han et al. [117] have proposed a new approach for the determination of Pb^{2+} based on fluorescence quenching of thiol-capped CdTe QDs. Under optimum conditions, the calibration graph was linear in the range of 2.0 μM to 0.1 mM of Pb^{2+} concentration, and a detection limit of 0.27 μM. Ag^+ and Cu^{2+} could also strongly quench the fluorescence signals of CdTe QDs Furthermore, other cations slightly interfere with the sensitivity and selectivity of the Pb^{2+} determination.

The surface of CdSe/ZnS QDs was capped with an organic compound incorporating a dithiolane anchoring group, an electron-rich indole, a 4-nitrophenylazophenoxy chromophore and a 1,3-oxazine ring in their skeleton [217]. The pH sensing is based on the 1,3-oxazine ring that opens upon addition of base or acid. The chemical stimulation generates a 4-nitrophenyl-azophenolate chromophore and is transduced into luminescence quenching due to PET from the QDs to the chromophores adsorbed on their surface. The probe can be used in pH range from 3 to 11. Nocera et al. [118] developed a ratiometric pH sensor based on CDSe/ZnS quantum dots and squaraine dye. A sensing approach was based on FRET between CdSe/ZnS QDs encapsulated within an amphiphilic polymer and a pH-sensitive squaraine dye conjugated on the cap surface (Fig. 11). As the absorption spectrum of the dye is pH dependent, so is the FRET efficiency, and the ratio of QDs to dye emission becomes a function of pH. Rogach et al. [119] reported the use of CdTe quantum dots capped by short-chain thiol molecules as pH probe which are pH-sensitive within the pH range of 4–6. The proposed sensors seem to be unappropriate for intracellular measurements since the physiological ions change the probe response. Authors [118, 119, 217] did not include the data regarding the effect of the ionic strength on those probes, which might possibly limit their use.

Lakowicz et al. [218] examined the steady-state and time-resolved luminescence spectral properties of two types of CdS nanoparticles. Polyphosphate-stabilized CdS, which displayed a longer wavelength red emission maximum and a zero anisotropy for all excitation wavelengths are mostly insensitive to dissolved oxygen but were quenched by iodide.

Zhang et al. [219] investigated the use of CdTe quantum dots modified with thioglycolic acid (TGA) for the determination of cationic surfactants. The applicability of TGA-CdTe QDs for this approach in real samples is questionable since

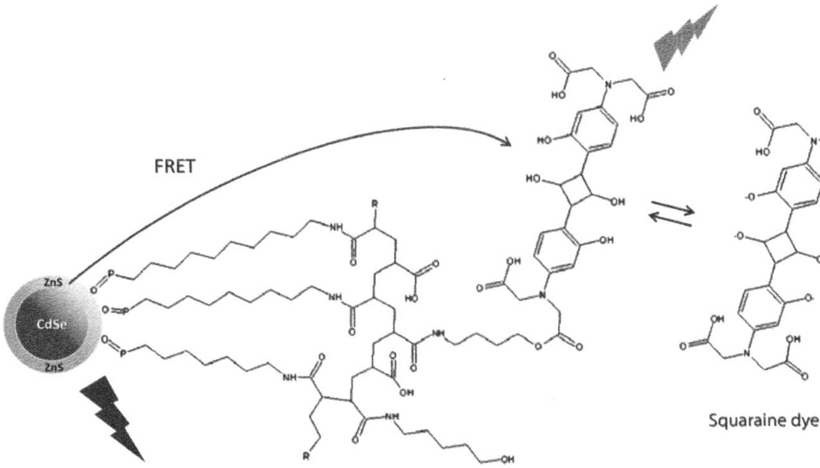

Fig. 11 A sensor constructed from a colloidal CdSe NC that is overcoated with an outer layer of ZnS. The native phosphine oxide ligands are encapsulated with an amphiphilic polymer upon which a pH-sensitive squaraine dye is conjugated. Upon excitation, the CdSe/ZnS nanocrystal may either fluoresce or transfer energy to the squaraine dye. The FRET efficiency is modulated by the environment as the dye's absorption profile is a function of pH. Consequently, the ratio of NC to dye emission becomes a function of environmental variables (Adapted from [118])

without the pre-treatment of the sample the metal ions significantly interfere in the determination procedure.

SdSe/ZnS quantum dots surface functionalized with Schiff base receptor for the determination of K^+ and Ca^{2+} was reported by Callan et al. [121]. However, Schiff base receptor itself demonstrated selectivity for Mg^{2+}. The attachment of it onto the nanoparticle surface alters the binding properties of it dramatically, although the general selectivity for alkali/alkali earth metal ions remains. Large enhancements in fluorescence were observed for both K^+ and Ca^{2+}, accompanied by a blue shift for K^+ (from $\lambda_{max} = 360$ nm to $\lambda_{max} = 320$ nm) and a slight red shift for Ca^{2+} (from $\lambda_{max} = 360$ nm to $\lambda_{max} = 370$ nm). Ca^{2+} causes significant interference in the measurement of K^+ and vice versa. No other ions had a major effect on the fluorescent intensity except Co(II) which resulted in a quenching of the fluorescence. Ca^{2+} could be measured accurately in the presence of up to four equivalents of K^+ whereas interference was observed in the measurement of K^+ when greater than 0.25 equivalents of Ca^{2+} was added.

The water-soluble luminescent CdSe quantum dots surface modified with triethanolamine (TEA) were prepared for the selective determination of nitric oxide in aqueous solutions [112]. The detection limit was 3.02×10^{-7} mol/l. The interference effect of some common interferents such as nitrite (NO_2^-), nitrate (NO_3^-), glucose and L-ascorbic acid on the detection of NO was negligible for the proposed system. The probe can be used for more than 3 months.

Yan et al. [116] presented the technique for sulfadiazine analysis based on the quenching of the fluorescence of CdS quantum dots modified by thioglycolic acid. Under optimum conditions, a good linear relationship between fluorescence intensity ratio of the system and concentration of sulfadiazine in the range from 12 μM to 2.13 mM can be achieved. Heavy metal ions such as Pb(II), Cu(II) can interfere with the detection.

A method for spironolactone analysis has been developed based on the quenching of the fluorescence of CdSe QDs by spironolactone [120]. In the optimum conditions, calibration graph was linear in the range 6.0–1,680 μM and the determinant detection limit is 0.48 μM. The present method was applied to determine spironolactone in spironolactone tablets. Therefore, the commonly used excipients in those tablets were tested for possible interference and showed no the fluorescence of CdSe probe.

3.2 *Sensors Based on Polymers and Sol-Gel Materials*

3.2.1 Nanosensors for pH

Determination of the pH has a critical importance for a wide range of applications, namely in the medical, environmental and biotechnological fields [26, 220–231]. The most common pH sensors are electrochemical devices (glass electrodes). Although they can be reliable analytical tools in many situations, the use of those for intracellular measurements is limited. In the field of biology and medicine, sensors able to monitor pH in real time are in great demand, as they can be used for the general understanding of biological processes and for biomedical diagnostic [232].

Recently, a lot of work has been done on optical pH-sensitive micro- [71, 73, 76, 232–234] and nano-sensors [38, 79, 80, 192, 235–237]. Here the latter will be discussed since they are less than 100 nm sized. Table 4 summarizes pH nanosensors.

Several sol-gel pH nanosensors were prepared by microemulsion procedure [38, 79, 80, 235–238] and one by Stöber [192]. Wang et al. [79, 235] designed pH sensors using the doping technique for the immobilization of indicators in sol-gel nanoparticles. Although leaching can be a serious problem, in this case it has not been tested. The pH dynamic range of sensing nanoparticles (60 ± 4 nm) is between 5.0 and 10.0 [235] and between 4.0 and 8.0 for nanosensors (30 ± 4 nm) based on Fluorescein indicator [79]. The response time (5 min) for both sensors is quite long compared to other pH nanosensors (Table 4), even compared to macro-sized pH optical sensors [26]. Different approaches can be used to prepare ratiometric sensors as shown in Fig. 12. Prasad et al. [236] designed ormosil nanoparticles (33 ± 6 nm) in which the ratiometric pH-responsive dye is covalently immobilized. Peng et al. [80] and Burns et al. [192] developed two-fluorophore-immobilized nanoparticles sensors. The first author used fluorescein

Table 4 Nanosensors/probes for pH

Material	Indicator	pK_a or dynamic range	Response time	Leaching	Operational lifetime	Effect of ionic strength	Ref.
Sol-gel (TEOS)	RITC	5.0–10.0	5 min	Not given	Not given	<5% Signal decrease	[235]
Sol-gel (TEOS)	Fluorescein	4–8	5 min	Not given	5 weeks	<4.5% Signal decrease	[79]
Sol-gel (ICTES)	NVP	4–8	Not given	Not given	>6 months	Not given	[236]
Sol-gel (TEOS, APTES)	FITC	4–7 0.05 pH units	1 s	30% Signal decrease	Not given	Not given	[80]
Sol-gel (TEOS, APTES)	FITC	6.4	Not given	No leaching	Not given	Not given	[192]
Polystyren	FITC	5.6	Few s	No leaching	>6 months	Change in pK_a	[237]
Polyacrylamide	AF	5.8–7.2	0.4 s	9% of the dye	Not given	Not given	[38]

TEOS tetraethoxysilane, *RITC* rhodamine-β-isothiocyanate, *ICTES* (3-isocyanatopropyl)triethoxysilane), *NVP* naphtalenylvinyl-pyridine, *APTES* aminopropyltriethoxysilane, *FITC* fluorescein isothiocyanate, *AF* aminofluorescein

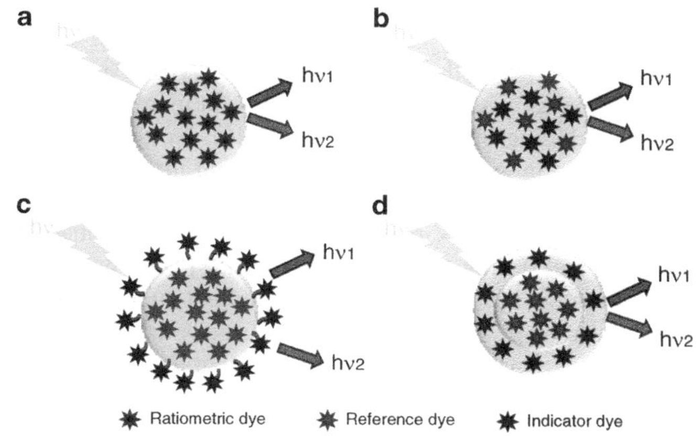

Fig. 12 Types of ratiometric approaches: (**a**) ratiometric indicator dye is immobilized in the nanoparticle; (**b**) indicator and reference dye are immobilized inside the nanoparticle; (**c**) reference dye is immobilized (covalently attached or doped) in the nanoparticle, and the indicator dye is covalently bound to the nanoparticle surface; (**d**) in the core-shell system, the reference dye is immobilized in the core and the indicator dye is immobilized in the shell

isothiocyanate as indicator dye that was covalently crosslinked with the network formed by the hydrolysis of TEOS and APTES, while tris(2,2′-bipyridyl) dichloro-ruthenium(II) hexahydrate (reference) was entrapped simultaneously inside the same nanosensors (42 nm) by electrostatic interaction. The second author synthesized a nanosensor (70 nm) via a modified Stöber method incorporating a reference (tetramethylrhodamine isothiocyanate) and an indicator dye (fluorescein isothiocyanate) covalently bound to the matrix in the core-shell architecture. In this case the pH sensor is composed of a shell of covalently bound indicator-dye molecules surrounding a core of sequestered, covalently bound reference-dye molecules. What is interesting is that although both authors used covalent immobilization, Peng et al. [80] reported extensive dye leakage. Although optical pH sensors are usually strongly dependent on the ionic strength [239], the studies [80, 192, 236] did not present this.

Two different approaches have been used for the development of polymeric ratiometric pH nanosensors [38, 237]. In the first case a reference dye (Rhodamine B derivate) and a pH sensitive dye (AF) were covalently embedded in polyacrylamide nanoparticles (50 nm) [38]. Allard et al. [237] presented a pH sensor based on polystyren nanoparticles, in the 20 nm range, with a hydrophobic reference dye (1,9-diphenylanthracene) embedded within the particle and the indicator dye (FITC) covalently attached to the surface. In this case the pH sensor may retain many of the drawbacks related to free fluorescent dyes because the indicator molecules are not protected from the cellular environment. It would be interesting to know, if the photostability of the covalently attached indicator to the nanoparticle surface is improved compared to free indicator.

3.2.2 Nanosensors for Oxygen

Oxygen is undoubtedly one of the most important analytes. Particularly, determination of dissolved oxygen is of utmost importance in many fields of science and technology [240] including biotechnology [241–246], biology [247, 248], marine science [249–253], and in medicine [254–257]. Recent interest in the methods for measuring dissolved oxygen concentration has been focused mainly on optical sensors, due to their advantages over conventional amperometric electrodes in that they are faster and do not consume oxygen [258]. The principle behind the operation of these sensors is the reduction in luminescent intensity as a consequence of oxygen quenching of the emitting state. The sensor optode, either in typical microparticles type [58, 73, 167, 259, 260] or in recent nanoparticle type [62, 72], consists of dye entrapped in a matrix with a high permeability to oxygen. Borisov et al. [62] showed that different polymer nanoparticles (size 44–55 nm), depending on their gas permeability, embedded with iridium (III) coumarin, could be used as oxygen nanosensors. However, the characterisation of these sensors should be included. Cywinski et al. [72] prepared ratiometric polystyrene-based oxygen nanosensors (20.9 $\pm$ 7.1 nm). Platinum(II)meso-tetra(pentafluorophenyl)porphine was used as an indicator dye whereas *N,N*-bis(1-hexylheptyl)perylene-3,4:9,10-bis-(dicarboximide) was used as a reference dye. Some additional tests, such as leaching, photostability, storage stability, reversibility should be done.

3.2.3 Nanosensors for Ions

Table 5 summarizes the recently developed ion nanosensors. Lead is one of the most toxic heavy metals, and it is particularly dangerous for its effects on children. Therefore, the determination of lead in low concentrations is important. Mancin et al. [78] developed a fluorescent nanosensor for the detection of Pb^{2+} ions. Silica core-shell nanoparticles were produced from 50 nm diameter "pure" silica cores and a 5 nm thick shell in which the indicator dansyl dye was covalently linked to silica network. Then the surface of the shell was functionalized with (mercaptopropyl) triethoxysilane. Surface thiols groups play a key role in the recognition of the analyte. Selectivity tests show that Cu^{2+} is a serious interferent. On the other hand, no effect is observed upon the addition of Zn^{2+}, Ca^{2+}, and Mg^{2+}, while Ni^{2+} and Co^{2+} produce much smaller quenching than Pb^{2+}.

Copper is an essential trace element that is required in various biological processes. Cu^{2+} is a very important element for hemopoiesis, metabolism, growth, and immune system [261]. On the other hand, excess copper concentrations are extremely toxic. The detection of trace amounts of copper ion is of increasing importance in light of its environmental and biomedical implications. Nanosensors for the detection of Cu^{2+} ions typically combine two components: a metal chelating molecule designed to bind the target ion selectively and a fluorophore as a readout system. The chelating molecules and the fluorophore are both covalently linked on to the surface of nanoparticle [188, 189] or the fluorophore is entrapped inside the

Table 5 Nanosensors/probes for ions

Analyte	Material	Indicator	LOD	Response time	Leaching	Operational lifetime	Interferences	Ref.
Pb^{2+}	Sol-gel (TEOS, MPS)	Dansyl-amide	1.1×10^{-6} M	Not given	Not given	>30 days	Cu^{2+}, Co^{2+}, Ni^{2+}	[78]
Cu^{2+}	Sol-gel	Dansyl-amide	Not given	Not given	Not given	Not given	Ni^{2+}	[188]
Cu^{2+}	Sol-gel	Dansyl-amide	3.0×10^{-8} M	Not given	Not given	Not given	Ni^{2+}	[189]
Cu^{2+}	Latex	BODIPY-ciklomat	5×10^{-9} M	Not given	No leaching	Not given	No	[169]
Cu^{2+}	Styren	DAP-PM567-ciklomat	20×10^{-9} M	Not given	No	>6 months	No	[170]
Cu^{2+}	Sol-gel	QIOEt	3.8×10^{-7} M	Not given	Not given	Not given	Zn^{2+}	[190]
Zn^{2+}	Polyacrylamide	Newport Green	4×10^{-6} M	<4 s	Intensity signal drop for 3% after 2 days	Not given	K^+	[75]
Zn^{2+}	Sol-gel (TEOS)	TSQ	2.6×10^{-7} M	Not given	Not given	Not given	Cu^{2+} Cd^{2+}	[81]
Fe^{3+}	Polyacrylamide	Alexa Fluor 448	1×10^{-6} M	10 min	Not given	Not given	Cu(I)	[74]
Mg^{2+}	Polyacrylamide	Coumarin 343	340×10^{-6} M	<4 s	Intensity signal drop for less than 11.8% after 24-h	Not given	pH	[281]
Ca^{2+}	Sol-gel (TEOS)	Calcein	7.48×10^{-8} M		Intensity signal drop for 17% after 12-h	Not given	pH	[77]

MPS (mercaptopropyl)triethoxysilane, *DAP* 9,10-diphenyl-anthracene, *PM567* pyrromethene, *TSQ* 6-methoxy-8-(*p*-toluensulfonamido)-quinoline, *QIOEt* *N*-(quinoline-8-yl)-2-(3-triethoxysilyl-propylamino)-acetamide nanoparticle [169, 170]. Tecilla at al. reported the use of two different metal chelating ligands, picolinamide [188] and (2-pyridinmethyl)-glycinamide [189], and the same fluorescent dye (danslyamide) for the surface modification of silica nanoparticles. The interference for both sensors is Ni^{2+}. Larpent et al. [169] described the concept of associating an efficient fluorescent reporter, a BODIPY derivative, and an excellent metal-chelating receptor, cyclam, in polymer nanoparticles. The hydrophobic dye is entrapped within the particle core and the receptor, covalently linked to the polymer backbone, is mainly located near the surface. The detection is based on the quenching of fluorescence by energy transfer from the dye to the receptor-target complex that occurs when the absorption band of the complex overlaps the emission band of the dye. In this case the Ni^{2+} ions did not interfere.

Core-shell type dual fluorescent nanoparticles (NPs) in the 16 nm diameter range with a selective ligand (cyclam) attached to the surface and two fluorophores, 9,10-diphenyl-anthracene (donor, D) and pyrromethene PM567 (acceptor, A), embedded within the polymer core were synthesized [170]. Dual-dye-doped fluorescent NPs permit the design of cascade FRET-mediated sensing devices. The two sequential FRET processes, the energy transfer from D to A, and quenching of the resulting sensitized emission of A by copper complexes that are formed at the NP surface, provide an efficient means to sense copper ions by monitoring the quenching of the acceptor dye upon irradiation at the donor excitation wavelength. This sensor is highly selective towards other divalent cations.

Huan et al. [190] have developed a Cu^{2+} sensor based on a fluorescent ligand (*N*-(quinoline-8-yl)-2-(3-triethoxysilyl-propylamino)-acetamide) which acts as both binding and readout system.

Zinc is the second most abundant trace element in humans behind iron. Most of the zinc found in the body is bound to proteins such as carbonic anhydrase or zinc finger proteins. Zinc has been implicated in Alzheimer's and Parkinson's disease [262, 263] and as a neuromodulator [264]. In addition, zinc has been found to perform structural functions in some enzymes and found to act as a Lewis acid in others [265]. The amount of zinc can range throughout the body from nanomolar concentrations in the cytosol of certain cells to millimolar concentrations in some neuronal vesicles [266]. However, the full extent of zinc's purpose in the body remains unclear. Therefore the development of nanosensors is required to further elucidate the role of zinc in vivo. Two different nanosensors for the detection of Zn^{2+} have been reported [75, 81]. Kopelman et al. [75] designed a polyacrylamide PEBBLE sensor, containing immobilized Newport Green as the indicator dye and Texas Red-dextran as the reference dye. The maximum response to zinc (percent increase in the "intensity ratio") is 50%, which is less than observed with the Newport Green dye in solution (which has an increase of approx. 250%). It is obvious that the matrix is affecting the sensitivity of the indicator dye. But on the other hand, the matrix is preventing the indicator dye to react to non-specific bindings of proteins, such as bovine serum albumin, which is a prime advantage of the PEBBLE sensors compared to free dye, especially for cellular analysis.

Mancin et al. [81] selected the 6-methoxy-8-(p-toluensulfonamido)-quinoline (TSQ) as the active unit for the preparation of Zn^{2+} sensing silica nanoparticles via the Stöber–Van Blaaderen procedure [267–272]. The Coumarin 5 was used as reference dye covalently linked to silica network. The sensor suffers from the Cu^{2+} interferences and, to minor extent, from Cd^{2+}. Iron is an integral metal ion in several physiological processes involving electron transfer and oxidation. Iron can be both beneficial and cytotoxic. Numerous enzymes use iron as a cofactor for hydroxylation, oxygen transport, DNA synthesis, hydrolysis, etc. [266]. Because of iron's vital role in a number of proteins, deficiency during the developmental stages can lead to irreversible loss of motor skills and has been linked to behavioral alterations in iron deficient rats [273–275]. Accumulation of iron in the central nervous system has been implicated in a number of diseases. Therefore the development of sensors that allow in situ measurements would be beneficial. A PEBBLE sensor based on Alexa Fluor 448 (indicator dye) and Texas Red (reference) immobilized in polyacrylamide nanoparticles for the detection of Fe^{3+} ions have been introduced [74]. What is surprising is that the reference dye is quenched by 20% in the presence of 1 μM copper.

Changes in intracellular ion concentrations accompany many processes in living cells, including transport, signalling, and enzyme function [276–280]. A magnesium-sensitive PEBBLE, which contains immobilized Coumarin 343 as the indicator dye, and Texas Red-dextran as a reference dye, has been reported [281]. The interference from calcium is a major complication of current strategies aimed at the measurement of magnesium concentrations in biological environments [282–284]. The results show that the Coumarin 343 is insensitive to Ca^{2+} ions. A fluorescence-based calcium nanosensor was described [77] that exploits silica nanoparticles (prepared by inverse microemulsion polymerization) doped with calcein as indicator dye for the determination of Ca^{2+} in blood serum. It is surprising that the effect of pH on the fluorescent intensity of nanosensor was not tested in the clinically important pH range (6.6–7.8) [285]. Some other "nano"sensors for the detection of different ions have been published [286–289] but are not commented here since the dimensions of those optrodes are bigger than 100 nm.

3.2.4 Nanosensors for Other Molecules

The characteristics of nanosensors for the detection of glucose, naproxen and Dipicolinic acid are summarized in Table 6. A nonenzymatic sensor for glucose based on amino-functionalized luminescent silica nanoparticles (LSNPs) doped with the europium(III) mixed complex, $Eu(TTA)_3phen$ with 2-thenoyltrifluoroacetone (TTA) and 1,10-phenanthroli-ne(phen) was reported [290]. Lapresta et al. [68] developed a fluorescent polyacrylamide nanosensor for the determination of non-steroidal anti-inflammatory drug naproxen. The fluorescentmonomer,(E)-4-[4-(dimethylamino)styryl]-1-[4-methacryloyloxymethyl) benzyl]-pyridinium chloride (mDMASP) was used as an indicator. Upon the immobilization of the indicator into a polymer nanoparticle the sensitivity and detection limit were reduced. Because of relatively high detection limit those nanosensors could only be used in wastewater or industrial routine tests.

Table 6 Nanosensors/probes for other molecules

Analyte	Material	Indicator	LOD	Response time	Leaching	Operational lifetime	Interferences	Ref.
Naproxen	Polyacryl-amide	mDMASP	0.5×10^{-3} M	Not given	Not given	Not given	cAMP	[68]
Glucose	Sol-gel (TEOS, APTES)	Eu complex	4.4×10^{-6} M	20 min	Not given	Not given	No in the 5% range of signal change	[290]
Dipicoli-nic acid	Sol-gel (TEOS, APTES)	Eu-EDTA complex	0.2×10^{-9} M	30 s	Not given	Not given	No	[291]
Dipicoli-nic acid	Sol-gel (TEOS, APTES)	Tb-EDTA complex	10.3×10^{-9} M	Not given	Not given	Not given	No	[292]

mDMASP (E)-4-[4-(dimethylamino)styryl]-1-[4-(methacryloyloxymethyl) benzyl]pyridinium chloride, *cAMP* adenosine 3:5-cyclic monophosphate, *EDTA* ethylenediamine tetraacetic acid, *APTES* aminoproplytriethoxysilane

For the detection of dipicolinic acid, a chemical marker for bacterial spores, silica nanosensors have been developed based on Eu-complex [291] and Tb-complex [292]. In both cases the EDTA ligand is covalently attached onto the surface of silica nanoparticles doped with reference dyes, such as fluorescein isothiocyanate (FITC) [291] or Ru-complex [292]. Interferences resulting from nonselective binding of aromatic ligands (benzoic acid, m-phthalic acid, o-phthalic acid) to Tb-complex should be tested. Moreover, it is surprising that a Ru-complex, a common indicator for oxygen sensors [293–295] has been used as reference but the possible effect of oxygen on its fluorescence properties has not been tested.

4 Future Trends of Nanomaterial-Based Optical Chemical Sensors

The development of nanoscale materials for optical chemical sensing applications has emerged as one of the most important research areas of interest over the past decades. Nanomaterials exhibit highly tunable size- and shape-dependent chemical and physical properties. They also show unique surface chemistry, thermal and electrical properties, high surface area and large pore volume per mass unit area. Because of their unique and advantageous features they can help to improve sensitivity, response time and detection limit of sensors. Although the zero dimensional materials are applied in sensor design, the factors, such as matrix, indicator and immobilization technique still need to be considered in order to achieve the desired sensor's characteristics. In other words, if the matter is only to reduce the size to less than 100 nm, this does not consequently lead to improved sensor's characteristics. The applications also decide the type and the size of nanomaterials that will be used in the nanosensors' development.

Although today's promising nanosensors have unprecedented sensitivity, the selectivity can still be improved. This can be achieved by functionalisation of nanomaterials (e.g. by modifying the surface or by doping atoms). Suitable control of the properties and response of nanostructures can lead to new devices and technologies. Novel methods of fabricating patterned nanostructures as well as new device concepts are therefore being constantly discovered. An important area, which is currently still in its early stage, but is expected to emerge in the near future, is represented by the in-vivo sensors. Nanosensors offer the possibility to be permanently ingested or injected into the living body, where they can act as reporters of in-vivo concentrations of key analytes. These devices would have a capability for sensing and transmitting data to an external data capture system. In this respect, probes encapsulated by biologically localized embedding (PEBBLEs) are sub-micron sized optical sensors specifically designed for minimally invasive analyte monitoring in single cells with applications for real-time analysis of drug, toxin, and environmental effects on cell function. Their protective polymer matrix protects the cell from the indicator dyes toxicity, prevents interference from biomolecules, and provides a phase for complex sensing schemes. However, there must be taken into account also the dark side of in-vivo sensors where the prime concern is the retention of nanomaterials in the body and the harmful effect in the long run. The toxicity is dependent on the nanoparticles chemical (charge on the particle) and physical (size of particle) properties, and also on the material biocompatibility. Thus, synthesizing particles that are biocompatible and are cytotoxic only at very high concentrations are being aimed at.

Many interesting concepts and ideas for designing NSs can be found in literature, but most of them do not fulfil all the requirements for practical use. But, as more and more types of smart nanomaterials with unique and tunable properties continue to be invented, increasing numbers of efficient and selective nanosensors are expected to emerge. Sensors capable of detecting very low analyte concentrations for non-invasive and non-degenerative analyses in a wealth of different promising applications will be produced. As well it is believed that the cost of individual sensors will be radically reduced and sensors will be more easily usable for the end-users. Moreover, other important challenges, such as development of fully engineered monitoring system requiring modules to collect the target molecules and bring them to the sensor, the nanosensor array to carry out detection, a mechanism to refresh and regenerate the nanosensor as needed, and data management capabilities to communicate any display the information as well as the sensors' long-term stability are expected to be fulfilled. An interesting task is the capability to distinguish among multiple chemical species in a sample stream, therefore the multi-target detection is possible through an emerged development of multi-analyte sensor arrays. In such system, hundreds, thousands or even millions of high-sensitivity nanoscale sensing elements would be widely distributed to simultaneously monitor a large number of chemical, biological and radiological analytes in a given application, with increased reliability, sensitivity, accuracy and selectivity.

Nanoscience is an interdisciplinary area covering physics, chemistry, biology, materials, and engineering, therefore the field of nanosensors will continue to involve mutual interactions and collaborations among chemists, physicists, materials scientists and engineers. Future discoveries and applications of nanoscale sensing devices will continue to evolve, greatly enhancing our daily lives and likely to benefit various fields: industrial sectors, including chemical and electronic industries, manufacturing, health care, medical practice, environmental protection, military and homeland defense, networking and communications.

References

1. Guozhong C (2005) Nanostructures & nanomaterials: synthesis, properties & applications. Imperial College Press, London
2. Göpel, W.: Sensors: a Comprehensive Survey, Fundamentals and General Aspects, vol. 1. VCH, Weinheim (1989)

3. Göpel, W.: Sensors: a Comprehensive Survey, Chemical and Biochemical Sensors Part 1, vol. 2. VCH, Weinheim (1991)
4. Göpel, W.: Sensors: a Comprehensive Survey, Chemical and Biochemical Sensors Part 2, vol. 3. VCH, Weinheim (1991)
5. Göpel, W.: Sensors: a Comprehensive Survey, Chemical and Biochemical Sensors Part 2, vol. 3. VCH, Weinheim (1990)
6. Göpel, W.: Sensors: a Comprehensive Survey, Optical Sensors, vol. 6. VCH, Weinheim (1991)
7. Göpel, W.: Micro- and nanosensor technology/trends in sensor markets. In: Sensors: a Comprehensive Survey, vol. 8. VCH, Weinheim (1995)
8. Pitkethly MJ (2004) Mater Today 7:20–29
9. Kalantar-Zadeh, K., Fry, B.: Inorganic nanotechnology enabled sensors. In: Nanotechnology-Enabled Sensors. Springer, USA (2008)
10. Hauptmann P (1993) Sensors: principles and applications. Prentice Hall, London
11. Shtykov SN, Rusanova TY (2008) Russ J Gen Chem 78:2521–2531
12. Schneider HJ (1991) Angew Chem Int Ed Engl 30:1417–1436
13. Reichardt C (1988) Solvents and solvent effects in organic chemistry. VCH Verlagsgesellschaft, Weinheim, FRG
14. Sellergren B, Shea KJ (1993) J Chromatogr A 654:17–28
15. Sellergren B (1994) J Chromatogr 673:133–141
16. Matsui J, Kato T, Takeuchi T, Suzuki M, Yokohama K, Tamiya E, Karube I (1993) Anal Chem 65:2223–2224
17. Ansell RJ, Kriz D, Mosbach K (1996) Curr Opin Biotechnol 7:89–94
18. Grundler P (2007) Chemical Sensors, an Introduction for Scientists and Engineers. Springer, Berlin Heidelberg
19. Demchenko AP (2009) Introduction to Fluorescence Sensing – Theoretical Aspects. Springer, Netherlands
20. Wolfbeis OS (1990) J Anal Chem 337:522–527
21. Nagl S, Wolfbeis OS (2008) Springer Ser Fluoresc 5:325–346
22. Blum LJ (1997) Bio- and Chemi-Luminescent Sensors. World Scientific, Singapore
23. Moreno-Bondi MC, Benito-Pena E (2006) Fundamentals of enzyme-based sensors (Ch. 16). In: Baldini F, Chester AN, Homola J, Martellucci S (eds) Optical Chemical Sensors. NATO Science Series II. Springer, Dordrecht
24. Galán-Vidal CA, Munoz J, Domínguez C, Alegret S (1995) Trends Anal Chem 14:225–231
25. Lobnik A (2006) Absorption-based sensors. In: Baldini F, Chester AN, Homola J, Martellucci S (eds) Optical Chemical Sensors. NATO Science Series II. Springer, Dordrecht
26. Jerónimo PCA, Araújo AN, Montenegro MCBSM (2007) Talanta 72:13–27
27. Lukowiak A, Strek W (2009) J Sol-Gel Sci Technol 50:201–215
28. Seitz WR (1988) Crit Rev Anal Chem 19:135–173
29. Yappert MC (1997) Chem Educ 1:1–10
30. Wolfbeis OS (2008) Anal Chem 80:4269–4283
31. Guilbault G (1990) Practical Fluorescence. Marcel Dekker Inc., New York
32. Szmacinski H, Lakowicz JR (1995) Sens Actuators B 29:16–24
33. Lippittsch ME, Draxler S (1993) Sens Actuators B 11:97–101
34. Lippitsch ME, Draxler S, Kieslinger D (1997) Sens Actuators B 38–39:96–102
35. Kostov Y, Harms P, Rao G (2001) Anal Biochem 297:105–108
36. Valledor M, Campo JC, Sánchez-Barragán I, Viera JC, Costa-Fernández JM, Sanz-Medel A (2006) Sens Actuators B 117:266–273
37. Buck SM, Koo Y-EL, Park E, Xu H, Philbert MA, Brasuel MA, Kopelman R (2004) Curr Opin Chem Biol 8:540–546
38. Sun H, Scharff-Poulsen AM, Gu H, Almdal K (2006) Chem Mater 18:3381–3384
39. Lakowicz JR (2006) Principles of Fluorescence Spectroscopy. Springer, USA
40. McDonagh C, Burke CS, MacCraith BD (2008) Chem Rev 108:400–422

41. Wolfbeis OS (1991) Fiber Optic Chemical Sensors and Biosensors. CRC, Boca Raton
42. Murković Steinberg I, Lobnik A, Mohr GJ, Wolfbeis OS (1996) Anal Chim Acta 334:125–132
43. Lobnik A, Majchin N, Niederreiter K, Uray G (2001) Sens Actuators B 74:200–206
44. Durkop A, Turel M, Lobnik A, Wolfbeis OS (2006) Anal Chim Acta 555:292–298
45. Turel M, Čajlakovic M, Austin E, Dakin JP, Uray G, Lobnik A (2008) Sens Actuators B 131:247–253
46. Turel M, Durkop A, Yegorova A, Scripinets Y, Lobnik A, Samic N (2009) Anal Chim Acta 644:53–60
47. Demchenko AP (2009) Introduction to Fluorescence Sensing – Design and Properties of Fluorescence Reporters. Springer, Netherlands
48. Han, J., Burgess, K.: Chem Rev. doi:10.1021/cr900249z (2009)
49. Lev O, Tsionsky M, Rabinovich L, Glezer V, Sampath S, Pankratov I, Gun J (1995) Anal Chem 67:22A–30A
50. Mohr GJ (2006) Polymers for optical sensors. In: Baldini F, Chester AN, Homola J, Martellucci S (eds) Optical Chemical Sensors, Part 1. Springer, Netherlands
51. Avnir D, Braun S, Lev O, Levy D, Ottolenghi M (2004) Organically doped sol-gel porous glasses: chemical sensors, enzymatic sensors, electrooptical materials, luminescent materials and photochromic materials. In: Klein LC (ed) Sol-Gel Optics: Processing and Applications. Springer, New York
52. Lobnik A, Oehme I, Murkovic I, Wolfbeis OS (1998) Anal Chim Acta 367:159–165
53. Oehme I, Prattes S, Wolfbeis OS, Mohr GJ (1998) Talanta 47:595–604
54. Lobnik A, Wolfbeis OS (2001) J Sol-Gel Sci Technol 20:303–311
55. Orellana G, Moreno-Bondi MC, Garcia-Fresnadillo D, Marazuela MD (2005) Springer Ser Chem Sens Biosens 3:189–225
56. Korent SM, Lobnik A, Mohr GJ (2007) Anal Bioanal Chem 387:2863–2870
57. Chu CS, Lo YL (2009) Sens Actuators B 143:205–210
58. Waich K, Mayr T, Klimant I (2008) Talanta 77:66–72
59. Borisov SM, Klimant I (2008) Analyst 133:1302–1307
60. Aylott JW (2003) Analyst 128:309–312
61. Riu J, Maroto A, Rius FX (2006) Talanta 69:288–301
62. Borisov SM, Mayr T, Mistlberger G, Waich K, Koren K, Chojnacki P, Klimant I (2009) Talanta 79:1322–1330
63. Buck SM, Xu H, Brausel M, Philbert MA, Kopelman R (2004) Talanta 63:41–59
64. Sounderya N, Zhang Y (2008) Recent Pat Biomed Eng 1:34–42
65. Nel A, Xia T, Mädler L, Li N (2006) Science 311:622–627
66. Shi J, Zhu Y, Zhang X, Baeyens WRG, Garca-Campana AM (2004) TrAC Trends Anal Chem 23:351–360
67. Asefa T, Duncan CT, Sharma KK (2009) Analyst 134:1980–1990
68. Lapresta-Fernandez A, Cywinski PJ, Moro AJ, Mohr GJ (2009) Anal Bioanal Chem 395:1821–1830
69. Zenkl G, Klimant I (2009) Microchim Acta 166:123–131
70. Hornig S, Schulz A, Mohr GJ, Heinze T (2009) J Photopolym Sci Technol 22:671–673
71. Borisov SM, Herrod DL, Klimant I (2009) Sens Actuators B 139:52–58
72. Cywinski PJ, Moro AJ, Stanca SE, Biskup C, Mohr GJ (2009) Sens Actuators B 135: 472–477
73. Borisov SM, Mayr T, Klimant I (2008) Anal Chem 80:573–582
74. Sumner JP, Kopelman R (2005) Analyst 130:528–533
75. Sumner JP, Aylott JW, Monson E, Kopelman R (2002) Analyst 127:11–16
76. Doussineau T, Trupp S, Mohr GJ (2009) J Colloid Interface Sci 339:266–270
77. Hun X, Zhang ZJ (2007) Microchim Acta 159:255–261
78. Arduini M, Mancin F, Tecilla P, Tonellato U (2007) Langmuir 23:8632–8636
79. Gao F, Tang LJ, Dai L, Wang L (2007) Spectrochim Acta A 67:517–521

80. Peng JF, He XX, Wang KM, Tan WH, Wang Y, Liu Y (2007) Anal Bioanal Chem 388: 645–654
81. Teolato P, Rampazzo E, Arduini M, Mancin F, Tecilla P, Tonellato U (2007) Chem Eur J 13:2238–2245
82. He L, Toh CS (2006) Anal Chim Acta 556:1–15
83. Tansil NC, Gao Z (2006) Nano Today 1:28–37
84. Vaseashta A, Dimova-Malinovska D (2006) Sci Technol Adv Mater 6:312–318
85. Davis F, Higson SPJ (2005) Biosens Bioelectron 21:1–20
86. James SW, Tatam RP (2006) J Opt A Pure Appl Opt 8:430–444
87. Jain PK, El-Sayed IH, El-Sayed MA (2007) Nano Today 2:18–29
88. Shi J, Zhu Y, Zhang X, Baeyens WRG, Garcia-Campana AM (2004) TrAC Trends Anal Chem 23:351–360
89. Costa-Fernandez JM, Pereiro R, Sanz-Medel A (2006) TrAC Trends Anal Chem 25:207–218
90. Yain J, Esteves CM, Smith JE, Wang K, He X, Wang L, Tan W (2007) Nano Today 2:44–50
91. Murphy CJ (2002) Anal Chem 74:520A–526A
92. Ashby MF, Ferreira PJ, Schodek DL (2009) Nanomaterials, Nanotechnologies and Design: an Introduction for Engineers and Architects. Butterworth & Heinemann, Amsterdam
93. Yu WW, Wang YA, Peng X (2003) Chem Mater 15:4300–4308
94. Gubin SP, Kataeva NA, Khomutov GB (2005) Russ Chem B 54:827–852
95. Mattoussi H, Radzilowski LH, Dabbousi DO, Thomas EL, Bawendi MG, Rubner MF (1998) J Appl Phys 83:7965–7974
96. Teranishi T, Nishida M, Kanehara M (2005) Chem Lett 34:1004–1005
97. Efros AL (1982) Sov Phys Semicond 16:772–775
98. Halperin WP (1986) Rev Mod Phys 58:533–606
99. Alivisatos AP (1996) Science 271:933–937
100. Frasco MF, Chaniotakis N (2009) Sensors 9:7266–7286
101. Costa-Fernandez, J.M., Pereiro, R., Sanz-Medel, A (2006) **25**, 207–218
102. Grecco HE, Lidke KA, Heintzmann R, Lidke DS, Spagnuolo C, Martinez OE, Jares-Erijman EA, Jovin TM (2004) Microsc Res Tech 65:169–179
103. Sapsford KE, Pons T, Medintz IL, Mattoussi H (2006) Sensors 6:925–953
104. Eychmuller A (2000) J Phys Chem B 104:6514–6528
105. Murray, C.B.: Synthesis and characterization of II–VI quantum dots and their assembly into 3D quantum dot superlattices. Ph.D. thesis, Massachusetts Institute of Technology, Cambridge, MA (1995)
106. Rossetti R, Ellison JL, Gibson JM, Brus LE (1984) J Phys Chem 80:4464–4469
107. Murray CB, Norris DJ, Bawendi MG (1993) J Am Chem Soc 115:8706–8715
108. Wang Y, Herron N (1987) J Phys Chem 91:257–260
109. Kuczynski JP, Milosavljevic BH, Thomas JK (1983) J Phys Chem 87:3368–3370
110. Yin Y, Alivisatos AP (2005) Nature 437:664–670
111. Wang J, Liang JG, Sheng ZH, Han HY (2009) Microchim Acta 167:281–287
112. Yan XQ, Shang ZB, Zhang Z, Wang Y, Jin WJ (2009) Luminescence 24:255–259
113. Koneswaran M, Narayanaswamy R (2009) Sens Actuators B 139:91–96
114. Callan JF, Mulrooney RC (2009) Phys Status Solidi C 4:920–923
115. Lai SJ, Chang XJ, Fu C (2009) Microchim Acta 165:39–44
116. Liu MM, Xu L, Cheng WQ, Zeng Y, Yan ZY (2008) Spectrochim Acta A 70:1198–1202
117. Wu HM, Liang JG, Han HY (2008) Microchim Acta 161:81–86
118. Snee PT, Somers RC, Nair G, Zimmer JP, Bawendi MG, Nocera DG (2006) J Am Chem Soc 128:13320–13321
119. Susha AS, Javier AM, Parak WJ, Rogach AL (2006) Colloids Surf A 281:40–43
120. Liang JG, Huang S, Zeng DY, He ZK, Ji XH, Ai XP, Yang HX (2006) Talanta 69:126–130
121. Singh N, Mulrooney RC, Kaur N, Callan JF (2009) J Fluoresc 19:777–782
122. Uskoković V, Košak A, Drofenik M (2006) Int J Appl Ceram Technol 3:134–143
123. Uskoković V, Košak A, Drofenik M (2006) Mater Lett 60:2620–2622

124. Wang JH, Wang HQ, Zhang HL, Li XQ, Hua XF, Huang ZL, Zhao YD (2007) Colloids Surf A 305:48–53
125. Zhang Y, Zhang H, Guo X, Wang H (2008) Microchem J 89:142–147
126. Dubois F, Mahler B, Dubertret B, Doris E, Mioskowski C (2007) J Am Chem Soc 129:482–483
127. Kamat PV (1993) Chem Rev 93:267–300
128. Schmid G (2004) Nanoparticles – from Theory to Application. Wiley-VCH Verlag GmbH & Co. KGaA, Weinheim
129. Weller H (1993) Angew Chem Int Ed Engl 32:41–53
130. Spanhel L, Haase M, Weller H (1987) J Am Chem Soc 109:5649–5655
131. Sooklal K, Cullum BS, Angel SM, Murphy CJ (1996) J Phys Chem 100:4551–4555
132. Dannhauser T, O'Neil M, Johansson K, Whitten D, McLendon GJ (1986) Phys Chem 90:6074–6076
133. O'Neil M, Marohn J, McLendon G (1990) J Phys Chem 94:4356–4363
134. Johansson K, Cowdery R, O'Neil M, Rehm J, McLendon G, Marchetti A, Whitten DG (1993) Isr J Chem 33:67–70
135. Fox MA, Dulay MT (1993) Chem Rev 93:341–357
136. Kumar A, Kumar S (1996) Chem Lett 7:711–712
137. Weller H, Koch U, Gutierrez M, Henglein A (1984) Phys Chem 88:649–656
138. Zhou HS, Honma I, Komiyama H, Haus JW (1993) J Phys Chem 97:895–901
139. Zhou HS, Sasahara H, Honma I, Komiyama H, Haus JW (1994) Chem Mater 6:1534–1541
140. Bedja I, Kamat PV (1995) J Phys Chem 99:9182–9188
141. Tian Y, Newton T, Kotov NA, Guldi DM, Fendler JH (1996) J Phys Chem 100:8927–8939
142. Mews A, Eychmuller A, Giersig M, Schooss D, Weller H (1994) J Phys Chem 98:934–941
143. Kamalov VF, Little R, Logunov SL, El-Sayed MA (1996) J Phys Chem 100:6381–6384
144. Li J, Zhang JZ (2009) Coord Chem Rev 253:3015–3041
145. Fujii H, Inata K, Ohtaki M, Eguchi K (2001) J Mater Sci 36:527–532
146. Cumberland SL, Berrettini MG, Javier A, Strouse GF (2003) Chem Mater 15:1047–1056
147. Manera MG, Spadavecchia J, Buso D, Fernandez C, Mattei G, Martucci A, Mulvaney P, Perez-Juste J, Rella R, Vasanelli L, Mazzoldi P (2008) Sens Actuators B 132:107–115
148. Diguna LJ, Shen Q, Sato A, Katayama K, Sawada T, Toyoda T (2007) Mater Sci Eng C Biomim Supramol Syst 27:1514–1520
149. Ge JP, Xu S, Zhuang J, Wang X, Peng Q, Li YD (2006) Inorg Chem 45:4922–4927
150. Youn HC, Baral S, Fendler HJ (1988) J Phys Chem 92:6320–6327
151. Henglein A (1989) Chem Rev 89:1861–1873
152. Henglein A (1995) Ber Bunsenges Phys Chem 99:903–913
153. Weller H, Eychmüller A (1995) Adv Photochem 20:165–216
154. Danek M, Jensen KF, Murray CB, Bawendi MG (1996) Chem Mater 8:173–180
155. RodriguezViejo J, Jensen KF, Mattoussi H, Michel J, Dabbousi BO, Bawendi MG (1997) Appl Phys Lett 70:2132–2134
156. Kuno M, Lee JK, Dabbousi BO, Mikulec FV, Bawendi MG (1997) J Chem Phys 106:9869–9882
157. Peng X, Schlamp MC, Kadavanich AV, Alivisatos AP (1997) J Am Chem Soc 119:7019–7029
158. Meisel D (1997) Charge transfer in nanoparticles. In: Kamat PV, Meisel D (eds) Semiconductor Nanoclusters. Elsevier Science BV, Amsterdam
159. Pavel FM (2004) J Dispers Sci Technol 25:1–16
160. Košak A, Makovec D, Drofenik M (2004) J Magn Magn Mater 272–276:1542–1544
161. Košak A, Makovec D, Drofenik M, Žnidaršič A (2004) J Eur Ceram Soc 24:959–962
162. Košak A, Makovec D, Drofenik M (2004) Mater Sci Forum 453–454:219–224
163. Košak A, Makovec D, Drofenik M (2004) Nanotechnology (Bristol) 15:S160–S166
164. Košak A, Makovec D, Drofenik M, Žnidaršič A (2005) J Magn Magn Mater 289:32–35
165. Košak A, Makovec D, Drofenik M (2005) J Metastable Nanocryst Mater 23:251–254

166. Higuchi T, Yabu H, Shimomura M (2006) Colloids Surf A 284–285:250–253
167. Borisov SM, Klimant I (2009) Microchim Acta 164:7–15
168. Borisov SM, Mayr T, Karasyov AA, Klimant I, Chojnacki P, Moser C, Nagl S, Schaeferling M, Stich M, Kocincova AS, Wolfbeis OS (2008) Springer Ser Fluoresc 4:431–463
169. Méallet-Renault, R., Pansu, R., Amigoni-Gerbier, S., Larpent, C.: Chem. Commun., 2344–2345 (2004)
170. Frigoli M, Ouadahi K, Larpent C (2009) Chem Eur J 15:8319–8330
171. Lobnik A (2000) Sol-Gels and Other Polymers used in Optical Chemical Sensors. SPIE, Bellingham
172. Lobnik A, Wolfbeis OS (1998) Analyst 123:2247–2250
173. Lobnik A, Wolfbeis OS (1998) Sens Actuators B 51:203–207
174. Lobnik A, Čajlaković M (2001) Sens Actuators B 74:194–199
175. Čajlaković M, Lobnik A, Werner T (2002) Anal Chim Acta 455:207–213
176. Murković Steinberg I, Lobnik A, Wolfbeis OS (2003) Sens Actuators B 90:230–235
177. Jain TK, Roy I, De TK, Maitra A (1998) J Am Chem Soc 120:11092–11095
178. Rossi LM, Shi L, Quina FH, Rosenzweig Z (2005) Langmuir 21:4277–4280
179. Shibata S, Taniguchi T, Yano T, Yamane M (1997) J Sol-Gel Sci Technol 10:263–268
180. Wang L, Tan W (2006) Nano Lett 6:84–88
181. Wang L, Yang C, Tan W (2005) Nano Lett 5:37–43
182. Bagwe RP, Yang C, Hilliard LR, Tan W (2004) Langmuir 20:8336–8342
183. Santra S, Zhang P, Wang K, Tapec R, Tan W (2001) Anal Chem 73:4988–4993
184. Zhao X, Tapec-Dytioco R, Tan W (2003) J Am Chem Soc 125:11474–11475
185. Yang HH, Qu HY, Lin P, Li SH, Ding MT, Xu JG (2003) Analyst 128:462–466
186. Santra S, Bagwe RP, Dutta D, Stanley JT, Walter GA, Tan W, Moudgul BM, Mericle RA (2005) Adv Mater 17:2165–2169
187. Schmidt J, Guesdon C, Schomäcker R (1999) J Nanopart Res 1:267–276
188. Brasola, E., Mancin, F., Ramazzo, E., Tecilla, P., Tonellato, U.: Chem. Commun., 3026–3027 (2003)
189. Ramazzo E, Brasola E, Marcuz S, Mancin F, Tecilla P, Tonellato U (2005) J Mater Chem 15:2687–2696
190. Zheng, J.N., Xiao, C., Fei, Q., Li, M., Wang, B.J., Feng, G.D., Yu, H.M., Huan, Y.F., Song, Z.G.: Nanotechnology. doi:10.1088/0957-4484/21/4/045501(2010)
191. Tapec R, Zhao XJJ, Tan WH (2002) J Nanosci Nanotechnol 2:405–409
192. Burns A, Sengupta P, Zedayko T, Baird B, Wiesner U (2006) Small 2:723–726
193. Wencel D, MacCraith BD, McDonagh C (2009) Sens Actuators B 139:208–213
194. Niu CG, Gui XQ, Zeng GM, Guan AL, Gao PF, Qin PZ (2005) Anal Bioanal Chem 383:349–357
195. Niu CG, Gui XQ, Zeng GM, Yuan XZ (2005) Analyst 130:1551–1556
196. Niu CG, Guan AL, Zeng GM, Liu YG, Huang GH, Gao PF, Gui XQ (2005) Anal Chim Acta 547:221–228
197. Park EJ, Reid KR, Tang W, Kennedy RT, Kopelman R (2005) J Mater Chem 15:2913–2919
198. Ge XD, Kostov Y, Rao G (2003) Biosens Bioelectron 18:857–865
199. Kermis HR, Kostov Y, Harms P, Rao G (2002) Biotechnol Prog 18:1047–1053
200. Xu Z, Rollins A, Alcala R, Marchant RE (1998) J Biomed Mater Res A 39:9–15
201. Song A, Parus S, Kopelman R (1997) Anal Chem 69:863–867
202. Chan WCW, Nie SM (1998) Science 281:2016–2018
203. Bruchez M, Moronne JM, Gin P, Weiss S, Alivisatos AP (1998) Science 281:2013–2016
204. Han MY, Gao XH, Su JZ, Nie SM, Nie SM (2001) Nat Biotechnol 19:631–635
205. Isarov AV, Chrysochoos J (1997) Langmuir 13:3142–3149
206. Chen Y, Rosenzweig Z (2002) Anal Chem 74:5132–5138
207. Fernandez-Arguelles MT, Jin WJ, Costa-Fernandez JM, Pereiro R, Sanz-Medel A (2005) Anal Chim Acta 549:20–25
208. Bo C, Ping Z (2005) Anal Bioanal Chem 381:986–992

209. Dong B, Cao L, Su G, Liu W, Qu H, Jiang D (2009) J Colloid Interface Sci 339:78–82
210. Liang JG, Xin-Ping A, He ZK, Pang DW (2004) Analyst 129:619–622
211. Jin WJ, Costa-Fernandez JM, Pereiro R, Sanz-Medel A (2004) Anal Chim Acta 522:1–8
212. Jin, W.J., Costa-Fernandez, J.M., Pereiro, R., Sanz-Medel, A (2005) Chem. Commun., 883–885
213. Moore DE, Patel K (2001) Langmuir 17:2541–2544
214. Li, J., Bao, D.S., Hong, X., Li, D., Li, J.H., Bai, Y.B., Li, T.J (2005) Colloids Surf. A Physicochem. Eng. Aspects, 257–258
215. Ruedas-Rama MJ, Hall EAH (2008) Anal Chem 80:8260–8268
216. Shang ZB, Wang Y, Jin WJ (2009) Talanta 78:364–369
217. Tomasulo M, Yildiz I, Kaanumalle SL, Raymo FM (2006) Langmuir 22:10284–10290
218. Lakowicz JR, Gryczynski I, Gryczynski Z, Murphy CJ (1999) J Phys Chem B 103: 7613–7620
219. Diao XL, Xia YS, Zhang TL, Li Y, Zhu CQ (2007) Anal Bioanal Chem 388:1191–1197
220. Dybko A, Wroblewski W, Rozniecka E, Poznisk K, Maciejewski J, Romaniuk R, Brozozka Z (1998) Sens Actuators B 51:208–213
221. Malins C, Glever HG, Keyes TE, Vos JG, Dressick WJ, MacCraith BD (2000) Sens Actuators B 67:89–95
222. Jorge PAS, Caldas P, Esteves da Silva JCG, Rosa CC, Oliva AG, Farahi F, Santos JL (2005) Fiber Integr Opt 24:201–225
223. Swindlehurst BR, Narayanaswamy R (2004) Optical sensing of pH in low ionic strength waters. In: Narayanaswamy R, Wolbeis OS (eds) Optical Sensors – Industrial Environmental and Diagnostic Applications. Springer, New York
224. Orellana G, Garcia-Fresnadillo D (2004) Environmental and industrial optosensing with tailored luminescent Ru(II) polypyridyl complexes in optical sensors. In: Narayanaswamy R, Wolbeis OS (eds) Optical Sensors – Industrial Environmental and Diagnostic Applications. Springer, New York
225. Fritzsche M, Barreiro CG, Hitzmann B, Scheper T (2007) Sens Actuators B 128:133–137
226. Seidel MP, DeGrandpre MD, Dickson AG (2008) Mar Chem 109:18–28
227. Brigo L, Carofiglio T, Fregonese C, Meneguzzi F, Mistura G, Natali M, Tonellato U (2008) Sens Actuators B 130:477–482
228. Ge FY, Chen LG (2008) J Fluoresc 18:741–747
229. Hakonen A, Hulth S (2008) Anal Chim Acta 606:63–71
230. Bradley M, Alexander L, Duncan K, Chennaoui M, Jones AC, Martin RMS (2008) Bioorg Med Chem Lett 18:313–317
231. Dong S, Luo M, Peng G, Cheng W (2008) Sens Actuators B 129:94–98
232. Doussineau T, Smaihi M, Mohr GJ (2009) Adv Funct Mater 19:117–122
233. Schulz A, Hornig S, Liebert T, Birckner E, Heinze T, Mohr GJ (2009) Org Biomol Chem 7:1884–1889
234. Hornig S, Biskup C, Grafe A, Wotschadlo J, Liebert T, Mohr GJ, Heinze T (2008) Soft Matter 4:1169–1172
235. Gao F, Wang L, Tang LJ, Zhu CQ (2005) Microchim Acta 152:131–135
236. Kim S, Pudavar HE, Prasad PN (2006) Chem Commun 19:2071–2073
237. Allard E, Larpent C (2008) J Polym Sci A Polym Chem 46:6206–6213
238. Welser K, Perera MDA, Aylott JW, Chan WC (2009) Chem Commun 43:6601–6603
239. Wolfbeis OS (2005) J Mater Chem 15:2657–2669
240. Papkovsky DB, O'Riordan TC (2005) J Fluoresc 15:569
241. Kostov Y, Harms P, Randers-Eichhorn L, Rao G (2001) Biotechnol Bioeng 72:346–352
242. Ge X, Hanson M, Shen H, Kostov Y, Brorson KA, Frey DD, Moreira AR, Rao G (2006) J Biotechnol 122:293–306
243. John GT, Klimant I, Wittmann C, Heinzle E (2003) Biotechnol Bioeng 81:829–836
244. Zanzotto A, Szita N, Boccazzi P, Lessard P, Sinskey AJ, Jensen KH (2004) Biotechnol Bioeng 87:243–254

245. Hanson MA, Ge X, Kostov Y, Brorson KA, Moreira AR, Rao G (2007) Biotechnol Bioeng 97:833–841
246. Mehta G, Mehta K, Sud D, Song JW, Bersano-Begey T, Futai N, Heo YS, Mycek MA, Linderman JJ, Takayama S (2007) Biomed Microdevices 9:123–134
247. Schmaelzlin E, van Dongen JT, Klimant I, Marmodee B, Steup M, Fisahn J, Geigenberger P, Loehmannsroeben HG (2005) Biophys J 89:1339–1345
248. Cao Y, Koo YEL, Kopelman R (2004) Analyst 129:745–750
249. Gouin JF, Baros F, Birot D, Andre JC (1997) Sens Actuators B 38–39:401–406
250. Klimant I, Meyer V, Kuhl M (1995) Limnol Oceanogr 40:1159–1165
251. Hasumoto H, Imazu T, Miura T, Kogure K (2006) J Oceanogr 62:99–103
252. Koenig B, Kohls O, Holst G, Glud RN, Kuehl M (2005) Mar Chem 97:262–276
253. Schroeder CR, Polerecky L, Klimant I (2007) Anal Chem 79:60–70
254. Meruva RC, Meyerhoff ME (1998) Biosens Bioelectron 13:201–212
255. Schmaelzlin E, Walz B, Klimant I, Schewe B, Loehmannsroeben HG (2006) Sens Actuators B 119:251–254
256. Babilas P, Liebsch G, Schacht V, Klimant I, Wolfbeis OS, Szeimies R-M, Abels C (2005) Microcirculation 12:477–487
257. Kimura S, Matsumoto K, Mineura K, Itoh T (2007) J Neurol Sci 258:60–68
258. McDonagh C, Bowe P, Mongey K, MacCraith BD (2002) J Non-Cryst Solids 306: 138–148
259. Koo YEL, Cao YF, Kopelman R, Brasuel M, Philbert MA (2004) Anal Chem 76:2498–2505
260. Borisov SM, Nuss G, Klimant I (2008) Anal Chem 80:9435–9442
261. Uauy R, Olivares M, Gonzalez M (1998) Am J Clin Nutr 67:952S–959S
262. Cuajungco MP, Lees GJ (1997) Neurobiol Dis 4:137–169
263. Frederickson CJ, Suh SW, Silva D, Thompson RB (2000) J Nutr 130:1471S–1483S
264. Choi DW, Koh JY (1998) Annu Rev Neurosci 21:347–375
265. Cunnane SC (1988) Zinc: Clinical and Biochemical Significance. CRC, Boca Raton
266. Silvia J, Williams R (1991) The Biological Chemistry of the Elements: the Inorganic Chemistry of Life. Claredon Press, Oxford
267. Stöber W, Fink A, Bohn E (1968) J Colloid Interface Sci 26:62–69
268. Van Blaaderen A, Kentgens APM (1992) J Non-Cryst Solids 149:161–178
269. Van Blaaderen A, Van Geest J, Vrij A (1992) J Colloid Interface Sci 154:481–501
270. Van Blaaderen A, Vrij A (1993) J Colloid Interface Sci 156:1–18
271. Van Blaaderen A, Vrij A (1992) Langmuir 8:2921–2931
272. Verhaegh NAM, Van Blaaderen A (1994) Langmuir 10:1427–1438
273. Burdo JR, Connor JR (2003) Biometals 16:63–75
274. Felt BT, Lozoff B (1996) J Nutr 126:693–701
275. Earley CJ, Connor JR, Beard JL, Malecki EA, Epstein DK, Allen RP (2000) Neurology 54:1698–1700
276. Golovina VA, Blaustein MP (1997) Science 275:1643–1648
277. Kennedy RT, Huang L, Aspenwall CAJ (1996) J Am Chem Soc 118:1795–1796
278. Nuccitelli R (1994) Methods in Cell Biology. Academic, San Diego
279. Ross WN (1993) Biophys J **64**, 1655–1656
280. Kotyk A, Slavik J (1989) Intracellular pH and Its Measurement. CRC, Boca Raton
281. Park EJ, Brasuel M, Behrend C, Philbert MA, Kopelman R (2003) Anal Chem 75:3784–3791
282. Gotoh H, Kajikawa M, Kato H, Suto K (1999) Brain Res 828:163–168
283. Cheng C, Reynolds IJ (2000) Neuroscience 95:973–979
284. Brocard JB, Rajdev S, Reynolds IJ (1993) Neuron 11:751–757
285. He H, Jenkins K, Lin C (2008) Anal Chem Acta 661:197–204
286. Ruedas-Rama MJ, Hall EAH (2006) Analyst 131:1282–1291
287. Graefe A, Stanca SE, Nietzsche S, Kubicova L, Beckert R, Biskup C, Mohr GJ (2008) Anal Chem 80:6526–6531
288. Clark HA, Kopelman R, Tjalkens R, Philbert MA (1999) Anal Chem 71:4837–4843

289. Brasuel M, Kopelman R, Miller TJ, Tjalkens R, Philbert MA (2001) Anal Chem 73: 2221–2228
290. Gao F, Luo FB, Chen XX, Yao W, Yin J, Yao Z, Wang L (2009) Talanta 80:202–206
291. Ai KL, Zhang BH, Lu LH (2009) Angew Chem Int Ed 48:304–308
292. Taylor KML, Lin WB (2009) J Mater Chem 19:6418–6422
293. Wu XD, Song LT, Li B, Liu YH (2010) J Lumin 130:374–379
294. Zhang HR, Li B, Lei BF, Li WL (2008) J Lumin 128:1331–1338
295. Li L, Gao XK, Lv BQ, Zhou ZD, Xiao D (2007) Sensor Lett 5:441–444

Zinc Oxide Nanostructured Thin Films: Preparation and Characterization

Mohamad Hafiz Mamat and Mohamad Rusop

Abstract The preparation and characterization of zinc oxide (ZnO) nanostructured thin films have been discussed. ZnO, a wide band gap semiconductor material, has proven to be of great interest for use in a lot of applications especially in electronics and green technology such as solar cells, sensors and light emitting devices. With wide applications by ZnO material, it is important to study its preparation and properties which could give the characteristics that suitable in applications. In this study, we prepared ZnO nanostructured thin films using economically viable and simple technique of sol-gel spin-coating technique. The effects of annealing temperature and precursor molar concentration on ZnO nanostructured properties are investigated. We characterized and discussed the morphology, crystallinity, optical and electrical properties of prepared ZnO nanostructured thin film.

1 Introduction

The semiconductor industry has grown very rapidly since the invention of semiconductor transistor by the scientists of Bell Labs in 1947. Research in semiconductor has also been progressively conducted as the semiconductor industry expanded over the years to fabricate more powerful devices which have faster

M.H. Mamat (✉)
NANO-ElecTronic Centre (NET), Faculty of Electrical Engineering, Universiti Teknologi MARA (UiTM), 40450 Shah Alam, Selangor, Malaysia
e-mail: hafiz_030@yahoo.com

M. Rusop
NANO-ElecTronic Centre (NET), Faculty of Electrical Engineering, Universiti Teknologi MARA (UiTM), 40450 Shah Alam, Selangor, Malaysia
NANO-SciTech Centre, Institute of Science, Universiti Teknologi MARA, 40450 Shah Alam, Selangor, Malaysia
e-mail: Rusop8@gmail.com

N. Yahya (ed.), *Carbon and Oxide Nanostructures,* Adv Struct Mater 5,
DOI 10.1007/8611_2010_23, © Springer-Verlag Berlin Heidelberg 2010,
Published online: 31 July 2010

operation speed, smaller size and lower power consumption [1]. Besides that, the research has also been focused to prepare high performance electronic devices using low cost materials and low preparation cost.

Since many years ago, silicon (Si) has dominated the semiconductor market and become preferred materials to be used in electronic devices [2]. The maturity of its fabrication technology also becomes a factor why it has been used widely in semiconductor industry. However, Si preparation is not easy and very costly which induce interest among researchers to find alternative materials for device applications [2]. Moreover, the indirect band gap properties of Si limited its usage to optoelectronic devices [3]. Thus, gallium arsenide (GaAs) has become a material of choice for electronic devices since the material is a direct semiconductor material and has properties such as high carrier mobility and effective carrier velocity. However, due to rapidness in the progress of semiconductor industry, the devices require more stability, durability and physical characteristic of the materials which is beyond the limits of Si and GaAs.

The need of new materials for application has led to the research in wide band gap semiconductor. The wide band gap semiconductor become very promising materials due to its inherent properties such as larger band gap, higher breakdown voltage, and higher electron mobility [4, 5]. The properties are very useful and suitable for the fabrication of high power, high temperature and short-wavelength electronic devices. Among the wide band gap materials which have been studied are magnesium oxide (MgO), tin oxide (SnO), titanium dioxide (TiO_2), and zinc oxide (ZnO).

ZnO materials received a lot of attentions since many years ago where it has already featured as the subject of numerous research papers as early as 1935 [6]. It has been useful in many industries such as paints, cosmetics, pharmaceuticals, plastic and rubber manufacturing and electronics as a result of its unique properties. ZnO has become valuable due to the properties such as ultra violet absorbance, electric conductivity, transparency, piezoelectricity and luminescence [7].

In electronics, ZnO is important because of its semiconducting properties which make ZnO become prospective candidate for optoelectronic device fabrication. ZnO thin films structure exhibits high conductivity, high carrier mobility, chemical stability and transparency in the visible range [8–10]. It is also non toxic and sustainable and could be prepared using low cost materials [11]. In light emitting technology, although gallium nitride (GaN) based materials have dominated in green, blue, white and ultra violet light emitting devices, ZnO enters the arena with several advantages. Among of the advantages are high exciton binding energy, ability to grow high quality single crystal substrate with low cost, and simple crystal growth technology [12]. In solar cell technology, due to its properties such as high transparency and conductivity, ZnO materials based dye sensitized solar cell has been developed and achieved efficiency as high as 5% [13, 14]. The ZnO based solar cells are expected to be an alternative candidate for Si based solar cell replacement due to its chemical and thermal stability including the stability against photo corrosion [15]. ZnO is also used as a various kind of sensors due to its properties. Pyroelectric, piezoelectric and other sensing properties enable ZnO to be used as thermal, pressure and gaseous sensors [16, 17].

2 Zinc Oxide Properties

Zinc oxide (ZnO) is II–VI compound semiconductor which has direct bandgap energy of approximately 3.2–3.4 eV at room temperature [18–22]. It has a melting point of 1,975°C to imply strong bonding and to suggest that ZnO is thermal and chemical resist materials [23, 24]. It is not completely soluble in water but could be dissolved in alkalis and acids. ZnO is known as zinc white and commonly available in white powder. Although under certain condition ZnO have been reported growth as p-type conductivity, naturally ZnO exhibit as n-type conductivity. The n-type conductivity might be caused by intrinsic defects such as zinc interstitials and oxygen vacancies [25]. The n-type conductivity of ZnO could be enhanced by introducing elements such as boron, aluminum, gallium and indium as dopant materials [26–29].

ZnO crystals are commonly found stable with hexagonal symmetry or wurtzite structure. ZnO has lattice parameters of a = 0.3249 nm and c = 0.5207 nm at 300 K which give c/a ratio of 1.602 to indicate that ZnO structure is closed with an ideal hexagonal closed packed structure (1.633) [24, 30]. Hexagonal (wurzite) structure ZnO consists of zinc (Zn) atoms which are tetrahedrally coordinated to four oxygen (O) atoms, where the Zn d-electrons hybridize with the O p-electrons [31]. There are four most common face terminations of wurtzite ZnO which are (0001), (000-1), (11-20) and (10-10). The (0001) is the polar Zn terminated plane and is the basal plane while (000-1) is the polar O terminated plane. The Zn terminated plane produce positively charged polar surface while the O terminated plane produce negatively charged polar surface [32]. The (11-20) and (10-10) are the non-polar planes which contain an equal number of Zn and O atoms. The tetrahedral coordination gives rise to polar symmetry along the hexagonal axis that is responsible to the properties of ZnO such as piezoelectricity and spontaneous polarization, and is also a key factor in crystal growth, etching and defect generation. The conduction band (CB) of ZnO arises from the Zn^{++} 4s orbital (symmetry $\Gamma 7$) and the upper valence bands (VB) from the O^{--} 2p states with an admixture of Zn^{++} 3d levels [31]. The valence band maxima and the conduction band minima occur at the same axis of Brillouin zone to indicate that ZnO is direct a semiconductor material.

ZnO is an important material which exhibit UV emission, UV absorption and visible light transparency [33–36]. As a direct and wide band gap semiconductor material, ZnO is considered as a good candidate for electronic devices application. The direct and wide bandgap properties make ZnO may have a high breakdown voltage, lower noise and could be operate at high power and high temperature [37]. The properties also allow ZnO to be used in light emitting devices application such as blue and ultraviolet light emission diode. High exciton binding energy of ZnO materials which is about 60 meV induces efficient light emission at room temperature [19]. The transparency properties allow ZnO materials to be used in solar cell technology as transparent electrode. The UV light absorption properties give a lot of benefits as a material in UV sensor and UV protection layer fabrications.

3 Preparation Methods of ZnO Thin Film

There are many techniques of ZnO preparation. Among them are molecular beam epitaxy (MBE), chemical vapor deposition (CVD), pulsed laser deposition (PLD), metal organic chemical vapor deposition (MOCVD), sputtering and sol-gel method [38–44]. MBE and MOCVD technique could produce high quality ZnO crystal but the systems are very expensive, complicated and require exclusive maintenance. PLD and sputtering also produce high quality ZnO crystal but require high quality target and vacuum chamber. Moreover, the techniques also require rigid experimental conditions such as cleaning, vacuum and controlled gas atmosphere [7]. CVD is a promising technique for ZnO preparation since it could prepare ZnO with good properties but need optimization in system for large scale deposition. The system also requires carrier gas such as argon (Ar) gas and high temperature to operate which could increase a cost for final product. Sol-gel method is a good preparation technique of ZnO since it could be done for large scale deposition, low temperature and relatively low cost compared other preparation technique [45]. The technique also could produce high quality ZnO for device application and expected to be used in industry for its cost effectiveness. However the technique requires optimization of solution preparation to obtain ZnO structure with good properties.

ZnO could be grown on wide range of substrate such as silicon (Si), gallium nitride (GaN), sapphire and glass [46, 47]. The growth ability on the glass substrate is very advantageous since the glass is a cheap substrate. The ability also indicates ZnO could be grown on amorphous and inexpensive substrate for research purpose and electronic devices fabrication. ZnO also could be prepared at relatively low temperature compared to other wide band gap materials which is benefited to manufacture electronic devices at lower cost.

3.1 Sol-Gel Method

The sol-gel process is a wet-chemical procedure in which a solution of a metal compounds or a suspension of very fine particles in a liquid (referred to as a sol) is converted into a semi-rigid mass (a gel) [48]. The sol-gel consists of mixture of solid materials suspended in a liquid solution. Gel occurs when the individual molecules form structures which then form molecules matrix network that are same with the formation of semiconductor crystals but without ordered spacing. Through this process, pores are formed between particles. The molecules matrix becomes large and dense depend on process, time and temperature. During thin films deposition, the parameters of deposition process such as deposition speed, deposition time, drying time and drying temperature will determine final material properties including structural, optical and electrical properties.

Sol-gel is attractive process because it offers low processing temperature, possibility of various forming process and high homogeneity and purity of resulting materials. The sol-gel process allows preparing materials in a wide variety of forms such as powders, thin films, microporous inorganic membranes, ceramic fibers and extremely porous materials (aerogels).

The sol-gel technology involved a few of processes which are hydrolysis, condensation, ageing, drying and densification [49]. Hydrolysis process occurs when the hydroxyl groups of water become attached to the metal atom by replacing the ligands in the precursors as shown in (1) [48]:

$$M(OR)_n + xH_2O \rightarrow M(OR)_{n-x}(OH)_x + xROH \tag{1}$$

The process where the hydroxyl groups merge to form metal-oxygen-metal bonds and release a water molecule is called as condensation. During the process, the chains and networks of particle are formed. The process of condensation may occur by oxolation and alcoxolation process. The oxolation process occurs when two hydroxide branches connected releasing water molecules as indicated in (2) [48]:

$$-M-OH + HO-M- \rightarrow -M-O-M- + H_2O \tag{2}$$

The process of when a hydroxide branch with an alkoxide branch connected and releasing an alcohol molecule is called alcoxolation. The process is shown in (3) [48]:

$$-M-OH + R-O-M- \rightarrow -M-O-M- + ROH \tag{3}$$

During ageing process where the solution is stirred at room temperature, both hydrolysis and condensation reactions are expected to occur to form a continuous solid network suspended in fluid phase. The process is called as polymerization.

The drying process is the process to evaporate solvent, water and other volatile components of the sol-gel materials that leads to the formation of an elastic or viscoelastic gel-like state. The process is very important in thin film preparation where the thickness of the film could be increased by the repetition of deposition and drying process. The process also produces stronger adhesion between the particles and the substrate.

Densification refers to the thermal treatment of the materials which collapse the open structure and reducing pores to form dense materials. The thermal treatment which is related to process of annealing will produce amorphous or crystalline phase of materials depend on applied energy and materials itself.

3.2 Sol-Gel in ZnO Thin Film Preparation

Sol-gel method has been widely used to prepare semiconductor solution materials such as titanium dioxide (TiO_2) and zinc oxide (ZnO). The method is well established due to its process which is simple, inexpensive, non-vacuum and low temperature to synthesized film. Moreover, the method has it merits including excellent control of the stoichiometry of precursor solutions, ease of compositional modifications, customizable microstructure, ease of introducing various functional groups, relatively low annealing temperatures and the possibility of coating deposition of large area substrates [50]. In our research, ZnO thin films have been prepared using three main materials which are zinc acetate dihydrate as a precursor or starting material, monoethanolamine (MEA) as a stabilizer and 2-methoxyethanol as a solvent.

The use of anhydrous zinc acetate, which avoids the introduction of large amounts of water to the sol-gel, enables the control over reactions within the sol-gel [21]. Monoethanolamine or other aminoethanols such as diethanolamine (DEA) and triethanolamine (TEA) act as bidentate ligands to Zn^{2+} to stabilize the solution against any precipitation thus producing clear solution for coating procedure [51]. Stabilizer plays very important role in the sol-gel process. It prevents uncontrollable growth of particles and also prevent particle from aggregation. Moreover, the stabilizer will control growth rate of particle thus controlling particle size. The other function of stabilizer is allowing particle solubility in various solvents.

MEA contain two functional segments which both plays a significant role as a complexing agent to govern the hydrolysis of zinc acetate dehydrate. MEA extends into the particle surface to lose their conformational entropy and to inhibit the particle agglomeration in polymer matrix by steric repulsion. MEA retarded condensation process but promote the formation of ZnO because of the presence of amine which increases the solution pH [52]. The use of MEA also induces a preferential growth along c-axis of ZnO which could avoid optical scattering in the thin films [36].

The process occurs for ZnO growth during sol-gel method has been described by literature [52]. In alcoholic solution, zinc acetate dihydrate will dissolve to form Zn^{2+}, CH_3COO^-, H^+ and OH^- where H^+ and OH^- come from dihydrate component. MEA, OH– and CH_3COO^- are competed each other to interact with Zn^{2+}. Zn^{2+} will interact with CH_3COO^- and OH^- to form zinc monoacetate ion ($ZnCH_3COO^+$) and zinc hydroxide ion ($ZnOH^+$) through the process of hydrolysis. Zn^{2+} reacts with MEA to form $ZnCH_2NOH^{2+}$. $ZnOH^+$ then go through for condensation process to produce $Zn(OH)_2$ and ZnO particle which then produce Zn-oxo-acetate colloidal particle. While $ZnCH_3COO^+$ will produce zinc oxo-acetato species ($ZnO_{1-X}(AcO)_{2X}$). Ageing process produces Zn-oxo-acetate colloidal particle through the process of condensation. The hydrolysis and condensation process are very slow due to the low quantities of water thus the size of ZnO particle could be controlled to be in nanometer size [51].

Spin-coating technique is a simple way to deposit thin film uniformly from precursor solution. An instrument for this technique is called as spin-coater as shown in Fig. 1. It is a device which is able to produce uniform thin films on the substrates at a constant spin speed for a given time. The spin coater has a sample

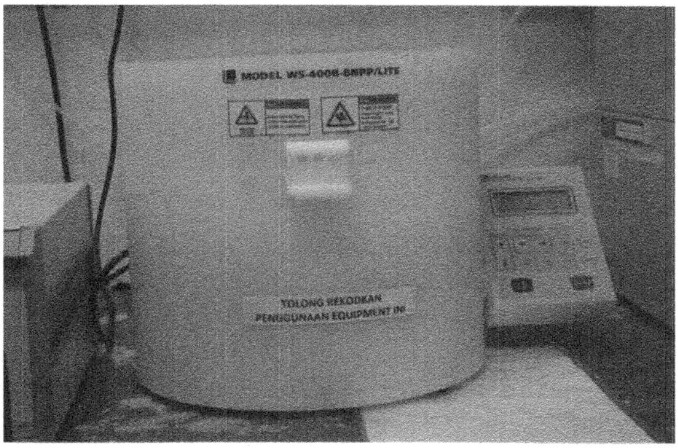

Fig. 1 Spin coater for ZnO thin film deposition process

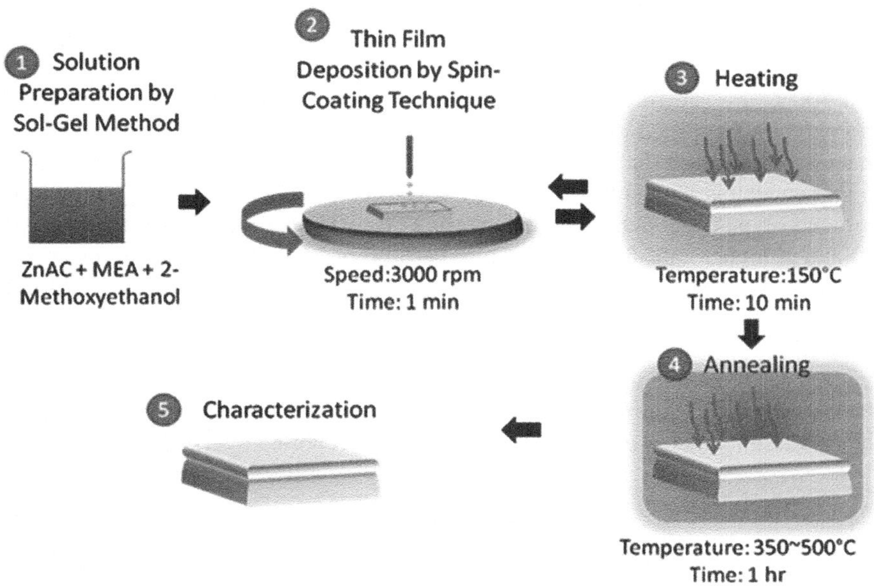

Fig. 2 Nanostructured ZnO thin film preparation by sol-gel spin-coating method

holder which use vacuum to hold the substrate during deposition process. The deposition process is done by dropping the solution onto the substrate which is accelerated at certain speed for certain time. The speed and deposition time are very important parameters which determine the thin film uniformity, morphology and thickness. An example of the thin film preparation process is illustrated in Fig. 2.

4 Effect of Annealing Process and Precursor Molar Concentration

4.1 Surface Morphology

In the sol-gel spin-coating method, there are many processing parameters that we need to consider in order to obtain good ZnO films. One of them is an annealing process. Annealing process plays very important role in a material growth mechanism. Generally, annealing process is a process to apply energy (usually thermal energy) on the materials to change its properties including electronic and structural properties. During the annealing process, due to the supplied energy, electron diffusion and atom rearrangement occurs to favorable energy position which improves the materials properties. There are many interesting characteristics of the materials that could be observed after the annealing process including an evolution of surface morphology and crystallization degree.

The dependence of annealing temperature on surface morphology could be investigated using field emission scanning electron microscope (FESEM). For example, FESEM images for ZnO thin films deposited on silicon substrate at 350 and 500°C are shown in Fig. 3. We can observe that the transformation in film morphology happened at the two of annealing temperatures. From the FESEM images, it could be understood that the ZnO film thin film become denser with higher annealing temperature up to 500°C. The condition could be originated from particles expansion during annealing process. The particles have tendency to expand at higher annealing temperature because of the energy supplied during the annealing process cause an atom rearrangement to form a crystalline structure. Since the thin film is deposited at room temperature using sol-gel spin-coating technique, the ZnO particles are mostly in amorphous structure. By annealing process, it favors the diffusion of atoms absorbed on the substrate and accelerates the migration of atoms to the favorable energy positions, resulting enhancement in the crystallinity and c-axis orientation as has been reported by M. Rusop et al. [53].

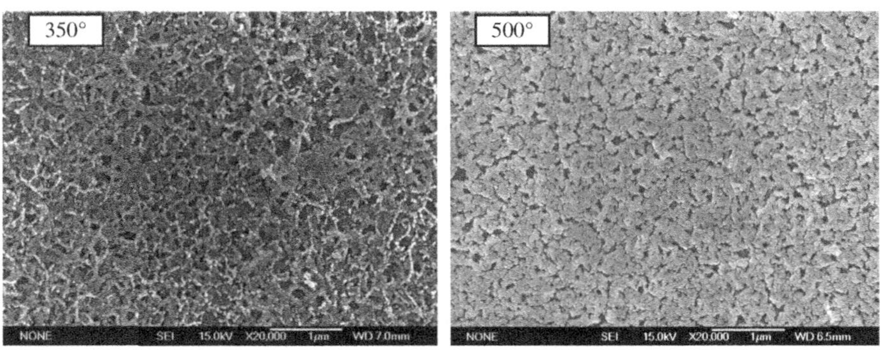

Fig. 3 SEM images of ZnO thin films annealed at 350 and 500°C

Other immediate effects of annealing process are porosity reduction and particle size increment. As the thin film become denser at higher annealing temperature, the porosity of thin film become lower. The particle expansion removed porosity from the thin films through space filling process between particles resulting denser thin film at higher annealing temperature. Through the migration of atoms to the favorable energy position, it produces larger ZnO particles when the thin film is annealed at higher annealing temperature. We can observe this transformation in the FESEM images whereby the average particle size of ZnO annealed at 350 and 500°C are measured to be 37 and 56 nm, respectively.

It is interesting to observe that the ZnO particles synthesized by sol-gel method are not in perfectly spherical shape. They exhibit some edges rather than perfect curved surfaces due to the difference in surface energy of crystallographic directions in the ZnO growth. In the wurtzite structure, the relative growth rate of each crystallographic plane differs somewhat according to the crystal orientation, so that it is difficult for crystalline ZnO to grow symmetrically into spherical particles [54].

4.2 X-Ray Diffraction (XRD) Spectra

The crystallinity of ZnO could be characterized by X-ray diffractometer (XRD). The degree of crystallization is determined based on diffracted incident X-ray beam intensity when it hit on the atoms which might have arrangement in random (amorphous) or regular pattern (crystalline) or both. A destructive interference occurs when the beam hits an amorphous structure while a constructive interference produced when the beam hits a crystalline structure. The latter interference gives a crystallization signal which is translated into XRD peak representing an orientation of crystal plane or orientation. For instance, we can observe the difference of crystallization degree for ZnO thin film annealed at 350 and 500°C. The XRD spectra for thin films annealed at 350 and 500°C are shown in Fig. 4. Both films exhibit polycrystalline structure that belongs to the hexagonal wurtzite type of ZnO. There are five orientation peaks that could be observed in the spectra, identified as (100), (002), (101), (102) and (110) orientation peaks. The spectra shows that the both films having preferential growth along c-axis or (002) plane but with the different peak intensity. The thin film annealed at 500°C has higher and sharper XRD peak, indicating an improvement in (002) orientation peak intensity compare than the thin film annealed at 350°C. The result indicates that the annealing process improved ZnO crystallinity which more atoms are moved to the favorable energy position in ZnO wurtzite structure.

Another important parameter that affected ZnO crystallinity is precursor molar concentration. The XRD patterns for ZnO thin films prepared at different precursor concentrations are shown in Fig. 5. Analysis of the XRD pattern of the films revealed that they have a polycrystalline structure. It is observed that all films have a (002) preferred orientation. The (002) diffraction peak for thin film prepared using 0.4 M solution is the highest compared to other thin films which indicates that

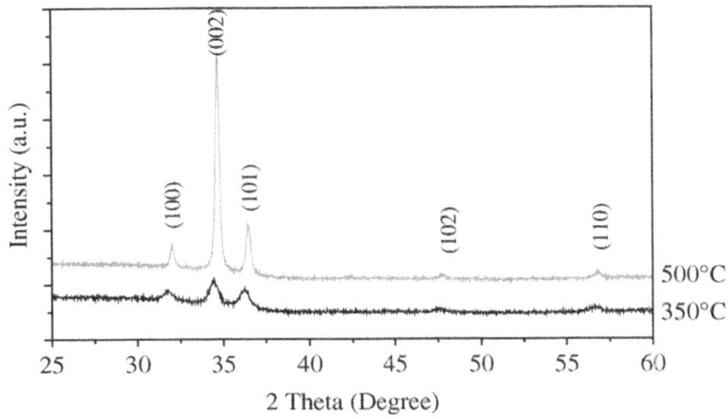

Fig. 4 XRD spectra of ZnO thin films annealed at 350 and 500°C

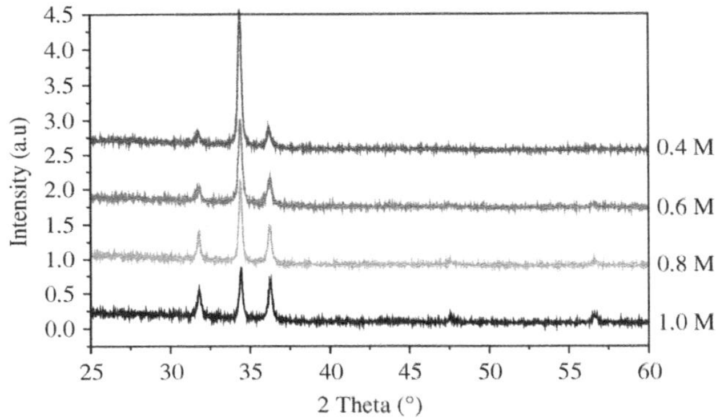

Fig. 5 XRD spectra of ZnO films prepared using different precursor molar concentrations

the film is strongly c-axis oriented. Increasing of solution concentration up to 1.0 M however led to decreasing of (002) orientation growth in the thin films but increased other orientations growth, namely (100), (101), (102), and (110). The increment of the other peaks also indicates the increment of the grain boundaries density in the thin films. The similar characteristic also reported by Y.-S. Kim et al. using isopropanol as solvent [36]. They reported that the (002) peak intensity reduces when the thin film is prepared at higher molar concentration up to 1.3 M.

To prepare high quality ZnO, the growth of ZnO must be oriented at its lowest free surface energy which is (002) plane or c-axis. The growth of material at its lowest free surface energy relieves the internal stress in the material which improves its properties such as optical and electrical properties. The growth along the lowest free surface energy plane also produces material close to an ideal crystal structure and with less defects. In most of electronic device fabrication, it is desirable since it could avoid optical scattering and enhance transparency properties [36]. For piezoelectric devices, the c-axis preferential thin film is important because the largest piezoelectric coupling constant for ZnO crystals is along the c-axis [39].The c-axis oriented films also has lower the resistivity due to shorter carrier path length in a c-plane and reduction in the scattering of carriers at grain boundaries and crystal defects which increase the apparent mobility of electrons [27].

4.3 Absorption Coefficient Spectra

Absorption coefficient gives information of material light absorbance capabilities at a certain light wavelength range at a certain thickness. It is very useful information to investigate the material optical absorption properties which suitable for a specific device or certain applications. For ZnO materials, the absorption coefficient spectrum is studied for some electronic device applications such as ultra-violet photoconductive sensor and dye-sensitized solar cell (DSSC). Figure 6 shows the absorption coefficient, α of ZnO thin films as a function of the wavelength at different annealing temperatures. The absorption coefficient has been calculated using thin film transmittance data which is obtained from UV–Vis–NIR spectrophotometer measurement. The value of absorption coefficient at respective light wavelength is obtained using Lambert's law as shown in following equation (4):

$$\alpha = \frac{1}{t} \ln\left(\frac{1}{T}\right), \tag{4}$$

where t is thin film thickness and T is the transmittance spectra of thin films. It is observed in the spectra that ZnO films exhibit low absorption in the visible and near infra-red (NIR) range. Contrastably, they exhibit very high absorption in the ultra-violet (UV) range. These behaviors of visible light transparency and UV light

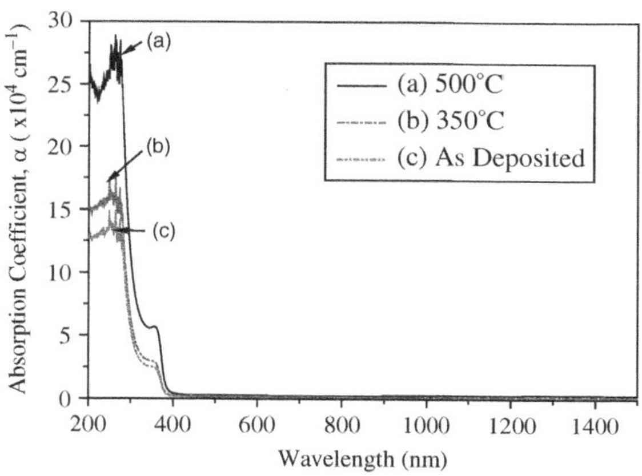

Fig. 6 Absorption coefficient of ZnO thin films at various annealing temperatures

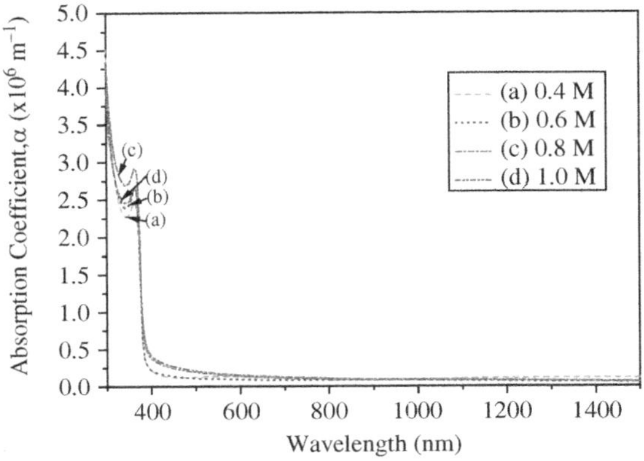

Fig. 7 Absorption coefficient of ZnO thin films prepared at different precursor molar concentrations

absorbability are beneficial for the optoelectronic and transparent electronic devices fabrication, and UV photoconductive sensor application, respectively. The UV absorption properties of ZnO thin films are due to its intrinsic optical band gap energy which is to be around 3.3 eV (~370 nm). The electron will have sufficient energy to jump from valence band to conduction band when the film is supplied with that amount of energy or higher (≥3.3 eV).

It clearly is seen in Fig. 6, the absorption coefficients of the thin films in the UV region improve with annealing temperatures. The improvement in UV absorption with annealing temperatures is very useful information especially in UV sensor fabrication process. By applying annealing treatment to the thin film, perhaps, the sensitivity or performance of the devices could be improved.

The absorption coefficient of ZnO thin films prepared at different precursor molar concentration is shown in Fig. 7. We can see that the absorption coefficient of the film at the UV region (<400 nm) is increased with molar concentration up to 0.8 M but reduce at 1.0 M. The increment in absorption coefficient with molar concentration might be due to the increment of particle size of ZnO with molar concentration or the increment of ZnO particles with molar concentration. This factor contributes to the improvement in UV light absorption properties. The reduction of UV light absorption properties for thin film prepared using 1.0 M precursor molar concentration might be due to a few reasons. Firstly, it is due to the increment of defects in the thin films as the growth along c-axis become lower at higher molar concentration. The grain boundaries effect which scatter the light as the ZnO growth along c-axis is reduced is also believed to be a factor in this phenomenon.

4.4 Photoluminescence (PL) Spectra

Photoluminescence (PL) measurement, usually, is conducted to investigate a light emission characteristic of the materials when irradiated with a light source. This photoluminescence property, however, only occurs if there is a radiative recombination between electron and hole which is normally found in a direct band gap semiconductor material. The light emission wavelength is depended on energy state difference during the recombination and thus providing some data about band gap energy and defect states in the materials. To explain this behavior on ZnO material, we have studied the PL properties of ZnO thin films prepared at different annealing temperatures and different precursor molar concentration.

The PL spectra of ZnO thin films annealed at different temperatures are presented in Fig. 8. The measurement was conducted at room temperature in the region of 350–600 nm using Xenon (Xe) lamp as a PL source with excitation of 325 nm. There are very significant difference in PL properties of as deposited thin films and annealed thin films. Two peaks centered at about 365 nm (3.40 eV) and 534 nm (2.34 eV) are observed in the PL spectrum for as deposited sample. However, PL spectra for annealed samples appear with more luminescence peaks where four detected main peaks are located at about 363 nm (3.41 eV), 394 nm (3.14 eV), 465 nm (2.67 eV) and 534 nm (2.32 eV).

The highest peak for as deposited ZnO thin film and annealed thin film which located at 363 and 365 nm, respectively, may be attributed to emission near bandgap from free exciton recombination. While, for the other emission peaks are probably originate from defect states in ZnO lattice [55]. It could be observed from the spectra that the intensity the UV emission increase with annealing temperature. This characteristic indicates that the crystallinity of the thin film improve at higher annealing temperature [36]. Other researcher reported that the increment in UV emission intensity also might be due to increment in carrier concentration which increases the recombination reaction [25]. For the emission peak at 394 nm or violet emission, it might be appear due to radiative defects from the interface traps in ZnO–ZnO boundaries since they are reported to locate about 0.3 eV below the conduction band in energy state position [56]. A broad emission peak at 465 nm or blue emission might be related to the surface defect in ZnO thin films, while a green emission at the peak around 534 nm might be associated to the recombination of a photogenerated hole with an electron that belongs to oxygen deficiency in the surface and sub-surface lattices of ZnO materials [55, 57, 58].

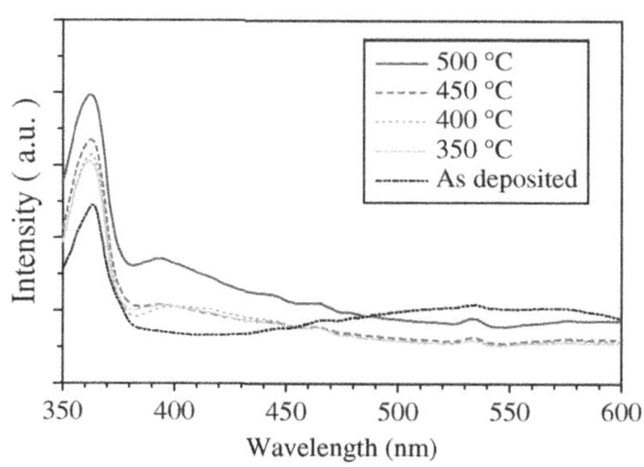

Fig. 8 Room temperature PL spectra of ZnO thin films on glass substrate annealed at different temperatures

Figure 9 shows photoluminescence (PL) spectra of ZnO thin films on glass substrates prepared using different precursor molar concentrations. The spectra indicate that all films emit UV emission. It could be observed in the spectra that UV peak intensity centered at 364 nm shows increment pattern at higher molar concentrations. According to H.S. Kang et al., it might be due to the increment of carrier concentration with molar concentration which contributes to the excitonic recombination [25]. The increment of carrier concentration is due to increment of Zn species in ZnO thin film with precursor molar concentrations. M. Dutta et al. reported that the size of ZnO particle increase when the molar concentration increase which increase the Zn^{2+} species in the ZnO thin film [59].

Visible emissions center at 392, 465 and 533 nm are significantly appear for thin film prepared at higher precursor molar concentration which indicates the increment of defects concentration in ZnO crystal. The defects such as oxygen vacancies, zinc interstitials, zinc vacancies, oxygen interstitials or oxygen antisites have been reported to be the origin of visible light emission [34, 55]. This behavior gives a conclusion that the thin films prepared using higher precursor molar concentration produce higher defect concentrations.

4.5 Current–Voltage (I–V) Spectra

Electrical properties of the semiconductor thin film sometimes represent the film uniformity, crystallinity and carrier concentration. Thus, electrical property of the thin film is considered one of the criteria that we need to study in order to have a quality thin film. There are a few measurement methods which we could use to characterize the electrical properties of thin film such as Hall Effect measurement and current–voltage (I–V) measurement. The I–V measurement is a most common method and essential which is widely used to study electrical conductivity of the thin film. Through I–V spectra, we can investigate the improvement or degradation of thin film electrical conductivity when deposited at certain parameters.

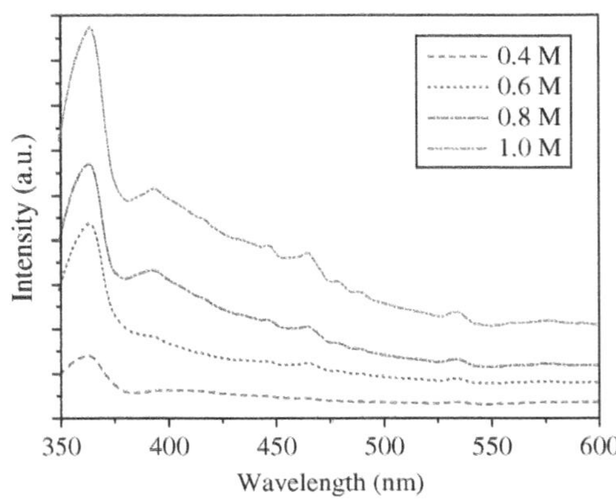

Fig. 9 PL spectra of ZnO thin films prepared using different precursor's molar concentration

For instance, Fig. 10 shows Current–Voltage (I–V) measurement spectra for ZnO thin films at different annealing temperature using gold (Au) as metal contacts. The results shows Au gives Ohmic contact to the ZnO thin films prepared in this study. It was found that the current intensity of the thin film at respective voltage increased with annealing temperatures.

Figure 11 shows electrical conductivity of ZnO thin films at different annealing temperatures calculated from I–V curve data. The result indicates the conductivity of ZnO thin films increased with annealing temperatures up to 500°C. The result suggested the improvement in electron mobility and electron concentration at higher annealing temperature to increase the conductivity of the thin films. During annealing process, rearrangement of atoms occurs to form fine crystalline structure. Energy supplied during annealing process favors the diffusion of atoms absorbed on the substrate and accelerates the migration of atoms to the favorable energy position [53]. The higher the annealing temperature, the higher will be energy supplied for the atoms rearrangement process. Thus, the crystallinity and c-axis orientation of ZnO improved with annealing temperatures which result in enhancement of electron mobility and electron concentration.

Current–Voltage (I–V) measurement results of ZnO thin films prepared using different precursor molar concentration are shown in Fig. 12. The results show all

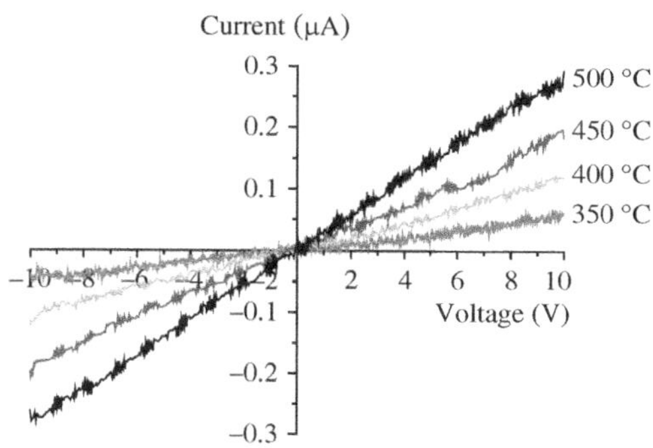

Fig. 10 I–V curve of ZnO thin films at different annealing temperatures using Au metal contact

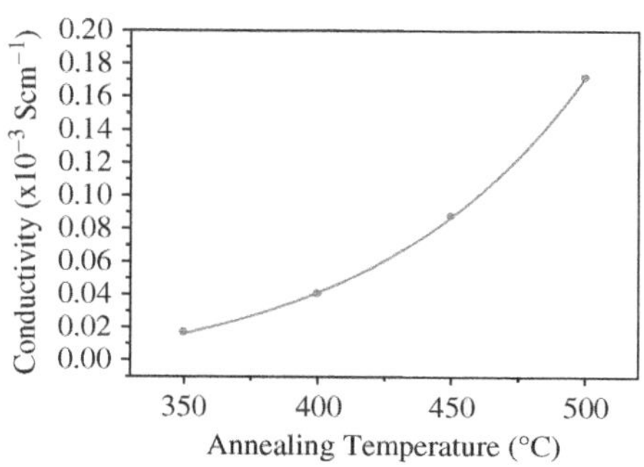

Fig. 11 Electrical conductivity of ZnO thin films with annealing temperatures

films exhibit Ohmic behavior with Au metal contact as linear I–V curve obtained from the measurement. It could also be seen that current value at respective voltage increased with molar concentration to show the increment of electron concentration at higher molar concentration.

The conductivity plots of ZnO thin films as the function of precursor solution concentrations is shown in Fig. 13. The result shows the conductivity is increased at higher molar concentration. This may attribute from the increasing of Zn species with higher solution concentration which increased carrier concentration in the thin films [60]. Higher solution concentration also increased particle size in the thin film which may increase the surface contact between particles and improve packing density in the thin films. This resulting reduction of oxygen adsorption in grain boundaries thus reducing carrier trapping phenomenon in the thin films to decrease thin films resistivity with higher solution concentration. The carrier traps could become electrically charged to create a potential energy barrier between ZnO particles that prevent carrier to move from one particle to another, leading in decrement in carrier mobility of ZnO thin films prepared using lower precursor molar concentration. Moreover, improvement in surface contact between particles as a result of particle size increment with molar concentration produces

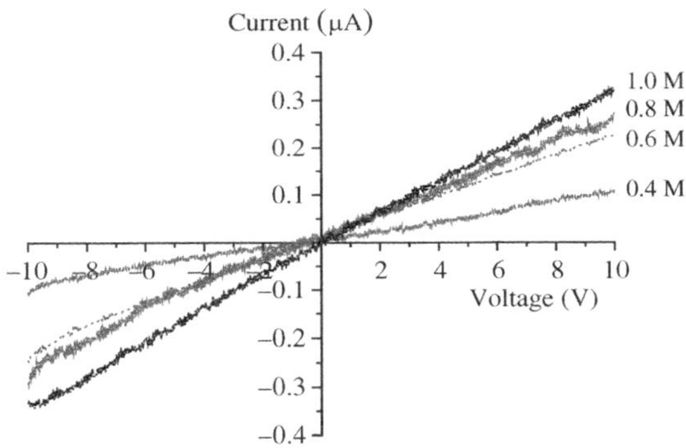

Fig. 12 I–V curve of ZnO thin films with precursor molar concentrations

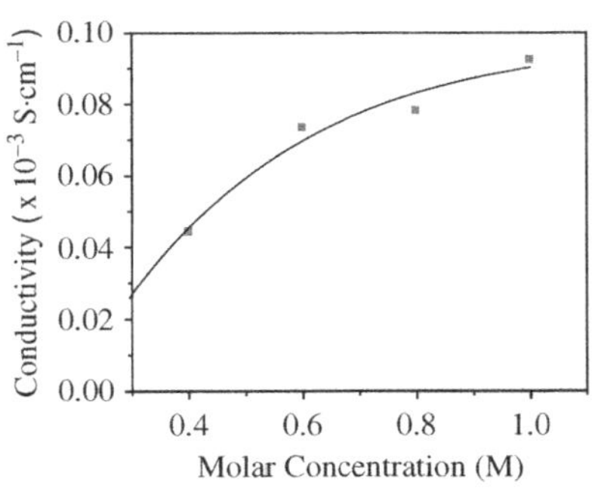

Fig. 13 Conductivity of ZnO thin films prepared using different molar concentrations

better electron mobility in the thin films to reduce resistivity at higher molar concentration [38].

5 Conclusion

This chapter provides an overview of the ZnO preparation and characterization, with particular attention given to ZnO thin film properties prepared using sol-gel spin-coating method. This preparation method is a promising technique which offers a lot of advantages such as low cost and large scale deposition possibility. There are many parameters in sol-gel processing which significantly affected ZnO thin film properties. For example, there are precursor molar concentration, deposition speed, pre-heating temperature and annealing temperature which could influence the thin film characteristic. This chapter has shown the effect of annealing temperature and precursor molar concentration on the ZnO properties including surface morphology, crystallinity, absorption coefficient, photoluminescence and electrical properties. Although there are many works remain to develop and optimize suitable ZnO preparation process, undoubtedly, the sol-gel spin-coating will become one of the established ZnO preparation methods in the semiconductor industry for electronic device and transparent electrode applications.

Acknowledgements The author would like to thank Universiti Teknologi MARA (UiTM) Malaysia and Jabatan Perkhidmatan Awam (JPA) Malaysia for the financial support. The author also would like to acknowledge NANO-SciTech Centre staff and members for their support during this research.

References

1. Lu, J.G., Chang, P., Fan, Z.: Mater. Sci. Eng. R **52**, 49–91 (2006)
2. Pfeiffer, M., Leo, K., Zhou, X., Huang, J.S., Hofmann, M., Werner, A., Blochwitz-Nimoth, J.: Org. Electron. **4**, 89 (2003)
3. Tsakalakos, L.: Mater. Sci. Eng. R **62**, 176 (2008)
4. Agarwal, K., Mani, S.S., Seshadri, S., Cassady, J.B., Sanger, P.A., Brandt, C.D., Saks, N.: Naval Res. Rev. **51**(1), 14–21 (1999)
5. Shenai, K., Scott, R.S., Baliga, B.J.: IEEE Trans. Electron Devices **36**(9), 1811–1823 (1989)
6. Yearian, H.J.: Phys. Rev. **48**, 631–639 (1935)
7. Jimenez Gonzalez, A.E., Soto Urueta, J.A.: Sol. Energy Mater. Sol. Cells **52**, 345–346 (1998)
8. Chopra, K.L., Major, S., Pandya, D.K.: Thin Solid Films **102**, 1–46 (1983)
9. Minami, T.: MRS Bull., 38–44 (2000)
10. Igasaki, Y., Saito, H.: J. Appl. Phys. **69**, 2190–2195 (1991)
11. Ismail, B., Abaab, M.A., Rezig, B.: Thin Solid Films **383**, 92 (2001)
12. Ozgur, U., Alivov, Y.I., Liu, C., Teke, A., Reshchikov, M.A., Doan, S., Avrutin, V., Cho, S.J., Morkoc, H.: J. Appl. Phys. **98**, 041301 (2005)
13. Zhang, Q., Chou, T.P., Russo, B., Jenekhe, S.A., Cao, G.: Angew. Chem. Int. Ed. **47**, 2402 (2008)

14. Ranga Rao, A., Dutta, V.: Nanotechnology **19**, 445712 (2008)
15. Kalyanasundaram, K., Grätzel, M.: Coord. Chem. Rev. **177**, 347 (1998)
16. Ryu, H.W., Park, B.S., Akbar, S.A., Lee, W.S., Hong, K.J., Seo, Y.J., Shin, D.C., Park, J.S., Choi, G.P.: Sens. Actuators B **96**, 717 (2003)
17. Karpina, V.A., Lazorenko, V.I., Lashkarev, C.V., Dobrowolski, V.D., Kopylova, L.I., Baturin, V.A., Pustovoytov, S.A., Karpenko, A.J., Eremin, S.A., Lytvyn, P.M., Ovsyannikov, V.P., Mazurenko, E.A.: Cryst. Res. Technol. **39**(11), 980–992 (2004)
18. Srinivasan, G., Kumar, J.: Cryst. Res. Technol. **41**(9), 893–895 (2006)
19. Lee, J.-H., Park, B.-O.: Mater. Sci. Eng. B **106**, 245 (2004)
20. Whangbo, S.W., Jang, H.K., Kim, S.G., Cho, M.H., Jeong, K., Whang, C.N.: J. Korean Phys. Soc. **37**(4), 456 (2000)
21. O'Brien, S., Koh, L.H.K., Crean, G.M.: Thin Solid Films **516**, 1391–1394 (2008)
22. Preetam Singh, Ashvani Kumar, Deepak, Davinder Kaur: J. Cryst. Growth **306**, 303 (2007)
23. Norton, D.P., Heo, Y.W., Ivill, M.P., Ip, K., Pearton, S.J., Chisholm, M.F., Steiner, T.: Mater. Today **7**, 35 (2004)
24. Olvera, M.D., Maldonado, A., Asomoza, R., Lira, M.M.: J. Mater. Sci. Mater. Electron. **11**(1), (2000)
25. Kang, H.S., Kim, G.H., Lim, S.H., Chang, H.W., Kim, J.H., Lee, S.Y.: Thin Solid Films **516**, 3147 (2008)
26. Pawar, B.N., Jadkar, S.R., Takwale, M.G.: J. Phys. Chem. Solids **66**, 1779 (2005)
27. Musat, V., Teixeira, B., Fortunato, E., Monteiro, R.C.C., Vilarinho, P.: Surf. Coating Technol. **180–181**, 659–661 (2004)
28. Park, S.-M., Ikegami, T., Ebihara, K.: Thin Solid Films **513**, 90 (2006)
29. Caglar, Y., Ilican, S., Caglar, M., Yakuphanoglu, F.: Spectrochim. Acta A **67**, 1113 (2007)
30. Steiner, T., Lu, Y., Zhong, J.: Semiconductor Nanostructure for Optoelectronic Applications, Chapter 6: Zinc Oxide Based Nanostructures. Artech House, Inc, Boston (2004)
31. Klingshirn, C., Hauschild, R., Priller, H., Decker, M., Zeller, J., Kalt, H.: Superlattice. Microst. **38**, 210 (2005)
32. Wang, Z.L.: J. Phys. Conf. Ser. **26**, 1 (2006)
33. Jung, E.S., Lee, J.Y., Kim, H.S.: J. Korean Phys. Soc. **47**, S482 (2005)
34. Sun, Y., George Ndifor-Angwafor, N., Jason Riley, D., Ashfold Michael, N.R.: Chem. Phys. Lett. **431**, 352–357 (2006)
35. Huang, B., Li, J., Wu, Y.-B., Guo, D.-H., Wu, S.-T.: Mater. Lett. **62**, 1316 (2008)
36. Kim, Y.-S., Tai, W.-P., Shu, S.-J.: Thin Solid Films **491**, 153–160 (2005)
37. Jin, Xu: MSc thesis, Virginia Commonwealth University, USA (2004)
38. Ogata, K., Sakurai, K., Fujita, S., Fujita, S., Matsushige, K.: J. Cryst. Growth **214–215**, 312 (2000)
39. Deng, H., Russell, J.J., Lamb, R.N., Jiang, B., Li, Y., Zhou, X.Y.: Thin Solid Films **458**, 43 (2004)
40. Gupta, V., Bhattacharya, P., Yuzuk, Y.I., Sreenivas, K., Katiyar, R.S.: J. Cryst. Growth **287**, 39 (2006)
41. Park, S.H., Seo, S.Y., Kim, S.H., Han, S.W.: J. Cryst. Growth **303**, 580 (2007)
42. Kang, S.J.: J. Korean Phys. Soc. **47**, S589 (2005)
43. Kuo, S.-Y., Chen, W.-C., Lai, F.-I., Cheng, C.-P., Kuo, H.-C., Wang, S.-C., Hsieh, W.-F.: J. Cryst. Growth **287**, 78 (2006)
44. Srinivasan, G., Gopalakrishnan, N., Yu, Y.S., Kesavamoorthy, R., Kumar, J.: Superlattice. Microst. **43**, 112 (2008)
45. Sagar, P., Kumar, M., Mehra, R.M.: Mater. Sci. (Poland) **23**(3), 686 (2005)
46. Look, D.C.: Mater. Sci. Eng. B **80**, 383–387 (2001)
47. Cho, M.W., Harada, C., Suzuki, H., Minegishi, T., Yao, T., Ko, H., Maeda, K., Nikura, I.: Superlattice. Microst. **38**, 349–363 (2005)
48. Yeping, Guo: PhD thesis, Universität des Saarlandes, State of Saarland, Germany (2002)

49. Wright John, D., Sommerddijk Nico, A.J.M.: Sol-Gel Materials Chemistry and Applications, p. 4. Taylor & Francis, Great Britain (2001)
50. Kumar, N., Kaur, R., Mehra, R.M.: J. Lumin. **126**, 784 (2007)
51. Fujihara, S., Sasaki, C., Kimura, T.: Appl. Surf. Sci. **180**, 347 (2001)
52. Znaidi, L., Soler Illia, G.J.A.A., Benyahia, S., Sanchez, C., Kanaev, A.V.: Thin Solid Films **428**, 258 (2003)
53. Rusop, M., Uma, K., Soga, T., Jimbo, T.: Mater. Sci. Eng. B **127**, 151 (2006)
54. Lee, S., Jeong, S., Kim, D., Park, B.K., Moon, J.: Superlattice. Microst. **42**, 361–368 (2007)
55. Jiang, P., Zhou, J.-J., Fang, H.-F., Wang, C.-Y., Wang, Z.L., Xie, S.-S.: Adv. Funct. Mater. **17**, 1309 (2007)
56. Cordaro, J.F., Shim, Y., May, J.E.: J. Appl. Phys. **60**, 4186 (1986)
57. Navale Shalaka, C., Gosavi, S.W., Mulla, I.S.: Talanta **75**, 1319 (2008)
58. Vanheusden, K., Warren, W.L., Seager, C.H., Tallant, D.K., Voigt, J.A., Gnade, B.E.: J. Appl. Phys. **79**, 7983 (1996)
59. Dutta, M., Mridha, S., Basak, D.: Appl. Surf. Sci. **254**, 2743–2747 (2008)
60. Goyal, D.J., Agashe, C.M., Marathe, B.R., Takwale, M.G., Bhide, V.G.: J. Mater. Sci. Lett. **11**, 708–710 (1992)

Superparamagnetic Nanoparticles

Boon Hoong Ong and Nisha Kumari Devaraj

Abstract Nanoscaled magnetic materials are great candidates for fundamental and applied research. 0D, 1D and 2D magnetic nanostructures have been extensively studied previously. One of the unique phenomena that only exists in nanoscaled magnetic structure (below a certain critical size) is superparamagnetism. In this chapter, various chemical synthesis methods to obtain superparamagnetic nanoparticles are compared. Strategies to prevent agglomeration of nanoparticles and the influent factors for nanoparticle synthesis are discussed. Three examples in biomedical application are introduced and a concluding remark for future synthesis approaches is highlighted.

1 Introduction

Superparamagnetic nanoparticles is a unique class of magnetic materials which is easily magnetized with an external magnetic field while not retaining remanence (zero remanence) in the absence of the magnetic field [1]. When the size of the magnetic nano particle decreases, thermal fluctuations can overcome its anisotropy barrier and lead to random flipping of the magnetic moment. Thus, the nanoparticles lose their stable magnetic order. The hysteresis loop of superparamagnetic nanoparticle is shown in Fig. 1. Extensive research on superparamagnetic nanoparticles has successfully demonstrated both function and performance adequacy for biomedical applications [2, 3]. Nanosized metal oxide particles are widely used for

B.H. Ong (✉) and N.K. Devaraj
Faculty of Engineering, Multimedia University, Jalan Multimedia, 63100 Cyberjaya, Selangor Darul Ehsan, Malaysia
e-mail: bhong@mmu.edu.my

N. Yahya (ed.), *Carbon and Oxide Nanostructures,* Adv Struct Mater 5,
DOI 10.1007/8611_2010_24, © Springer-Verlag Berlin Heidelberg 2010,
Published online: 13 August 2010

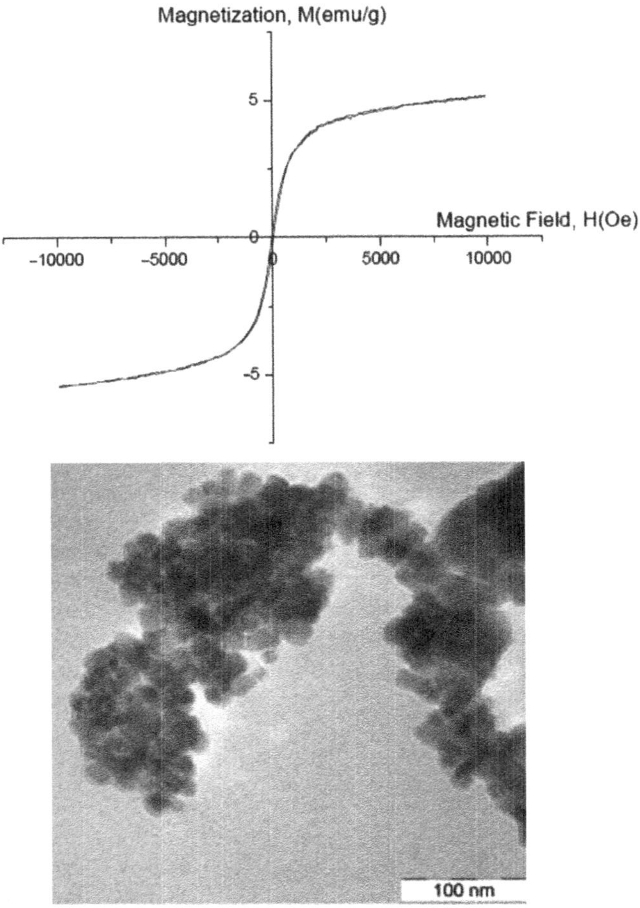

Fig. 1 Hysteresis loop of superparamagnetic iron oxide nanoparticle with size of 14.25 ± 0.9 nm

biomedical applications [4]. Examples of metal oxides include iron oxide [5, 6], cobalt oxide [7, 8], and nickel oxide [9, 10]. For biomedical applications, the nanosized metal oxide should have size comparable to size of virus (20–500 nm), protein (5–50 nm), and gene (2 nm wide and 10–100 nm long). Besides that, the particle must be able to obey Coulomb's law and can be manipulated by an external magnetic field gradient. Large surface of the particles can be modified to attach biological agents. Properties such as size monodispersity, magnetization stability, non-toxicity, biocompatibility and injectability are to make sure that the particles are not harmful and stable enough to be used in human body or in-vivo purposes. This chapter reviews recent progress on the synthesis methods with emphasis on the role of surfactant and also some other synthesis influent parameters. Biomedical applications of superparamagnetic nanoparticles are highlighted at the end of this chapter.

2 Magnetic Nanoparticles Synthesis Methods

The synthesis technique of nanoparticles is very important in determining the size and shape of the particles, their size distribution, surface chemistry, degree of structural defects or impurities as well as the distribution of such defects within the particle. These factors influence the magnetic properties of the particles. This section summarizes some of the most common methods used by researchers to prepare magnetic nanoparticles. The advantages and disadvantages of these methods are as outlined in Table 1.

2.1 Aerosol/Vapour (Pyrolysis) Method

This method has been reported for the preparation of metal and metal oxide materials [11]. The precursors are liquified to form aerosol/vapour which solidify through evaporation of solvent or through heating [11, 17]. This may be followed by further reaction with chemicals. The solids will lastly sinter to form particles [17].

2.2 Gas Deposition Method

The reactants are in gaseous form. This method is useful for the synthesis of thin films. The gaseous species will react with each other and decompose on a substrate surface to form continuous films at high temperatures [11].

2.3 Transferred Arc Plasma Induced Gas Phase Condensation Method

Nagar et al. [13] reported this method for the preparation of magnetite (Fe_3O_4) nanoparticles. First, plasma is initiated between the anode and cathode terminals in a stainless steel reaction chamber, and is made to impinge on a metal iron anode. Due to the thermal energy transferred into the anode, vapourization of the iron anode occurs, followed by oxidation in a controlled environment. The particles are formed through nucleation and growth and condense on the inner walls of the chamber. The powder is obtained by scraping the chamber walls [13].

2.4 Sol-Gel Method

Oxide and oxide based-hybrids can be easily prepared using the sol-gel technique. Hydrolysis and condensation of precursors are involved in this method. Catalysts

Table 1 Characteristics of common magnetic nanoparticles synthesis methods

Synthesis methods	Advantages	Disadvantages	References
Aerosol/vapour (pyrolysis) method	High production rate	• Large aggregates are formed • Difficult to introduce stabilization mechanism during synthesis process	[11, 12]
Gas deposition method	Useful for protective coatings and thin film deposition	Require very high temperatures	[12]
Transferred arc plasma induced gas phase condensation method	• Nanocrystalline particles can be produced on a large scale (10–20 g/h) • Reproducible and consistent	• Nanoparticles obtained are of multiple phases. Need to control temperature and oxygen pressure to achieve single phase • Particles have wide size distribution	[13]
Sol-gel method	• Particles of desired shape and length can be synthesized, useful for making hybrid nanoparticles • Particles obtained have high surface area	• Products usually contain sol-gel matrix components at their surfaces • Only particles with sizes <20 nm can be synthesized • Precursors involved could be expensive, flammable, viscous and dangerous to handle	[12, 14, 15]
Co-precipitation	• Simple and economical • Large quantities can be synthesized • Easy to tune particle properties • Does not require sophisticated equipments • Particles with high surface area are obtained	• Without surface coating, particles tend to agglomerate • Difficult to obtain narrow size distribution • Only particles with sizes between 5–12 nm can be synthesized • Final product may contain high traces of impurities	[12, 14, 15]
Microemulsion (or reverse micelle synthesis)	• Particles with uniform properties are obtained • Size of the particles can be modulated	• Surfactants are difficult to remove • Only small quantities of nanoparticles can be synthesized • Extensively agglomerated particles are generated • Nanoparticles obtained are poorly crystalline	[12, 16]
Polyol process	• Particles with well-defined shape and size can be obtained by controlling the kinetics of precipitation • Average size can also be adjusted by seeding the reactive medium with foreign particles (heterogeneous nucleation)	• Liquid polyols usually have high boiling points of around 200–250°C, thus increasing cost of production as well as raising safety concerns	[17]

(*continued*)

Table 1 (continued)

Synthesis methods	Advantages	Disadvantages	References
Electrochemical synthesis	• Size can be controlled by adjusting the electrooxidation current density (*i*) or potential (E) of the system • Aggregation can be prevented by use of surfactant • Particles with sizes >20 nm can be obtained • Since it is performed in aqueous solutions, the particles obtained are hydrophilic (soluble in water and polar solvents) and their surface can be easily modified for biomedical applications	• Need to adjust the distance between the electrodes and also the potential applied so that the desired material can be obtained • Impurities from the electrodes may be present in the final product	[14]
Pechini (citrate method)	• Components utilized are easy to handle and inexpensive • Flexible method	• Residual carbon contamination • Particles obtained have lower surface area as compared to those obtained by precipitation and sol-gel methods	[15]
Gel-to-crystalline method	• No need for calcination treatment at higher temperatures to produce final product • No expensive reactants are required		[18]

are often added to promote hydrolysis and condensation reactions [11]. The following steps (Fig. 2) are entailed in the sol-gel technique [19].

2.5 Co-Precipitation

Involves mixing of multiple chemicals simultaneously under vigorous stirring. The nanoparticles are formed through nucleation when the concentration of constituent species reaches critical supersaturation. A large number of nuclei will be formed. This will be followed almost concurrently by growth [17] and other secondary processes such as Ostwald ripening may occur, depending on the preparation parameters [19].

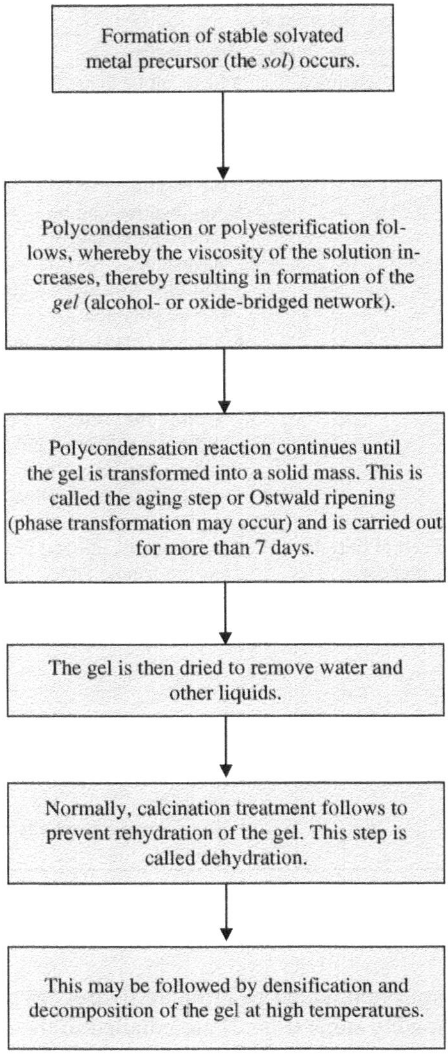

Fig. 2 Steps involved in the sol-gel synthesis method

2.6 *Microemulsion (or Reverse Micelle Synthesis)*

Synthesis is achieved through confining the reactants in a restricted space (inside micelles, formed by dissolving surfactant or polymer in a solvent). The particles stop growing when the reactants are fully consumed [11]. The particles are stabilized by a monolayer of the surfactant [20].

2.7 Polyol Process

Liquid polyol (such as ethylene glycol, trimethylene glycol or tetraethylene glycol) [21] is used as the solvent for precursor solution, as the reducing agent and sometimes as the complexing agent for metallic precursor [17]. The solution is stirred and heated to the boiling point of polyol. Formation of intermediate phases as well as reduction of metal ions is followed by nucleation and growth processes, thus resulting in the generation of particles [21].

2.8 Electrochemical Synthesis

The synthesis is performed by using suitable cathode and anode materials in the presence of an electrolyte. The particles are formed when a voltage potential and current density are applied to the system due to electron transfer between the electrodes and the electrolyte [14].

2.9 Pechini (Citrate Method)

This is also known as the citrate method. Synthesis is performed by combining a metal precursor with water, citric acid and a polyhydroxyalcohol, such as ethylene glycol. The mixture is then heated to remove the water, thus leaving behind some viscous oil. A solid resin is obtained by heating the viscous oil to a temperature that polymerizes the citric acid and ethylene glycol. Finally, the resin is calcined at temperature $>500^{\circ}C$ to burn off the polymer matrix, hence resulting in porous particles [15].

2.10 Gel-to-Crystalline Method

This method was used by Ozkaya et al. [18] to prepare Fe_3O_4 nanoparticles. The synthesis is performed by adding a strong base to ferrous chloride solution to form a metal hydroxide gel (polymeric chains forming an entangled network in which the solvent is entrapped). Continuously in-fluxing the solvent breaks the gel network and creates small crystalline iron oxide regions at around $100^{\circ}C$ [18].

Physical methods such as aerosol/vapour pyrolysis and gas deposition methods are not usually employed due to the high cost involved and sophisticated equipment requirement. On the other hand, through wet chemical techniques such as co-precipitation, water-in-oil (W/O) microemulsion (or reverse micelle synthesis), polyol, high temperature decomposition of organic precursors, electrochemical method and sol-gel method [17], the particle size, chemical composition, morphology [18] and sometimes even their shapes can be controlled by varying the experimental parameters. In addition, chemical approach is also simpler, more efficient, economical [12] and can produce large quantities of the final product [20].

One of the most common chemical techniques in the synthesis of magnetic nanoparticles is co-precipitation. This method entails the addition of two types of solutions (with different ionic charges and molar ratio, containing the salt of the metal to be synthesized as nanoparticles) to an alkaline base under vigorous stirring to precipitate the nanoparticles. The advantage of using this method is that the average size of the nanoparticles can be varied by controlling the pH of the medium, reaction temperature, molar ratio of the salts, concentration of the base, ionic strength of the aqueous medium [22] and the reaction time [20].

However, a major weakness arising from the co-precipitation technique is the agglomeration of the particles [23], thus resulting in a larger particle size [12] and limited control over size distribution [22]. This is highly unacceptable in industrial and biomedical applications as any slight diversion in the physical, magnetic or chemical properties of the nanoparticles can adversely affect their efficacy. Mono-dispersed nanoparticles are required for various applications because properties of the clusters formed by agglomerated nanoparticles differ from their individual counterparts and are also not biocompatible [24].

Several researchers have come up with different preparation methods to identify the optimum technique that can produce magnetite nanoparticles with uniform size, shape and chemical composition. Zhu et al. [25] used a novel method called precipitation with forced mixing to synthesize the magnetite nanoparticles. Results indicated that the particles synthesized via this method had sizes ranging from 6 to 13 nm with uniform distribution. Suber et al. [26] explored the synthesis of iron oxide particles, tubes and fibrils within the pores of nanoporous polycarbonate and alumina membranes. According to Kim et al. [27], controlled co-precipitation by spraying the iron salts instead of dropping them through the typical pipette method is relatively simple compared to other methods and offers good control over particle properties.

Park et al. [28] developed a method for the ultra-large scale synthesis of monodisperse nanoparticles. Not only does their synthesis technique enable the production of 40 g of magnetite nanoparticles in a single batch of reaction, but it also uses inexpensive and environmentally friendly reactants. In addition, the particle size may be adjusted by changing the experimental parameters. Their technique is also very economical, simple as well as general and can be repeated for other types of transition metal oxides. In their research, metal-oleate precursors were first prepared by reacting metal chlorides and sodium oleate. The metal-oleate precursors were then subjected to thermal decomposition in high boiling solvent to produce monodisperse nanoparticles. Simply by controlling the aging period and temperature, the size of the nanoparticles could be varied. By setting up reactors in parallel, multi-kilograms of monodisperse nanoparticles can be prepared.

On the other hand, Wang et al. [29] developed a general strategy for the synthesis of nanocrystals. In their study, uniform noble metal nanocrystals were produced through the reduction of noble metal ions by ethanol at a temperature of

20–200°C under hydrothermal or atmospheric conditions. The reaction involved in the synthesis of the nanocrystals is a liquid–solid-solution (LSS) process whereby the reduction of the metal ions takes place at the interfaces of metal-linoleate (solid), ethanol–linoleic acid liquid phase (liquid) and water–ethanol solutions (solution). This LSS phase transfer and separation process can be used to produce semiconducting, magnetic, dielectric and fluorescent nanoparticles. The prepared nanocrystals can also be dispersed in non-polar solvents to form homogenous colloidal solutions that usually remain stable for months.

Some of the other methods used to synthesise iron oxide nanoparticles are [30]:

1. Sonochemical decomposition of iron pentacarbonyl
2. Thermal decomposition of iron complexes followed by oxidation

Through these two methods, monodispersed nanoparticles with sizes 3–20 nm for magnetite and 4–16 nm for maghemite were obtained by varying preparation parameters.

However, most organometallic precursors used in some of these synthesis processes are toxic and expensive. In addition, the decomposition temperature is very high, typically 250–320°C. Wen et al. [31] reported the preparation of monodispersed magnetite nanoparticles at a mild temperature of 74°C by using cheap and non-toxic precursors. Their methodology overcomes the disadvantages of, and at the same time, combines the advantages of chemical co-precipitation and thermal decomposition methods. As such, their method is cost-effective, able to produce monodispersed nanoparticles at a large scale, involves low synthesis temperature and low toxicity. They utilized ferric chloride hexahydrate, ferrous chloride tetrahydrate and sodium oleate, which were mixed in toluene/ethanol/water mixture solvent and were refluxed at 74°C to obtain monodispersed magnetite nanoparticles. The particles were mostly well-dispersed with an average size of 4–5 nm. However, the M_s value obtained was 23.6 emu/g, which is significantly lower than the bulk magnetite value of 92 emu/g.

In summary, results from various literatures show that the synthesis process determines the particle size and size distribution. This subsequently has an effect on the magnetic property of the sample.

3 Stabilization of Magnetic Nanoparticles

There are two ways to prevent agglomeration of nanoparticles: steric stabilization and electrostatic stabilization [32]. Steric stabilization entails coating the particles with organic or inorganic molecules during or after the synthesis process way to avoid the aggregation and oxidation of the particles [12]. The latter approach involves introducing surface charges on the particles to provide electrostatic repulsion between the particles [32]. For this, usually the particles have to precipitated at a pH far away from the point of zero charge (PZC) of the material so that the adsorption of hydroxyl ions (or protons) onto the surface of the particles will be increased [33]. In the pH range of 5–9, stable magnetite colloidal solutions cannot be obtained without the use of surfactants due to neutralization of the surface charges [34].

For steric stabilization, the suitable coating material has to be chosen carefully according to the application of the nanoparticles. For example, the coating material has to be non-toxic and biocompatible for biomedical applications. An ideal surface coating should fulfill the following requirements [35]:

- Should prevent the nanoparticles from aggregation during long-term storage
- Soluble in water
- Can retain functionalities of the nanomaterials
- Able to maintain biocompatibility of the particles so that they may interact with their biological targets

Magnetic nanoparticles are usually coated with polymer molecules [12]. Petri-Fink et al. [36] co-precipitated ferrous and ferric chloride solutions with concentrated ammonia and later coated the magnetite particles with polyvinyl alcohol (PVA). Liao et al. [37] treated the surface of the co-precipitated Fe_3O_4 particles with polyacrylic acid (PAA).

Agglomeration can also be prevented by adding in a suitable surfactant pre- or post-synthesis. Kim et al. [38] utilised sodium oleate as surfactant to coat the co-precipitated magnetite nanoparticles. Poddar et al. [39] and Horak et al. [40] coated the Fe_3O_4 nanoparticles prepared by co-precipitation with oleic acid for surface treatment and stabilization. Oleic acid is a surfactant which contains surface charges, thus providing repulsion between the particles. On the other hand, Morales et al. [41] stabilized their magnetite nanoparticles using oleic acid and Pluronic® block copolymer. There are numerous other coating materials used for co-precipitated Fe_3O_4 nanoparticles by various researchers, as follows:

- Kim et al. [42] – polyoxyethylene (10) oleyl ether
- Zhang et al. [43] – poly(methacrylic acid) (PMAA)
- Urban et al. [44] – lauric acid
- Fu et al. [45] – lauric acid and decanoic acid

The particles can also be coated with starches, proteins, polyelectrolytes and non-ionic detergents [46]. Some of the different polymers or molecules which may be used to stabilize the nanoparticles and also for other biological applications are as shown in Table 2 [12].

Results from these researches indicated that coating the nanoparticles did provide stabilization and was effective in preventing agglomeration and further growth of the nanoparticles. Coated nanoparticles can be dispersed in organic solvents, thus expanding their applications. However, coated particles tend to exhibit lower saturation magnetization values as compared to uncoated particles [12]. In addition, the surfactant coating on the surface of the nanoparticles has to be removed by means of purifiers or by washing them with alcoholic solution before being used for most applications [47]. The purity of the nanoparticles is also affected when coated [33]. Residual surfactant remaining on the surface of the nanoparticles may also

Table 2 Different coating materials and their advantages

Polymers/molecules	Advantages
Polyethylene glycol (PEG)	Improves the biocompatibility and blood circulation time due to non-covalent immobilization
Dextran	Stabilizes the colloidal solution (nanoparticles dispersed in solvent) and enhances blood circulation time
Polyvinylpyrrolidone (PVP)	Stabilizes the colloidal solution and enhances blood circulation time
Fatty acids	Enhances colloidal stability and provides functional carboxyl groups
Polyvinyl alcohol (PVA)	Prevents agglomeration of particles
Polyacrylic acid	Enhances stability and biocompatibility of particles, aids in bioadhesion
Polypeptides	Useful for cell biology, for example cell targeting
Phosphorylcholine	Stabilizes colloidal solution
Poly(D. L – lactide)	Biocompatible and low cytotoxicity
Poly(*N*-isopropylacryl-amide) (PolyNIPAAM)	Useful for cell separation and drug delivery
Chitosan	Natural linear polymer used in agriculture, medicine, food, biotechnology, textiles, water treatment and polymers; Biocompatible and hydrophilic, widely used as non-viral gene delivery system
Gelatin	Natural polymer used as gelling agent, emulsifier; biocompatible and hydrophilic

hinder their surface modification and the use of toxic surfactants might affect their biocompatibility [46].

Furthermore, if nanoparticles are coated with surfactants with long hydrocarbon chains, they become hydrophobic and cannot be dispersed in water or other polar solvents, thus limiting their potential for biomedical applications [47]. However, Wang et al. [47] have come up with a technique of coating the particles with bilayer surfactants. They prepared the nanoparticles using co-precipitation method and subsequently coated the particles with sodium oleic as the inner surfactant layer and sodium dodecyl benzene sulfonate as the outer surfactant layer. With these bilayer surfactants, they were able to stabilize the nanoparticles and at the same time, obtain hydrophilic nanoparticles.

In order to protect the nanoparticles against oxidation and environmental contamination, the synthesis process is usually carried out in a controlled and oxygen-free environment by passing an inert gas, such as nitrogen or argon, through the reaction medium.

4 Effects of Varying Synthesis Parameters

The properties of the magnetite nanoparticles synthesized via the co-precipitation technique depend on the type of iron salts used, the pH value and the ionic strength of the aqueous medium [12]. It is indeed interesting to investigate the effects of the preparation parameters on the physical, chemical and magnetic properties of the nanoparticles. For example, Hsin et al. [48] investigated the effects of the preparation parameters such as the type of iron salt used, pH value of the solution, drying temperature and the feeding rate of the aqueous solution of the iron salt on the properties of the iron oxide prepared by the precipitation method. Kim et al. [48] reported the synthesis and coating of superparamagnetic monodispersed iron oxide nanoparticles (SPION) by the controlled co-precipitation method under different pH and concentration of the NaOH solution.

Zhu et al. [25] meanwhile investigated the effect of varying the molar ratio of Fe^{2+} to Fe^{3+} salt solutions. Tang et al. [49] and Cheng et al. [23] used tetramethylammonium hydroxide (TMAOH) as the alkaline base instead of NH_4OH or NaOH. Kouassi et al. [50] compared the difference between utilizing NH_4OH and NaOH as the alkaline base in terms of the percentage of yield of the nanoparticles. Mehta et al. [51] and Urban et al. [44] co-precipitated $FeSO_4$ and $FeCl_3$ instead of the common protocol involving ferrous and ferric chlorides. Some of the effects of varying the synthesis parameters on the properties of the nanoparticles are as summarized below.

4.1 pH

The size decreases as pH of the reaction medium increases (medium becomes more alkaline) and vice-versa [38]. The decrease in the size of the particles as they are precipitated at a pH far away from PZC (point of zero charge) of materials is due to the increased adsorption of hydroxyl (or proton ions if medium is acidic) onto the particle surface [33]. Thus, the particles develop surface charges which causes repulsion, thereby preventing further growth and resulting in the decreased average size.

4.2 Reaction Temperature

The size increases as temperature increases because reaction rate is increased [20, 52] and the growth of particles is promoted at higher temperatures. The morphology of the particles may also be altered from spherical (at 30°C) to cubic (at 90°C) [53]. Elevated reaction temperature can yield nanoparticles with crystalline structure [40]. However, particle growth at elevated temperatures does lead to diversification of particle sizes and wide size distribution [54].

4.3 Reaction Time

Size increases as time increases because more monomeric species are generated [20]. As a result of increased reaction time, secondary processes such as Ostwald ripening and agglomeration may dominate, hence increasing the average particle size.

4.4 Concentration of Alkaline Solution (Base)

Size increases as concentration of base is increased [38] and this is because of higher concentration of growth species.

4.5 Molar Ratio of Fe^{3+}/Fe^{2+} Salts

Varying the ratio from 1 to 2 decreases the average particle size and saturation magnetization value [47]. Using different ratios of 1, 2, 10 and 20 further decreased the particle size and saturation magnetization as the ratio is increased. Also, increased oxidation of magnetite was observed as hematite peaks were detected from the XRD profiles [27].

4.6 Type of Base

Using tetramethylammonium hydroxide (TMAOH) instead of NH_4OH or NaOH resulted in good stability for nanoparticles in dispersed solvent between pH 7 and 13 but at a lower pH of 5, the particles settled [23]. Using TMAOH as the base results the particles being coated with hydroxide (OH^-) anions, which in turn attracted the tetramethylammonium, $N(CH_3)_4^+$ cations to form a diffuse shell around each particle, subsequently creating repulsion between them [55]. Using NaOH resulted in higher yield (73%) of magnetite nanoparticles compared to using NH_4OH (66% yield) [51]. Utilizing urea of hydrazine as base can result in little or no impurities in the final product, as compared to high amount of impurities by using NaOH or NH_4OH [15].

4.7 Iron Salts Addition Method

Using the controlled drop method by a novel piezoelectric as compared to the typical pipette drop method (with uncontrolled droplet sizes and drop times) resulted in smaller particles with narrower size distribution [56]. Meanwhile, by using the spray-coprecipitation method, magnetite peaks from XRD profiles were better compared to the typical pipette drop method [51].

4.8 Aging Period

Size increased as the aging period increased due to Ostwald Ripening (bigger particles growing at the expense of smaller ones) [11].

4.9 Concentration of Fe^{2+} Salt

Increasing the concentration increased the particle size and saturation magnetization values. Oxidation of magnetite was reduced [51].

By varying the materials and experimental conditions, nanoparticles with different properties can be obtained. Therefore, by identifying the suitable preparation parameters, the properties of the nanoparticles can be further optimized to suit the intended applications.

5 Applications of Superparamagnetic Nanoparticles

A superparamagnetic material is characterized by zero remanence and zero coercivity. This opens up the possibility of controlling the movement and functionality of the superparamagnetic nanoparticles in the body of an organism by manipulating the applied external field. Thus, these particles are being considered for biomedical applications, namely:

- Contrast enhancement agent for Magnetic Resonance Imaging (MRI)
- Hyperthermia agents
- Site-specific drug delivery

5.1 Contrast Enhancement Agent for Magnetic Resonance Imaging (MRI)

Magnetic Resonance Imaging (MRI) is a diagnostic technique used to image various organs and tissues in the human body to detect various kinds of malignant tumours and diseases. In this method, a pulse of radio frequency will hit the hydrogen protons in molecules of water and the change in their magnetization is measured. Different tissues will have different proton reactions, thus producing pictures of anatomical structures [57]. The presence of magnetic nanoparticles in body tissues has enabled larger signals to be obtained from an MRI scanner [57]. The nanoparticles used are usually coated with dextran, polyethylene glycol (PEG), polyvinyl alcohol (PVA) and polysaccharide to provide colloidal stability and biocompatibility [58].

However, the reliability of magnetic nanoparticles as contrast agents for MRI is predominantly influenced by their sizes. Nanoparticles with size of 30 nm can be collected by the lymph nodes while those below 10 nm aren't easily recognizable. Smaller particles can last longer in the body before being collected by the reticulo-endothelial cells [59]. In general, to image different parts of the body, different sized nanoparticles are needed [58]:

- Liver and spleen imaging – ~150 nm
- Lymph node imaging – ~30 nm
- Cancer imaging – 80–150 nm
- Macrophage imaging – 15–30 nm

Two examples of superparamagnetic magnetite nanoparticles currently on the market are Endorem® (magnetite particles with 150 nm diameter) and Lumirem® (silicon-coated iron oxide nanoparticles with 300 nm diameter) [57]. These contrast agents are being used for spleen and liver disease detection and for gastro-intestinal tract imaging, respectively. Sinerem®, another brand of magnetite particles with a diameter of 30 nm, is being utilized for tumour detection [57].

5.2 Hyperthermia Agents

Hyperthermia is a technique where tumours or other growths are destroyed by high temperature of the affected body part. Cancer cells are normally killed at temperatures above 41–42°C [17]. Superparamagnetic nanoparticles can be injected into the affected body part and controlled with an alternating magnetic field. The applied magnetic field will induce currents in the particles, thereby producing heat [57]. The amount of heat produced is directly proportional to the particle size and is also influenced by its shape. The heat produced is usually sufficient to kill tumour cells [57]. However, the level of heating must be below the material's Curie temperature so that their magnetic properties are not affected.

For hyperthermia applications, the nanoparticles are usually coated with biological molecules such as folic acid to increase their uptake by cancer cells. The frequency of the alternating magnetic field should be in between 100 and 1,000 kHz to avoid adverse effects to healthy cells in the body [17]. The optimal size of the particle appears to be between 20 and 50 nm for heating applications [54].

5.3 Site-Specific Drug Delivery

The use of chemotherapy in conventional cancer treatment has resulted in the destruction of relatively healthy surrounding cells due to lack of site-target specificity and sensitivity in detecting and attacking cancerous cells [59]. Also, high doses of chemicals are needed to produce the desired effect [17]. These drawbacks can be overcome by using magnetic nanoparticles with drugs or antibodies attached to them. The illustration of this hypothetical concept is given in Fig. 3. Magnetic nanoparticles with drugs attached to them are injected into the bloodstream and are attracted into tissues by an external magnetic field [60]. The medication can then be released locally. This allows the reduction in the dosage of the medication and also minimizes any adverse effects to surrounding healthy cells.

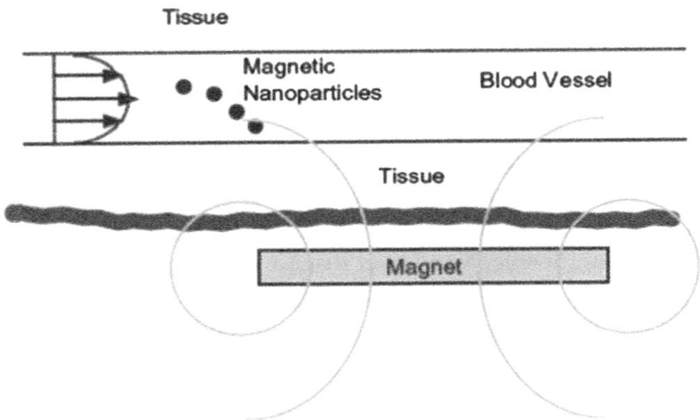

Fig. 3 A hypothetical diagram depicting a site-specific drug delivery system [60]. (Figure 3 is taken from [60])

The surface of the nanoparticles is commonly functionalized with genetic materials, drugs and proteins [46]. Some of the related key issues to be considered are the particle size, surface property and strength of drug-to-particle binding [17].

In summary, for biomedical applications, the nanoparticles must have the following characteristics [17, 46]:

- Must be made of non-toxic and non-immunogenic materials
- Size must be small enough so that it may remain in circulation after being injected into the body and also to pass through the capillary systems of various organs and tissues
- Must have high magnetization in order for the movement to be controlled by a magnetic field and be immobilized close to the affected tissue or area
- Must be superparamagnetic at room temperature to prevent their agglomeration within the body and subsequently, avoid possible blockage of blood vessels
- Must be stable in water at neutral pH and physiological conditions

6 Conclusion

In this chapter, we have reviewed and compared various synthesis methods to fabricate superparamagnetic nanoparticles. Issues such as size control, and stability are also discussed. In order to make it more reliable and environmental friendly, this dimension-dependent material needs precise synthetic chemistry, including rational synthetic strategies, functional performance, and green chemistry principles as proposed by Peng [61]. In the future, the design-by-purpose approaches will dominate the synthesis of functional superparamagnetic nanoparticles based on their niche applications.

References

1. Bean, C.P., Livingston, J.D.: Superparamagnetism. J. Appl. Phys. **30**, S120 (1959)
2. Jeong, U., Teng, X., Wang, Y., Yang, H., Xia, Y.: Superparamagnetic colloids: controlled synthesis and niche applications. Adv. Mater. **19**, 33–60 (2007)
3. Lu, A.-H., Salabas, E.L., Schuth, F.: Magnetic nanoparticles: synthesis, protection, functionalization, and application. Angew. Chem. **46**, 1222–1244 (2007)
4. Park, J., Joo, J., Kwon, S.G., Jang, Y., Hyeon, T.: Synthesis of monodisperse spherical nanocrystals. Angew. Chem. **46**, 4630–4660 (2007)
5. Rebolledo, A.F., Miguel, O.B., Marco, J.F., Tartaj, P.: A facile synthetic route for the preparation of superparamagnetic iron oxide nanorods and nanorices with tunable surface functionality. Adv. Mater. **20**, 1760–1765 (2008)
6. Tannenbaum, R., Reich, S., Flenniken, C.L., Goldberg, E.P.: Shape control of iron oxide nanoclusters in polymeric media. Adv. Mater. **14**, 1402–1405 (2002)
7. Sun, S., Murray, C.B.: Synthesis of monodisperse cobalt nanocrystals and their assembly into magnetic superlattices (invited). J. Appl. Phys. **85**, 4325 (1999)
8. Tomou, A., Gournis, D., Panagiotopoulos, I., Huang, Y., Hadjipanayis, G.C., Kooi, B.J.: Weak ferromagnetism and exchange biasing in cobalt oxide nanoparticle systems. J. Appl. Phys. **99**, 123915 (2006)
9. Richardson, J.T., Yiagas, D.I., Turk, B., Forster, K., Twigg, M.V.: Origin of superparamagnetism in nickel oxide. J. Appl. Phys. **70**, 6977 (1991)
10. Khadar, M.A., Biju, V., Inoue, A.: Effect of finite size on the magnetization behavior of nanostructured nickel oxide. Mater. Res. Bull. **38**, 1341–1349 (2003)
11. Cao, G.: Nanostructures & Nanomaterials: Synthesis, Properties & Applications. Imperial College Press, London (2004)
12. Gupta, A.K., Gupta, M.: Synthesis and surface engineering of iron oxide nanoparticles for biomedical applications. Biomaterials **26**, 3995–4021 (2005)
13. Nagar, H., Kulkarni, N.V., Karmakar, S., Sahoo, B., Banerjee, I., Chaudhari, P.S., Pasricha, R., Das, A.K., Bhoraskar, S.V., Date, S.K., Keune, W.: MÖssbauer spectroscopic investigations of nanophase iron oxides synthesised by thermal plasma route. Mater. Characterisation **59**, 1215–1220 (2008)
14. Cabrera, L., Gutierrez, S., Menendez, N., Morales, M.P., Herrasti, P.: Magnetite nanoparticles: electrochemical synthesis and characterisation. Electrochim. Acta **53**, 3436–3441 (2008)
15. Bluthardt, C., Fink, C., Flick, K., Hagemeyer, A., Schlichter, M., Volpe, A.J.: Aqueous synthesis of high surface area metal oxides. Catal. Today **137**, 132–143 (2008)
16. Lee, Y., Lee, J., Bae, C.J., Park, J.-G., Noh, H.-J., Park, J.-H., Hyeon, T.: Large-scale synthesis of uniform and crystalline magnetite using reverse micelles as nanoreactors under reflux conditions. Adv. Funct. Mater. **15**(3), 503–509 (2005)
17. Tartaj, P., Puerto Morales, M.D., Veintemillas-Verdaguer, S., Gonzalez-Carreno, T., Serna, C.J.: The preparation of magnetic nanoparticles for applications in biomedicine. J. Phys. D Appl. Phys. **36**, R182–R197 (2003)
18. Ozkaya, T., Toprak, M.S., Baykal, A., Kavas, H., Köseoglu, Y., Aktas, B.: Synthesis of Fe_3O_4 nanoparticles at 100°C and its magnetic characterisation. J. Alloys Compd. 472(1–2), 18–23 (2009)
19. Cushing, B.L., Kolesnichenko, V.L., O'Connor, C.J.: Recent advances in the liquid-phase syntheses of inorganic nanoparticles. Chem. Rev. **104**(9), 3893–3946 (2004)
20. Hyeon, T.: Chemical synthesis of magnetic nanoparticles. Chem. Commun. **8**, 927–934 (2003)
21. Hinotsu, T., Jeyadevan, B., Chinnasamy, C.N., Shinoda, K., Tohji, K.: Size and structure control of magnetic nanoparticles by using a modified polyol process. J. Appl. Phys. **95**(11), 7477–7479 (2004)
22. LaConte, L., Nitin, N., Bao, G.: Magnetic nanoparticle probes. Mater. Today **8**(Suppl. 1), 32–38 (2005)

23. Cheng, F.-Y., Su, C.-H., Yang, Y.-S., Yeh, C.-S., Tsai, C.-Y., Wu, C.-L., Wu, M.-T., Shieh, D.-B.: Characterisation of aqueous dispersions of Fe_3O_4 nanoparticles and their biomedical applications. Biomaterials **26**, 729–738 (2005)
24. Zhang, Y., Sun, C., Kohler, N., Zhang, M.: Self-assembled coatings on individual monodisperse magnetite nanoparticles for efficient intracellular uptake. Biomed. Microdevices **6**(1), 33–40 (2004)
25. Zhu, Y., Wu, Q.: Synthesis of magnetite nanoparticles by precipitation with forced mixing. J. Nanopart. Res. **1**, 393–396 (1999)
26. Suber, L., Imperatori, P., Ausanio, G., Fabbri, F., Hofmeister, H.: Synthesis, morphology and magnetic characterisation of iron oxide nanowires and nanotubes. J. Phys. Chem. B **109**, 7103–7109 (2005)
27. Kim, D.-H., Kim, K.-N., Kim, K.-M., Shim, I.-B., Lee, M.-H., Lee, Y.-K.: Tuning of magnetite nanoparticles to hyperthermic thermoseed by controlled spray method. J. Mater. Sci. **41**, 7279–7282 (2006)
28. Park, J., An, K., Hwang, Y., Park, J.-G., Noh, H.-J., Kim, J.-Y., Park, J.-H., Hwang, N.-M., Hyeon, T.: Ultra-large-scale syntheses of monodisperse nanocrystals. Nat. Mater. **3**, 891–895 (2004)
29. Wang, X., Zhuang, J., Peng, Q., Li, Y.: A general strategy for nanocrystal synthesis. Nature **437**, 121–124 (2005)
30. Osaka, T., Matsunaga, T., Nakanishi, T., Arakaki, A., Niwa, D., Iida, H.: Synthesis of magnetic nanoparticles and their application to bioassays. Anal. Bioanal. Chem. **384**, 593–600 (2006)
31. Wen, X., Yang, J., He, B., Gu, Z.: Preparation of monodisperse magnetite nanoparticles under mild conditions. Curr. Appl. Phys. **8**, 535–541 (2008)
32. Lopez-Perez, J.A., Lopez-Quintela, M.A., Mira, J., Rivas, J.: Preparation of magnetic fluids with particles obtained in microemulsions. IEEE Trans. Magn. **33**(5), 4359–4362 (1997)
33. Vayssieres, L.: On the design of advanced metal oxide nanoparticles. Int. J. Nanotechnol. **1**(1–2), 1–41 (2004)
34. Massart, R.: Preparation of aqueous magnetic liquids in alkaline and acidic media. IEEE Trans. Magn. **17**(2), 1247–1248 (1981)
35. Gao, J., Xu, B.: Applications of nanomaterials inside cells. Nano Today **4**, 37–51 (2009)
36. Petri-Fink, A., Chastellain, M., Juillerat-Jeanneret, L., Ferrari, A., Hofmann, H.: Development of functionalised superparamagnetic iron oxide nanoparticles for interaction with human cancer cells. Biomaterials **26**, 2685–2694 (2005)
37. Liao, M.H., Chen, D.H.: Preparation of characterisation of a novel magnetic nano-adsorbent. J. Mater. Chem. **12**, 3654–3659 (2002)
38. Kim, D.K., Zhang, Y., Voit, W., Rao, K.V., Muhammed, M.: Synthesis and characterisation of surfactant-coated superparamagnetic monodispersed iron oxide nanoparticles. J. Magn. Magn. Mater. **225**, 30–36 (2001)
39. Poddar, P., Telem-Shafir, T., Fried, T., Markovich, G.: Dipolar interactions in two- and three-dimensional magnetic nanoparticle arrays. Phys. Rev. B **66**, 060403-1–060403-4 (2002)
40. Horak, D., Lednicky, F., Petrovsky, E., Kapicka, A.: Magnetic characteristics of ferrimagnetic microspheres prepared by dispersion polymerization. Macromol. Mater. Eng. **289**, 341–348 (2004)
41. Morales, M.A., Jain, T.K., Labhasetwar, V., Leslie-Pelecky, D.L.: Magnetic studies of iron oxide nanoparticles coated with oleic acid and Pluronic block copolymer. J. Appl. Phys. **97**, 10Q905-1–10Q905-3 (2005)
42. Kim, D.K., Zhang, Y., Voit, W., Rao, K.V., Kehr, J., Bjelke, B., Muhammed, M.: Superparamagnetic iron oxide nanoparticles for bio-medical applications. Scripta Mater. **44**, 1713–1717 (2001)
43. Zhang, H., Wang, R., Zhang, G., Yang, B.: A covalently attached film based on poly (methacrylic acid)-capped Fe_3O_4 nanoparticles. Thin Solid Films **429**, 167–173 (2003)

44. Urban, P., Idzikowski, B., Kostyrya, S., Andrzejewski, B.: Grains size distribution and thermal stability of surfactant stabilised Fe_3O_4-based magnetic fluid. Czech. J. Phys. **54**, D683–D686 (2004)
45. Fu, L., Dravid, V.P., Johnson, D.L.: Self-assembled (SA) bilayer molecular coating on magnetic nanoparticles. Appl. Surf. Sci. **181**, 173–178 (2001)
46. Teja, A.S., Koh, P.-Y.: Synthesis, properties, and applications of magnetic iron oxide nanoparticles. Prog. Cryst. Growth Characterisation Mater. 55, 22–45 (2009)
47. Wang, X., Zhang, C., Wang, X., Gu, H.: The study on magnetite particles coated with bilayer surfactants. Appl. Surf. Sci. **253**, 7516–7521 (2007)
48. Hsin, Y.L., Yu, W.C., Wei, J.W.: Preparation of nanosized iron oxide and its application in low temperature CO oxidation. J. Nanopart. Res. **7**, 249–263 (2005)
49. Tang, J., Myers, M., Bosnick, K.A., Brus, L.E.: Magnetite Fe_3O_4 nanocrystals: spectroscopic observation of aqueous oxidation kinetics. J. Phys. Chem. B **107**, 7501–7506 (2003)
50. Kouassi, G.K., Irudayaraj, J., McCarty, G.: Activity of glucose oxidase functionalised onto magnetic nanoparticles. Biomagn. Res. Technol. **3**(1), 1–10 (2005)
51. Mehta, R.V., Upadhyay, R.V., Charles, S.W., Ramchand, C.N.: Direct binding of protein to magnetic particles. Biotechnol. Tech. **11**(7), 493–496 (1997)
52. Wu, J.H., Ko, S.P., Liu, H.L., Jung, M.-H., Lee, J.H., Ju, J.-S., Kim, Y.K.: Sub 5 nm Fe_3O_4 nanocrystals via coprecipitation method. Colloids Surf. A Physicochem. Eng. Asp. **313–314**, 268–272 (2008)
53. Liu, Z.L., Liu, Y.J., Yao, K.L., Ding, Z.H., Tao, J., Wang, X.: Synthesis and magnetic properties of Fe_3O_4 nanoparticles. J. Mater. Synth. Process. **10**(2), 82–87 (2002)
54. Mürbe, J., Rechtenbach, A., Töpfer, J.: Synthesis and physical characterisation of magnetite nanoparticles for biomedical applications. Mater. Chem. Phys. **110**, 426–433 (2008)
55. Berger, P., Adelman, N.B., Beckman, K.J., Campbell, D.J., Ellis, A.B., Lisensky, G.C.: Preparation and properties of an aqueous ferrofluid. J. Chem. Educ. **76**(7), 943–948 (1999)
56. Jeong, J.-R., Shin, S.-C., Lee, S.-J., Kim, J.-D.: Magnetic properties of superparamagnetic γ-Fe_2O_3 nanoparticles prepared by coprecipitation technique. J. Magn. Magn. Mater. **286**, 5–9 (2005)
57. Berry, C.C., Curtis, A.S.G.: Functionalisation of magnetic nanoparticles for applications in biomedicine. J. Phys. D Appl. Phys. **36**, R198–R206 (2003)
58. Thorek, D.L.J., Chen, A.K., Czupryna, J., Tsourkas, A.: Superparamagnetic iron oxide nanoparticle probes for molecular imaging. Ann. Biomed. Eng. **34**(1), 23–38 (2006)
59. Guiden, D., Dravid, V., Su, M., Fu, L.: Smart therapy: the multivariate potentials of iron oxide nanoparticles in drug delivery. Nanoscape **1**, 71–75 (2004)
60. Pankhurst, Q.A., Connolly, J., Jones, S.K., Dobson, J.: Applications of magnetic nanoparticles in biomedicine. J. Phys. D Appl. Phys. **36**, R167–R181 (2003)
61. Peng, X.: An essay on synthetic chemistry of colloidal nanocrystals. Nano Res. **2**(6), 425–447 (2009)

Ammonia Synthesis

Noorhana Yahya, Poppy Puspitasari, Krzysztof Koziol, and Pavia Guiseppe

Abstract Ammonia production is a very energy- and capital-intensive industry as it requires high temperature (400–500°C) and also high pressure (150–300 bar) for its daily operations. Two moles of ammonia are obtained by reacting one mole of nitrogen and three moles of hydrogen gases in the presence of conventional catalyst which is magnetite. The process to produce ammonia is known as Haber–Bosch process which was developed and patented by Fritz Haber and Carl Bosch in 1916. Since then more work on ammonia production was carried out with the aim to obtain higher ammonia yield. Catalytic reaction giving emphasis on types of catalysts was reviewed in this chapter. Different catalysts synthesis methods and their characterisations were also reviewed A variety of microreactors were proposed by different authors and some patent fillings have been described. A new method to synthesize ammonia at room temperature and ambient pressure were described. The reaction was done in 1 T magnetic field. This work offers the ammonia producers a potential contender in the market place.

1 Introduction

Ammonia is a very important feedstock for urea synthesis which is an important ingredient as a fertilizer. The ammonia synthesis discovery is marked as an important discovery due to our ability to synthesize synthetic fertilizer. Due to the

N. Yahya (✉) and P. Puspitasari
Department of Fundamental and Applied Sciences, Universiti Teknologi PETRONAS, Bandar Seri Iskandar, 31750 Tronoh, Perak, Malaysia
e-mail: noorhana_yahya@petronas.com.my, povopivi@gmail.com

K. Koziol
Department of Materials Science and Metallurgy, University of Cambridge, Pembroke Street, Cambridge CB2 3QZ, UK
e-mail: kk292@cam.ac.uk

P. Guiseppe
Carl Zeiss NTS GmbH, Carl-Zeiss-Straße 56, 73447 Oberkochen, Germany

N. Yahya (ed.), *Carbon and Oxide Nanostructures*, Adv Struct Mater 5,
DOI 10.1007/8611_2010_25, © Springer-Verlag Berlin Heidelberg 2010,
Published online: 3 August 2010

Table 1 Ammonia production in some Asian countries [1]

(Thousand metric tons contains nitrogen)

Country	2003	2004	2005	2006	2007
Malaysia	910	843	920	950	960
India	10,048	10,718	10,800	10,900	11,000
Indonesia	4,250	4,120	4,400	4,300	4,400
Iran	1,115	1,088	1,020	1,020	2,000
Japan	1,061	1,101	1,083	1,091	1,090
Pakistan	2,357	2,114	2,114	2,200	2,250
Saudi Arabia	1,743	1,726	1,780	2,000	2,600

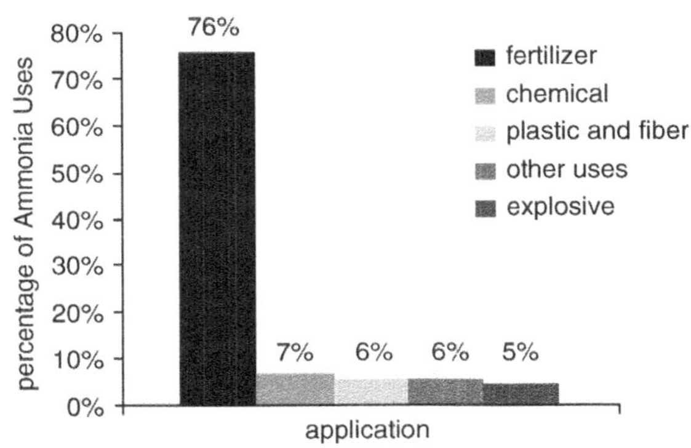

Fig. 1 Applications of Ammonia in the industry [1]

importance of ammonia particularly as a fertilizer in the agriculture industry, the production rate has been extensively increased each year. Large-scaled ammonia production plants exists world widely. Table 1 shows the ammonia production for some Asian countries in years 2003–2007 [1].

Based on Fig. 1, ammonia-based fertilizers have been seen as the key factor in developing and promoting agriculture. In Malaysia, two ammonia production plants were built located at Gurun, Kedah, PETRONAS Fertilizer (Kedah) Sdn. Bhd. (PFK) and at Kerteh, Terengganu, PETRONAS Ammonia Sdn. Bhd. (PASB).

2 Ammonia Synthesis

Ammonia is a molecule comprising of one nitrogen atom and three hydrogen atoms with chemical formula of NH_3. It is a very energy and capital intensive industry. Ammonia can be produced by reacting hydrogen and nitrogen gasses in high temperature and high pressure environment in the presence of iron catalysts. Fritz Haber and Carl Bosch made the first commercial high-pressure synthesis of ammonia.

2.1 *Catalytic Reaction*

Activation of catalyst is an initial process that has to be performed before the reaction takes place. It involves the reduction of catalyst from metal oxide to metallic state in hydrogen as the reducing gas. The purpose of reduction is to eliminate oxygen in order to allow the electron pairing or sharing between the reactant atoms and partially filled d-orbital of a reduced metal. The overall chemical reaction is as follows [2]:

$$Fe_3O_4 + 4H_2 \rightarrow 3Fe + 4H_2O \quad (1)$$

The mechanism of reduction involves two main processes which occur separately. The first process is generation of reducing electrons via a reaction of hydrogen and oxygen, and followed by the transfer of these electrons to the species to be reduced. This process will continuously occur until the metallic state of catalyst is formed. The mechanism is illustrated as follows [2]:

$$O^{2-} + H_2 \rightarrow H_2O + 2\text{electrons} \quad (2)$$

$$Fe_3O_4 + Fe^{2+} + 2\text{electrons} \rightarrow 4FeO \quad (3)$$

2.1.1 Adsorption Process

Adsorption process in ammonia synthesis is an essential process where the reactant molecules in this case the hydrogen and nitrogen gasses adhere to the surface of the metal catalyst. It is categorized into two parts which are physical (physisorption) and chemical (chemisorption). When these reactant gases are flowed in the reaction medium, they will be firstly adsorbed to the metal surfaces by physisorption. At this stage, it involves a Van der Waals interaction between the reactants and metal surfaces. The enthalpy change due to adsorption (ΔH_{ads}) is 5–50 kJ/mole [3]. This enthalpy change is insufficient for the bond breaking to occur. Hence physisorbed molecules are retained. Multilayer adsorption would then occur. Multilayer adsorption is a phenomenon where the adsorption surface allows more than one layer of molecules. It should be noted that not all adsorbed molecules are in direct contact with the surface of the metal. Because of this, the chemisorption is performed to initiate the dissociation of molecules so that the next process can be completed. Figure 2 reveals the chemisorption mechanism on the iron metal surface [3, 4].

Chemical adsorption is a process of sharing or exchanging electrons between the reactants and the partially occupied d-orbital of iron forming chemical bond. The chemisorption enthalpy is 50–500 kJ/mole which is greater than for physisorption [4].Theoretically when the nitrogen molecules are chemically adsorb, the triple bond which ties two nitrogen atoms together is favoured to donate their electron to

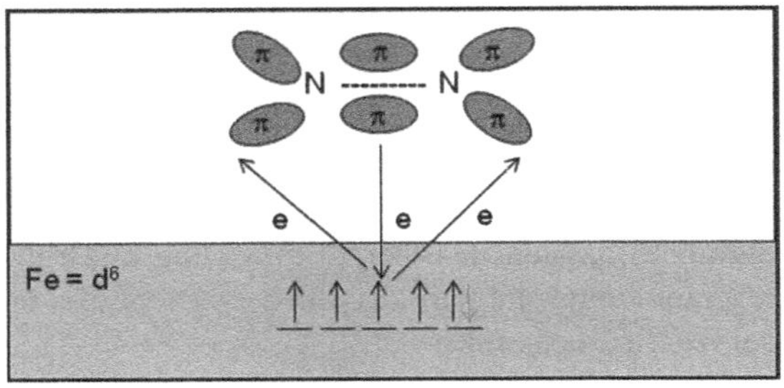

Fig. 2 Mechanism of chemisorption on the iron metal surface

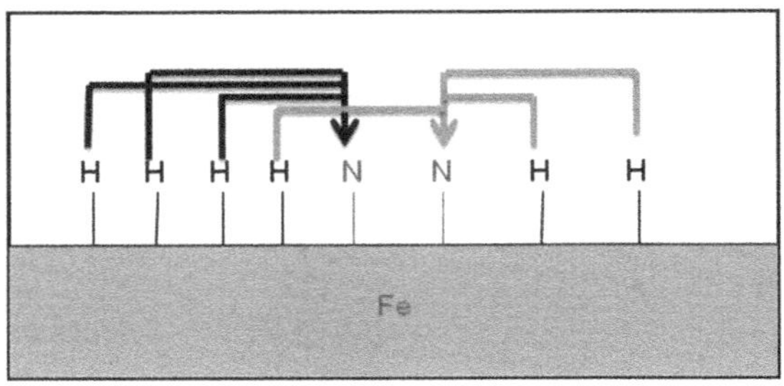

Fig. 3 Mechanism of migration on the iron metal surface

the partially occupied d-orbital of iron metal. This is usually referred as sigma (σ) donation which forms a strong chemical bond which resulting in shorter distance between the adsorb nitrogen molecules and iron metal surfaces. Iron metal has four unpaired electron (d^6) in d-orbital and needs four additional electrons to be stabilized. Because of this, the excess electron will be donated back to the empty pie (π^*) orbital's of nitrogen molecules. This process is known as π-back donation. In consequence, the N≡N bond is lengthened and weakens. Thus, the nitrogen molecules will be torn apart or rather separated and fragmented on the iron metal surface to form monolayer adsorption. At this stage, all the adsorbed atoms are in contact with the surface layer of the adsorbent [3, 4]. Figure 3 shows the mechanism of migration process occurs on the iron metal surfaces.

The migration process involves the movement of chemisorbed single atoms and reacts among each others to form the product in this case ammonia. It happens because the bonding between chemisorbed molecules and the iron metal surfaces provide thermodynamic driving force for the release of atoms so that the migration can be accomplished [5].

2.1.2 Desorption

Figure 4 shows how the mechanism of desorption process occurs on the iron metal surfaces.

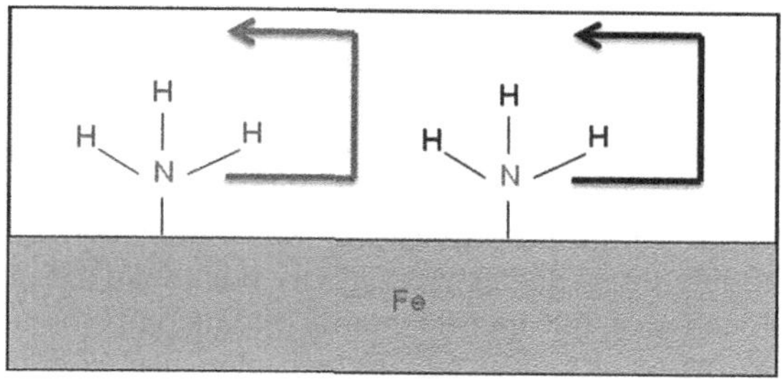

Fig. 4 Mechanism of desorption on the iron metal surface

When the product is formed, the intermediate strength of surface bonds allow desorption of products (NH_3). It leaves free active sites for the incoming hydrogen and nitrogen reactants molecules to be adsorbed and react on the iron metal surfaces. Using nanosize catalyst is attractive due to its high surface area leading to its active site for physi- and chemi-sorption process.

Understanding of adsorption and desorption process is utmost important for catalytic reaction from the viewpoint of fundamental science in the ammonia production. It should be noted that the first commercial ammonia plant was built in Oppau, Germany and has the production capacity of 30 tons/day [6]. This plant was set up by a German chemical giant, Badashe Analine and Soda Fabrils (BASF).

3 Catalyst for Ammonia Synthesis

This part reviews types of catalysts for ammonia synthesis by several authors. Traditionally and conventionally magnetite (Fe_3O_4) is used as catalyst for ammonia synthesis. The magnetite is promoted with oxides which are not likely to be reduced. Typically the promoters are aluminium, potassium and calcium [7]. In short, magnetite is known as a precursor due to the easy cation substitution of Al^{3+} for Fe^{3+}, hence homogeneous distribution of aluminium in solid can be obtained [8]. Recently wustite was proposed as a new precursor for industrial processes (ammonia synthesis catalyst) [9]. It was reported that higher quantity (30% of yield) of ammonia was synthesized using wustite as catalyst, compared to magnetite. Higher efficiency of Fe surface sites on the calcium oxide was able to stabilize wustite structure down to room temperature. It was also reported that wustite seems to be favourable to magnetite due to its ability to be reduced while thermoresistancy and mechanical strength are fully retained [9].

Based on the extensive kinetic studies done by several research groups in the past decades, ruthenium-based catalyst has been a promising candidate to replace magnetite for ammonia synthesis. The ruthenium-based catalyst is favoured owing to its long term stability and activity. It is also able to produce high ammonia yield in low pressure and low temperature conditions [10]. Mesoporous Ru/MgO catalyst prepared by sol gel method for ammonia synthesis was also studied [11]. It was found that specific surface area and metal dispersion were increased when increasing Ru carbonyl complex concentration. Highest ammonia formation rate was consequently observed on magnesia supported ruthenium Ru/MgO (Ru: 7.1 wt.%) with high surface (290 m^2/g) [11]. Ruthenium promoted by potassium metal has been proposed as a very active catalyst for non conventional ammonia synthesis which may be performed under atmospheric pressure [12]. Ruthenium supported on γ-Al_2O_3 which was modified with KOH was tested as ammonia synthesis catalyst under atmospheric pressure. Ruthenium derived from $Ru_3(CO)_{12}$ exhibited up to five times higher productivity then the corresponding $RuCl_3$ precursors. It was found that the highest hourly productivity of ammonia was obtained with 8% Ru/Al_2O_3–KOH at 623 K and atmospheric pressure [13]. High ammonia yield approximately 40–50% was successfully produced using Ru/C catalyst. The pressure and temperature conditions are 370–400°C and 50–100 atm respectively. The extremely high cost of ruthenium is the major drawback [14]. On top of the price of ruthenium, it was reported that carbon tends to react with H_2 to produce methane, (CH_4) during the catalytic activity [15]. It is well known that at different temperatures both iron and ruthenium are good catalysts for ammonia synthesis. Mutual influence in various proportions on the two active catalysts; ruthenium and iron was reported. It was observed that though ruthenium is very active at optimum temperature (573–623 K), the presence of iron had resulted in very low activity, indicating negative synergism of iron and ruthenium [16]. Ammonia synthesis over Ru/C catalysts with different carbon supports namely activated carbon fiber (ACF), activated carbon (AC), and carbon molecular sieve (CMS) was studied. The ammonia synthesis was done in 350–450°C and 3.0 MPa in a microreactor. It was found that Ru – Ba/ACF gave the highest turnover-frequency (TOF) value (0.089 s^{-1}) which is due to high purity and electronic conductivity of ACF [17]. Structure sensitivity of ruthenium catalysts supported on the graphitised carbon was also done. Ammonia synthesis studies have revealed that the reaction rates (400°C, 63 bar, 8.5% NH_3 or 400°C, 90 bar, 11.5% NH_3) expressed in terms of TOF had increased versus particle size regardless of the type of promoter [14].

Recently, a new type of nanocatalyst, namely $Mn_{0.8}Zn_{0.2}Fe_2O_4$ was proposed by Yahya et al. [18]. The catalyst was synthesis using sol gel method and was reacted under electromagnetic (EM) induction. The ammonia yield was found to be high due to the synergism of Mn, and Fe metal (which was reduced in hydrogen gas) and the EM induction.

Besides catalysts there are also two other parameters that could assist in the ammonia yield which are given as follows:

1. Promoter
2. Support

It is well known that metal crystallites especially in nanoscale range have high surface area as well as surface energy. Thus, the agglomeration is favoured to overcome these side effects which eventually lead to the formation of bigger crystallite size. The catalytic activity would occur infrequently due to less surface area available for the reactants to be adsorbed a process which was described in Sect. 2.1.1. In order to ensure the catalyst works efficiently, small amount of chemical additive or promoter is often introduced. This promoter usually comes with two categories which are textural or physical and electronic.

One of the best solutions to prevent agglomeration problem is by attaching metal crystallites on the carrier or supporter. Carrier which possesses great features such as excellent thermal stability and high surface area; is highly required. Alumina, γ-Al_2O_3 is known as typical carrier for ammonia synthesis catalyst. It has high surface area (100–300 m^2/g) and also capable to withstand at higher temperature environment. Moreover, higher degree of dispersion of metal crystallites on carrier is also desired which could induce the catalytic activity. This can be accomplished via the synthesis approach [19].

Electronic promoter is another vital component for catalyst. It is usually doped in relatively small amount. However, excessive doping may inhibit the catalytic activity as it can largely cover the surface of metal crystallites. Potash or potassium hydroxide, KOH is an example of promoter for ammonia synthesis catalyst. Potassium, K^+ behaves as electron donor who donates it electrons directly into d-orbital of iron. The continuous donation will create a high electron density region of iron. Unfortunately, this phenomenon will lead to the destabilization of iron. Thus, the surplus electrons will be than transferred to the π antibonding orbitals (π^*) in nitrogen molecules. The occupation of antibonding orbital makes the N≡N elongate and weaker. As a consequence, the N≡N bond cleavage is accelerated [20, 21].

The catalysts and supported/unsupported used in ammonia synthesis are summarized as follows (Table 2):

Table 2 Type of catalyst for ammonia synthesis

Catalysts/Ref.	Support	Yield NH_3	Temperature	Pressure
Ruthenium [14]	Graphitised Carbon	8.5%	400°C	63 bar
Ruthenium [14]	Graphitised Carbon	11.5%	400°C	90 bar
Ruthenium [13]	Active carbon	40–50%	370–400°C	50–100 atm
Ruthenium – Iron [16]	Active carbon	15 mol NH_3/hr/g.cat	623 K	Atmospheric pressure
Ruthenium [17]	Barium	Nil	350–450°C	3.0 MPa
Ruthenium [22]	Barium hexaaluminate (BHA)	5,426 μmol NH_3/hr/g.cat	653 K	1.1 MPa
Ruthenium [23]	CNT and MgO	4,453 mol NH3/hr/g.cat	673 K	0.2 MPa
NH_4ReO_4/Al_2O_3 [24]	Rhenium	0.2 mol NH_3/hr/g.cat	873 K	3 MPa
Iron cobalt [25]	Carbon	7.0 μmol/g/s	673 K	10 bar
Molybdenum Nitride [26]	KNO_3 or $CsNO_3$	15.0 mmol NH_3/hr/g.cat	673 K	3.1 MPa
K-C-Fe [27]	Al_2O_3	0.48% vol	350°C	Atmospheric pressure
Wustite A301 [28]	Al_2O_3	27% vol	425°C	15 MPa
Polyacrylonitrile (PAN) [29]	Active carbon	0.98% vol	673 K	Atmospheric pressure
$Mn_{0.8}Zn_{0.2}Fe_2O_4$ [18]	MWCNT	24.9%	28°C	Atmospheric pressure
$Mn_{0.8}Zn_{0.2}Fe_2O_4$	Nil	46%	28°C	Atmospheric pressure

3.1 Synthesis and Characterization of Catalysts

There are different types of synthesis methods on catalyst for ammonia production. Generally, there is a growing need to produce nanosized catalyst for the synthesis of ammonia. The synthesis methods are as stated below [30–34]:

1. Sol gel
2. Co-Precipitation
3. Hydrothermal

This part describes on the $Mn_{0.8}Zn_{0.2}Fe_2O_4$ catalysts which was synthesized using sol gel method. Phase, morphology and elemental analysis are crucial for us to understand and relate with the catalytic activities. A representative of single phase for $Mn_{0.8}Zn_{0.2}Fe_2O_4$ is characterized by X-Ray Diffraction (XRD) and the morphology is characterized by Field Emission Electron Scanning Microscope (FESEM) and Transmission Electron Microscope (TEM). Table 3 shows the samples name for $Mn_{0.8}Zn_{0.2}Fe_2O_4$ sintered at three different temperatures.

X-ray diffraction (XRD) analysis was done for all the samples (Table 3) using Philips X-Ray Diffractometer having CuKα, with $\lambda = 1.5418$ Å. The scanning angles (θ) were performed from 10° to 80° and the speed of the counter was 0.02°/2θ per min. The X-Ray Diffraction (XRD) patterns are shown in Fig. 5 for sol gel technique after annealing 700, 800, and 900°C for 4 h. The unit cell size and geometry were resolved from the angular positions of the diffraction peaks, whereas arrangement of atoms within the unit cell is associated with the relative intensities of these peaks. The pattern shows the evolution of crystallization of manganese zinc ferrite powders with the increasing annealing temperature. The highest peak intensity is observed for (311) plane at 2Ø which corresponds to the $MnFe_2O_4$ [30]. This signified that the annealing process had caused the atom to move form its own lattice completely. It is also speculated that Zn^{2+} ions had occupied the tetrahedral site, Mn^{2+} ions occupied the octahedral site and the Fe^{3+} ions were distributed over both sites [30]. Powder sintered at 700°C shows a mixture of hematite and ferrite phase.

Table 4 shows the XRD analysis identifies the value of full width half maximum (FWHM), d-spacing and crystallite size for the (311) plane. Scherer equation is used to determine the diameter of manganese zinc ferrite nano-particles by using XRD results.

Table 3 Sample name for $Mn_{0.8}Zn_{0.2}Fe_2O_4$ sintered at five different temperatures

Material	Sample name
$Mn_{0.2}Zn_{0.8}Fe_2O_4$ sintered at 700°C	700 SG
$Mn_{0.2}Zn_{0.8}Fe_2O_4$ sintered at 800°C	800 SG
$Mn_{0.2}Zn_{0.8}Fe_2O_4$ sintered at 900°C	900 SG

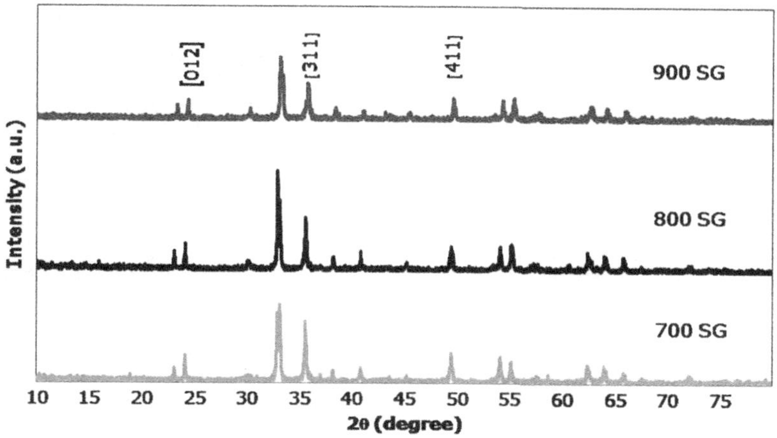

Fig. 5 X-Ray diffraction pattern for $Mn_{0.2}Zn_{0.8}Fe_2O_4$ annealed at 700, 800, and 900°C for 4 h

Table 4 Intensity, FWHM, d-spacing, crystallite size of $Mn_{0.8}Zn_{0.2}Fe_2O_4$ prepared using the sol gel technique, sintered at 700, 800, and 900°C for 4 h

Samples	X-ray diffraction (correspond to [311] peaks)					
	FWHM	d-spacing (Å)	Crystallite size (nm)	a	b	c
700 SG	0.4	2.7	28.8	8.4915	8.4915	8.4915
800 SG	0.3	2.7	31.2	8.4915	8.4915	8.4915
900 SG	0.2	2.7	37.2	8.3820	8.3820	8.3820

The Scherer equation is:

$$D = \frac{K\lambda}{\beta \cos \theta} \tag{4}$$

where K = 0.9, λ = wavelength, β = FWHM 2θ (Full width half maximum) and θ = maximum peak.

The FWHM values at 700, 800 and 900°C are 0.4, 0.3, 0.2, respectively. Hence, the crystallite size is increase while the value of FWHM is decrease. Referring to a, b, c values, all samples exhibit cubic structure since a = b = c.

Figure 6a describes nanofiber-like $Mn_{0.8}Zn_{0.2}Fe_2O_4$. The diameter of the rods is in the range of 62–72 nm. The $Mn_{0.8}Zn_{0.2}Fe_2O_4$ samples sintered at 800 and 900°C exhibit spherical-like morphologies (Fig. 6b, c) in particular are very unique, and would be an added advantage during reaction due to its large surface area.

Activation of catalyst for ammonia synthesis is an initial process before reaction process begins. This process requires reduction of catalysts from magnetite (Fe_3O_4)

Fig. 6 FESEM morphology of $Mn_{0.8}Zn_{0.2}Fe_2O_4$ at different sintering temperature (**a**) 700°C, (**b**) 800°C and (**c**) 900°C

to hematite (Fe_2O_3) and finally to wustite (FeO) before it was reduced to metal iron, in the ammonia synthesis process. The reducing gas is hydrogen. The reduction of metal oxide is done to eliminate oxygen gas to allow the electron pairing between the reactant atoms and partially filled d-orbital of the reduced metal, in this case iron. The overall chemical reaction is as follows [31]:

$$Fe_3O_4 + 4H_2 \rightarrow 3Fe + 4H_2O \tag{5}$$

A comparative behaviour of Fe_2O_3, ZnO and $ZnFe_2O_4$ by TPR technique was done [35]. It was found that $ZnFe_2O_4$ is easier to be reduced comparing to its counterpart Fe_2O_3. When $ZnFe_2O_4$ is reduced by hydrogen gas in TPR, three reduction peaks which correspond to Fe_3O_4, FeO, and Fe kinetics of reduction was also done on supported nanoparticles, silica-zirconia of iron oxide. It was found that two well defined and narrow temperature programmed reduction (TPR) peaks were observed. It was:

$$Fe_2O_3 \rightarrow FeO \tag{6}$$

$$\text{(I)}\quad Fe_2O_3 \rightarrow Fe(400°C) \tag{7}$$

$$\text{(II)}\quad FeO \rightarrow Fe(800 - 900°C) \tag{8}$$

The temperature of the second-step-peak increased with the zirconia content in the support due to the strong interaction of the iron oxide with the zirconia support, hence very high activation energy values were observed [32]. The disproportionation reaction wustite $\Leftrightarrow$ magnetite + iron, makes simple wustite reduction FeO $\rightarrow$ Fe a much more complicated process. In the case of thermodynamically forced FeO disproportionation, the closed packed cubic network of the oxygen sub-lattice does not change during wustite to magnetite transformation. However, the metallic iron phase formation requires temperature activated diffusion of iron atom into the hematite into the iron metallic phase, which can occur at 380°C [33]. $MnZnFe_2O_4$ was synthesized as a new type of catalyst to produce ammonia [18]. The catalysts were reduced in hydrogen and TPR results are presented below (Fig. 7).

TPR experiment was conducted for $Mn_{0.8}Zn_{0.2}Fe_2O_4$ (Fig. 7). Experiment was carried out at a heating rate of 5°C/min. The reactive gas composition is hydrogen (5 vol%) in nitrogen. The flow rate was fixed at 20 ccm/min. The total reactive gas consumption during TPR analysis was measured. The TPR measurement was carried out following activation after cooling the sample in nitrogen flow to 40°C. Sample was then held at 1,000°C for 10 min. The TPR experiment was performed at temperature 800°C.

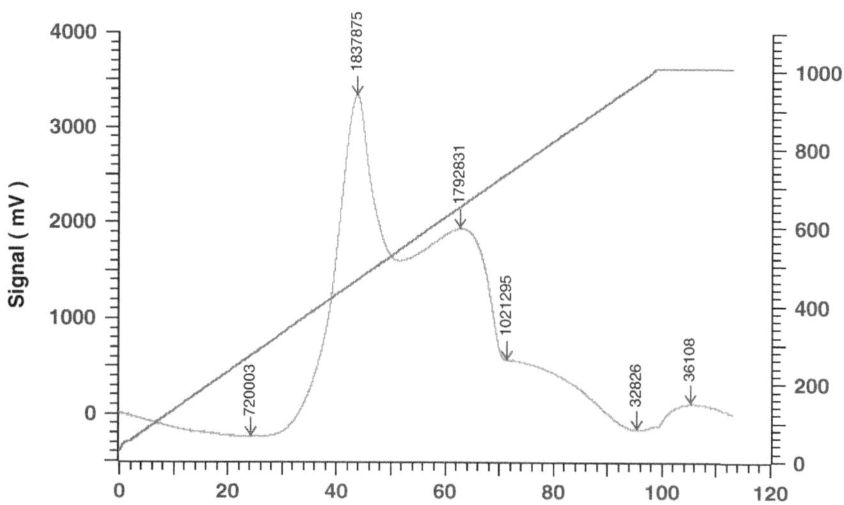

Fig. 7 Temperature program reduction (TPR) profile for $Mn_{0.2}Zn_{0.8}Fe_2O_4$ annealed at 700°C for 4 h

Table 5 Data TPR for $Mn_{0.8}Zn_{0.2}Fe_2O_4$ sintered at 700°C

Temperature (°C)	% Hydrogen	mVs	μmol/g
273	13.23	720003.28	625.33194
463	33.78	1837875.24	1596.21780
650	32.95	1792831.35	1557.09670
736	18.77	1021295.31	887.00787
1007	0.66	36108.31	31.36052

Referring to Fig. 7, the first reduction occurs at 273°C for Fe_2O_3 to Fe_3O_4. It has been reported that the reduction for Fe_2O_3 to Fe_3O_4 happened at 350°C [33]. Other researchers found that it occurred at 370°C [36]. The second reduction peak for $Mn_{0.8}Zn_{0.2}Fe_2O_4$ is at 463°C. It can be speculated as reduction peak to MnO and FeO. It was reported that the second reduction peak occurred at 500–600°C and at 540°C ferrite will reduce into FeO and MnO [37]. The third reduction peak is at 736°C which is the reduction temperature for oxide to the metallic state. It was reported that above 613°C the metallic iron was formed [34]. On the other hand other researchers reported the full reduction happened at above 650°C [37]. $Mn_{0.8}Zn_{0.2}Fe_2O_4$ needs less energy activation to reduce to Mn and Fe metallic because the present of Mn that facilitates the reduction of Fe_2O_3 [34, 36, 37].

Table 5 shows that the temperature at 463 and 650°C, percentage of hydrogen consumption was at 33.78% and 32.95%, respectively. While, at temperature above 1,000°C, it was expected that all reduction process are completed with 0.66% hydrogen consumed.

4 Microreactor

Studies on chemical reactors in a smaller scale have recently shown an increased interest due to its smallness, safety and energy efficiency [38, 39]. The microreactors have several advantages compared to the larger scale reactors. The large surface to volume ratio by a few orders of magnitude comparing to the industrial reactors is able to enhance heat or mass transfer. Several studies have been conducted to develop microreactors for chemical-catalytic reactions. Microreactor technology (MRT) was reviewed in context of discovery, development and commercialization of catalytic systems [39]. Issues relating to mathematical modelling of microreactor using fundamental principles transport-kinetic equations which are related to mass, heat, species, momentums kinetics, etc. and the appropriate boundary conditions for a specific application were proposed. Development of a structured aluminium-anodized alumina microreactor that exhibits high catalytic activity for decomposition of anhydrous ammonia to nitrogen and hydrogen gas was done [40]. Ammonia conversion of 99% at 600°C in a volume of 0.35 cm^3 was obtained using enhanced surface area of the anodized aluminum support [41]. A microreactor with induced electromagnetic field was designed and developed

Fig. 8 Helmholtz coil and ammonia microreactor connected [18]

for the ammonia synthesis [18]. Figure 8 shows ammonia yield was attributed to the $MnFe_2O_4$ nanocatalyst which was performed by using Helmholtz coil to align spins of the catalysts this method has produced ammonia gas when reaction was done in EM field (produced by Helmholtz coil). Specific microreactor for ammonia synthesis has been built by a group of researcher in Republic of Korea. They had used an integrated microammonia analysis system (IMAAS) with microreactor for their analysis on the reaction. Their project involves the microreactor with a size of 5×5 cm and thickness of 1 mm and it was fabricated using typical and simple microelectromechanical systems (MEMS). This resulted in an increasing temperature from 308 to 318 K for the reaction rates in their experiment. The reaction of this 318 K in temperature was complete in less than 1 min. Reaction rates also increased with ammonia concentration from 4.1 to 33.0 ppm [42]. There are many designs for microreactor have been proposed. Figure 9 shows the Y-Shape Microreactor which was connected to Helmholtz coil as displayed at Fig. 8.

Design and fabrication of reactor are done using AUTOCAD 3D (Fig. 10) and AUTOCAD 2D (Fig. 11). It should be noted that prior to this, studies on heating element and pressure relating to temperature and volume were initiated. An accurate control on temperature, pressure and the magnetic induction using current and voltage was done. Subsequently, fabrication of the chamber which has a heating element, two inlets and one outlet, a pressure gauge, and a magnetizer was done. The prototype was fabricated and shown in Fig. 12.

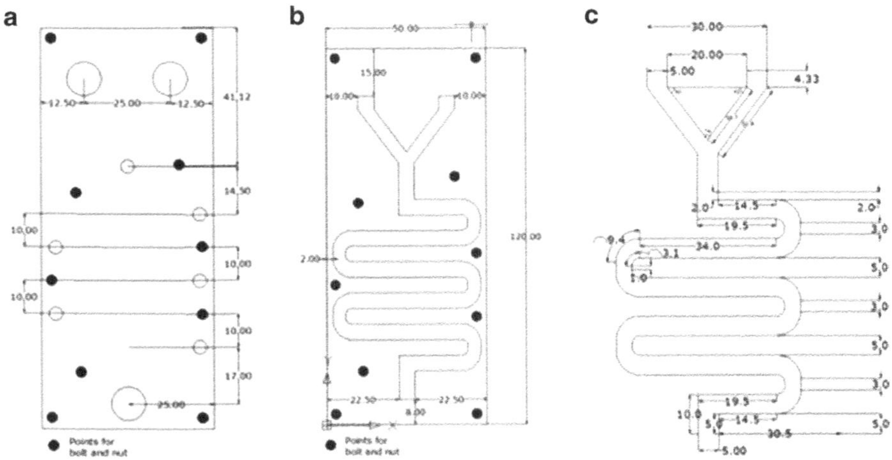

Fig. 9 Y-Shape ammonia microreactor [43]

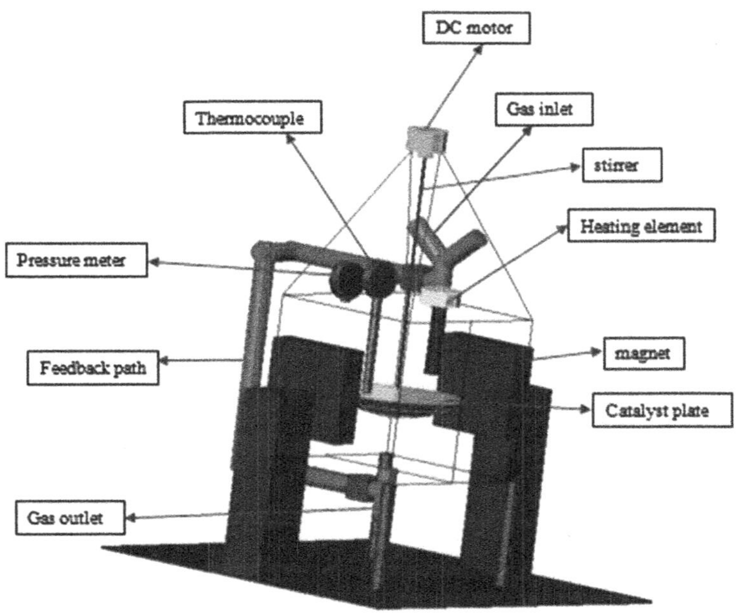

Fig. 10 Ammonia microreactor [44]

This part discusses on the microreactor which was designed at Universiti Teknologi PETRONAS and the catalytic reaction which was also conducted in-house. The catalysts were reacted in a chamber. Hydrogen gas was flowed to

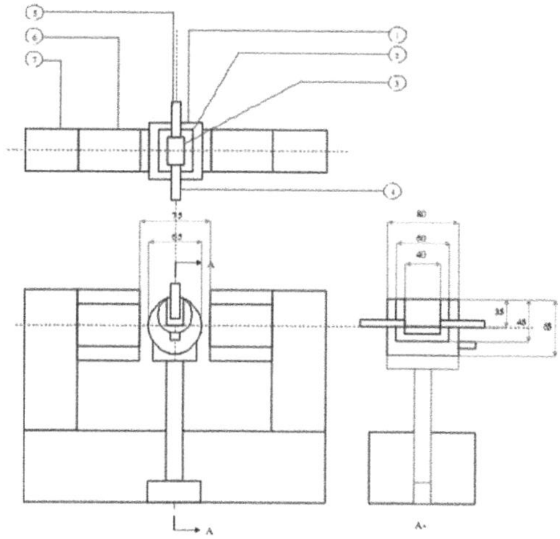

No.	Part
1	Jacket
2	Chamber
3	Catalyst Bed
4	Inlet
5	Outlet
6	Magnet
7	Stand

Fig. 11 Design of the ammonia reactor using AUTOCAD (2D) software

Fig. 12 Ammonia synthesis reactor with magnetic induction

ensure that the ferrites based catalyst is reduced. After 1 h of running the hydrogen gas, the magnetizer was activated to magnetize the catalysts. Nitrogen gas was subsequently flowed for about 2 h. The product was collected in a glass beaker filled with 0.1 M HCl to react with the NH_4OH (ammonia hydroxide) according to the Kjeldahl Method [45, 46]. Prior to this, TPR studies were done to understand the reduction profile of the MnZn Fe_2O_4 catalyst.

Figure 13 shows that ammonia yield (24%) has been obtain by using nanocatalyst $Mn_{0.8}Zn_{0.2}Fe_2O_4$ synthesized by sol gel method.

4.1 Ammonia Production Patents

In this part we discuss on the ammonia production patent that was filed (in United States). In particular, several studies on reactor for ammonia synthesis have been

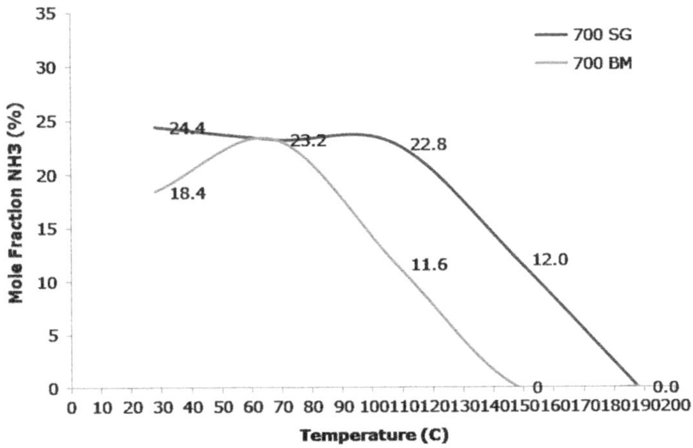

Fig. 13 Reaction rate of ammonia yield with magnetic induction

patented (Table 6). The first US patent for ammonia production was made by Haber and Bosch, U.S patent 1202995 [47] back in 1916. Since then, more patents of ammonia synthesis have been filed. One of the patents was been made by Wright et al. [47] (U.S patent 3721532). He invented a system of apparatus and process for synthesizing ammonia which includes first and second catalytic synthesis. The system has converters with heat exchanger interposed and operatively connected to the inlet and outlet of one converter. The inlet of the other converter permits a feed gas stream to be passed into the heat exchanger. The heat exchanger is connected to a partially synthesized gas stream passing from the outlet of the first converter. There was also a support platform for the converters and the heat exchanger. This process needs pressure range of 1,440–4,550 psig and temperature range of 400–1,100F. Rosa [48] (U.S patent 4107277) filed another patent that described the improvement of production of ammonia which employs high pressure electrolyzer to produce hydrogen. This invention claimed that the ammonia synthesis was done by eliminating mechanical compressor. It also utilizes the product of high pressure oxygen for refrigeration purposes needed in the process and combines hydrogen and nitrogen in a high pressure reactor. Meanwhile, the steam formed in the cooling of the reactor was used to generate electricity to activate the electrolyzer to produce hydrogen and oxygen. The pressure employed for this process was 200 atm and the temperature range is 80–360 K. Another U.S patent (US 4148866) that introduced low energy synthesis process for producing ammonia [49]. One aspect worth to mention is that the ammonia synthesis has been carried out at low pressures between 20 and 100 atmospheres. The ammonia in the effluent was then recovered by scrubbing with water absorption system to form water–ammonia mixture. The recovery of the ammonia from the water–ammonia mixture was done by distillation using the heat recovered from the raw synthesis gas. Ammonia product was refrigerated in an ammonia absorption system and the same distillation system was used to reconstitute the absorbent

Table 6 Summarize of ammonia production patents

Patent filling	Title of invention	Short summary	Ref.
3721532	Ammonia synthesis system	A system of apparatus and process for synthesizing ammonia, which includes first and second catalytic synthesis converter and heat exchanger. Pressure condition = 1,440–4,550 psig, temperature condition = 400–1,100F	[47]
4107277	Process for production of ammonia	Production of ammonia by the synthesis of hydrogen and nitrogen employs a high pressure electrolyzer and eliminate the main compressor. Pressure condition = 200 atm, temperature condition = 80–360 K	[48]
4148866	Low energy ammonia synthesis process	The improvement of ammonia production wherein the hydrogen and nitrogen reacted at pressure between 20 and 100 atm, temperature condition = 315–424°C	[49]
4479925	Preparation of ammonia synthesis	The production of ammonia from normally gaseous hydrocarbons with reduced fuel by deletion of the fired primary reformer. Pressure condition = 25 50 bars, temperature condition = 450–700°C	[50]
4695442	Ammonia synthesis process	The production of hydrogen and in particular to the production of purified ammonia synthesis gas from a raw gas. Pressure condition = 40–120 bar, temperature condition = 300–450°C. Ratio H_2/N_2 = 2.7–3.0	[51]
4792441	Ammonia synthesis	The integrated process from reforming process. The quantity of unreacted methane from the primary reforming introduced into the secondary reformer with the quantity of air to the ultimate result (H_2/N_2 3:1). Pressure condition = 1,100–1,200 psig, temperature condition = 1,680–1,720F	[52]

and to purify the ammonia for both the water absorption system and the refrigeration system, respectively. Shires et al. (U.S patent 4479925) describes ammonia synthesis which was produced by natural gas nitrogen and hydrogen. These gases were feed and steam to an endothermic catalytic conversion zone operated under primary reforming conditions to produce primary reformed gas [50]. This condition was then introduced with air to an adiabatic, catalytic conversion zone operated under autothermal steam reforming conditions to produce raw and hot, ammonia gas. In turn, it was passed to the endothermic catalytic conversion zone in an indirect heat exchanger. This process needs high temperature condition from 700 to 900°C and low pressure from 25 to 50 bars. Since the ammonia plant produces ammonia from raw gas, industrial plant is facing another problem on removing various gases from its mixtures. A study of adsorption properties of gases that involved over a narrow range of raw gas composition has been filed [51]. The content of medium boiling point gases, especially nitrogen, was in balance with the contents of hydrogen and carbon dioxide. Hydrogen and nitrogen are needed to produce ammonia gas at high percentage hydrogen recovery without excessive adsorption bed volume and with little or without external purge gas. This process operates at 350–500°C and pressure range of 25–50 bars. An integrated process was presented whereby the total raw hydrocarbon charged is separated for a reforming process. The reforming process requires a small portion of the total raw hydrocarbon into the primary steam reforming operation. Larger portion of it are being charged directly to the secondary reformer, together with the unconverted methane which was reacted with oxygen-enriched air of controlled composition [52]. This invention needs temperature range from 1,680 to 1,720F and pressure range from 1,100 to 1,200 psig. Another new concept of microreactor has been proposed by the main author of this chapter. This new microreactor induced magnetic field to a cylindrical chamber (Fig. 12). The magnetic field is used to align the electrons spin in spinel based nanocatalyst which in turn will enhance the catalytic activity to produce the high yield of ammonia (46%) as compared to the industrial plant (14%). The ammonia synthesis for this process requires room temperature (28°C) and ambient pressure.

5 Conclusion

In this chapter some aspects ammonia production issues were briefly described. Ultimately catalytic reaction was scrutinized. Issues on reactor, reaction conditions, catalyst, and yield of ammonia were elaborated. New microreactor and new nanocatalyst $Mn_{0.8}Zn_{0.2}Fe_2O_4$ are proposed (Table 6).

References

1. Nachtergaele, F., Brinman, R.B.: ftp://ftp.fao.org/agl/agll/docs/fertusemalaysia November 13 (2009)
2. Clark, J.: http://www.chemguide.co.uk/physical/equilibria/haber.html
3. Roebuck, C.M.: Exeel HSC Publisher, pp. 35–41 (2008)
4. Whitten, K.N., Davis, K.E., Peck, M.L., Stanley, G.G.: Thomson Brookscole, 7, pp. 813–720 (2003)
5. Somojai, G.A.: Wiley-Interscience, 1, pp. 452–454 (1994)
6. Modak, J.M.: Haber process for ammonia synthesis. Resonance, 69–77 (2002)
7. Tamaru, K.: In: Jennings, J.R. (ed.) Plenum Press, New York, Chapter 1 (1991)
8. Garbassi, F., Fagherazzi, G., Calcaterra, M.: J. Catal. **26**, 338 (1972)
9. Pernicone, J., Ferrero, F., Rosetti, Z., Forni, L., Canton, P., Rinello, P., Fagherazzi, G., Signorot, M., Pinna, J.: Appl. Catal. A Gen. **25**, 121–129 (2003)
10. Hinrichsen, O.: Catal. Today **53**, 177–188 (1999)
11. Juan, I., Masahiro, I., Yoshiro, K., Makoto, S., Ken-ichi, M.: Catal. Commun. **8**, 941–944 (2007)
12. Atsumu, O.: Acc. Chem. Res. **14**, 16 (1981)
13. Pietro, M., Giancarto, A., Giovanni, P., Giuseppe, S.: Appl. Catal. A Gen. **123**, 145–159 (1995)
14. Wioletta, R.P., Elzbieta, M., Darius, S., Zbigniew, K.: J. Catal. **231**, 11–19 (2008)

15. Mgrdichian, L.: http://www.bnl.gov/bnlweb. August 18 (2008)
16. Mahapatra, H., Kalyuzhunaya, E.S., Yunusov, S.M., Shur, V.B.: Recent Adv Basic Appl. Asp. Ind. Catal. Stud. Surf. Sci. Catal. **115**, 267 (1998)
17. Changhiang, L., Zhaobin, W., Qin, X., Can, L.: Appl. Catal. A Gen. **208**, 193–201 (2001)
18. Yahya, N., Puspitasari, P., Koziol, K., Zabidi, N.A., Othman, M.F.: Int. J. Eng. **10**, 1 (2010)
19. Bartholomew, C.H., Farrauto, R.J.: Fundamental of Industrial Catalytic Process, 2nd edn. Wiley Interscience, Weinheim (2006)
20. Hagen, J.: Industrial Catalysis: a Practical Approach, 2nd edn, pp. 189–192. Wiley-Interscience, Weinheim (2006)
21. Whitten, K.N., Davis, R.E., Peck, M.L., Stanley, G.G.: General Chemistry, 7th edn, pp. 713–720. Thomson Brookscole, Belmont (2003)
22. Zhixiong, Y., Koji, I., Ken-ichi, A., Toshihide, B.: J. Catal. **251**, 321–331 (2007)
23. Qing-Chi, X., Jing-Dong, L., Jun, L., Xian-Zhu, F., Zhen-Wei, Y., Wei-Ming, G., Dai-Wei, L.: J. Mol. Catal. A Chem. **259**, 218–222 (2006)
24. Ryoichi, K., Hiroyuki, E., Martin, M., Ken-ichi, A.: Appl. Catal. A Gen. **246**, 311–322 (2003)
25. Stefan, H., Rasmus, B., Rasmus, F., Claus, J.H.J., Herman, T., Teunissen, Chorkendorff: J. Catal. **214**, 327–335 (2003)
26. Ryoichi, K., Ken-ichi, A.: Appl. Catal. A Gen. **218**, 121–128 (2001)
27. Agnieszka, P.J., Barbara, M., Antoni, W.M.: Appl. Catal. A Gen. **141**, 1–16 (1996)
28. Hua-Zhang, L., Xiao-Nian, L., Zhang-Neng, H.: Appl. Catal. A Gen. **142**, 209–222 (1996)
29. Morawski, A.W., Lendzion-Bielufi, Z., Katucki, K., Kaleficzuk, R.J.: Appl. Catal. A Gen. **156**, 19–27 (1997)
30. Ali, A.M., Shaterian, H.R., Stuart, H.T., Graham, J.H.: Catal. Lett. **87**, 3–4 (2003)
31. Haznan, A., Byoung, S.A., Chang, S.K., Kye, S.Y.: Ind. Eng. Chem. Res. **46**, 7936–7941 (2007)
32. Wataru, U., Kenzo, O., Damien, V., Tokio, H., Youhei, K.: Catal. Surv. Jpn. **6**, 1–2 (2002)
33. Wataru, U., Ning, F.C., Kenzo, O.: Chem. Commun., 517–518 (1999)
34. Yucai, Y., Ping, Y., Tao, L., Jiang, Z., Dangdong, Z.: Indian J. Chem. **46a**, 1270–1272 (2007)
35. Meisheng, L., Wenkai, K., Kechang, X.: J. Nat. Gas Chem. **18**, 110–113 (2009)
36. Oliveira, L.C.A., Fabris, J.D., Rios, R.R.V.A., Mussel, W.N., Lago, R.M.: Appl. Catal. A Gen. **259**, 253–259 (2004)
37. Flavia, G.D., Bibiana, P.B., Luis, E.C., Cristina, R., Miguel, A.C., Jose, A.O.: Appl. Catal. B Environ. **92**, 194–201 (2009)
38. Bulbul, A.: Thesis, National University of Singapore
39. Patrick, L.M., David, J.Q., James, F.R.: Department of Chemical and Natural Gas Engineering (2007)
40. Ganley, J.C., Seebauer, E.G., Masel, R.I.: J. Power Sources **137**, 53–61 (2004)
41. Franz, T., Cedric, H., Phillip, R.V.R.: Chem. Eng. J. **135S**, S309–S316 (2008)
42. Joon-Shik, P., Kwang, B.P., Kyu-Shik, S., Hyo, D.P., Min-Chan, K., Jeong-Rim, K., Sang-Joon, P., Young-Hwa, S.: Sen. Actuators B Chem. **117**, 516–522 (2006)
43. Othman, M.F.: Microreactor design for ammonia synthesis. Thesis, Universiti Teknologi PETRONAS (2009)
44. Tarmizi, I.A.: Design of microreactor for ammonia synthesis. Thesis, Universiti Teknologi PETRONAS (2009)
45. Yunusov, S.M., Kalyuzhnaya, E.S., Moroz, B.L., Ivanova, A.S., Reshtenko, T.V., Avdeeva, L.B., Likholobov, V.A., Shur, V.B.: J. Mol. Catal. A Chem. **219**, 149–153 (2004)
46. Labconco: An Industry Service Publication
47. Lee, E.W.: Ammonia synthesis system. U.S. Patent 3721532, 20 March 1973
48. Aldo VdR: Process for production of ammonia. U.S Patent 4107277, 15 August 1978
49. Colman, L.B.: Low energy ammonia synthesis process. U.S Patent 4148866, 10 April 1979
50. Philip, J.S.: Preparation of ammonia synthesis gas. U.S Patent 4479925, 30 October 1984
51. Alwyn, P.: Ammonia synthesis process. U.S Patent 4695442, 22 September 1987
52. Shoou, I.W.: Ammonia synthesis. U.S Patent 4792441, 20 December 1988

The manufacturer's authorised representative in the EU is Springer Nature Customer Service Centre GmbH, Europaplatz 3, 69115 Heidelberg, Germany. If you have any concerns regarding our products, please contact ProductSafety@springernature.com

Printed and bound by CPI Group (UK) Ltd, Croydon, CR0 4YY
22/09/2026
02220985-0008